AF788839

# Current Developments in Plant Genetics and Breeding

# Current Developments in Plant Genetics and Breeding

Edited by **Kiara Woods**

SYRAWOOD
PUBLISHING HOUSE
New York

Published by Syrawood Publishing House,
750 Third Avenue, 9th Floor,
New York, NY 10017, USA
www.syrawoodpublishinghouse.com

**Current Developments in Plant Genetics and Breeding**
Edited by Kiara Woods

© 2016 Syrawood Publishing House

International Standard Book Number: 978-1-68286-260-5 (Hardback)

This book contains information obtained from authentic and highly regarded sources. Copyright for all individual chapters remain with the respective authors as indicated. All chapters are published with permission under the Creative Commons Attribution License or equivalent. A wide variety of references are listed. Permission and sources are indicated; for detailed attributions, please refer to the permissions page and list of contributors. Reasonable efforts have been made to publish reliable data and information, but the authors, editors and publisher cannot assume any responsibility for the validity of all materials or the consequences of their use.

The publisher's policy is to use permanent paper from mills that operate a sustainable forestry policy. Furthermore, the publisher ensures that the text paper and cover boards used have met acceptable environmental accreditation standards.

**Trademark Notice:** Registered trademark of products or corporate names are used only for explanation and identification without intent to infringe.

Printed in China.

# Contents

# Preface

Plant genetics and breeding has been widely practiced to develop innovative breeding methodologies and discover important genes and varieties. The text focuses upon theoretical and conceptual practices that bridge the gap between fundamental researches and their implementation in plant breeding. This book discusses the variations in plant DNAs and how such aspects can be utilized for breeding a superior quality of plants and crops. It also delves into phytopathology and disease management in plants. It brings forth revolutionary researches in the discipline from across the globe. It will be helpful to researches, experts and interested readers in this field.

This book is the end result of constructive efforts and intensive research done by experts in this field. The aim of this book is to enlighten the readers with recent information in this area of research. The information provided in this profound book would serve as a valuable reference to students and researchers in this field.

At the end, I would like to thank all the authors for devoting their precious time and providing their valuable contributions to this book. I would also like to express my gratitude to my fellow colleagues who encouraged me throughout the process.

**Editor**

# 1

# Effect of inter and intra row spacing on seed tuber yield and yield components of potato (*Solanum tuberosum* L.) at Ofla Woreda, Northern Ethiopia

**Harnet Abrha[1] , Derbew Belew[2] and Gebremedhin Woldegiorgis[3]**

[1]Alamata Agricultural Research Center, Alamata, Ethiopia.
[2]Jimma University, Jimma, Ethiopia.
[3]Holleta Agricultural Research Center, Addis Ababa, Ethiopia.

**Farmers in southern zone of Tigray are using different spacing below or above the national recommendation depending on the purpose of planting either for seed tuber or consumption due to lack of recommended plant spacing. This study was therefore conducted with the objective of determining the best inter and intra-row spacing for optimum tuber seed yield and quality of potato seed tuber at Ofla Wereda, Northern Ethiopia. Four different intra-row (20, 25, 30 and 35 cm) and inter-row (65, 70, 75 and 80 cm) spacing were used in the experiment. The result reveals that inter and intra-row spacing significantly ($p<0.001$) affected seed tuber yield $ha^{-1}$, the maximum seed tuber yield (36.89 and 37.54 ton $ha^{-1}$) was recorded at 65 and 20 cm inter and intra-row spacing, respectively. From this study, it can be concluded that the narrow spacing (20 and 65 cm intra and inter-row spacing) produced higher seed tuber yield per hectare than other spacings. Thus, potato (Jalenie variety) growers in the study area can benefit if they use this narrow spacing (20 and 65 cm intra and inter-row spacing).**

**Key words:** Potato, intra-row spacing, inter-row spacing, seed tuber yield.

## INTRODUCTION

Potato (*Solanum tuberosum* L.) originated from the high Andes of South America and was first cultivated in the vicinity of Lake Titicaca near the present border of Peru and Bolivia (Horton, 1987). In terms of quantity produced and consumed worldwide, potato is the most important vegetable crop. It is one of the most important food crops in the world; it produces more energy and protein per unit area and unit of time than most other major food crops.

The potato crop was introduced to Ethiopia around 1858 by Schimper, a German botanist (Pankhurst, 1964). Among African countries, Ethiopia has possibly the greatest potential for potato production; 70% of its arable land mainly in highland areas above 1500 m is believed to be suitable for potato. Since the highlands are also home to almost 90% of Ethiopia's population, the potato could play a key role in ensuring national food security

(FAO, 2008). However, the current area cropped with potato is about 0.16 million hectares and the national average yield is about 7.2 t/ha, which is very low as compared to the world's average production of 16.8 t/ha (Adane et al., 2010). The crop yield in Ethiopia is lower than that of most potato producing countries in Africa like South Africa and Egypt, which produce 34.0 and 24.8 t/ha, respectively (FAO, 2008).

Many diverse and complex biotic, abiotic and human factors have contributed to the existing low productivity of potato. Some of the production constraints which have contributed to the limited production or expansion of potato in Ethiopia include shortages of good quality seed tubers of improved cultivars, disease and pests, and lack of appropriate agronomic practices including optimum plant density, planting date, soil moisture, row planting, depth of planting, ridging and fertility status (Berga et al., 1994).

The optimization of plant density is one of the most important subjects of potato production management, because it affects seed cost, plant development, yield and quality of the crop (Bussan et al., 2007). The yield of seed potato can be maximized at higher plant population (closer spacing) or by regulating the number of stems per unit area and to certain extent by removing the haulm earlier during the maturity (O'Brien and Allen, 2009). Rahemi et al. (2005) reported that the effect of intra-row spacing on yield of potatoes was significant especially at 20 cm intra-row spacing, which showed 36.85% yield increment as compared to 30 cm intra-row spacing. Intra-row distance of 20 cm increased total tuber number and weight, and tuber weight per plant and the marginal return rate increased by 13% when intra-row distance decreased from 35 to 25 cm. EARO (2004) also determined that there is a little difference in yield between intra-row spacing of 25 and 30 cm for all varieties released so far in Ethiopia and the 30 cm intra-row and 75 cm inter-row spacing accepted as standard.

Farmers in the study area (Sothern zone of Tigray) are using different spacing below or above the national recommendation depending on the purpose of planting either for consumption or for seed tuber due to lack of recommended inter and intra-row spacing. Hence, it is important to maintain appropriate plant population per unit area to have high yield, marketable size and good quality of seed tuber. Even though different research is done in different parts of the country on potato plant density, the condition is not studied in Ofla Woreda, Southern Zone of Tigray. This study was therefore conducted to determine the best inter and intra-row spacing for optimum tuber seed yield and quality of potato seed tuber at Ofla Wereda, Northern Ethiopia.

## MATERIALS AND METHODS

The experiment was conducted in 2011/2012 under irrigation condition in Southern Zone of Tigray, Ofla Woreda at Hashenge Kebele, on farmer's field. The experimental site is located at an elevation of 2500 m above sea level. Maximum and minimum temperature ranges from 22.57and 6.8°C, respectively. The mean annual rainfall of the area is 806.5 mm. The major soils include clay (28%), loam (57%) and sandy (15%) with a pH of 6.8 (BoARD, 2009).

The Woreda is classified into three agro-ecological zones, namely, highland, midland and lowland. The midland covers the largest part which accounts about 42% of the total 133, 296 ha while both the highland and lowland covers 29%. The average land holding in the Woreda is about 0.5 ha per household and estimated total population of 132,491 (BoARD, 2009).

Different local and improved potato varieties are being grown in the area. Among the improved variety, Jalenie is growing widely and has got acceptance by farmers due to its high yielding ability and acceptability by consumers.

The experiment was laid out in 4 x 4 factorial arrangements using a Randomized Complete block design (RCBD) with three replications and two factors, which consisted of four different intra-row spacing: 20, 25, 30 and 35 cm, and four different inter-row spacing: 65, 70, 75 and 80 cm. Each plot contain four rows with different plot size of (3.15 x 3.2, 3.15 x 3, 3.15 x 2.8, 3.15 x 2.6) and different number of plants per row which includes 15, 12, 10, 9 plants for 20, 25, 30 and 35 cm intra row spacing, respectively. A foot path of 0.5 and one meter was left between plots and blocks, respectively.

The collected data on different growth stage was analyzed by using SAS Computer software version 9.0 (SAS Institute Inc., 2008).

## RESULTS AND DISCUSSION

### Leaf area index

Intra-row spacing showed a very highly significant ($P<0.001$) effect on leaf area index. However, the effect of inter-row spacing and interaction showed no significant difference in leaf area index (Figure 1). The result revealed that significantly the highest leaf area index (3.21) was recorded at 20 cm intra-row spacing, and this could be due to high number of haulms per unit area. Whereas the lowest (2.32) leaf area index was recorded from 35 cm intra-row spacing and it is statistically difference from the other three (30, 25 and 20 cm) intra-row spacings.

This result is in agreement with the findings of Ronald (2005) and Tamiru (2005) who reported that the highest density increased leaf area index, possibly indicating potential partitioning of assimilates for vegetative growth.

### Total tuber seed yield (t/ha)

The effect of inter-row and intra-row spacing showed a very highly significant ($P<0.001$) differences on total tuber yield $ha^{-1}$ (Table 1). However, the interaction effect was non-significant ($P>0.05$). The highest yield (36.89 t/ha) was obtained from 65 cm inter-row spacing, whereas the lowest (31.87 t/ha) yield was recorded at 80 cm inter-row spacing.

Regarding the intra-row spacing, the higher total yield

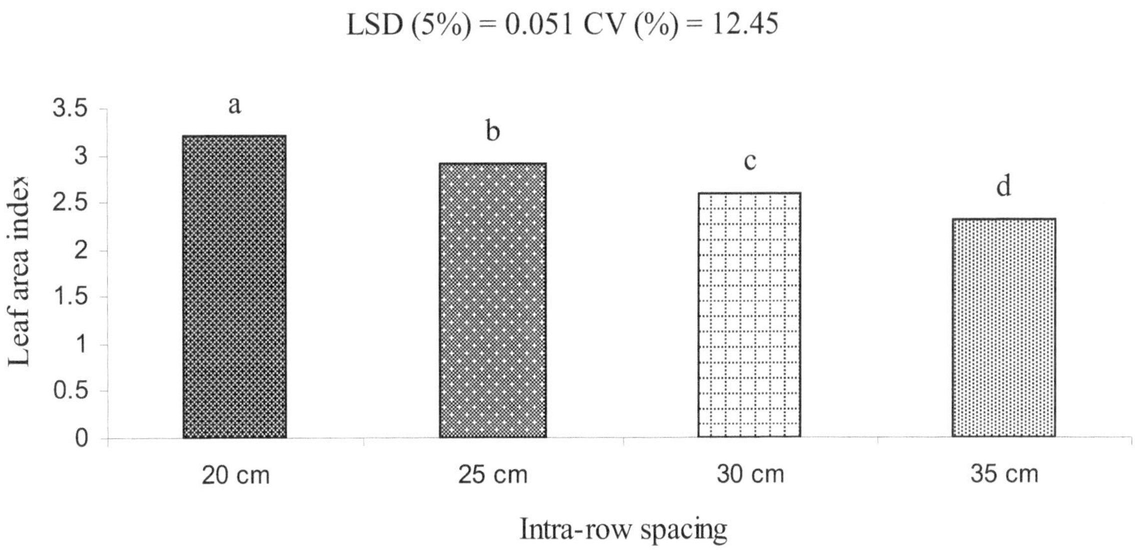


**Figure 1.** Means for the effect of intra-row spacing on leaf area index.

**Table 1.** Means for the effect of inter and intra-row spacing on total tuber yield and marketable tuber seed yield per hectare

| Treatment | Total tuber seed yield (t/ ha) | Marketable seed tuber yield (t/ ha) |
|---|---|---|
| **Intra-row spacings (cm)** | | |
| 20 | 37.54[a] | 35.89[a] |
| 25 | 35.75[b] | 34.49[b] |
| 30 | 35.61[b] | 34.66[b] |
| 35 | 29.38[c] | 28.65[c] |
| **Inter-row spacings (cm)** | | |
| 65 | 36.89[a] | 35.09[a] |
| 70 | 35.33[b] | 33.86[b] |
| 75 | 34.18[b] | 33.32[b] |
| 80 | 31.87[c] | 31.42[c] |
| LSD (5%) | 1.18 | 1.18 |
| CV (%) | 11.25 | 10.31 |

Means followed by the same letter within the same column are not significantly different at 5% level of significance.

per hectare (37.54 t/ha) was obtained from 20 cm intra-row spacing. As intra-row spacing increased from 20 to 35 cm, total tuber yield decreased from 37.54 to 29.38 t/ha. Intra-row spacing of 35 cm showed lower total tuber yield (29.38 t/ha) and it was significantly different from the three levels. It was clearly evident from the results that the yield of seed tuber per hectare was increased with decreasing plant spacing.

The increased yield was attributed to more tubers produced at the higher plant population per hectare although average tuber size was decreased because of increased inter-plant competition at closely spaced plants leading to more unmarketable tuber yield. At closer spacing there is high number of plants per unit area which brings about an increased ground cover that enables more light interception, consequently influencing photosynthesis. It is therefore, very likely that substantial increases in rate of land coverage and thereby tuber yield could be achieved by dramatically increasing the stem density per unit area.

The present result agrees with the findings of Zabihi et al. (2011) who reported that plant density in potato affects some of the important plant traits such as total yield, tuber size distribution and tuber quality. Increase in plant

**Table 2.** Means for the effect of inter and intra-row spacing on total and marketable seed tuber number of tuber $ha^{-1}$.

| Treatment | Total number of seed tuber per hectare | Number of marketable seed tuber per hectare |
|---|---|---|
| **Intra-row spacing(cm)** | | |
| 20 | 558174$^{a}$ | 501651$^{a}$ |
| 25 | 486858$^{b}$ | 445568$^{b}$ |
| 30 | 455014$^{bc}$ | 423513$^{bc}$ |
| 35 | 430311$^{c}$ | 395106$^{c}$ |
| **Inter-row spacing (cm)** | | |
| 65 | 532865$^{a}$ | 485144$^{a}$ |
| 70 | 496599$^{a}$ | 455026$^{a}$ |
| 75 | 453307$^{b}$ | 411315$^{b}$ |
| 80 | 447586$^{b}$ | 414352$^{b}$ |
| LSD (5%) | 37587.6 | 37667.7 |
| CV (%) | 15.57 | 15.61 |

Means followed by the same letter within the same column are not significantly different at 5% level of significance.

density led to decrease in mean tuber weight but number of tubers and yield per unit area were increased. In contrast, Berga et al. (1994) reported that wider row width by wider in-row distance (80 x 40 cm) gave the highest yield (34 t/ha) and the 60 x 20 treatment gave the lowest yield (22.2 t/ha).

### Marketable seed tuber yield (t/ha)

The data concerning marketable yield as influenced by planting density is presented in Table 1. Inter and intra-row spacing showed a very highly significant (P<0.001) effect on marketable yield. Significantly maximum marketable yield (35.89 and 35.09 t/ha) was obtained at a 20 and 65 cm intra and inter-row spacing, respectively. While the lowest marketable yield (28.65 and 31.42 t/ha) was obtained at the wider spacing (35 cm intra and 80 cm inter-row spacing, respectively). However the interaction effect did not show significant difference on marketable yield per hectare.

The highest marketable yield recorded at closer spacing is attributed to more tubers produced at the higher plant population per hectare. The present result agreed with the findings of many authors (Stoffella and Bryan, 1988; Khalafalla, 2001) regarding plant density effect on marketability of the crop. Close spacing of 15-25 cm was reported to give better proportion of marketable yield than wider spacing of 35 cm.

### Total number of tubers per hectare

The results of total number of tuber ($ha^{-1}$) as influenced by inter and intra-row spacing is presented in Table 2. Inter and intra-row spacing had very highly significantly (P<0.001) affected total number of tuber per ha. Significantly maximum total number of tuber per hectare (532,865) was recorded at 65 cm inter-row spacing. While the lowest number of tuber per hectare (447,586) was obtained at wider spacing (80 cm) inter-row spacing.

As far as the intra-row spacing is concerned, significantly maximum total number of tuber per hectare (558,174) was obtained from 20 cm spacing. Whereas the lowest total number of tuber per hectare (430,311) was obtained at 35 cm spacing. Total tuber number per hectare was increased with closer spacing. The highest number of tuber at closer spacing is due to high number of plants per unit area. Rahemi et al. (2005) reported that intra-row distance of 20 cm increased total tuber number and weight per unit area.

### Marketable seed tuber number per hectare

Marketable tuber number (000's $ha^{-1}$) as influenced by inter-row and intra-row spacing is presented in Table 2. Inter and Intra-row spacing had very highly significant (P<0.001) effect on marketable tuber number per hectare. However, the interaction effect had no significant (P>0.05) effect on marketable tuber number per hectare. Maximum marketable tuber number (485,144 and 501,651) was obtained at 65 and 20 cm inter and intra-row spacing respectively, while the result recorded at 20 cm intra-row spacing was significantly different from the other intra-row spacings. The lowest number of marketable tuber per hectare (411,315 and 395,106) was obtained at 80 cm inter and 35 cm intra-row spacing, respectively. Among the inter-row spacings, statistically, the same results were obtained from 65 and 70 cm, which scored the highest marketable tuber number per hectare, 485,144 and 455,026, respectively.

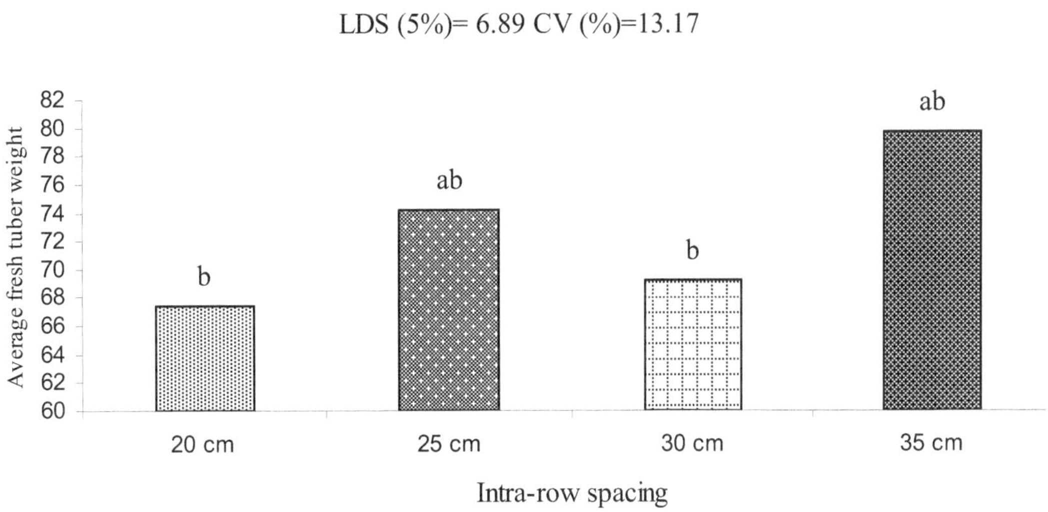


**Figure 2.** Means for the effect of intra-row spacing on average fresh tuber weight.

Related study was reported by Burton (1989); wider spacing may produce few tubers as it gave rise to few stems that could lead to high number and possibly misshapen tuber while, closer spacing improved quality and saleable yield.

## Average fresh tuber weight (g)

Intra-row spacing showed highly significant (P<0.01) difference on average fresh tuber weight per plant (Figure 2). However, the main effects of inter-row spacing and its interaction with intra-row spacing had no significant (P>0.05) difference on average fresh tuber weight. The maximum mean tuber weight (79.68 g) was recorded at 35 cm intra-row spacing but not statistically different with 25 cm intra-row spacing. The smallest average fresh tuber weight (67.3 g) was recorded at 20 cm intra-row spacing. However, it was not significantly different from 25 and 30 cm intra row spacing for the values of (74.24 and 69.16 g, respectively).

Increase in density probably increased competition between and within plants and hence, leads to decrease in availability of nutrients to each plant and consequently, resulted in decline of mean tuber weight. This result is in line with that of Ali (1997), who found higher average fruit weight at wider spacing as compared to closer spacing. Berga and Caesar (1990) also reported that stem number per plant and tuber number per plant are positively related, however, average tuber weight increased with wider spacing.

## Tuber size category

Intra-row spacing had shown highly significant (P<0.01) effect on number of tubers graded less than 20 mm (Table 3). Maximum (9.96%) less than 20 mm number was recorded at intra-row spacing of 20 cm. However, it was not significantly different from 25 cm intra-row spacing. While, the lowest (6.629%) was at 35 cm. Intra-row spacing also showed a very highly significant (P<0.001) effect on weight of tubers graded less than 20 mm. Significantly maximum (0.74%) less than 20 mm weight was recorded at intra-row spacing of 20 cm. It was significantly different from the other intra-row spacings. However, the effect of inter-row spacing and interaction effect had no significant (P>0.05) difference for number and weight of tubers graded less than 20 mm.

Intra-row spacing also showed very highly significant (P<0.001) effect on tubers graded greater than 50 mm in terms of number and weight. Significantly maximum (23.74 number and 52.91 weight percent) greater than 50 mm graded tuber was recorded at 35 cm intra-row spacing. While, the lowest (18.50 number and 42.30 weight percent) was recorded at 20 cm intra-row spacing. Inter-row spacing showed highly significant (P<0.01) effect on tuber graded 30-40 mm weight.

The highest (17.14 percent) tubers graded 30-40 mm weight was recorded at 65 cm inter-row spacing. The results of this investigation clearly indicated that the level of intra-row spacing largely affected potato tuber size distribution. Thus, based on market and consumers' demand, it is possible to produce either seed potato or ware potato of required size through the selection of appropriate planting density (intra-row spacing).

The present result is in agreement with the findings of Wiersema (1987) who reported that at higher stem density, the tuber produced will remain smaller than at lower stem densities. Khajehpour (2006) also reported that increase in plant density decreases mean tuber size probably because of plant nutrient elements reduction, increase in interspecies competition and large number of tubers produced by high numbers of stems. Generally,

**Table 3.** Means for the effect of intra-row spacing on tuber size category.

| Intra-row spacing (cm) | Weight of tubers graded less than 20 mm (%) | Number of tubers graded less than 20 mm (%) | Weight of tubers graded greater than 50 mm (%) | Number of tubers graded greater than 50 mm (%) |
|---|---|---|---|---|
| 20 | 0.7335[a] | 9.961[a] | 42.30[c] | 18.50[b] |
| 25 | 0.7066[b] | 8.121[ab] | 44.21[c] | 19.02[b] |
| 30 | 0.6808[c] | 7.485[b] | 49.06[b] | 22.00[a] |
| 35 | 0.5005[d] | 6.629[b] | 52.91[a] | 23.74[a] |
| LSD (5%) | 0.005 | 1.904 | 3.401 | 2.419 |
| CV (%) | 14.10 | 25.2 | 12.12 | 17.55 |

Means followed by the same letter within the same column are not significantly different at 5% level of significance.

the result of this study indicates that tuber size category is influenced mainly by intra-row spacing rather than inter-row spacing.

## Summary and conclusion

The result of this study demonstrated that yield per unit area is influenced by the different level of inter and intra-row spacing. From this study, it can be concluded that the narrow spacing (20 and 65 cm intra and inter-row spacing) produced higher seed tuber yield and marketable yield per hectare than other spacings. Thus, potato (Jalenie variety) growers in the study area (southern zone of Tigray) can benefit if they use this narrow spacing (20 and 65 cm intra and inter-row spacing).

## Conflict of Interests

The author(s) have not declared any conflict of interests.

## ACKNOWLEDGEMENTS

This research was conducted in partial fulfillment of the M.Sc. degree at Jimma University by the first author. Funding was provided by the Rural Capacity Building project.

### REFERENCES

Adane H, Miranda PMM, Agajie T, Willemien JML, Alfons OL, Admasu T, Paul CS (2010). Analysis of Seed Potato Systems in Ethiopia. Am. J. Potato Res. 87(6):537-552.

Ali N (1997). Sesamum Research in Pakistan. In: sesame and safflower status and potentials. Althertoni, J. and J.Rudich, 1986. The Tomato Crop. Chapman and Hall, London, U.K. J. Food Sci. 15:842-859.

Berga L, Caesar K (1990). Relationships between the number of main stems and yield components of potato (*Solanum tuberosum* L.) as influenced by different day- lengths. Potato Res. 33:257-267.

Berga L, Gebremedhin GW, Teressa J, Bereke T (1994). Potato agronomic research in Ethiopia in: Horticulture research and development in Ethiopia. Proceeding of the 2[nd] national Horticulture workshop. 1-3 December 1992, Addis Ababa, Ethiopia. Herath, E, and Lemma Dessalgne (Eds.), pp. 101-119. IAR/FAO. Addis Ababa.

BoARD (Bureau of Agriculture and Rural Development) (2009). Annual report, Bureau of Agriculture and Rural Development. Ofla, Tigray, Ethiopia. Unpublished document.

Burton WG (1989). The Potato. 3rd ed. Longman Publisher Ltd., African Crop Science Conference Proceedings, Tanzania 8:1207-1210.

Bussan AJ, Mitchell PD, Copas ME, Drilias MJ (2007). Evaluation of the effect of density on potato yield and tuber size distribution. Crop Sci. 47:2462-2472.

EARO (Ethiopia Agriculture Research Organization), 2004. Directory of released crop varieties and their recommended cultural practices: Ethiopian Agricultural Research Organization, Addis Ababa.

FAO (2008). Production year book. Food and Agriculture Organization of the United Nations, Rome, Italy.

Horton D (1987). Potatoes: production, marketing and for developing countries. West view press (Boulder), IT Publications (London), p. 243.

Khajehpour M (2006). Production of industrial plants. Jehad-e-Daneshgahi Isfahan press. Isfahan. Iran. 580: ISBN 961-6122-63-9.

Khalafalla AM (2001). Effect of Plant Density and Seed Size on Growth and Yield of Solanum Potato in Khartoum State, Sudan. Afr. Crop Sci. J. 9(1):77-82.

O'Brien PJ, Allen EJ (2009). Effects of date of planting, date of harvesting and seed rate on yield of seed potato crops. The Journal of Agricultural Science, Cambridge University Farm, Huntingdon Road, Girton, Cambridge, UK. 118:289-300.

Pankhurst R (1964). Notes for a history of Ethiopian agriculture. Ethiopian Observer. 7:210-240.

Rahemi A, Hasanpour A, Mansoori B, Zakerin A, Taghavi TS (2005). The effects of intra-row spacing and n fertilizer on the yield of two foreign potato cultivars in Iran. Int. J. Agric. Biol. 7(5):705-707.

Zabihi-e-Mahmoodabad R, Jamaati-e-Somarin S, Khayatnezhad M, Gholamin R (2011). Correlation of Tuber Yield Whit Yield Components of Potato Affected by Nitrogen Application Rate in Different Plant Density. Advances in Environmental Biology, 5(1) Islamic Azad University, Ardabil, Iran. pp. 131-135.

Ronald A (2005). Effect of pre-harvest management on yield, process quality, and disease development in Russet Burbank potatoes. M.Sc. Thesis, the University of Manitoba Winnipeg. p. 174.

SAS Institute Inc. 2008. SAS/STAT. 9.2 User's Guide. Cary, NC: SAS Institute Inc. USA.

Stoffella PJ, Bryan HH (1988). Plant population influences growth and yield of bell pepper. J. Am. Soc. Hortic. Sci. 113:835-839.

Tamiru H (2005). Effect of plant population and harvesting time on tuber Yield of potato (*Solanum tuberosum* L.). Ethiopian J. Biol. Sci. 4(1):1-9.

Wiersema SG (1987). Effect of stem density on potato production. Technical information bulletin 1. International Potato Center (CIP), Lima, Peru. (3[rd] revised ed.). pp. 1-16.

# Effect of seed-borne fungi on germination and seedling vigour of watermelon (*Citrullus lanatus* thumb)

**S. T. Anjorin[1] and M. Mohammed[2]**

[1]Department of Crop Science, University of Abuja, Abuja, Nigeria.
[2]Department of Crop Production, Federal University of Technology, Minna, Niger State, Nigeria.

**The effect of *Mucor racemosus* and *Rhizopus nigricans* inoculation (g and 0.1 g $L^{-1}$ distilled water) and a seed dressing fungicide Seedplus® (1.25 g 0.5 $kg^{-1}$ seeds) on germination and seedling vigour of watermelon (cv. Chaliston gray) was investigated. The experiment involved a complete randomized design (CRD) with three replicates. It was confirmed that the combined inocula with higher density caused significantly poor germination and low seedling vigour ($P \leq 0.05$) than those seeds inoculated with lower inocula density. Higher germination percentage and seedling vigour were observed with single fungus inoculation than with the two fungi inoculation. Dressing of the fungi-infected seeds with Seedplus® 30 WS significantly improved germination percentage and seedling vigour of watermelon ($P \leq 0.05$) but not as high as that of the control.**

**Key words:** *Mucor racemosus*, *Rhizopus nigricans,* Seedplus.

## INTRODUCTION

In Nigeria, the largest production of watermelon (*Citrullus lanatus* Thumb Family Cucurbitaceae) comes from the northern part of the country, where the suitable agro-ecology is found (IITA, 2007). Watermelon is cultivated extensively for its pleasant-tasting nature. Its fruits are mostly threatened by some pathogenic fungi which are seed-borne and also soil-borne such as *Mucor* spp. and *Rhizopus* spp. (Pamela and Tom, 2006). The effects of such fungi on the seedlings include poor germination, low seedling vigour and even complete failure of seedlings. These usually result in low yield and low income arising from poor yield quantity and quality.

Although much work has been done on investigating and discovery of fungal diseases that affect the fruits, there are no substantial report on the effect of pathogenic fungi on the seeds and seedling of watermelon. *Rhizopus stolonifer*, a type of black mould, has a wide host range and can affect over 300 plant species including fruits, vegetables and ornamentals (Farr et al., 2007). *Mucor* spp. which is a soil-borne pathogen may infect the fruit and stem of several plants such as pears and apple (Michailides, and Spotts, 1990). This study assessed the effect of *Mucor racemosus* and *Rhizopus nigricans* individually and in combination, and confirmed the efficacy of Seedplus® fungicide on germination and seedling of infected watermelon seeds.

**Table 1.** Effects of inoculum density and types of pathogenic mould on germination percentage of watermelon.

| Treatment | Germination percentage | | | |
|---|---|---|---|---|
| | 5DAS (%) | 6DAS (%) | 7DAS (%) | 8DAS (%) |
| *M. racemosus* + *R. nigricans* (g $L^{-1}$) | 46.67$^{b*}$ | 46.67$^{c}$ | 46.67$^{c}$ | 46. 67$^{c}$ |
| *M. racemosus* + *R. nigricans* (0.1g$L^{-1}$) | 60.00$^{ab}$ | 60.00$^{bc}$ | 60.00$^{bc}$ | 60. 00$^{bc}$ |
| *M. racemosus* (g $L^{-1}$) | 86.67$^{a}$ | 93.33$^{a}$ | 93.33$^{a}$ | 93.33$^{a}$ |
| *R. nigricans* (g $L^{-1}$) | 73.33$^{ab}$ | 73.33$^{abc}$ | 73.33$^{abc}$ | 73.33$^{ab}$ |
| *M. racemosus* (0.1 g $L^{-1}$) | 86.67$^{a}$ | 93.33$^{a}$ | 93.33$^{a}$ | 93.33$^{a}$ |
| *R. nigricans* (0.1 g $L^{-1}$) | 86.67$^{a}$ | 93.33$^{a}$ | 93.33$^{a}$ | 93.33$^{a}$ |
| *M. racemosus* + *R. nigricans* (g $L^{-1}$) + Seedplus® | 78.00$^{ab}$ | 80.00$^{ab}$ | 80.00$^{ab}$ | 80.00$^{ab}$ |
| Control | 88.33$^{a}$ | 93.33$^{a}$ | 93.33$^{a}$ | 93.33$^{a}$ |

*Within each column, means followed by the same letter(s) are not significantly different (P ≤ 0.05) according to Duncan Test. **DAS, Days after sowing.

## MATERIALS AND METHODS

### Collection of seed sample

Watermelon seeds (cv Chaliston gray commonly grown in Nigeria) were obtained from Gwadabe market, Minna, Nigeria (Lat. 4° 30$^1$ N of the Equator and Long. 10° 30$^1$ E) where most farmers procure their planting seeds.

### Inoculum production and identification

Confirmed strains of *R. nigricans* and *M. racemosus* by CABI Biosciences Identification Services (IMI 392668 & 392668) were maintained on cowpea seeds kept in Biochemistry Laboratory. The infected seeds samples were aseptically placed in 90 mm diameter Petri dishes containing 15 ml each of an autoclaved Potato Dextrose Agar, (PDA, Oxoid) added with 0.05 g $l^{-1}$ chloramphenicol. It was incubated in lamina hood at 28°C and examined from 2-3 days in order to obtain the pure culture of the inoculum. On the third day after incubation, hyphal fragments inoculum was prepared by flooding the surface of the agar slant with sterile distilled water and gently scraping the surface of the sporing surface with a loop. Hyphal structures which are germ tubes were at least five times as long as the diameter of the spores. The resulting suspension with spores was then filtered off through sterile gauze. To reconfirm the identity of the fungi, mycelia speck from each colony were aseptically placed on a slide, stained with lactophenol blue, covered with slips and viewed under microscope (40x). The identification was accomplished using fungi catalogue in the Microbiology Department in F.U.T, Minna.

### Inoculum quantification and inoculation

The hyphal strands of each fungus were diluted by adding sterile distilled water 1 g and 0.1g $l^{-1}$ water to obtain the working suspensions. One hundred seeds (5 g) were pre-inoculated with 2 ml of each of the hyphal suspension concentrations by soaking for 20 min, that is, 0.4 L $kg^{-1}$ seeds. The treatments included seeds pre-inoculated with high and low density of *M. racemosus* and *R. nigricans* (g 1 $L^{-1}$) and 0.1g $L^{-1}$ distilled water as follows:

*M. racemosus* (g $L^{-1}$ ) + *R. nigricans* g $L^{-1}$; *M. racemosus* (0.1 g $L^{-1}$) + *R. nigricans* (0.1 g $L^{-1}$) *M. racemosus* (g $L^{-1}$ ), *R. nigricans* (g $L^{-1}$), *M. racemosus* (0.1 g $L^{-1}$), *R. nigricans* (0.1 g $L^{-1}$) and *M. racemosus* g $L^{-1}$+ *R. nigricans* (g $L^{-1}$) plus (10% imidacloprid + 10% metalaxyl + 10% carbendazim 2.5 g $kg^{-1}$ seeds (Seedplus® 30 WS, Jiangsu Flag Industry Co., Ltd, Nianjing, China) and the control with uninoculated seeds.

A completely randomized experimental design with three replicates was involved. Ten seeds each were placed in 24 Petri dishes containing three layers of blotters moistened with 10 ml distilled water, set to germinate at 28 ± 2°C in the incubator and observed for eleven days for germination, fungal infection and seedling vigour.

Germination percentage, length of root and length of plumule were recorded daily as from the fifth to eleventh day. Germination percentage, length of radicle and length of plumule were recorded daily as from 5 to 11 days after sowing (DAS). Vigour index was calculated at 8 DAS according to Randahawa et al. (1985) and modified as follows:

$$VI = (PL + RL) \times GP$$

Where VI = Vigour index; PL = plumule length (cm); RL = radicle length (cm); GP = germination percentage (%).

Data collected were subjected to analysis of variance (ANOVA) and means were separated with Duncan multiple range test (DMRT) and *P* values <0.05 were considered statistically significant.

## RESULTS

At five days after sowing (DAS), the germination percentage of watermelon seeds inoculated with high inoculum density (*M. racemosus* + *R. nigricans* (g $L^{-1}$) had the lowest percentage germination (46.67%) (Table 1).

The highest germination percentage was observed in the control (88.33%) and this was significantly different (P ≤ 0.05) from the seeds applied with *M. racemosus* + *R. nigricans* (g$L^{-1}$). At 6, 7 and 8 DAS, germination percentage of seeds applied with *M. racemosus* + *R. nigricans* (g $L^{-1}$ or 0.1g$L^{-1}$) was significantly lower (P≤ 0.05) than that of other treatments (Table 2).

At 5 DAS, length of plumule (2.45 cm) from the seeds applied with *M. racemosus* + *R. nigricans* (g $L^{-1}$) was significantly lower (P≤ 0.05) than those applied with *M. racemosus* + *R. nigricans* (0.1g $L^{-1}$). The plumule length

**Table 2.** Effect of inoculum density and types of pathogenic mould on plumule length of watermelon seeds.

| Treatment | Plumule length (cm) | | | |
|---|---|---|---|---|
| | 5 DAS** | 6DAS | 7DAS | 8DAS |
| *M. racemosus* + *R. nigricans* (g L$^{-1}$) | 3.45$^{b*}$ | 3.55$^{b}$ | 4.27$^{b}$ | 4.50$^{c}$ |
| *M. racemosus* + *R. nigricans* (0.1 gL$^{-1}$) | 4.78$^{ab}$ | 4. 88$^{ab}$ | 7.30$^{a}$ | 7.86$^{bc}$ |
| *M. racemosus* (g L$^{-1}$) | 6.48$^{a}$ | 6.48$^{a}$ | 9.07$^{a}$ | 11.26$^{ab}$ |
| *R. nigricans* (g L$^{-1}$) | 5.72$^{a}$ | 5.72$^{a}$ | 8.40$^{a}$ | 9.50$^{ab}$ |
| *M. racemosus* (0.1 g L$^{-1}$) | 6.60$^{a}$ | 6.70$^{a}$ | 10.03$^{a}$ | 11.10$^{ab}$ |
| *R. nigricans* (0.1 g L$^{-1}$) | 5.45$^{a}$ | 5.45$^{a}$ | 7.77$^{a}$ | 10.90$^{ab}$ |
| *M. racemosus* + *R. nigricans* (g L$^{-1}$) + Seedplus® | 5.33$^{a}$ | 5.33$^{a}$ | 10.97$^{a}$ | 11.33$^{ab}$ |
| Control | 6.75$^{a}$ | 8.95$^{a}$ | 11.50$^{a}$ | 12.50$^{a}$ |

*Within each column, means followed by the same letter(s) are not significantly different (P ≤ 0.05) according to Duncan Test. **DAS, Days after sowing.

**Table 3.** Effects of inoculum density and types of pathogenic mould on radicle length vigour indices of watermelon seeds.

| Treatment | Radicle length (cm) | | | | V.I*** (x100) |
|---|---|---|---|---|---|
| | 5 DAS** | 6 DAS | 7 DAS | 8DAS | 8DAS |
| *M. racemosus* + *R. nigricans* (gL$^{-1}$) | 3.17$^{a*}$ | 4.47$^{a}$ | 4.57$^{a}$ | 4.83$^{a}$ | 0.39$^{d}$ |
| *M. racemosus* + *R. nigricans* (0.1 gL$^{-1}$) | 4.40$^{a}$ | 5.90$^{ab}$ | 6.37$^{ab}$ | 6.87$^{ab}$ | 0.89$^{c}$ |
| *M. racemosus* (g L$^{-1}$) | 5.25$^{b}$ | 7.00$^{b}$ | 7.40$^{b}$ | 7.93$^{ab}$ | 1.53$^{ab}$ |
| *R. nigricans* (g L$^{-1}$) | 4.50$^{a}$ | 6.38$^{b}$ | 6.47$^{ab}$ | 7.23$^{a}$ | 1.84$^{a}$ |
| *M. racemosus* (0.1 g L$^{-1}$) | 5.67$^{b}$ | 7.48$^{b}$ | 7.96$^{b}$ | 8.70$^{b}$ | 1.21$^{bc}$ |
| *R. nigricans* (0.1 g L$^{-1}$) | 6.37$^{b}$ | 5.48$^{ab}$ | 6.53$^{ab}$ | 7.20$^{a b}$ | 1.68$^{ab}$ |
| *M. racemosus* + *R. nigricans*(g L$^{-1}$)+ Seedplus® | 5.38$^{b}$ | 7.38$^{b}$ | 8.30$^{b}$ | 8.68$^{b}$ | 1.60$^{ab}$ |
| Control | 6.38 $^{b}$ | 7.10$^{b}$ | 8.06$^{b}$ | 8.70$^{b}$ | 1.99$^{a}$ |

*Within each column, means followed by the same letter(s) are not significantly different (P ≤ 0.05) according to Duncan Test. **DAS, Days after sowing, ***V.I. = vigour index.

of seeds in the control was the highest (6.75 cm) but was only significantly higher (P≤ 0.05) than for seeds applied with *M. racemosus* + *R. nigricans* (g L$^{-1}$). At 6 DAS, seeds applied with *M. racemosus* + *R. nigricans* (g L$^{-1}$) had the least plumule length (3.55 cm) and this was significantly lower (P≤ 0.05) than for all other treatments except those applied with *M. racemosus* + *R. nigricans* (0.1g L$^{-1}$). At 7 and 8 DAS, seeds in the control had the highest plumule length (11.50 and 12.50 cm, respectively). This was significantly higher (P≤ 0.05) than for seeds applied with *M. racemosus* + *R. nigricans* (g L$^{-1}$) (Table 3).

At 5 DAS, radicle growth was highest in the control but not significantly different (P≤ 0.05) from fungi inoculated seeds applied with Seedplus®. This was significantly higher (P≤ 0.05) than for seeds applied with *M. racemosus* + *R. nigricans* (g L$^{-1}$) and *R. nigricans* only (g L$^{-1}$). At 6 and 7 DAS, radicle length of seeds applied with *M. racemosus* + *R. nigrican s* (g L$^{-1}$) was significantly lower (P≤ 0.05) than for those applied with *M. racemosus* only (0.1 or g L$^{-1}$), Seedplus® and the control. At 8 DAS, seeds applied with *M. racemosus* + *R. nigricans* (g L$^{-1}$) had the least radicle length (4.83 cm) and was significantly lower (P≤ 0.05) than those applied with *M. racemosus* (0.1 g L$^{-1}$), Seedplus and the control. Seedling vigour index was significantly lower (P≤ 0.05) in seeds applied with *M. racemosus* + *R. nigricans* (g L$^{-1}$) than for those applied with lower inoculum density (0.1 g L$^{-1}$). Highest vigour index was obtained in the control (1.99) and this was not significantly different (P≤ 0.05) from the inoculated seeds with the fungi but with Seedplus® treatment.

## DISCUSSION

Water melon seeds inoculated with the fungi without fungicide treatment in this study exhibited some pathogenic symptoms such as root rot. Moss and Smith (2006) earlier reported that pathogenic seed-borne fungi include *R. nigricans*, *Mucor* spp. and *Fusarium oxysporum*. Mehrotra and Aggarwal (2003) reported that such fungi could seriously retard seed germination through softening and necrosis of tissues. They also confirmed the association of seed-borne fungi with seed viability, wilting of plants and stem flaccidity. Incidences of *R. nigricans* and several other pathogenic seed-borne fungi on seeds have been reported by Leslie et al. (2005) and Anjorin et al. (2008). The factors influencing the development of seed-borne fungi include the moisture content of the seed, prevailing temperature, storage period and degree of seed invasion with the pathogen). Others are level of host genetic resistance, activities of insects and mites and amount of foreign materials in the seed lot (Miller and Trenholen, 1994).

The inhibition of radicle and plumule growth especially by seeds applied with high inoculum density led to lower germination percentage of up to 50% (Linn and Ehret, 1991; Gilbert and Tekauz, 1995; Menzies et al., 1996). Pathogenenic fungi may only be present at such low density such that their inoculum potential is low. Thus infected plants may not show significant symptoms of infection even though the pathogen is present in their cell or tissue (Sanogo and Moorman, 1993). Higher inoculum density can overcome the plant's defense mechanisms and cause death. However, the actual level of inoculum needed to overcome the host's defenses would vary with the environmental conditions (Paternotte, 1992). Whether this is because of an increased severity of infection caused by increased number of primary infections at few sites on the roots, or an increased number of infection sites along the roots, is not known. It may be that the seeds applied with low inoculum density are able to overcome the low level of infections, and once the defence responses are activated the plants are able to tolerate the infections that occurred from secondary inoculum produced from primary infections (Menzies et al., 1996). A strong linear relationship between pathogen inoculum density and growth and yield parameters monitored by Sanogo and Moorman (1993) are good indications of how the increasing density of the pathogen inoculum increases the stress on the host plants.

The relatively higher radicle growth observed on the fungal inoculated seeds applied with imidacloprid, metalaxyl and carbendazim (Seedplus®) and also in the control was in line with the submission of Paternotte (1992) on disease development of *Pythium* in glass house cucumber. Seeds treated with Seedplus® recorded relatively high percentage germination. This could be due to the systemic action of Seedplus® which is strongly inhibitory to hyphal development and fungal spore germination (Maynard and Hopkins, 1999). This supports the fact that this seed dressing fungicide was effective in the control of the fungi infection (Freshpatents.com, 2005).

## Conclusion

This study confirms that *M. racemosus* and *R. nigricans* were capable of causing considerable low germination and seedling vigour in watermelon. It was also confirmed that the higher the inoculum density especially when the fungi are combined, the greater the pathogenic capacity on watermelon seeds. Thus, infected seeds with low inoculum density had relatively higher germination percentage, plumule and radicle length and seedling vigour than those applied with high inoculum density. Better germination and seedling vigour was observed when only one fungus was involved than when two fungi were involved. Hence inoculum level on seeds should be reduced to the tolerable level through the use of seed dressing fungicide such as imidacloprid, metalaxyl and carbendazim (Seedplus®) before sowing. This could greatly improve radicle and plumule development, germination and vigour indices of watermelon seeds. Farmers should preferably sow watermelon seeds that are clean, fungicide-dressed or fungal-resistant.

## Conflict of Interests

The author(s) have not declared any conflict of interests.

### REFERENCES


Anjorin ST, Hussaini AM, Titilayo A, Ibrahim K (2008). Effects of *Fusarium verticilloides*, its metabolites and neem leaf extract on germination and vigour indices of maize (*Zea mays* L.). Afr. J. Biotech. 7(14):2402-2406.

Farr DF, Rossman AY, Palm ME, McCray EB (2008). Fungal Databases, Systematic Botany & Mycology Laboratory, ARS, USDA. Retrieved May 3, from http://nt.ars-grin.gov/fungaldatabases/

Freshpatents.com (2005). http:www.freshpatents.com/aqueous. composition for seed treatment and use. NFDC, TVA, Muscle Shoals, Alabama. Bull. Y-69:46-62.

Gilbert J, Tekauz A (1995). Effects of *Phaeosphaeria nodorum*-induced seed shriveling on subsequent wheat emergence and plant growth. Euphytica 82:9-16.

IITA (2007). Watermelon production in Nigeria. A practical guide pp. 3-15.

Leslie JF, Zeller KA, Lampecht SC, Rheeder JP, Marasa WFO (2005). Toxicity, pathogenic and differentiation of five spp. of *Fusarium* from sorghum and millet. Phytopathology 95:275-285.

Linn WC, Ehret DL (1991). Nutrient concentration and fruit thinning effect on shelf life of long English Cucumber. Hortsci. 26:1299-1300.

Maynard DN, Hopkins DL (1999). Watermelon fruits disorders. Hort.Technol. 9(2): 155-161.

Mehrotra RS, Aggarwal A (2003). Plant pathology. Teta Mc Graw-hill Publishing Company New Delhi. Second edition. p.254.

Menzies JG, Ehret DL, Stan S (1996). Effects of inoculum density of *Pythiumaphanidermatum* on the growth and yield Cucumber plants grown in recirculating nutrient film culture. Can. J. Plant Pathol. 18:50-54.

Michailides TJ, Spotts RA (1990). Postharvest diseases of pome and stone fruits caused by *Mucor piriformis* in the Pacific Northwest and California. Plant Dis. 74:537-543.

Miller JD, Trenholen HL (1994). Mycotoxins in grain compounds. Eagan Press USA. pp. 541.

Moss MO, Smith JE (2006). Mycotoxins: Formulation, Analysis and Significance. John Willey and Sons. Chichester, Britain. pp.143.

Pamela R, Tom K (2006). Plant Disease Management Guide-V3-55. one of a series of the department of Plant Pathology, University of Florida.

Randahawa HS, Sharma HL, Kaur J, Dalhiwal AS (1985). Effects of fungicide on seed germination and seed mycofloras of wheat under different storage conditions. Pestic. 2:36-38.

Sanogo S, Moorman GW (1993). Transmission and control of *Phytium aphanidermatum* in an ebb and flow sub irrigation system. Plant Dis. 77:287-290.

# Effect of growing media on seed germination and seedling growth of papaya cv. 'Red lady'

**R. L. Bhardwaj**

Krishi Vigyan Kendra, Sirohi, Agriculture University Jodhpur-342304, Rajasthan, India.

**This study was carried out to explore the effect of growing media on seed germination and seedling growth of papaya (*Carrica papaya*). The research was conducted at the model nursery of Krishi Vigyan Kendra, Sirohi during two successive seasons from July to August 2008-2009 and 2009-2010. This research was arranged in a complete randomized design with nine treatment combinations and three replications. The treatments were combination between types of media with level of cocopeat. Each treatment has 100 poly bags. The results showed that the medium of vermicompost + sand + pond soil (1:1:1) with 2 cm cocopeat in top of the poly bags ($T_9$) gave maximum speed of emergence (277.6 and 709.09), highest germination percent (95.27 and 90.15%), highest seed vigour (91.97 and 86.69), maximum germination index (7.15 and 7.22), germination value (17.33 and 33.83) and least time required for imbibition (9.45 and 9.30 days) and minimum germination period (3.70 and 2.75 days), respectively in both years of experimentation. This medium was also found to be the best medium for the growth of papaya seedlings as it gave the highest parameters in term of seedling height (23.43 and 22.67 cm), leaf area (349.33 and 329.20 $cm^2$), number of leaves (10.02 and 9.67), stem girth (3.16 and 3.48 mm), number of roots (17.20 and 16.17), root length (10.20 and 9.67 cm), production of total biomass (5.02 and 4.77 g/plant) and least root/shoot ratio (0.22 and 0.20). This treatment also significantly reduces the seedling mortality and produce maximum healthy seedlings (92.23 and 93.15%) in minimum days (35.33 and 35.15) with highest net profit (Rs. 3493.30/1000 seedling and Rs. 3448.00/ 1000 seedling) and B:C ratio (1.85 and 1.84) of seedlings, in both years (2008-09 and 2009-10), respectively.**

**Key words:** Seedling, plant growth, cocopeat, pond soil, vermicompost, B:C ratio.

## INTRODUCTION

Use of suitable growing media or substrates is essential for production of quality horticultural crops. It directly affects the development and later maintenance of the extensive functional rooting system. A good growing medium would provides sufficient anchorage or support to the plant, serves as reservoir for nutrients and water, allow oxygen diffusion to the roots and permit gaseous exchange between the roots and atmosphere outside the root substrate (Abad et al., 2002). Nursery potting media influence quality of seedlings produced (Agbo and

Omaliko, 2006). The quality of seedling obtained from a nursery influences re-establishment in the field and the eventual productivity of an orchard (Baiyeri, 2006). Papaya is an important fruit crop which is propagated by seeds only. Seed germination is affected by many factors, which include type of substrate used, environmental factors such as oxygen, water, temperature and for some plant species, light (Hartmann et al., 2001). The germination of seed of papaya (*Carrica papaya*) is frequently reported to be slow, erratic and is incomplete (Chako and Lange, 1966). Red lady is choicest variety of papaya grower due to hermaphrodite nature and prolonged self life of fruits. But the seed cost of this variety is very high (Rs. 2.0 lakh $kg^{-1}$) so, increasing germination percent and producing more healthy seedling is a challenge for papaya growers. The germination of papaya (*C. papaya*) cv. Red lady seeds faces certain problems in germination and has high seedling mortality due to damping off disease in nursery stage. Initial mortality and incomplete germination is also one of the causes of reduced survival percent of papaya plants. In heavy soil without enough drainage, the development of root system is suppressed and plants are more susceptible to soil borne diseases (Beattie and White, 1992). The papaya seed is endosed within a gelatinous sarcotesta (aril or outer seed coat which is formed from the outer integument). Growing media also plays important role for seed germination. Growing medium not only acts as a growing place but also as a source of nutrient for plant growth. Media composition used influences the quality of seedling (Wilson et al., 2001). Generally, media for fruit crop seedling are composed of soil, organic matter, pond soil and sand. The pond soil is usually used as a basic medium because it is cheapest and easy to procure. Supplementing of the sand is aimed to make media more porous while the organic matter (FYM and vermicompost) is added so as to enrich adequate nutrients for the seedling. There is better relationship between the manure and rooting rather than conventional soil mix and less susceptibility of the seedling to soil borne pests and diseases (Akanbi et al., 2002). Several studies on growth media had been conducted on the various fruit commodities by previous researchers. The best growth of mangosteen seedling was reached on soil medium as compared to the other media (Jawal et al., 1998). Baiyeri (2003) mentioned that the best seedling qualities of African breadfruit (*Treculia africana* Decne) were obtained when grown in medium formulated with top soil + poultry manure + river sand in 1:2:3 (v/v/v) ratios. Humic acids (vermicompost) applied in the medium increased plant height, leaf area and dry weight of peppers, tomatoes and marigold (Arancon et al., 2004). Ratna et al. (2006) working on banana cv. Raja Serai proved that soil and sand medium was the most suitable medium for shoot and leaf growth of this banana. *Uapaca kirkiana* Müell Arg. planted on the medium comprising 75% forest soil and 25% sawdust produced the tallest seedlings, larger root collar diameter and higher survival at 10 months after planting (Mhango et al., 2008).

Cocopeat is an agricultural by-product obtained after the extraction of fiber from the coconut husk (Abad et al., 2002). As a growing medium, cocopeat can be used to produce a number of crop species with acceptable quality in the tropics (Yahya and Mohklas, 1999; Yau and Murphy, 2000). Cocopeat is considered as a good growing media component with acceptable pH, electrical conductivity and other chemical attributes (Abad et al., 2002). Cocopeat has good physical properties, high total pore space, high water content, low shrinkage, low bulk density and slow biodegradation (Evans et al., 1996; Prasad, 1997). The results of many experiments revealed that cocopeat used alone, or as a component of soil medium, is suitable for roses (Blom 1999), gerbera (Labeke and Dambre, 1998), many potted plants (De Kreij and Leeuven, 2001; Treder and Nowak, 2002) and also for vegetables. Due to usually high initial level of potassium and sodium in cocopeat, the fertilization program should be adjusted carefully to plant requirements.

Keeping in view the influence of media in germination and seedling growth of papaya, the present investigation was carried out to study the effect of different media viz. sand, pond soil, FYM, vermicompost and cocopeat on seed germination, seedling growth and vigour of papaya seedlings.

## MATERIALS AND METHODS

### Seed material and treatment

Seed germination and seedling growth experiments of papaya were carried out at NHM Model nursery of Krishi Vigyan Kendra-Sirohi (Rajasthan) during two successive season from July to August, 2008-09 and 2009-10. Experimental treatments comprised of nine treatment combinations consisting of different combination of growth media and cocopeat filling at the top of seedling poly bags (10 x 12 cm) namely, $T_1$- Sand + pond soil (1:1) without cocopeat, $T_2$- Sand + pond soil (1:1) with 1 cm cocopeat, $T_3$- Sand + pond soil (1:1) with 2 cm cocopeat, $T_4$- FYM + Sand + pond soil (1:1:1) without cocopeat, $T_5$- FYM + Sand + pond soil (1:1:1) with 1 cm cocopeat, $T_6$- FYM + Sand + pond soil (1:1:1) with 2 cm cocopeat, $T_7$- Vermicompost + sand + pond soil (1:1:1) without cocopeat, $T_8$- Vermicompost + sand + pond soil (1:1:1) with 1 cm cocopeat, $T_9$- Vermicompost + sand + pond soil (1:1:1) with 2 cm cocopeat. The seed sowing was done in month of July about 1 cm deep in different media as per treatments. The poly bags were irrigated immediately after seed sowing and repeated every day till the final emergence. After the completion of germination, the bags were irrigated once in 2 days.

### Experimental design and measured parameters

For seed germination and seedling growth experiments, treatments of the experiment were conducted in complete randomized design with three replications. Each treatment was composed of 100 polybags seedlings. All the observation on germination parameters were recorded at the time of germination and growth parameter at the time of transplanting (45 days after seed sowing) from 100 seeds for germination parameter and randomly selected 10 seedling for growth parameters. Data on germination was recorded from the first germination until no further germination at two days interval. The imbibition period, number of days from sowing to commencement of germination, was recorded for all studied treatments. The rate of emergence (RE) was calculated according to Islam et al. (2009) using the following formula:

$$\text{Rate of emergence} = \frac{\text{No. of seedlings emerged 5 days after sowing}}{\text{No. of seedlings emerged 15 days after sowing}} \times 100$$

The germination percentage was calculated as the percent of germinating seeds starting from the first germination to no further germination. Germination percentage was calculated by number of germinated seedling divided by the total number of seeds sown in poly bags and multiplied by 100. The germination period was calculated as the difference between initial and final emergence (number of days) recorded. Seed vigour was calculated by total number of healthy seedling divided by the number of total seedlings and multiplied by 100. The germination index was calculated as described in the Association of Official Seed Analysis (1983) by the following formula:

$$\text{Germination index} = \frac{\text{No. of germinating seeds}}{\text{Days of first count}} + \frac{\text{No. of germinating seeds}}{\text{Days of final or last count}}$$

The germination value (GV) was calculated according to Hossain et al. (2005) by the following formula: Germination value = $(\Sigma\ DGs/N) \times GP/10$. Where (GP) is the germination percentage at the end of experiment, (DG) is the daily germination speed obtained by dividing the cumulative germination percentage by the number of days since sowing, ($\Sigma$ DGs) is the total germination obtained by adding every DGs value obtained from the daily counts, (N) is the total number of daily counts starting from the first germination and (10) is constant. Counting of number of leaves was done at the end of experiment (45 days after seed sowing) when the true leaves have emerged. Stem girth was measured 1 cm from the base of the stem using vernier caliper. Plant height was measured from poly bag top soil surface upto the highest leaf tip by straightening all leaves. Leaf area was calculated by the leaves traced on a graph paper. Number of roots, root length was measured by destructive method of uprooting the plants and taking measurement by standard method. Stem and root were weighed to record stem, root fresh weight, root/shoot ratio, and total weight per plant (g) was recorded at time of transplanting. Survival percent (after transplanting in main field) was recorded by following formula:

$$\text{Survival percent} = \frac{\text{Total survival transplanted plants}}{\text{Total transplanted plants}} \times 100$$

The net return was calculated by subtracting cost of each treatment from the gross return and benefit: cost ratio = Gross income/cost of seedling production. All data was subjected to analysis of variance (ANOVA) to determine significant differences and comparison of means at significant level of 5%.

## RESULTS AND DISCUSSION

The results showed that growing media and cocopeat had beneficial effect on seed germination and growth of papaya seedling.

### Seed germination parameters

Seed germination parameters of papaya (*C. papaya*) as affected by growing media and use of cocopeat are presented in Table 1. The treatment $T_9$ was found to be best followed by $T_8$ with regard to germination behaviour as these media have suitable physical properties and good water holding capacity that supports the germination of papaya seeds (Table 1). Coir dust when mixed with organic manure is the best media as coir dust has good physical characteristics (Garcia and Daverede, 1994) and also successfully tested as a growing medium in ornamentals (Van Holm, 1993). Germination started at the 9.45 and 9.30 days after sowing on vermicompost medium with 2 cm cocopeat ($T_9$) for both year of experimentation, respectively. Germination continued until the 25.72 and 22.03 days from sowing where no further germination was noticed in both year of experimentation, respectively. For both year of experimentation, the maximum speed of emergence (277.6 and 709.0), highest germination percent (95.27 and 90.15%), highest seed vigour (91.97 and 86.69), maximum germination index (7.15 and 7.22), germination value (17.33 and 33.83) and least time required for imbibition (9.45 and 9.30 days) and minimum germination period (3.70 and 2.75 days) were obtained in vermicompost + sand + pond soil (1:1:1) with 2 cm filling with cocopeat of seedling poly bags ($T_9$) in both years of experimentation, respectively. The sand + pond soil (1:1) without cocopeat showed the least results in most cases. The vermicompost medium with 2 cm cocopeat allowed increased germination parameters from the beginning to the end of experiment as compared to other media

**Table 1.** Effect of seedling growing media and cocopeat on the germination parameters of papaya (*C. papaya*).

| Treatment | Imbibition period | | Speed of emergence | | Germination (%) | | Germination period | | Seed vigour | | Germination index | | Germination value | |
|---|---|---|---|---|---|---|---|---|---|---|---|---|---|---|
| | 2008-2009 | 2009-2010 | 2008-2009 | 2009-2010 | 2008-2009 | 2009-2010 | 2008-2009 | 2009-2010 | 2008-2009 | 2009-2010 | 2008-2009 | 2009-2010 | 2008-2009 | 2009-2010 |
| $T_1$ | 16.67 | 14.67 | 129.51 | 125.22 | 61.70 | 57.67 | 9.05 | 7.35 | 52.80 | 57.67 | 2.00 | 2.73 | 1.66 | 2.21 |
| $T_2$ | 14.60 | 12.60 | 133.12 | 205.39 | 69.14 | 66.50 | 7.35 | 6.60 | 62.72 | 65.60 | 2.81 | 3.46 | 3.20 | 3.79 |
| $T_3$ | 12.35 | 11.75 | 166.74 | 285.68 | 79.36 | 81.30 | 6.50 | 5.25 | 73.78 | 77.69 | 3.70 | 4.97 | 5.64 | 8.02 |
| $T_4$ | 16.45 | 14.25 | 152.02 | 156.21 | 69.45 | 72.13 | 7.90 | 6.65 | 62.08 | 67.30 | 2.94 | 3.53 | 2.53 | 4.20 |
| $T_5$ | 13.55 | 13.62 | 188.06 | 366.19 | 79.70 | 83.40 | 5.65 | 4.10 | 73.18 | 80.33 | 4.23 | 4.78 | 5.59 | 9.84 |
| $T_6$ | 11.75 | 11.42 | 206.08 | 445.04 | 84.65 | 86.65 | 4.72 | 3.17 | 80.22 | 85.39 | 5.30 | 5.60 | 9.01 | 15.92 |
| $T_7$ | 14.30 | 13.38 | 231.47 | 273.54 | 80.33 | 74.90 | 6.95 | 5.35 | 73.92 | 73.60 | 3.86 | 3.51 | 4.44 | 5.59 |
| $T_8$ | 11.13 | 12.12 | 257.91 | 603.16 | 90.13 | 85.00 | 4.55 | 3.60 | 84.90 | 81.62 | 5.69 | 6.29 | 10.35 | 18.64 |
| $T_9$ | 9.45 | 9.30 | 277.65 | 709.09 | 95.27 | 90.15 | 3.70 | 2.75 | 91.97 | 86.69 | 7.15 | 7.22 | 17.33 | 33.83 |
| SEM± | 0.377 | 0.618 | 5.874 | 12.833 | 1.830 | 2.173 | 0.264 | 0.265 | 1.804 | 1.853 | 0.224 | 0.136 | 0.278 | 0.278 |
| CD at 5% | 1.117 | 1.829 | 17.385 | 37.979 | 5.416 | 6.432 | 0.783 | 0.786 | 5.338 | 5.484 | 0.665 | 0.404 | 0.822 | 0.834 |

combination in both years of experimentation. The reason for the best performance of pond soil and vermicompost are high organic matter content which increases the water and nutrient holding capacity of the medium, which improve the water utilization capacity of plant. Joiner and Nell (1982) found similar results in peat + perlite mixture for Aglaonema and Dieffenbachia. Vermicompost is reported to have bioactive principles which are considered to be beneficial for root growth and this has been hypothesized to result in greater root initiation, higher germination, increased biomass, enhanced growth and development (Bachman and Metzger, 2008) and also balanced composition of nutrients (Zaller, 2007). The higher available well decomposed organic matter (vermicompost) may preserve soil humidity, increase nutrient content and improve soil structure which increase water absorption and maintains the cell turgidity, cell elongation and increase respiration at optimum level, leading to favourable seed sprouting. Vermicompost mixed with pond soil affects properties of soil physics, chemistry and biology, since organic matter acts as glue for soil aggregate and source of soil nutrient (Soepardi, 1983). Vermicompost granules may develop soil aggregate and it's granulating. Soil aggregation will improve permeability and airflow in the polybags. Vermicompost and pond soil (due to high organic matter) may decrease fluctuation of soil temperature. Further, seed germination and root growth becomes easier to the particular depth so that plant grows well and may absorb more water and nutrient (Jo, 1990). Organic matter may also improve nutrient availability and improve phosphorus absorption (Karama and Manwan, 1990). All these factors are favourable for seed germination and ultimate by increase seed germination percent, speed of emergence, seed vigour, germination index, germination value and reduce imbibition period. Combined application of vermicompost and cocopeat in the treatment $T_9$ showed significant effect on germination, seedling growth and plant biomass probably due to the synergistic combination of both factors in improving physical condition of the media and nutritional factors (Sahni et al., 2008).

## Seedling growth and development parameters

Data presented in Tables 2 and 3 show growth and development of papaya seedling as significantly affected by growing media and cocopeat. Significant differences were observed among the different treatments with regard to seedling growth characters and maximum number of leaves was observed in $T_9$ treatment (10.02 and 9.67) which was at par with $T_6$ treatment (9.20 and 9.00), respectively, in both years (Table 2). Maximum seedling girth (3.16 and 3.48 mm), highest seedling height (23.43 and 22.67 cm), largest leaf area (349.33 and 329.20 $cm^2$), longest root length (10.20 and 9.67 cm) and highest fresh weight of plants (5.02 and 4.77 g) were recorded in $T_9$ treatment in both years of experimentation, respectively. Similarly maximum number of roots per plant was also higher in $T_9$ treatment (17.20 and

**Table 2.** Effect of seedling growing media and cocopeat on the growth parameters of papaya (*C. papaya*) at 45 days after seed sowing.

| Treatment | Number of leaves | | Stem girth (mm) | | Seedling height (cm) | | Leaf area ($cm^2$) | | Number of roots | | Root length (cm) | | Fresh weight of plants (g) | |
|---|---|---|---|---|---|---|---|---|---|---|---|---|---|---|
| | 2008-2009 | 2009-2010 | 2008-09 | 2009-2010 | 2008-2009 | 2009-2010 | 2008-2009 | 2009-10 | 2008-2009 | 2009-10 | 2008-2009 | 2009-2010 | 2008-2009 | 2009-2010 |
| $T_1$ | 4.00 | 3.07 | 0.96 | 1.28 | 7.20 | 9.33 | 18.30 | 15.14 | 5.00 | 6.60 | 3.01 | 3.90 | 0.60 | 0.63 |
| $T_2$ | 6.05 | 5.13 | 1.28 | 1.60 | 8.45 | 10.33 | 33.35 | 30.30 | 9.45 | 10.35 | 3.55 | 4.21 | 0.80 | 0.79 |
| $T_3$ | 7.11 | 8.25 | 1.92 | 2.24 | 10.45 | 12.33 | 51.23 | 50.37 | 12.03 | 12.30 | 6.02 | 5.45 | 1.00 | 1.02 |
| $T_4$ | 7.23 | 6.00 | 1.28 | 1.80 | 9.05 | 9.17 | 41.25 | 59.45 | 7.10 | 7.83 | 5.25 | 4.90 | 0.72 | 0.65 |
| $T_5$ | 8.25 | 8.15 | 2.24 | 2.44 | 12.20 | 12.17 | 84.31 | 99.00 | 11.13 | 9.69 | 7.37 | 6.47 | 1.20 | 0.80 |
| $T_6$ | 9.20 | 9.00 | 2.87 | 2.76 | 17.45 | 17.17 | 134.00 | 134.30 | 12.17 | 12.35 | 7.60 | 6.94 | 1.65 | 1.23 |
| $T_7$ | 8.13 | 6.63 | 1.89 | 2.21 | 16.43 | 11.67 | 126.17 | 149.12 | 13.80 | 12.00 | 8.15 | 7.19 | 2.67 | 2.17 |
| $T_8$ | 9.02 | 7.60 | 2.53 | 2.84 | 20.18 | 19.67 | 216.25 | 249.25 | 16.83 | 15.20 | 9.45 | 8.58 | 3.82 | 3.30 |
| $T_9$ | 10.02 | 9.67 | 3.16 | 3.48 | 23.43 | 22.67 | 349.17 | 329.20 | 17.20 | 16.17 | 10.20 | 9.67 | 5.02 | 4.77 |
| SEm± | 0.318 | 0.386 | 0.083 | 0.119 | 0.422 | 0.448 | 3.457 | 3.457 | 0.581 | 0.536 | 0.218 | 0.196 | 0.046 | 0.051 |
| CD at 5% | 0.941 | 1.142 | 0.245 | 0.353 | 1.251 | 1.326 | 10.232 | 10.232 | 1.721 | 1.587 | 0.647 | 0.580 | 0.137 | 0.152 |

**Table 3.** Effect of seedling growing media and cocopeat on the growth parameters, survival percent, net return and B : C ratio of papaya (*C. papaya*).

| Treatment | Fresh weight of shoot (g) | | Fresh weight of root (g) | | Survival per cent | | Root /Shoot ratio | | Days required for gaining transplanting size of seedling | | Net return (Rs./1000 seedlings) | | B : C ratio | |
|---|---|---|---|---|---|---|---|---|---|---|---|---|---|---|
| | 2008-2009 | 2009-2010 | 2008-2009 | 2009-2010 | 2008-09 | 2009-2010 | 2008-2009 | 2009-2010 | 2008-2009 | 2009-2010 | 2008-2009 | 2009-2010 | 2008-2009 | 2009-2010 |
| $T_1$ | 0.39 | 0.39 | 0.21 | 0.23 | 77.30 | 78.12 | 0.55 | 0.59 | 47.35 | 45.30 | 805.00 | 755.0 | 1.17 | 1.15 |
| $T_2$ | 0.53 | 0.51 | 0.27 | 0.27 | 80.37 | 82.14 | 0.50 | 0.54 | 45.30 | 42.35 | 1005.0 | 955.0 | 1.19 | 1.18 |
| $T_3$ | 0.69 | 0.68 | 0.31 | 0.34 | 84.35 | 86.00 | 0.45 | 0.50 | 42.50 | 41.20 | 1905.0 | 1855.0 | 1.35 | 1.34 |
| $T_4$ | 0.54 | 0.47 | 0.18 | 0.18 | 81.80 | 82.60 | 0.33 | 0.37 | 43.60 | 40.60 | 1093.0 | 1043.0 | 1.20 | 1.19 |
| $T_5$ | 0.95 | 0.61 | 0.27 | 0.19 | 84.88 | 86.65 | 0.28 | 0.31 | 40.50 | 39.50 | 1893.0 | 1843.0 | 1.34 | 1.33 |
| $T_6$ | 1.27 | 0.96 | 0.35 | 0.27 | 86.86 | 90.61 | 0.27 | 0.28 | 38.58 | 37.62 | 2393.0 | 2343.0 | 1.41 | 1.40 |
| $T_7$ | 2.12 | 1.74 | 0.50 | 0.43 | 87.20 | 84.30 | 0.24 | 0.25 | 40.25 | 40.30 | 2193.3 | 2148.0 | 1.65 | 1.64 |
| $T_8$ | 3.11 | 2.77 | 0.71 | 0.60 | 90.25 | 89.35 | 0.23 | 0.22 | 38.30 | 38.35 | 2993.3 | 2948.0 | 1.78 | 1.77 |
| $T_9$ | 4.12 | 3.98 | 0.89 | 0.78 | 92.23 | 93.15 | 0.22 | 0.20 | 35.33 | 35.15 | 3493.3 | 3448.0 | 1.85 | 1.84 |
| SEm± | 0.057 | 0.057 | 0.016 | 0.016 | 1.633 | 1.570 | 0.009 | 0.009 | 1.006 | 1.006 | 60.324 | 45.057 | 0.066 | 0.066 |
| CD at 5% | 0.170 | 1.170 | 0.048 | 0.048 | 4.835 | 4.649 | 0.029 | 0.028 | 2.979 | 2.979 | 178.53 | 133.34 | 0.198 | 0.198 |

16.17) which was at par with $T_8$ treatment (16.83 and 15.20). Highest fresh weight of shoot (4.12 and 3.98 g), fresh weight of roots (0.89 and 0.78 g) and least root/shoot ratio (0.22 and 0.20) was also reported in $T_9$ treatment in both years of experimentation (2008-09 and 2009-10), respectively. Manure (vermicompost) provides adequate nutrients and enhances both the physical properties and the water holding capacity (Soegiman, 1982). Similar result was also reported by Supriyanto et al. (1990) working on orange seedling where media containing manure produced growth and roots better than those containing sawdust and rice hulk. Purbiati et al. (1994) proved that soil + manure (1:1) was the best medium for the growth components of salacca cv. Pondoh and Bali. Merrow (1995) found similar results in sedge peat medium for *Ravenea rivularis*. Combined application of vermicompost and cocopeat in the treatment $T_9$ showed significant effect on seedling growth parameters and plant biomass probably due to the synergistic combination of both factors in improving the physical conditions of the media and nutritional factors (Sahni et al., 2008). This result is akin to the findings of Campos Mota et al. (2009) and Abirami et al. (2010) who suggested that since coir dust is low in nutrients when mixed with vermicompost, provides a better growth medium for plant establishment. However, the air filled porosity (AFP), easily available water (EAW) and aeration of vermicompost and FYM were not at the recommended level which in turn limit the root growth and lowered the water holding capacity. Therefore, the medium with vermicompost and cocopeat is more suitable than vermicompost alone because of the better physical properties and enhanced nutrient level.

This treatment combination was also helpful in reducing damping off disease in seedling due to proper aeration in root zone of the seedling and produce highest survival percent of seedling (92.23 and 93.15%, respectively) which was at par with $T_8$ treatment (90.25 and 93.15%, respectively). Because of the better physical properties and enhanced nutrient level in $T_9$ treatment, growth of seedling are too fast and minimum days required for gaining transplanting size is 35.33 and 35.15 days which is at par with $T_6$ (37.62 days) in second year (2009-10) of experimentation (Table 3). Vermicompost with cocopeat may improve soil porosity, water content, pore of drainage, soil permeability and water availability, whereas weight of soil may decrease. This may develop soil aggregation, and moreover it improves permeability and air flow in the soil, this type of condition sharply reduce damping off disease in nursery stage and provide support to fast growth of the seedling due to availability of better nutrition with water and air in root zone of the seedling ultimately, the seedling gain transplanting size very soon in this treatment combination than other treatments. It seems that good physical and biological conditions in cocopeat and vermicompost had positive effect on root development, which is helpful in increased survival percent of seedling in main field after transplanting. Beneficial effect of cocopeat on root system was observed on nutmeg seedling by Abirami et al. (2010), *Osteospermum* cuttings by Nowak (2004), salvia, viola by Pickering (1997) and *Impatiens* by Smith (1995).

Application of vermicompost : pond soil : sand (1:1:1) with 2 cm cocopeat media ($T_9$) for preparation of papaya seedling proved profitable and showed maximum net return (Rs. 3493.30/1000 seedlings and Rs. 3448.00/1000 seedlings) and benefit : cost ratio (1.85 and 1.84) for the first and second year of experimentation, respectively due to higher germination percent and survival percent obtained (Tables 1 and 3). This treatment was significantly superior to the rest of the treatments during both year but benefit : cost ratio was *at par* with $T_8$ treatment.

## Conflict of Interests

The author(s) have not declared any conflict of interests.

## Conclusion

In conclusion, presented results showed that vermicompost and cocopeat, due to suitable physical, chemical and biological properties could be used successfully in preparation of papaya seedling. On the basis of results obtained from this study, it is concluded that growing media significantly influenced the germination, growth and development parameters of papaya seedling in which medium of vermicompost + pond soil + sand (1:1:1) with 2 cm cocopeat filling of poly bags was the best media since the germination, seedling growth and development parameters were higher than those on the other media. The overall results revealed that media supplemented with cocopeat gave higher parameters of germination, growth and development of papaya seedling as compared to media without cocopeat. Therefore, this result suggested that vermicompost, pond soil and sand with cocopeat should be used as growing media for higher germination percent quickening of papaya seedling growth and earn more profit by sale of seedlings.

## REFERENCES

Abad M, Noguera P, Puchades R, Maquieira A, Noguera V (2002). Physico-chemical and chemical properties of some coconut dusts for use as a peat substitute for containerized ornamental plants. Biores. Technol. 82:241-245.

Abirami K, Rema J, Mathew PA, Srinivasan V, Hamza S (2010). Effect of different propagation media on seed germination, seedling growth and vigour of nutmeg (*Myristica fragrans* Houtt.). J. Med. Plants Res. 4:2054-2058.

Agbo CU, Omaliko CM (2006). Initiation and growth of shoots of *Gongronema latifolia* Benth stem cuttings in different rooting media. Afr. J. Biotechnol. 5:425-428.

AOSA (1983). Seed vigour testing handbook contribution No. 32 to handbook on seed testing. Association of Official Seed Analysis.

Arancon NQ, Lee S, Edwards CA, Atiyeh R (2004). Effect of humic acids derived from cattle, food and paper-waste vermicomposts on growth of greenhouse plants. *Pedobiologia* 47:741-744.

Baiyeri KP (2003). Evaluation of nursery media for seedling emergence and early seedling growth of two tropical tree species. Moor J. Agric. Res. 4:60-65.

Baiyeri KP (2006). Seedling emergence and growth of pawpaw (*Carica papaya*) grown under different coloured shade polyethylene. Int. Agrophys. 20: In press.

Beattie DJ, White JW (1992). Lilium–hybrids and species. In: De Hertogh A.A. and Le Nard M. (eds.), The Physiology of Flower Bulbs. Elsevier, Amsterdam.

Blom TJ (1999). Coco coir versus granulated roockwool and 'arching' versus traditional harvesting of roses in a recirculating system. Acta Hortic. 481:503-507.

Campos Mota L, Van Meeteren U, Blok C (2009). Comparison of physical properties of vermicompost from paper mill sludge and green compost as substitutes for peat based potting media. Acta Hortic. 819:227-234.

Chako EK, Singh RN (1966). Studies on the longevity of Papaya, Phalsa, guava and Mango seeds. *Pro.* Int. Seed Test. Assoc. 36:147-158.

De Kreij C, Leeuven GJL (2001). Growth of pot plants in treated coir dust as compared to peat. Commune Soil Sci. Plant Analy. 32:2255-2265.

Evans MR, Konduru S, Stamps RH (1996). Source variation in physical and chemical properties of coconut coir dust. HortiSci. 31:965-967.

Garcia M, Daverede C (1994). Dust from coir fibres: New substrate for soilless culture. PHM Revue Horticole 348:7-12.

Hartmann HT, Kester DE, Davies FT, Geneve RL (2001). Plant propagation; Principles and Practices 7th Edition, Prentice Hall Publishers, New Jersey.

Hossain MA, Arefin MK, Khan BM, Rahman MA (2005). Effect of seed treatments on germination and seedling growth attributes of horitaki (*Terminalia chebula* Retz.) in the nursery. Res. J. Agric. Biol. Sci. 1:135-141.

Islam AK, Anuar N, Yaakob Z (2009). Effect of genotypes and pre-sowing treatment on seed germination behavior of jatropha. Asian J. Plant Sci. 8:433-439.

Jawal AS, Hadiati S, Susiloadi A, Indriyani NLP (1998). Pengaruh Media Tumbuh terhadap Pertumbuhan Semai Manggis (*Garcinia mangostana* L.). J. Stigma 6:213-218.

Jo I S (1990). Effect of organic fertilizer on soil physical properties and plant growth. Paper presented at seminar on "The use of organic fertilizers in crop production", at Suwon, South Korea, 18-24 June. 1990. pp. 25.

Joiner JN, Nell TA (1982). Fired montmalillonite clay as a propagation and hydroponic growing medium. Proceed. Florida State Hortic. Soc. 93:214-215.

Karama AS, Manwan I (1990). Penggunaan pupuk organic pada tanaman pangan. Makalah pada Lokakarya Nasional Efisiensi Penggunaan Pupuk. Cisarua Bogor, 12-13 November 1990, pp. 44.

Labeke Van MC, Dambre Van P (1998). Gerbera cultivation on coir with recalculating of the nutrient solution: a comparison with roockwool culture. Acta Hortic. 458:357-362.

Mhango J, Akinnifesi FK, Mng'omba SA, Sileshi G (2008). Effect of growing medium on early growth and survival of *Uapaca kirkiana* Müell Arg. seedlings in Malawi. Afr. J. Biotechnol. 7:2197-2202.

Nowak J (2004). The effect of rooting media and $CO_2$ enrichment, P - nutrition and micorrhizal inoculation on rooting and growth of *Osteospermum*. Acta Hortic. 644:589-593.

Pickering JS (1997). An alternative to peat. The Garden 122:428-429.

Prasad M (1997). Physical, chemical and biological properties of coir dust. Acta Hortic. 450:21-29.

Purbiati T, Ernawanto QD, Soemarsono SR (1994). Pengaruh Komposisi Media Tumbuh and Ukuran Pot terhadap Keberhasilan and Pertumbuhan Tunas Anakan Salak yang Diperbanyak secara Vegetatif. Penel. Hortic. 6:1-12.

Ratna TE, Awaludin H, Sutanto A (2006). Pengaruh Media terhadap Pertumbuhan Bibit Pisang Susu Asal Bonggol di Sambelia, Lombok Timur NTB. J. Hortikultura 1:15-22.

Sahni S, Sarma BK, Singh DP, Singh HB, Singh KP (2008). Vermicompost enhances performance of plant growth promoting rhizobacteria in *Cicer arietinum* rhizosphere against *Sclerotium rolfsii.* Crop Prot. 27:369-376.

Smith C (1995). Coir: a viable alternative to peat for potting. Hortic. 4:25-28.

Soegiman (1982). Ilmu tanah. Terjemahan dari. The nature and Properties of Soils. Buckman and Brady. Bhatara Karya Aksara. Jakarta. 788 hal.

Soepardi G (1983). Sifat dan cirri tanah. Department Ilmu-ilmu Tanah, IPB. Bogor.

Supriyanto A, Ernawanto QD, dan Setiono (1990). Media Tumbuh Pembibitan Jeruk. Penelitian Hortikultura 5:1-8.

Treder J, Nowak J (2002). Zastosowanie podłoży kokosowych w uprawie roślin rabatowych. Zesz. Probl. Post. Nauk Rol. 485:335-358 (in Polish with English abstract).

Van Holm L (1993). Coir as a growing medium 7th floricultural symposium, October 11, Institute of fundamental studies: Hantana, Kandy, Srilanka, pp. 1-23.

Wilson SB, Stoffella PJ, Graetz DA (2001). Use of compost as a media amendment for containerized production of two subtropical perennials. J. Environ. Hortic. 19:37-42.

Yahya A, Safie H, Mohklas MS (1999). Growth and flowering responses of potted chrysanthemum in a coir dust-based medium to different rates of slow released-fertilizer. J. Trop. Agric. Food Sci. 27:39-46.

Yau PY, Murphy RJ (2000). Biodegraded cocopeat as a horticultural substrate. Acta Hortic. 517:275-278.

Zaller JG (2007). Vermicompost as a substitute for peat in potting media: Effects on germination, Biomass allocation, Yields and fruit quality of three tomato varieties. Sci. Hortic. 112:191-199.

# 4

# Efficacy of *Moringa oleifera* as a phytoextraction plant in the remediation of heavy metals polluted soil

**Amadi, N. and Tanee, F. B. G.**

Department of Plant Science and Biotechnology, College of Natural and Applied Sciences, University of Port Harcourt, Port Harcourt, Rivers State, Nigeria.

**The efficacy of *Moringa Oleifera*, for the phytoextraction of heavy metals (Zn, Fe and Pb) in tropical soil was investigated. A randomized complete block design (RCBD) consisting of 3 blocks was used. Each block (A, B and C) was polluted with 100 mg Zn, Pb and Fe, respectively. Each block was subdivided into treatment and control of 5 replicates each. Two seedlings of *M. Oleifera* were planted in each treatment plot (phytoextraction) while no planting was done in the control. Zn, Fe and Pb, were analyzed for the soil and plant parts. Results show that *M. oleifera* was able to extract and accumulate high concentration of these metals (Zn, Fe and Pb) in their roots and shoots. The percentage reductions of the heavy metals in the soil were Zn (20.1%), Fe (22.7%) and Pb (64%) after 3 months of phytoextraction. The transfer factor was in the order of Zn > Fe > Pb while translocation factors indicate that metals were largely retained in roots. On the basis of the result obtained, *M. oleifera*, can be classified as hyperaccumulator of some measured heavy metals and therefore, it is suitable for phytoextraction of heavy metal (Zn, Fe and Pb) contaminated soils.**

**Key words:** Heavy metal, contamination, *Moringa oleifera*, phytoextraction, translocation factor, transfer factor.

## INTRODUCTION

Heavy metals are natural constituents of the earth's crust. Heavy metals occur naturally in the ecosystem with large variations in concentrations. In modern times, anthropogenic sources of heavy metals have been introduced to the ecosystem. Heavy metals have an effect on the stability of colloids. Colloids are sensitive to change of ion or heavy metal concentration and it leads to violation of colloid stability and subsequent disintegration (Lane and Morel, 2010).

Heavy metal can occur in any ecosystem- terrestrial or aquatic and when it occurs it affects all components of the ecosystem (biotic and abiotic). The effects range from physical and chemical contaminations of soil, air and water to deleterious impacts on flora and fauna.

Some metals such as Fe, Mn, Zn, Cu, Mg, Mo, and Ni are essential for plant growth in low concentration. However, high concentrations of these heavy metals in soil can negatively affect crop growth, as these metals interfere with metabolic functions in plants such as inhibition of photosynthesis, and respiration and degeneration of main cell organelles, even leading to death of plants (Garbisu and Alkorta, 2003; Schmidt, 2003; Schwartz et al., 2003). Soil contamination with heavy metals may also cause changes in the composition of soil microbial community, adversely affecting soil characteristics (Giller, 1998; Kozdrój and Van-Elsas, 2001).

Due to the negative effects of these heavy metals on the ecosystem especially at high concentration, it is imperative to apply some remedial measures. Several methods are already being used to clean up the environment from these kinds of contaminants, but most of them are costly and far away from their optimum performance. Conventionally, remediation of heavy-metal-contaminated soils involves either onsite management or excavation and subsequent disposal to a landfill site. This method of disposal solely shifts the contamination problem elsewhere along with the hazards associated with transportation of contaminated soil and migration of contaminants from landfill into an adjacent environment. Soil washing for removing contaminated soil is an alternative way to excavation and disposal to landfill. This method is very costly and produces a residue rich in heavy metals, which will require further treatment. Moreover, these physico-chemical technologies used for soil remediation render the land usage as a medium for plant growth, as they remove all biological activities (Gaur and Adholeya, 2004).

Recent concerns regarding the environmental contamination have initiated the development of appropriate technologies to assess the presence and mobility of metals in soil (Shtangeeva and Laiho, 2004). Presently, phytoextraction which is a form of phytoremediation has become an effective and affordable technological solution used to remove metal pollutants from contaminated soil. This technology is environmental friendly and potentially cost effective. Many species of plants have been successful in absorbing contaminants such as lead, cadmium, chromium, arsenic and various radionuclides from soils. One of phytoremediation categories, phytoextraction, can be used to remove heavy metals from soil using its ability to uptake metals (Fe, Mn, Zn, Cu, Mg, Mo and Ni) which are essential for plant growth.

This study attempts to determine the uptake and accumulation of heavy metals in different parts of the studied plant. It is expected that the findings obtained from this study will provide us with a clue on the efficacy of *M.oleifera* in the phytoextraction of heavy metal in the tropical soil.

## MATERIALS AND METHODS

This study was carried out at the Ecological Centre, University of Port Harcourt. It is located in the Niger Delta area of Nigeria on geographical coordinates: latitude 4° 65′ N and longitude 7° 5′ E. The climate condition of the area is characterized by high temperature, high rainfall and high relative humidity.

A randomized complete block design (RCBD) was used for the experiment. Weighing balance (Setra 480S USA) calibrated in kilogram, was used to weigh 5 kg of soil into 40 polythene bags. The polythene bags were arranged in three blocks (designated as A, B and C) of 10 replicates each. Each block was polluted with 100 mg of a particular heavy metal solution. That is, block A was polluted with 100 mg of Zn; block B was with polluted 100 mg of Fe and block C was polluted with 100 mg of Pb.

The soil was thoroughly mixed in the bag to enhance the harmonization of the metal solution with the soil. The bags were allowed to stand for two weeks. After the two weeks of post-pollution treatment, each block (treatment) was subdivided into two subsets of five replicates each.

Two seedlings of *M. oleifera* were transplanted from the nursery into each bag of one subset for each block while the other subset acted as the control for each block with no planting. This was done for all the blocks (A, B and C). It was also ensured that the plants were of the same sizes. Watering was done three times a week. The quantity of water used for watering the planting bags was 50 cl per bag. Weeding was done when the need arose. The experiment lasted for 12 weeks.

Soil samples were collected from each bag after pollution (but before the plants were transplanted into them). This was done to ascertain the level of these metals present in the soil before phytoextraction. Soil sample collections were also done at the termination of the experiment. The heavy metals (Zn, Fe and Pb) were analyzed for the soil samples.

At the end of the 12 week period, the plants were carefully harvested from each bag. The shoots were separated from the root by cutting. All samples collected from each treatment were taken to the laboratory immediately for analysis. The plants parts and soil were analyzed for their heavy metal (Pb, Fe and Zn) content. This was done in accordance with their treatment. The plant concentration factor and translocation factor were also calculated.

Soil samples collected were analyzed by air-drying and all clods and clumps removed. Dried soil was sieved using 2 mm sieve to remove coarse particles before analysis. One gram (1 g) of the dried soil sample was placed in 100 ml beaker and; 3 ml of perchloric acid and 5 ml of nitric acid were added. The mixture was allowed to stand for 15 min before digestion by gently heating at low temperature on a hot plate and allowed to cool for 5 min. The digest was then filtered into 50 ml standard flask. The filtrate was analyzed for heavy metal using atomic absorption spectrophotometer (AAS).

Plant samples were analyzed by first rinsing with distilled water and oven-dried at 100°C for 48 h. The plant materials were ground to fine powder. One gram of the powder was digested as described above and analyzed for heavy metals AAS. Transfer factor or plant concentration factor (PCF) using metal concentration in the extracts of soils and plants were calculated. The plants concentration factor (PCF) was calculated as follows:

$$PCF = C_{plant}/C_{soil}$$

Where, $C_{plant}$ and $C_{soil}$ represents the heavy metal concentration in extracts of plants and soils respectively (Cui et al., 2005).

Translocation factor (TF) or mobilization ratio (Barman et al., 2000; Gupta et al., 2008) using the formula below.

$$TF = \frac{\text{concentration in shoot}}{\text{concentration in root}}$$

From the data generated, means and standard error mean (SEM) were calculated. The data obtained were also subjected to analysis of variance (ANOVA) using Microsoft Office Excel and least significant difference (LSD) was calculated using SPSS data analysis package (2007 version 9.1).

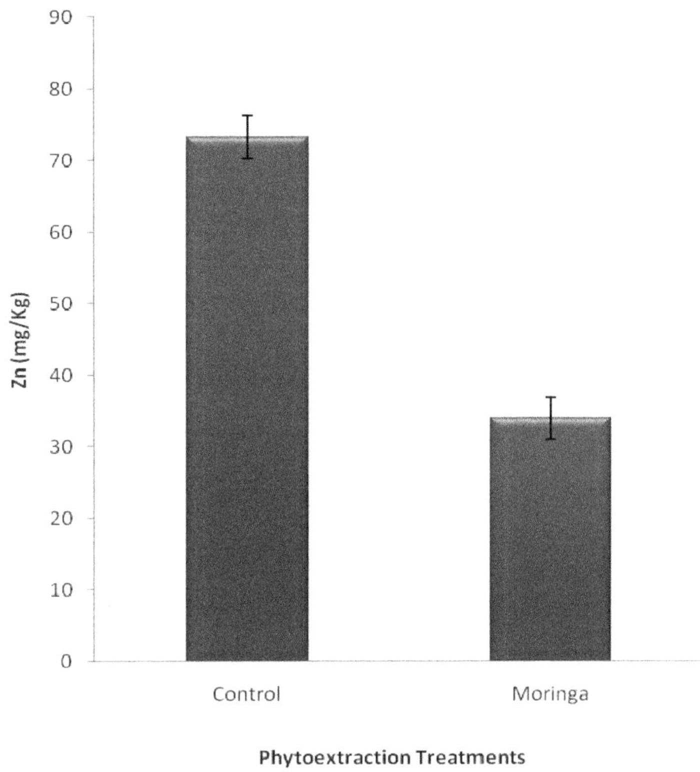


**Figure 1.** Soil Zn concentrations in phytoextraction treatment.

## RESULTS

Figures 1 to 3 show the efficacy of the test plant (*M. oleifera*) for the phytoextraction of the heavy metal polluted soil. It was observed that there was a significant reduction in soil Zn in the polluted soil phytoextracted with the test plant species as compared to the control soil (no planting) (Figure 1). Similar results were observed in the Fe (Figure 2) and Pb (Figure 3) polluted soil phytoremediated with the test plant. These reductions were found to be significantly different (p = 0.05) from the control.

It was also observed that the reduction of the heavy metals in the soil differed considerably from metal to metal. Results show that the percentage reduction of Pb, Zn and Fe were 64, 20.1 and 22.7%, respectively.

Figures 4-6 showed the concentrations of the metals in both the above-ground and the below-ground parts of the test plant. Figure 4 shows the concentration of Zn in plant parts (roots and shoots) while Figures 5 and 6 shows the concentration of Fe and Pb in the plant parts of the test plant, respectively. It was observed that *M. oleifera* retained more of the heavy metals in its root than in the above-ground part (shoot). In other words, the concentration of Zn in plant's roots was significantly (*P = 0.05*) higher than the shoots. *M. oleifera* accumulation of zinc was significantly higher in root than shoot. Similar results were obtained for the concentration of Fe and Pb in the plant's above-ground (shoot) and below-ground (root) parts.

The calculated result (Figure 7) showed the transfer values from soil to plant. It was observed that the transfer factors for the three (3) heavy metals studied were below one (1). The result indicates that transfer factor for heavy metals in the test plant species were in the orders Zn > Fe >Pb treatments. Reduction in transfer factor was lowest in Pb and highest in Zn polluted soil.

The translocation factor which is an indication of the ratio of heavy metal retention in the shoot over the root showed that the values for the 3 heavy metals studied were below one (1). This also buttress the fact that the phytoextracted plant (*M. oleifera*) accumulated and retained heavy metals (especially Fe, Zn and Pb) more in the root system that the shoot system (Figure 8). The order of translocation factor was: Zn > Pb > Fe.

## DISCUSSION

It is a well established fact that some plants can extract heavy metals from the soil or water and accumulate or retain them in their body parts (roots, stems or leaves). This uptake and accumulation of metal ions by plants differ from plant to plant and in varying degrees. The bioavailability of the metals can also influence their uptake by plants which are in turn determined by both

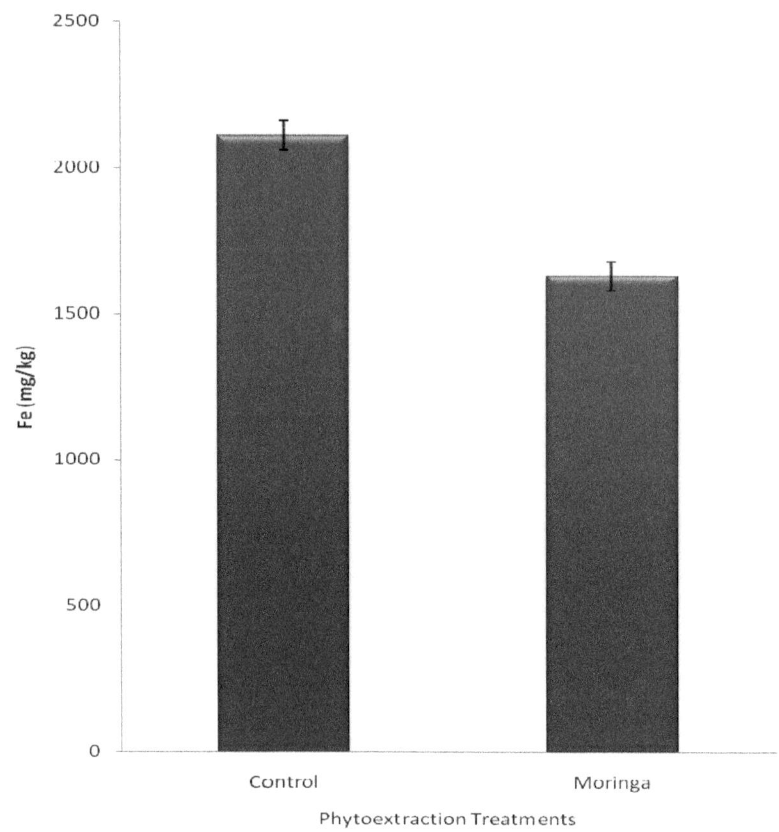


**Figure 2.** Soil Fe concentrations in phytoextraction treatment.

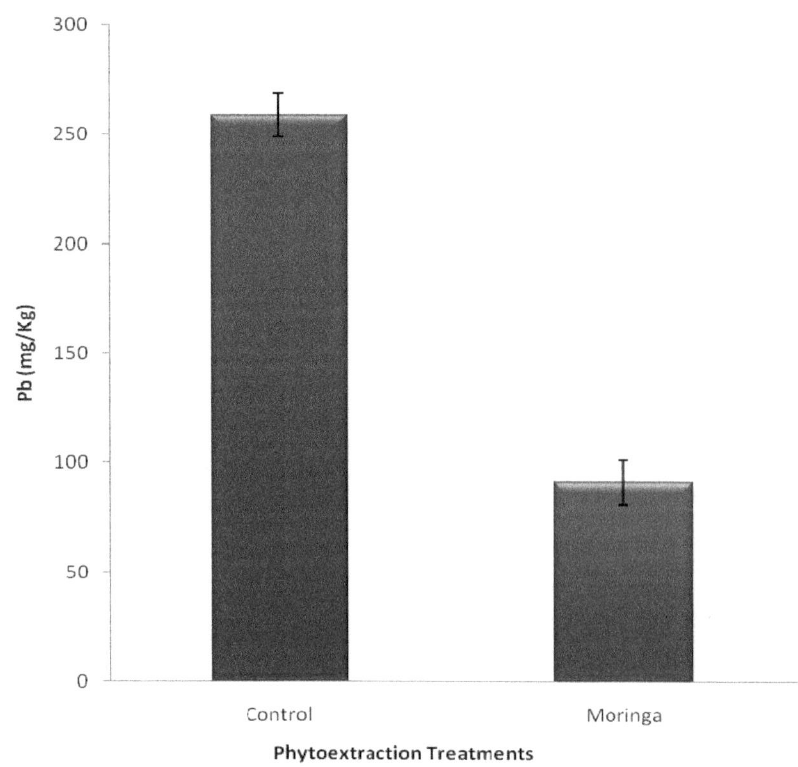


**Figure 3.** Soil Pb concentrations in phytoextraction treatment.

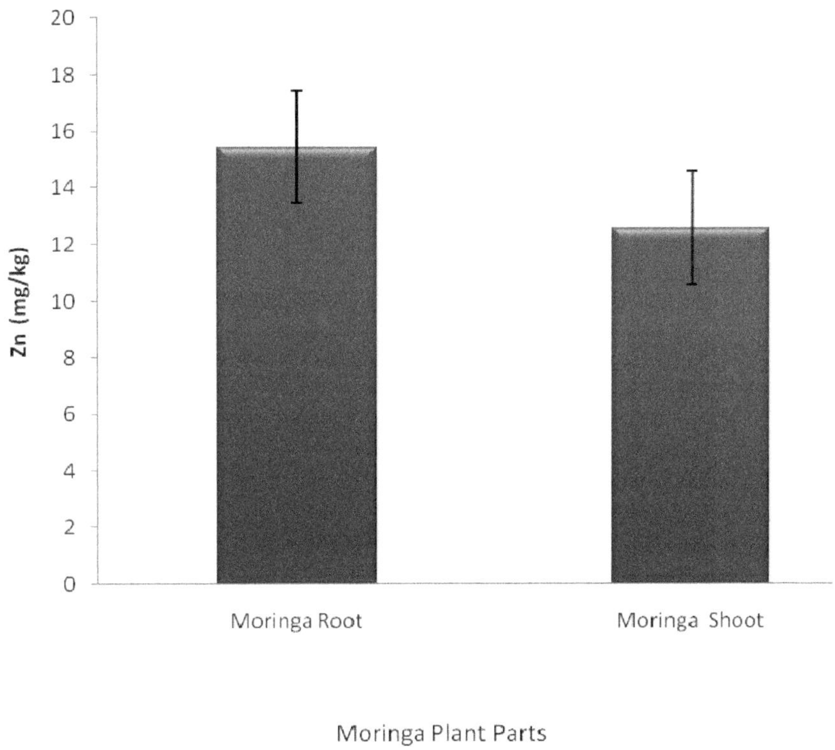


**Figure 4.** Zn concentration in plants part.

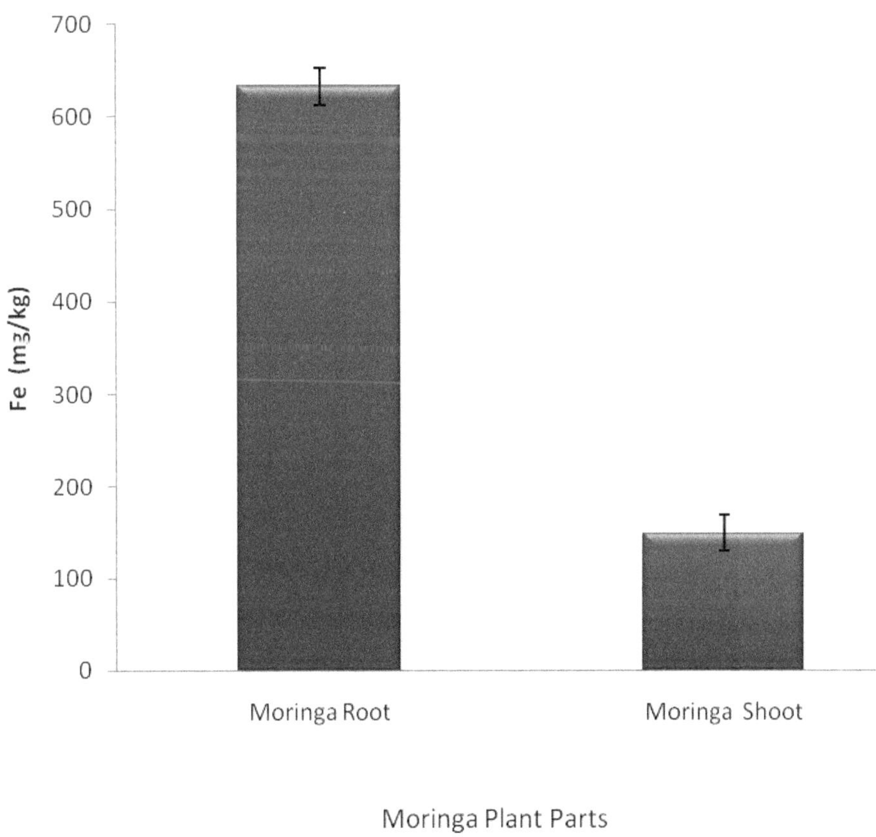


**Figure 5.** Fe concentration in plants parts.

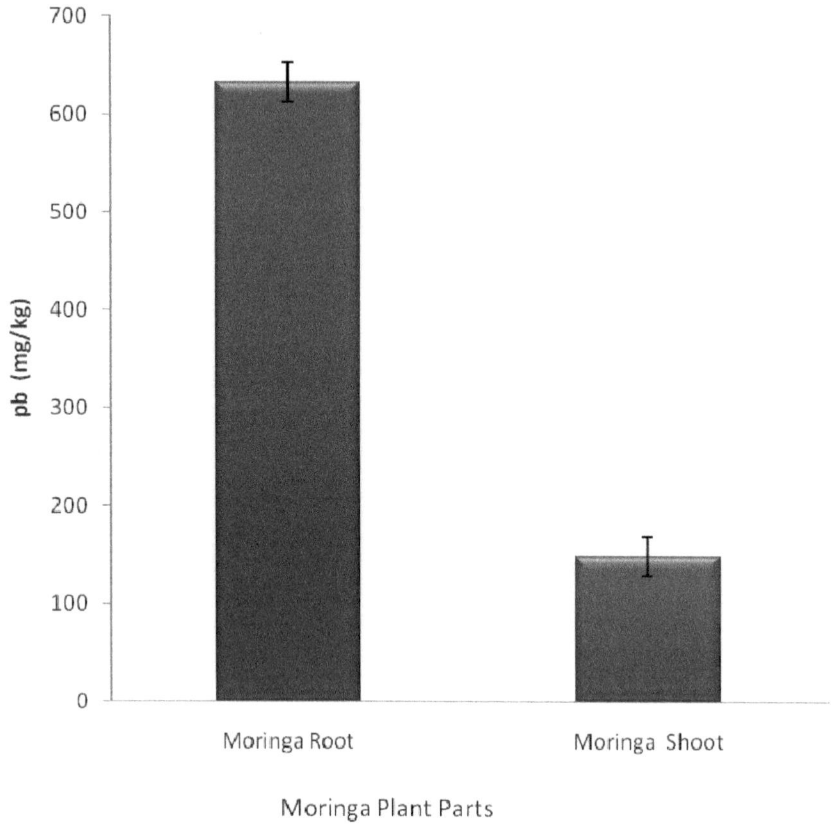


**Figure 6.** Pb concentration in plants parts.

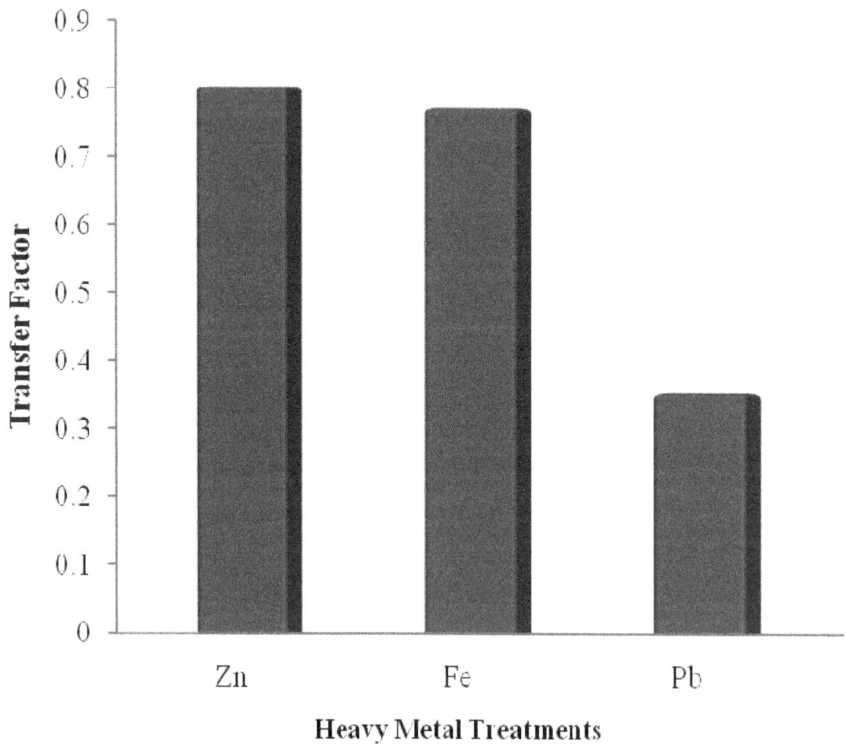


**Figure 7.** Transfer factor of different metals.

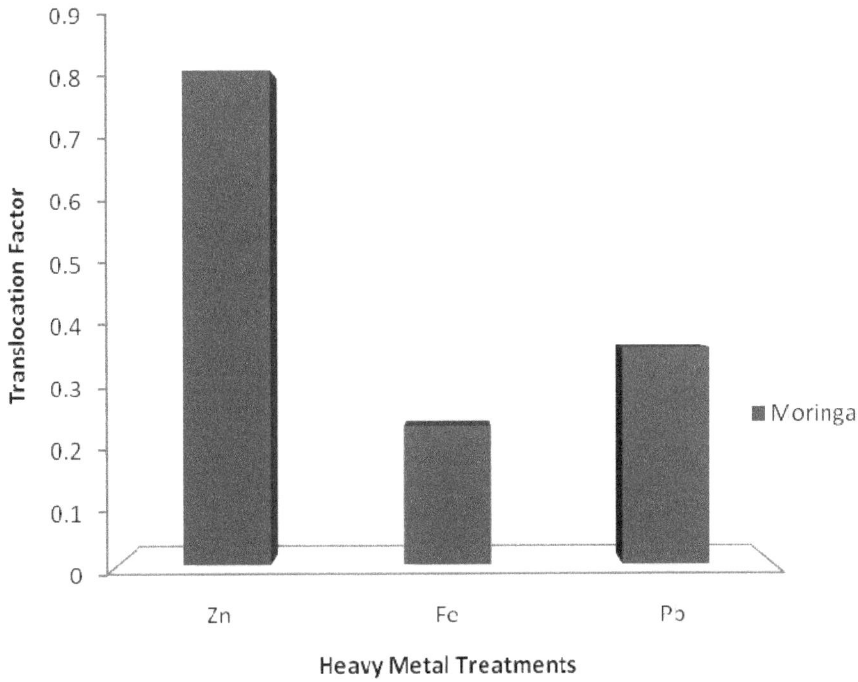


**Figure 8.** Translocation factors of different metals.

external (soil-associated) and internal (plant-associated) factors.

Results showed that *M. oleifera* plant has the ability to phytoextract heavy metals especially Fe, Zn and Pb and accumulate them in their body parts (below-ground and above-ground). Although, the strategy employed by *M. oleifera* plant for the uptake of heavy metal varies among heavy metal treatments as shown in the result in which the percentage reduction was in the order of Pb (64%) > Fe (22.7%) > Zn (20.1%). This is understandable since some plant exhibit selective absorption of metal ions from soil in which a particular plant may prefer a particular metal ion to another. Fleming and Parle (1977) reported that the uptake of heavy metal varies widely depending on the plant species being studied. It was long recognized that the uptake of metals by plants were plant species dependent (Adriano, 1986). Some plants release specific metal-chelating or reducing compounds into the rhizosphere to aid the absorption when availability of these micronutrients is low (Marschner, 1995).

Results in Figures 4, 5 and 6 show that the accumulation/retention of heavy metals (Zn, Fe and Pb) was more in root system than shoot system. This suggests that plant have different ability in accumulating particular heavy metals in different parts of their body. The accumulation of heavy metal contents in root and shoot varies from species to species (Salt et al., 1995). This is understandable since the breaking up of soil aggregates is a physical effect of root tips pushing through soil as the root tips grow. Fungi associated with some plant roots (mycorrhizae) can also influence the chemical conditions within the soil, which could account for the accumulation of heavy metal more in root. Similar result has also been reported by Trusby (2003). This also suggests that a number of factors (including anatomical, biochemical and physiological factors) might contribute to heavy metal accumulation in the root than shoot. Although, Salt et al. (1995) reported a contrary observation of high accumulation of heavy metals in the upper vegetative parts of some plants.

The analysis of transfer factor indicates that the uptake capabilities of heavy metals from soil to *M. oleifera* is in the order of Zn > Fe > Pb. This could be attributed to the plant species used. Zurera et al. (1987) have reported that the mobility of metals from soil to plants is a function of the physical and chemical properties of the soil and of plants species, and is altered by innumerable environmental and human factors.

Translocation factor, the ratio of shoot to root metals, indicates internal metal transportation. The data presented indicate that metals accumulated by the test plant species (*M. oleifera*) were largely retained in roots, as shown by TF values (<1). Translocation factor (TF>1) showed that translocation of metals effectively was made to the shoot from root (Baker and Brooks, 1989; Fayiga and Ma, 2006).

## Conclusion

It can be concluded that the ability of a plant species to phytoextract and accumulate heavy metals varies among the heavy metals. There were drastic reductions in the heavy metal contents of soil phytoremediated with *M. oleifera*. Moreover, the level of accumulation of these metals studied (Zn, Fe and Pb) was more pronounced in roots than shoots. Therefore, *M. oleifera* is capable of phytoextracting these metals (Zn, Fe and Pb), but their accumulation rate is dependent on a particular metal.

## Conflict of Interests

The author(s) have declared that there is no conflict of interests.

### REFERENCES

Adriano DC (1986). Utilization and disposal of fly ash and other coal residues in terrestrial ecosystems: A Review. J. Environ. Qual. 99:333-345.

Baker JM, Brooks RR (1989). Terrestrials higher plants which hyper accumulate metallic elements. A review of their distribution, ecology and phytochemistry. *Biorecovery* 1: 81-26Barman SC, Sahu RK, Bhargava SK, Chatterjee C (2000). Distribution of heavy metals in wheat, mustard and weed grains irrigated with industrial effluents. Bull. Environ. Contam. Toxicol. 64:489-496.

Cui C (2005). Junk-greedy greens: phytoremediation as a new option for soil decontamination. Asian Pac. J. Environ. Biol. 2:60-67.

Fayiga AQ, Ma LQ (2006). Using phosphate rock to immobilize metals in soils and increase arsenic uptake in *Pteris vittata*. Sci. Total Environ. 359:17-25.

Fleming G, Parle J (1977). Heavy metals in soils, herbage and vegetables from an industrialized areaWegt of Dublin City. Irish J. Agric. Res. 16:48.

Garbisu C, Alkorta I (2003). Basic concepts on heavy metal soil bioremediation. Eur. J. Min. Process Environ. Prot. 3(1):58-66.

Gaur A, Adholeya A (2004). Prospects of arbuscular mycorrhizal fungi in phytoremediation of heavy metal contaminated soils. Curr. Sci. 86:528-534.

Giller KE (1998). Toxicity of heavy metals to microorganism and microbial processes in agricultural soils: A Review. Soil Biol. Biochem. 30:10-11.

Gupta S, Nayek S, Saha RN, Satpati S (2008). Assement of heavy metal accumulation in macrophyte, agricultural soil and crop plants adjacent to discharge zone of sponge iron factory. Environ. Geol. 55:731-739.

Kozdrój J, Val Elsa T (2001). Structural diversity of microbial communities in arable soils of a heavily industrialized area determined by PCR-DGGE finger printing and FAME profiling. Appl. Soil Ecol. 43:197-212.

Lane TW, Morel FM (2010). A biological function for cadmium in marine diatoms. Proc. Natl. Acad. Sci. U.S.A. 97:4627-463.

Marschner H (1995). Mineral nutrition of higher plants. 2nd Ed. Academic Press Inc., LondonSalt DE, Smith RD, Raskin L (1995). Phytoremediation: a novel strategy for removal of toxic metals from the environment using plants. Biotechnol. 13:468-474.

Schmidt U (2003). Enhancing phytoextraction: The effect of chemical soil manipulation on mobility, plant accumulation and leaching of heavy metals. 87:212-229.

Shtangeeva I, Laiho H (2004). Improvement of phytoremediation effects with help of different fertilizers. Soil Sci. Plant Nutr. 50:885-889.

Schwartz C, Echevarria G, Morel J (2003). Phytoextraction of cadmium with *Th laspi caerulescens*. Plant Soil 249:27-35.

Trusby P (2003). Impact of heavy metals on forest trees from mining areas. Proceedings of the International Conference on Mining and the Environment III, October 24, 2005, Sudbury, Ontario, Canada.

Zurera G, Estrada B, Rincon F, Pozo R (1987). Lead and cadmium contamination levels in edible vegetables. Bull. Environ. Contamin. Toxicol. 38:805-812.

# 5

# Approaches for the control of banana *Xanthomonas* wilt in East and Central Africa

**J. Kubiriba and W. K. Tushemereirwe**

National Agricultural Research Organization, P.O. Box 7065, Kampala, Uganda

**The livelihoods of millions of banana farmers in East and Central Africa have been threatened by the devastating epidemic of banana *Xanthomonas* wilt (BXW) caused by *Xanthomonas campestris* pv. *musacearum*. The disease is widespread and has been reported in East and Central Africa. The economic impact of the disease emanates from loss of yield and cost of control measures. All banana cultivars grown in the region are susceptible to BXW and no source of resistance has been identified. Cultural interventions are therefore the most practical and recommended practices for BXW control. A number of approaches: farmer field schools, community action, going public and top-down conventional approaches (traning of trainers, mass media and posters) have been used to mobilize stakeholders in Kenya, Uganda, Rwanda, Tanzania and Democratic Republic of Congo. Different approaches have been used in different countries with varied success in controlling BXW. This review discusses approaches, which have been successfully used in the management of BXW in some areas of East and Central Africa with a view of evaluating them individually or in combination under local conditions. This hopefully will help scale-out the promising approaches and contribute to more effective BXW control in the region.**

**Key words:** Banana *Xanthomonas* wilt, disease management approaches, Eastern and Central Africa.

## INTRODUCTION

The communities of the Great Lakes of Eastern and Central Africa have traditionally depended on a perennial banana cropping system for food and income (Karamura et al., 2008). This slow-changing farming system has ensured regional food security for many centuries. The system has come under stress due to progressive decrease in farm size, land fallow periods and production. This is mainly due to biotic threats particularly weevils, nematodes, fungal and viral diseases. On farm, such pests and diseases were traditionally managed by exploiting exisiting differences in cultivar tolerance. Although there was reduced productivity, farmers maintained reasonable levels of food and income security. However, following the arrival of banana *Xanthomonas* wilt (BXW) caused by *Xanthomonas campestris* pv. *Musacearum,* in the great lakes region in 2001, entire crop holdings were wiped out in some areas where highly susceptible genotypes were dominating the farming systems ((Tushemereiwe et al., 2004, Karamura et al., 2006). Different approaches have been used across the region to control BXW with varied success. For example, in the major banana growing areas in Uganda, BXW incidence remained below 5% for about 4 years (Tushemereirwe et al., 2006). In addition, in areas around the lake Victoria zone of Tanzania and Northern Rwanda, the spread of BXW was relatively reduced. Successful control of disease and pest epidemics is only possible if effective technologies are deployed together with institutional approaches that efficiently mobilise stakeholders. This review documents technological and institutional approaches that contributed to successful control of BXW in some areas, which may be validated and scaled out in the region.

## Technical aspects of BXW control

There is no resistance to BXW so far reported among the local *Musa* germplasm in East and Central Africa (Ssekiwoko et al., 2006; Ndungo, unpublished). Therefore, the control strategies for BXW are based on cultural practices. However, the design and successful implementation of disease control programme based on cultural practices requires a clear understanding of the etiology and epidemiology of the disease (Thresh, 1980). Furthermore, critical factors that influence pathogen reproduction, survival, propagation, selection and evolution of more virulent strains must be understood (Jeger, 2004). This review provides a detail of the technologies used for control of BXW together with underlying epidemiological logic, which the farmers or other recipient stakeholders need in order to make informed decisions.

### *Prevention of introduction of BXW into uninfected fields*

Introduction of pathogens to new locations and fields may be attributed to the use of pre-infected planting material or movement of vectors from remote infected sources (Cabaleiro and Segura, 1997). Currently, there are no quick screening techniques in the region for the planting materials (Karamura et al., 2008). Therefore, in locations where bananas have completely been cleared especially by BXW, strategies for large scale production of affordable clean planting materials are urgently needed to supplement the limited macro and micro-propagation technology already available.

Early removal of the malebud is an effective means of preventing the spread of BXW by insect vectors (Tinzaara et al., 2006). Removal of malebuds is recommended for both affected and unaffected fields. Using a forked stick, the malebud is removed as soon as the last cluster on banana flower is formed, that is, at about two weeks after emergence of the inflorescence. A forked stick rather than the commonly used cutting tool should be used because the stick can only contaminate should the broken bud. As such, there is no exchange of plant sap between banana plants on which the forked stick has been used and therefore no risk of spreading BXW between the plants. This technique is a preventive measure that effectively blocks the spread of BXW infection via insect vectors from within and without field.

Some banana cultivars do not have persistent neuter flowers. Other cultivars have persistent or semi-persistent bracts which do not easily break and thus escape insect vectored BXW spread. In a few extreme cases, deployment of cultivars such as scambia which do not have a male bud, but are tested for juice production, yield and farmer acceptability, may reduce the cost of removal of male buds and therefore the control of BXW cost effectively.

### *Management of initial inoculum from the banana field*

Once BXW infection has been detected in a banana field, infected plants are the source of inoculum. In addition to removal of malebuds, farmers are advised to remove infected plants and disinfect tools that have had contact with infected plants (Tushemereirwe et al., 2004). Some farmers rouge the whole stool once they find an affected plant on them. Others remove the only infected plants from the affected stool. At the beginning of the BXW epidemic in Uganda, rouging the whole affected stool was recommended (Tushemereirwe et al., 2004). Overtime the recommendation changed to single stem removal (that is, removing infected plants only from the stool). This was after it became evident that the process of rouging whole stools was extremely labour intensive and it was possible to only remove infected plants from the affected stool and save the remaining uninfected plants from disease infection on the same stool. This was possible because most BXW infection starts from the upper parts of the plant on the inflorescence and takes some time to get to the lower parts of the plant (Ssekiwoko et al., 2006). If the infected plant is carefully removed from the stool before infection gets to the corm, it is possible to save other plants on the stool from BXW infection (Turyagyenda et al., 2006). Single stem removal has also been evaluated and found to be effective in control of BXW in Kenya and DRC (Kubiriba et al., 2014). Furthermore, after cutting infected banana plants, farmers may leave banana refuse on the ground. It had been previously recommended to burry all banana refuse (Tushemereirwe et al., 2004). However, no *Xcm* was isolated from refuse of infected banana plants after 5 days even when they were left on the ground (Mwebaze et al., (2006). As long as rotting conditions are created, Xcm seems to compete poorly with other organisms involved in the rotting process of the banana refuse. This suggests that formerly affected fields can be replanted after 3months without risk of reinfection from old inoculum. However, a minimum of 6 months have been recommended to completely remove the risk of reinfection in previously infected fields. It is impotant to note that the survival of the bacterial is dependent on ecological conditions of different areas therefore it may be necessary to validate this recomendation before its application.

### *Management of incubating BXW infection*

In tool-mediated infection, movement of the disease within the plant is dependent on the site of infection and this can be the roots, leaves, the inflorescence or the stem. During the incubation period, suspending the use of cutting tools in infected fields was key in the control of BXW (Tushemereirwe et al., 2006). The epidemiological basis for this practice is that, when a plant is infected with BXW, it shows symptoms within 3 months. Meanwhile, if infected plants are continuously removed, the number of infected plants will gradually reduce until no more infected plants are identified. During the 3 months incubation period, if pruning is discontinued, *Xcm* will not spread from infected plants with incubating BXW infection to unaffected plants. As a result, the disease outbreak is then effectively controlled. However, all preventive measures have to be applied continuously to prevent reintroduction of the bacteria.

### *Disinfection of tools*

Disinfection of cutting tools used for pruning trash off the plants or after removing infected plants limits mechanical spread of BXW. The Xcm was detected on cutting tools kept at room temperature up to 22 days (Buregyeya, 2010). During that period, BXW is spread from the tools to banana plants up to 18 days. This implies that if traders who collect bananas from fields that are hundreds of kilometres apart do not disinfect the tools between plants and farms, they can spread BXW over long distances. Cutting tools may be disinfected by burning directly on a fire flame or cleaning with commercial JIK (sodium hypochloride).

## Institutional approaches for stakeholder mobilisation

### *Conventional approaches*

Every stakeholder in the banana value chain should be engaged in the battle against BXW. Subsistence farmers, who form the majority of farmers in the region, have substantial difficulty in managing plant diseases (Sherwood, 1997). This is partly because they cannot see the organisms that cause the plant diseases (Nelson et al., 2001). It is therefore important to package the message in a clear and concise manner stating the epidemiological underpinnings, negative impact of failure to implement the intervention as well as the benefits of such interventions to the individual farmer and the larger community. The choice and design of the communication strategies should take into account the dynamics of the target area such as literacy levels, numbers of radio receivers and availability of extension support. The strategies would include training of trainers, posters, pamphlets, leaflets, brochures and billboards designed to suit the levels of literacy and other socio-economic dynamics in the target community. Whereas this approach, which is widely used in the region is instrumental in swiftly raising awareness of stakeholders about the disease across countries, it is ineffective in triggering actions to control the disease (Tushemereirwe et al., 2006).

## Participatory approaches

In the conventional approaches of stakeholder mobilization, pest and disease epidemic control programmes begin with generation of technologies and most often stop at dissemination of information to the farming communities, hoping that the recipients will use the information to control the epidemics (Hawkins et al., 2009). The awareness campaign deployed in Uganda was successful with more than 85% of the banana farmer knowing how to identify BXW, how it spreads and how it is controlled (Bagamba et al., 2006). However, only 30% of the farmers were undertaking BXW control (Tushemereirwe et al., 2006). It is possible to improve on the proportion of farmers that adopt control measures at community level using participatory approaches.

### *Participatory development communication*

Participatory development communication (PDC) is the use of communication to facilitate community participation in a development initiative such as control of BXW (Bessette, 2004). Deploying PDC facilitates dialogue among the different stakeholders around a common problem has the aim of developing and implementing an action plan to solve the problem. The approach follows the following steps:

1. Members of the community and their leaders are brought together to discuss the BXW problem with facilitation of a PDC resource person.
2. The community members explore and discuss alternative ways of controlling BXW and identify the appropriate practices they can implement.
3. The community members develop an action plan to tackle the BXW problem.
4. The community agrees on the mechanism for implementing the formulated action plan.
5. The action plan is executed with various stakeholders in the community given roles and responsibilities to see

that the agreed action plan is properly implemented in time and space.
6. The community also agrees on how to monitor and evaluate activities for disease control as part of the action plan.
7. The community agrees on how to share successful outcomes.

The key output of the above process is the formulation and implementation of an action plan for the control of BXW (Table 1). The action plans may be at community level, sub-county, district and national levels. In locations where PDC approaches were religiously implemented, the incidence of BXW was drastically reduced (Figure 1). In the communities of Lake zone, Tanzania, PDC or its modification was also used with some success in controlling BXW (Mgenzi et al., 2006).

### *Farmer field schools (FFS)*

Farmer field schools empower farmers to learn in an informal setting within their own environment. The FFSs are schools without walls where groups of farmers meet weekly with facilitators. It is a participatory method of learning, technology development and dissemination based on adult learning principles. The approach uses experiential learning and a group approach to facilitate farmers in making decisions, solving problems and learning new techniques. The concept is based on farmer participatory discovery approach for improving decision-making capacity of farming communities and stimulating local innovation for sustainable agriculture. The school involves 25 – 30 farmers. The group dynamics contribute towards team building and organizational skills. The FFS approach has successfully been used in Kenya, Uganda and Tanzania (Davis, 2008)

### *Farmer field schools with BXW control*

The FFS approach was employed in controlling BXW in Uganda between 2006 and 2008. The process started with training of trainers (TOT). In this TOT, it was emphasized to the facilitators that:

1. It is not possible to improve banana production with their farmers' groups unless BXW was eradicated. It was agreed that BXW is enemy number one to the banana crop; which should be eradicated quickly to allow the FFS time to deal with other banana constraints.
2. The cultural package for control of BXW is effective, giving experiences of successful control.
3. The farmers should know how to identify infected plants and how BXW is spread through Agroecosystem analysis (AESA). This then would help them to explain the logic behind the recommended practices, not only for enhanced adoption but also modification of the practices where applicable.

The FFS were thereafter established in 50 subcounties in five banana growing districts of Mbarara, Kiboga, Mukono, Kamuli and Lira. By April 2008, BXW had been significantly reduced in all the groups of participating farmers (Figure 2). In subcounties or districts hosting FFS, other farmers soon started controlling BXW thus spreading the efforts. In areas such as Lira district, where control of BXW had been difficult for the national disease control efforts, FFS were islands of effective control in a sea of no compliance in BXW control. The FFS approach is invaluable where some recommended technologies may not work in new locations and therefore must first be tested, validated and adapted locally. The FFS approach has potential benefits because it is cheap, flexible, incorporates farmers' aspirations and empowers them. In addition, FFS approach has an in-built monitoring and evaluation mechanism and it can be integrated with other research/development approaches.

## DISCUSSION

Effective disease control is possible if effective technologies are generated and promoted together with institutional approaches that effectively mobilise stakeholders who use the technologies (Kubiriba et al., 2012). The cultural practices reviewed in this paper were formulated with information generated from epidemiological studies in Uganda with participation of various partnerships from both local and International Research Institutes and Universities (Tushemereirwe et al., 2006). Other stakeholders from Government and Non-Government Organisations involved in rural development also contributed to the planning, generation and promotion of these technologies. In some locations, farmers sometimes modified recommended practices to suit their conditions (Bagamba et al., 2006). Experiments were established to scientifically validate such technologies and valuably improved development of appropriate technology e.g. single stem removal (Ssekiwoko et al., 2006). The technologies for the control of BXW proved effective in all areas where they have diligently and consistently been deployed (Tushemereirwe et al., 2006). Although most of the recommended practices may have region-wide applicability, most of them were ecology specific and may therefore need to be validated for customization to the local ecological conditions.

Accessibility to improved technologies for disease control influences a farmers' knowledge and their decision to control plant diseases (Sherwood, 1997). The farmers should be provided with correct information that enables them to make more effective action decisions. Bentley and Thiele (1999) showed that too much information would cause needless confusion, while oversimplification of concepts would not be effective in improving disease

**Table 1.** Action plan executed by communities to control BXW in Rugando, Mbarara District.

| Goal | Objectives | Activities | Time frame | | | | Target | Resp. Person | Budget unit cost | | |
|---|---|---|---|---|---|---|---|---|---|---|---|
| | | | March - June 2005 | July – September 2005 | October - December 2005 | January - June 2006 | | | Item | Amount | Source of funds |
| To eradicate BBW from Rugando Sub-county | Create awareness about BXW in Rugando | Conduct sensitization workshops | X | | | | Technical staff/local leaders | DPC & DAO | Facilitation<br>Stationery<br>Announce ments | 100,000=<br>20,000=<br>50,000= | District |
| | | Training farmers | X | X | X | X | Community | AO & SC leaders | Fuel<br>SDA<br>Stationary<br>Posters | 60,000=<br>60,000=<br>15,000= | NGOs<br>S/county |
| | Mobilize and organize farmers for community action for effective control of BXW | Formulate and enforce community bye-laws - village meetings | X | X | | | Affected villages | AO & Community leaders, TF | Fuel<br>SDA<br>Stationary | 50,000=<br>40,000=<br>10,000= | S/county |
| | | Organize farmers to effect BBW control – village meetings | X | X | | | Affected villages | AO & Community leaders, TF | Fuel<br>SDA<br>JIK<br>Posters | 100,000=<br>60,000=<br>9,000= | S/county<br>APEP |
| | Effectively supervise and monitor the BXW control activities | Formulation of Taskforces | X | | | | Affected villages | AO & community leaders, TF | Fuel<br>SDA<br>Stationary<br>Posters | 100,000=<br>30,000=<br>15,000= | S/County |
| | | Supervise implementation of control measures | X | X | X | | Affected villages | AO & community leaders, TF | Fuel<br>SDA<br>Stationary | 60,000=<br>40,000=<br>10,000= | S/County<br>APEP<br>District |
| | | Tracking new outbreaks/ infections | X | X | X | X | All villages | Community | - | - | - |
| | Ensure sharing of information | Organize exchange visits | | | X | X | Community and other stakeholder s | Community | Fuel<br>SDA<br>Stationary | 100,000=<br>80,000=<br>20,000= | NGOs<br>S/County |
| | | | | | | | | | **Total** | **884,000=** | |

DPC = District Production Coordinator; DAO = District Agricultural Officer; AO = Agricultural Officer; TF = Taskforce; NGOs = Non-governmental Organization; S/county = Sub-county. The action plan was totally facilitated by funds locally mobilized. To date no plant infected with BXW has been sighted in the Rugando sub-county, Mbarara for over 12 months.

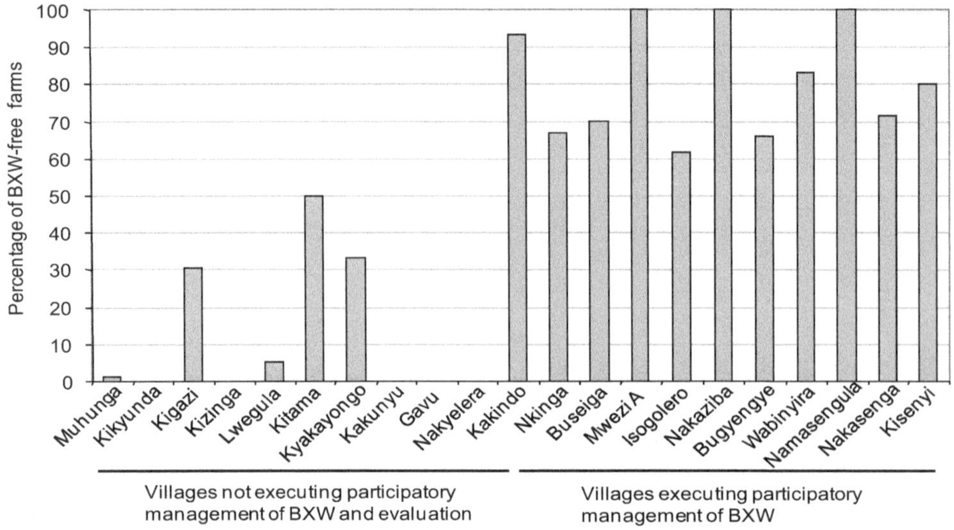


**Figure 1.** Success of participatory management of banana *Xanthomonas* wilt within one year in different villages in Uganda.

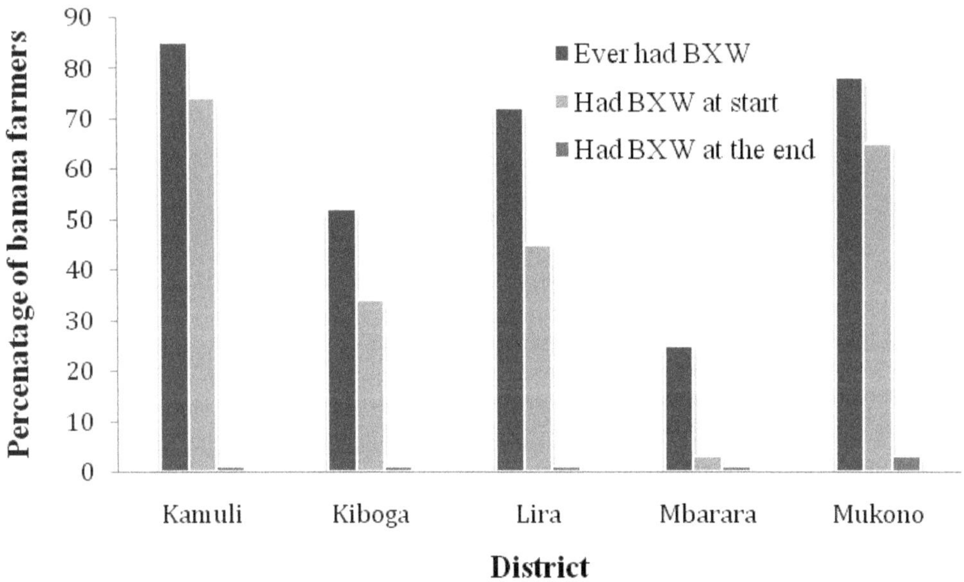


**Figure 2.** Level of BXW control in communities hosting farmer field schools in Uganda.

management. Research and extension agents too often use scientific terminology-laden approaches in information dissemination; they do not attempt to explain the disease life cycles to farmers, thereby underestimating the farmers contribution as key to unlocking the mystery of the disease. This problem was circumvented by involvement of various stakeholders (including the end users) in packaging information and selecting dissemination pathways (Ngambeki et al., 2006). Following the massive and aggressive awareness campaigns in Uganda, over 80% of the banana farmers knew BXW - its diagnosis, spread and means of control (Muhangi et al., 2006). Information materials developed from this process have been revised and further translated into different languages including Swahili and French used in all the 6 ECA countries (Karamura et al., 2006). The impact of using such information should be assessed to enable more appropriate modifications where necessary.

Despite the fact that most banana farmers knew how to control BXW from the awareness campaigns, only 30% utilized the knowledge to effect BXW control (Tushemereirwe et al., 2006). In order to address this problem, participatory approaches reviewed in this paper were used. The key elements of participatory approaches included getting communities together, facilitating them to formulate and implement action plans, mobilizing political and other leaders to support community efforts and ensuring their effective monitoring and evaluation were all important BXW control.

As a result, it was possible to control BXW to below 5% in important banana growing areas in Uganda for over 3 years while more than 90% of the farmers participating in FFS controlled the disease in their fields and their communities (Kubiriba et al., 2012). In future, the stakeholders in ECA who may use these participatory approaches needed to continuously evaluate their effectiveness and to make the necessary modifications to suit end user situation without compromising the key principles.

## CONCLUSION

A number of management options and approaches for the control of BXW have been explored. However, no single management option can effectively control the disease. Successful control of BXW and possibly other epidemics is only possible by deploying effective technologies together with participatory approaches that effectively mobilize stakeholders for the control of the epidemic. While generation of technologies that effectively control BXW is mainly the role of researchers, other elements for successful control of BXW are roles of different stakeholders including local leaders (political and opinion leaders), extension officers (both governmental and non-governmental organizations) and farmers.

The East and Central African regional partners under the Crop Crisis Control Project (C3P), Association for strengthening Agricultural Research in Eastern and Central Africa (ASARECA) project used these approaches to control BXW in 6 countries: Uganda, Tanzania, Kenya, Rwanda, Burundi and Democratic Republic of Congo. A remarkable reduction in the incidence of BXW has been recorded in most of the pilot areas. Approaches that have registered successful disease control need to be carefully evaluated and scaled out in order to effectively control BXW in the region. This will only be possible with the support of all key stakeholders including policy makers at regional and national levels.

## Conflict of Interests

The author(s) have not declared any conflict of interests.

### REFERENCES


Bagamba F, Kikulwe E, Tushemereirwe WK, Ngambeki D, Muhangi J, Kagezi GH, Green S (2006). Awareness of banana bacterial wilt control in Uganda: Farmers' perspective. Afr. Crop Sci. J. 14:(2):157-164.

Bentley JW, Thiele G (1999). Bibliography: Farmer knowledge and management of crop disease. Agric. Hum. Values 16:75-81.

Bessette G (2004). Involving the community: A Guide to participatory Development Communication. IDRC. 143pp.

Blomme G, Mpiira S, Ssemakadde R, Mukasa H (2005a). Controlling banana Xanthomonas wilt through de-budding. Info Musa 14(1):46.

Buregyeya (2010). Long distance spread of banana bacterial wilt. Msc. Thesis. Makerere University, Kampala Uganda.

Cabaleiro C, Segura A (1997). Field transmission of grapevine leafroll associated virus (GLRa-V-3) by mealybug, *Planococcus citri.* Plant Dis. 81:283-287.

Davis KNE (2008). Developing methodology for assessing the impact of farmer field schools in East Africa. Proceedings of the 24$^{th}$ Annual meeting of Association the International Agricultural and Extension Education (AIAEE). 4:93-99.

Hawkins R, Heemskerk W, Booth R, Maatman A, Adekunle AA (2009). Integrated Agricultural Research for Development (IAR4D). A concept paper prepared for Forum for Agricultural Research in Africa (FARA) and Sub-Saharan Africa Challenge Programme (SSA CP). http://www.fara-africa.org/.

Jeger MJ (2004). Analysis of disease progress as a basis for evaluating disease management practices. Ann. Rev. Phytopathol. 42:61-82.

Karamura EB, Osiru M, Blomme G, Lusty C, Picq C (2006). Developing a regional Strategy to address the outbreak of Banana *Xanthomonas* wilt in East and Central Africa: Proceedings of the Banana *Xanthomonas* wilt regional preparedness and strategy development workshop held in Kampala, Uganda - 14-18 February 2005. International Network for the Improvement of Banana and Plantain, Montpellier, France

Karamura EB, Turyagyenda FL, Tinzaara W, Blomme G, Ssekiwoko F, Eden-Green S, Molina A, Markmham R (2008). *Xanthomanas* wilt of banana in East and Central Africa. Diagnostic and management guide. Bioversity International, Uganda.

Kubiriba J, Karamura EB, Jogo W, Tushemereirwe WK, Tinzaara W (2012). Community mobilization: A key to effective control of banana Xanthomonas wilt. J. Dev. Agric. Econ. 45(5):125-131.

Kubiriba J, Muthomi J, Ndungo V, Kwach J, Erima R, Rwomushana I, Tushemereirwe W, Opio F (2014). Strategies for rehabilitation of banana fields infested with *Xanthomonas campestris* pv. *musacrearum.* J. Crop Protect. 3(1):21-29.

Mwebaze JM, Tusiime G, Tushemereirwe WK, Kubiriba J (2006). The survival of *Xanthomonas campestris* pv. *musacearum* in soil and plant debris. Afr. Crop Sci. J. 14:(2)121-128.

Muhangi J, Nankinga C, Tushemereirwe WK, Rutherfold M, Ragama P, Nowakunda K, Abeyasekera S (2006). Impact of awareness campaigns for banana bacterial wilt in Uganda. Afr. Crop Sci. J. 14:(2)175-183.

Ndungo V, Eden-Green S, Blomme G, Crozier J, Smith J (2005). Presence of banana *Xanthomonas* wilt (*Xanthomonas campestris* pv. *musacearum*) in the Democratic Republic of Congo (DRC). Plant Pathol. 55:294.

Nelson R, Mardt C, Orrego R, Ortiz O, Fredrix M, Tenorio J, Vien NV (2001). Working with resource poor farmers to manage plant diseases. Plant Dis. 85:684-695.

Ngambeki D, Tushemereirwe WK, Okaasai O (2006). Awareness of banana bacterial wilt control in Uganda. Community Leaders' perspective. Afr. Crop Sci. J. 14:(2):165-174.

Sherwood SG (1997). Little things mean a lot: Working with Central American farmers to address the mystery of plant diseases. Agric. Hum. Values 14:181-189.

Ssekiwoko F, Tushemereirwe WK, Batte M, Ragama PE, Kumakech A (2006). Reaction of banana germplasm to inoculation with *Xanthomonas campestris* pv. *Musacearum.* Afr. Crop Sci. J. 14:(2):151-156.

Thresh JM (1980). The origins and epidemiology of some important plant virus diseases. Appl. Biol. 5:1-65.

Tinzaara W, Gold CS, Ssekiwoko F, Bandyopadhyay R, Abera A, Eden-Green SJ (2006). Role of insects in the transmission of banana bacterial wilt. Afr. Crop Sci. J. 14:(2)105-110.

Tushemereirwe W, Kangire A, Ssekiwoko F, Offord LC, Crozier J, Ba E, Rutherford M, Smith JJ (2004). First report of *Xanthomonas campestris* pv. *musacearum* on banana in Uganda. Plant Pathol. 53:802.

Tushemereirwe WK, Okaasai O, Kubiriba J, Nankinga C, Muhangi J, Odoi N, Opio F (2006). Status of banana bacterial wilt in Uganda. Afr. Crop Sci. J. 14(2):73-82.

# Salt spray as a micro-environmental factor affecting the growth of *Commelina maritima* L. at Lekki Beach, Nigeria

**Otitoloju Kekere[1] and Joseph Femi Bamidele[2]**

[1]Department of Plant Science and Biotechnology, Adekunle Ajasin University, Akungba Akoko, Ondo State Nigeria.
[2]Department of Plant Biology and Biotechnology, University of Benin, Benin City, Edo State, Nigeria.

***Commelina maritima*** **L. is a dominant perennial halophyte that is restricted to sandy beaches where it plays major ecological roles in Southern Nigeria. This study examines the response of this plant to saltwater sprays, a factor that affects the growth of coastal plants. Plants were sprayed with seawater twice per week (2/week), four times per week (4/week) or six times per week (6/week) while control was sprayed without seawater (de-mineralized water) six times per week (0/wk). Survival, growth and biomass allocation of the plants were determined. Salt spray did not affect plant survival but significantly ($p \leq 0.05$) decreased number of leaves, shoot length, stem girth, leaf area and root growth. Relative growth rate and number of branches were not significantly ($p \geq 0.05$) affected by salt spray. Sea spray significantly ($p \leq 0.05$) reduced fresh and dry mass of plant parts, total biomass and leaf total chlorophyll when compared with the control. Root : shoot ratio increased significantly ($p \leq 0.05$) under seawater treatment as the shoot growth was more negatively affected than root growth. Relatively more biomass was allocated to the root than shoot in seawater-treated plants. Salt spray increased shoot ash content and negatively affected plant organic content.** ***C. maritima*** **can be classified as a salt spray-sensitive plant. Salt spray is a micro-environmental factor affecting its survival and growth, thus influencing its distribution in strandline.**

**Key words:** Strandline, survival, growth, biomass allocation, Commelinaceae, distribution.

## INTRODUCTION

Strandlines are created along the high water mark on a range of habitats including coastal vegetated shingle, coastal sand dunes and coastal salt marshes (Lee and Ignaciuk, 1985). Several natural and anthropogenic factors affect the vegetation of strandlines (Owen et al., 2001; Peter et al., 2003; Pichler and Oberhuber, 2007). Studies on coastal plant communities in many parts of the world have found that air-borne saltwater spray is an important natural selective abiotic factor (Boyce, 1954; Cartica and Quinn, 1980; Sykes and Wilson, 1988; Hesp,

1991). Salt-sensitive species can be eliminated entirely from areas with high salt spray so that plants adapted to salt spray grow close to the ocean and are replaced by less salt-resistant plants farther inland (Cheplick and White, 2002). It has been found that natural salt spray accumulation on plants in the field can cause water stress, leaf necrosis, growth reduction and death depending on the frequency and severity of storms and winds that carry aerial salt drift inland (Griffiths and Orians, 2004; Griffiths, 2006).

Strandline community composition is structured by the effects of salt spray, and plants typical of coastal beaches are salt spray tolerant, which gives them a competitive advantage over others (Tominaga and Ueki, 1991; Greaver and Sternberg, 2007; De Vos et al., 2010). To date, the possible influence of natural processes on coastal vegetation has largely focused on soil salinity while air-borne salinity has been largely ignored.

*Commelina maritima* L. commonly called white mouth dayflower belongs to the family Commelinaceae. It is a fleshy, creeping, almost glabrous herb with blue flowers, in coastal sand bar vegetation. It is perennial and monocotyledonous with succulent stem and swollen nodes often mucilaginous (Hutchinson et al., 1986). This plant provides essential habitat, spawning sites and food for many diverse species of terrestrial and marine wildlife. The plant is able to withstand periodic coastal disturbances and it is particularly important on exposed shores, where its population can act as precursors to sand dunes. It helps in storm protection and controls shoreline erosion. It helps in shoreline stabilization by enhancing organic and moisture content; pioneer plants can then become established and sand dune formation is initiated (Williamson, 2005). *C. maritima* is widely distributed in West Tropical Africa where it is found a short distance from sea on the sandy beaches (Hutchinson et al., 1968). This plant species is dominant, wide spread and restricted to strandline habitat in Nigeria. We hypothesize that salt spray is a micro-environmental factor affecting its growth. This is because strand vegetation is affected by environmental factors such as soil salinity, sand burial, drought, waterlogging, nutrient deficiency, trampling and other anthropogenic activities. Each factor is usually considered as a micro-environmental factor among the various environmental conditions that affect strand vegetation (Barbour, 1978). It was reported that the death of *Impereta cylindrica* and *Miscanthus sinensis* transplanted to beach was as a result of salt spray, inhibiting them from becoming established on the front dunes. They concluded that salt spray is among the non-negligible factors controlling the distribution of plants in sand dune vegetation (Ogura and Yura, 2007). Plant's ability to survive in the strandline is largely influenced by salt spray. Since *C. maritima* thrives exclusively in the environment, salt spray is a natural abiotic factor that influences its growth in the strandline (Barbour, 1978). We experimentally manipulated salt spray in laboratory studies and examined its influence on the survival, growth and biomass allocation of the plant species. The results are used to discuss its tolerance and some adaptations for survival.

## MATERIALS AND METHODS

Stem cuttings of *C. maritima* collected from Lekki Beach in Lagos, Nigeria were used to raise uniform plants in perforated plastic pots filled with 2:1 ratio of washed river sand to top soil (Khan et al., 2000a, b). Seawater was collected off the shore, the source of the air-borne salts and stored in a plastic keg at room temperature. Salinity of the seawater was 30.00 ppt and later measurements with conductivity meter did not change with short time storage. Plants were randomly assigned to either salt-spray treatments or control. On Mondays and Thursdays, plants were sprayed with seawater twice per week (one seawater spray on each of the two days), four times per week (two seawater sprays on each of the two days) or six times per week (three seawater sprays on each of the two days) while control were sprayed without seawater six times per week (three de-ionized water prays on each of the two days) following the methods of Cheplick and Demetri (1999), Cheplick and White (2002) and De Vos et al. (2010). Individual plant was sprayed to run-off with all the aerial parts equally exposed. The accumulated salt onto the shoot for 1, 2 and 3 sprays was estimated following the method described by Cheplick and Demetri (1999), which equaled on average 4, 8 and 12 mg NaCl $dm^{-2}$ leaf area $day^{-1}$. These fall within the levels found in the natural habitat of beach plants (Barbour et al., 1985; Griffiths, 2006). A plant mist bottle held about 20 cm from the shoot was used for spraying. Pots were arranged on the Greenhouse bench of the Plant Science and Biotechnology Department, Adekunle Ajasin University, Akungba Akoko, Ondo State Nigeria (Lat. 7° 28$^1$N, Long. 5° 44$^1$ E). Seawater was applied as a fine mist from a spray bottle held 20 cm from the side of each shoot. Plants in all treatments were repositioned after each saltwater treatment. Salt that might possibly deposit on soil during misting was flushed out weekly with water. There were 6 single-plant replicates per treatment in a completely randomized experimental design. Prior to beginning the salt spray treatments, five plants were harvested to determine the initial growth parameters and successive measurements were taken at an interval of two weeks.

Percentage survival and growth parameters were recorded. Shoot length was measured from the soil level to the terminal bud and leaf area was obtained with a leaf area meter. Stem girth was measured at about 10 cm from stem base using model 0-200 mm digital caliper. Number of leaves and lateral branches of individual plants were counted manually. The experiment was terminated at 12 weeks after initiation of treatment. Plants were harvest, major roots were counted and their lengths measured. The harvested plants were separated into leaves, stems and roots and their fresh mass determined. Dry mass was obtained after oven-drying to constant mass at 80°C. Root dry mass was divided by that of the shoot to obtain root : shoot ratio while total biomass of all plant parts was weighed. Relative growth rate was calculated using the formula, RGR (relative growth rate) = (ln mass2-ln mass1)/time.

Total chlorophyll was extracted in 80% (v/v) aqueous acetone and absorption was read in a spectrophotometer at 645 and 663 nm (Arnon, 1949). Each sample of the dried plant materials was ground to fine powder using Philips model blender. Percentage ash content was determined by weighing the plant sample (powder) in a Pyrex beaker and placed in muffle furnace set at 500°C. Ash and organic mass components were also determined. The soil physico-chemical properties were analyzed in the Central Laboratory of The National Institute for Oil Palm Research (NIFOR), Nigeria, following the standard methods of Association of Official Analytical Chemists

**Table 1.** The physico-chemical properties of soil used for planting.

| Physico-chemical parameter | Value |
|---|---|
| pH | 5.48 |
| % C | 3.67 |
| N (ppm) | 20.42 |
| P (ppm) | 3.56 |
| K (meg/100 g) | 3.56 |
| Ca (meg/100 g) | 2.32 |
| Mg (meg/100 g) | 2.60 |
| CEC (meg/100 g) | 8.2 |
| Sand (%) | 80.68 |
| Silt (%) | 12.06 |
| Clay (%) | 8.36 |

(AOAC, 1985). Data were subjected to single factor analysis of variance and means were separated with Duncan's MRT using SPSS version 17.0 (SPSS Inc., Chicago, IL, USA) at $P \leq 0.05$.

## RESULTS

Table 1 shows the physico-chemical properties of the soil used for planting. The ratio 2:1 of washed river sand to top soil is typical of sandy beach soil in strandline. The sandy soil supported the plant growth while the top soil was for nutrient supply. The soil was low in nutrients in comparison with nutrient composition of agricultural soils. Damage caused by salt spray include leaf browning, chlorosis and necrosis. Salt spray led to leaf firing and leaf injury which resulted in early leaf senescence and defoliation. Leaf death and defoliation in plants sprayed with seawater resulted in leaf number reduction. Leaf number decreased with increase in level of seawater (Figure 1). One-way ANOVA however revealed that reduction in leaf number became significant ($p \leq 0.05$) only at 4/week and 6/week when compared with the control treatment.

Table 2 shows the influence of salt spray on the growth parameters of *C. maritime*. Generally, plant growth was inhibited by salt spray but there was no plant mortality. Leaf area decreased significantly ($p \leq 0.05$) in plants sprayed with seawater relative to the control. At the highest level of salt application, leaf area was reduced by as high as 53.45%. In plants exposed to salt spray, root growth was negatively affected. This is evidenced by a significantly ($p \leq 0.05$) reduced root length and root number under seawater treatment as compared to the control. There was however no statistical difference ($p \geq 0.05$) when different levels of salt treatments were compared. Plants sprayed with seawater exhibited a significantly reduced ($p \leq 0.05$) stem girth in comparison with those exposed to de-ionized water. Relative growth rate likewise decreased with increase in salt spray but there was no significant difference ($p \geq 0.05$) between control and seawater-treated plants. Total biomass and chlorophyll content of plants subjected to salt treatment were reduced significantly ($p \leq 0.05$) as compared to the control. Root : shoot ratio increased significantly ($p \leq 0.05$) under seawater treatments relative to de-mineralized water treatment. This revealed that shoot growth was more negatively affected by salt spray than root growth. Shoot length of *C. maritima* treated with seawater decreased progressively with increasing salt level. This became significant ($p \leq 0.05$) from eight weeks after treatment. At the end of the investigation, shoot length was reduced by 37.14% at 6/week relative to the 0/week treatment (Figure 2). However, plants sprayed with seawater had reduced number of branches but there was but there was no significant difference ($p \geq 0.05$) between the control and those treated with salt spray (Table 2).

Fresh and dry mass of plant parts declined progressively with increasing salt spray levels. One-way ANOVA showed that fresh and dry mass were significantly lower ($p \leq 0.05$) under salt sprays than in the control treatment (Table 3). Figure 3 shows the biomass allocation pattern of *C. maritime* under different spray treatments. Relatively more organic biomass was allocated to the root than the shoot in plants subjected to salt spray. This was due to salt accumulation on the shoot following foliar spray with seawater. This led to an increase in the shoot ash content thus having a negative effect on percentage organic content. Organic biomass allocation was not affected by salt treatment in roots since they were not in direct contact with salt. When compared with the control, shoot ash mass was significantly increased ($p \leq 0.05$) by salt sprays. However, there was no significant difference ($p \geq 0.05$) when root ash mass was compared between control and seawater treatments (Figure 4). Organic mass was significantly ($p \leq 0.05$) reduced both in the shoot and root due to the influence of salt deposition on shoot surfaces (Figure 5).

## DISCUSSION

The ratio 2:1 washed river sand to top soil has been reported to have properties of beach soil where strand plants naturally grow (Khan et al., 2000a, b). Although, the soil was low in nutrients, strand plants have low nutrient requirements for growth (Lee and Ignaciuk, 1985). Since no negative effect was observed in control plants, the observed effect on the plant was not due to nutrient unavailability but salt spray. The changes in plant morphology due to salt sprays are consistent with the previous studies demonstrating damage to plants as a result of salt spray accumulation (Griffiths, 2006). Necrosis on *Myrica pensylvanica* leaves decreased as distance from the dune crest increased (Griffiths and Orians, 2003). Sea spray likewise resulted in leaf necrotic

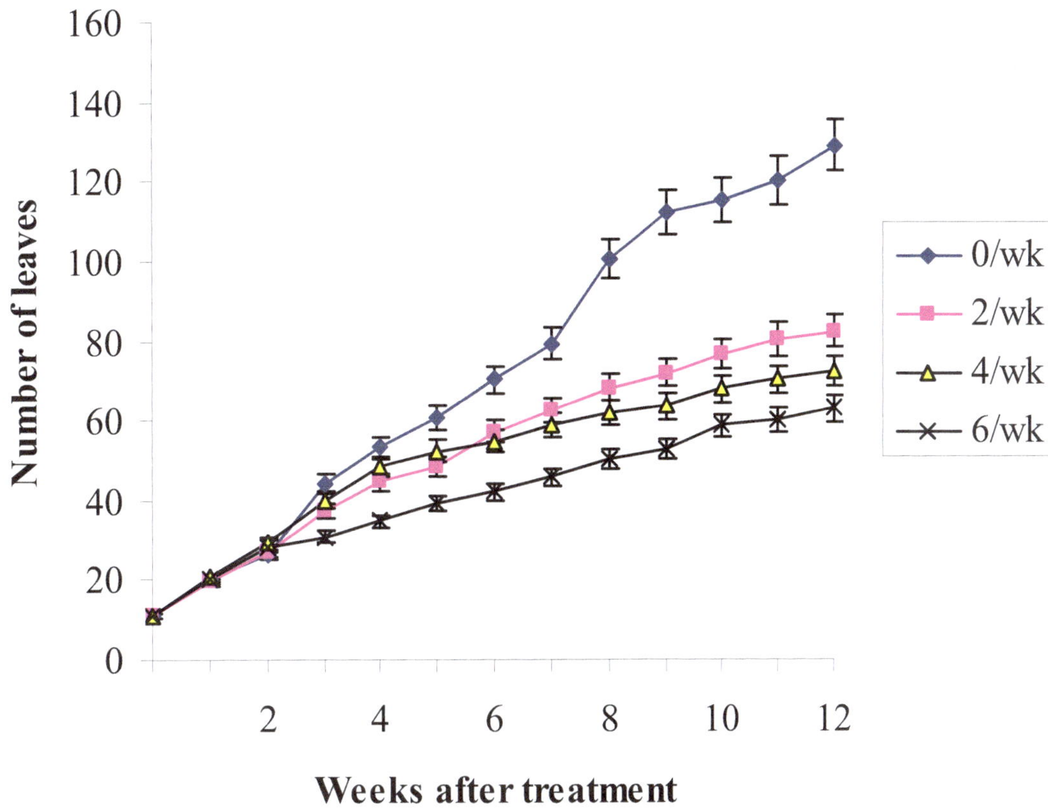


**Figure 1.** Effect of salt sprays on the leaf number of *C. maritima*, taken at 2-week interval for 12 weeks. Each value is a mean ± standard error of 6 determinations. 0/week = sprayed without seawater (de-mineralized water) six times per week, 2/week = sprayed with seawater twice per week, 4/week = sprayed with seawater four times per week, 6/week = sprayed with seawater six times per week.

damage in seedlings of *Pinus rigida* (Griffiths and Orians, 2004), *Solidago puberula*, *Solidago rugosa*, *Gaylussacia baccata* and *Quercus ilicifolia* (Griffiths and Orians, 2003) and in some coastal plant species (Griffiths et al., 2006). Kim et al. (2004) also observed browning symptoms followed by defoliation after the foliar spray of 3% NaCl in apple, pear, peach and grape trees. These could result in a decrease in net photosynthesis that might be expressed in reduced growth (Griffiths and Orians, 2004).

In coastal areas, salt spray often influences the distributions of species (Cheplick and Demetri, 1999; Griffiths and Orians, 2003; Ogura and Yura, 2007). Complete survival of plants under salt sprays conforms to the findings of Gagne and Houle (2002) on *Leymus mollis* and De Vos et al. (2010) on *Crambe maritima*. It is however in contrast with that of Gagne and Houle (2002) who reported a significantly reduced seedling survival in *Honckenya peploides*.

Growth reduction in *C. maritima* in this investigation confirms the report that salt spray usually leads to growth reduction (Tominaga and Ueki, 1991; Griffiths, 2006). Salt spray caused growth reduction in *Leymus mollis* (Gagne and Houle, 2002) and *M. pensylvanica* (Griffiths and Orians, 2003). Reduction in root elongation has been reported on many coastal plant species (Griffiths and Orians, 2004; Griffiths et al., 2006). The reduction in plant shoot length as obtained in this study affirms the report that salt spray inhibits the shoot elongation in coastal plant species (Griffiths et al., 2006). This could be one mechanism through which the characteristic dwarf stature of strand vegetation is maintained (Griffiths and Orians, 2003).

Reduction in number of branches agrees with the previous work of Cheplick and Demetri (1999) who recorded reduced number of tillers in *Triplasis purpurea* sprayed with salt relative to control plants. Reduction in leaf number in plants sprayed with seawater likewise supports the earlier report on *Scaevola sericea* seedlings by Goldstein et al. (1996) and De Vos et al. (2010) on *Crambe maritima*. Reduced leaf number in this study was due to leaf firing leading to leaf death and defoliation. Leaf area reduction is in conformity with that of Ogura and Yura (2007) on *Miscanthus sinensis* and *Imperata cylindrica* and in *Pinus rigida* (Griffiths and Orians, 2004). Leaf length was also reduced in *Crambe maritima* seedlings 13 weeks after spraying with salt water. Reduced leaf size was as a result of reduction in leaf area expansion and hence reduction of light interception. Reduction in leaf area also provided reduced area for water loss through transpiration and changes in

**Table 2.** Effect of salt sprays on leaf total chlorophyll and growth parameters of *C. maritima*.

| No. of spray (s) per week | Survival (%) | Stem Girth (cm) | Leaf area ($cm^2$) | Number of branches | Total biomass ($gplant^{-1}$) | Root: Shoot | RGR ($gg^{-1}d^{-1}$) | Total chlorophyll ($mgg^{-1}$) | Root number | Root length (cm) |
|---|---|---|---|---|---|---|---|---|---|---|
| 0 | 100 | $0.40^{a}$ | $5.04^{a}$ | $22.72^{a}$ | $20.67^{a}$ | 1.34 | $0.08^{a}$ | $3.24^{a}$ | $67.43^{a}$ | $44.71^{a}$ |
| 2 | 100 | $0.31^{b}$ | $2.41^{b}$ | $19.55^{a}$ | $13.45^{b}$ | 1.95 | $0.07^{a}$ | $1.43^{b}$ | $49.87^{b}$ | $32.04^{b}$ |
| 4 | 100 | $0.31^{b}$ | $2.36^{b}$ | $16.89^{a}$ | $11.50^{b}$ | 2.15 | $0.07^{a}$ | $1.71^{b}$ | $47.65^{b}$ | $34.11^{b}$ |
| 6 | 100 | $0.28^{b}$ | $2.54^{b}$ | $16.87^{ab}$ | $8.47^{bc}$ | 1.87 | $0.07^{a}$ | $1.68^{b}$ | $48.55^{b}$ | $34.67^{b}$ |

RGR = Relative growth rate. Each value is a mean of 6 replicates taken at 12 weeks after initiation of treatments. Means with the same letter (in superscript) in the same column are not significantly different at $P \geq 0.05$.

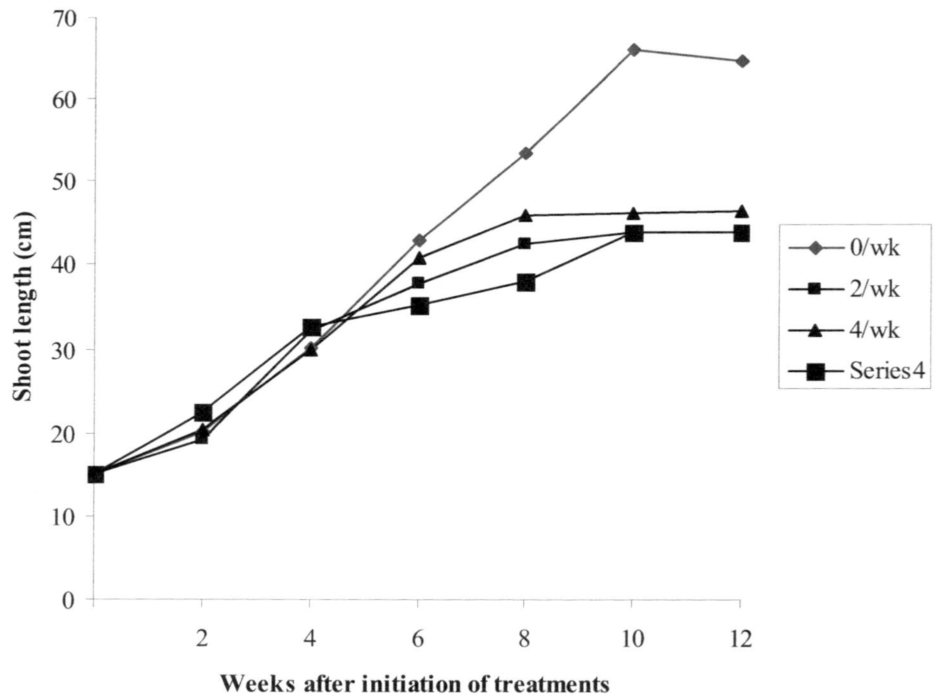


**Figure 2.** Effect of salt sprays on the shoot length (cm) of *C. maritima*, taken at 2-week interval for 12 weeks. Each value is a mean of 6 replicates. 0/week = sprayed without seawater (de-mineralized water) six times per week, 2/week = sprayed with seawater twice per week, 4/week = sprayed with seawater four times per week, 6/week = sprayed with seawater six times per week.

**Table 3.** Effect of salt sprays on the fresh and dry mass of plant parts in *C. maritima*.

| Parameter | Plant part | No of spray(s)per week | | | |
|---|---|---|---|---|---|
| | | 0 | 2 | 4 | 6 |
| Fresh mass (g) | Leaf | 5.10$^{a}$ | 2.56$^{b}$ | 2.11$^{b}$ | 1.25$^{bc}$ |
| | Stem | 10.03$^{a}$ | 5.94$^{b}$ | 4.70b | 5.16$^{b}$ |
| | Root | 14.87$^{a}$ | 11.14$^{a}$ | 9.76$^{a}$ | 7.38$^{c}$ |
| Dry mass (g) | Leaf | 3.77 | 2.04 | 1.67 | 1.01 |
| | Stem | 5.07 | 2.53 | 1.97 | 1.96 |
| | Root | 11.88 | 8.91 | 7.81 | 5.56 |

Each value is a mean of 6 replicates, taken at 12 weeks after initiation of treatments. Means with the same letter (in superscript) in the same row are not significantly different at $P \geq 0.05$. 0/week = sprayed without seawater (de-mineralized water) six times per week, 2/week = sprayed with seawater twice per week, 4/week = sprayed with seawater four times per week, 6/week = sprayed with seawater six times per week.

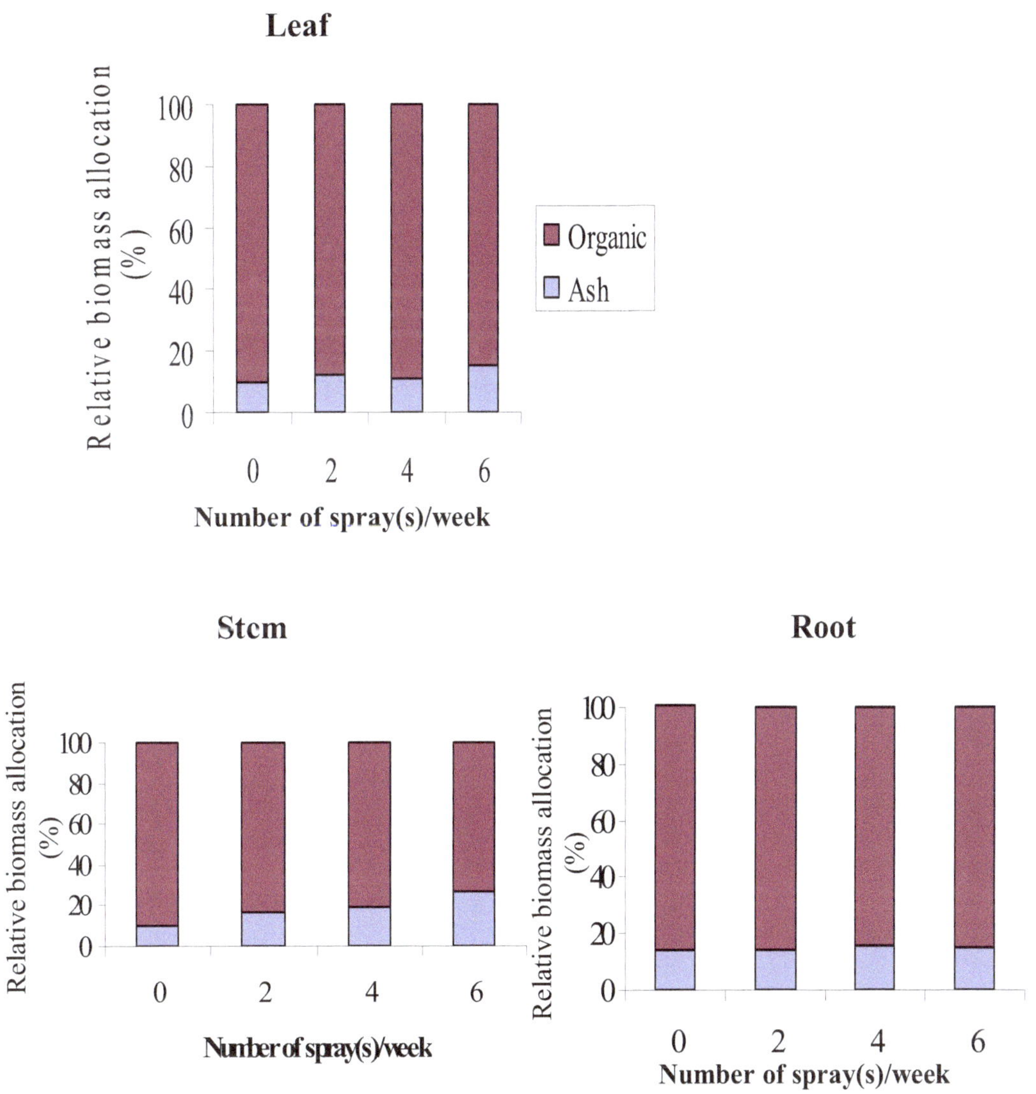


**Figure 3.** Effect of salt sprays on the relative biomass allocation (%) of *C. maritima*. Each bar represents mean of six determinations, taken at 12 weeks after initiation of treatments.

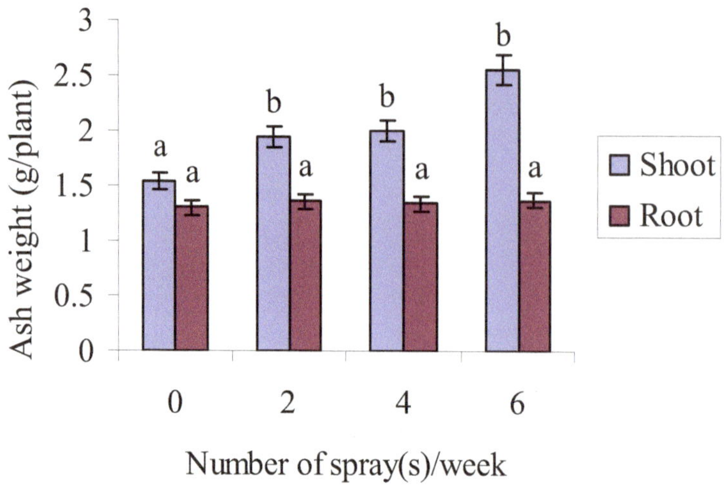


**Figure 4.** Effect of salt sprays on the ash mass (g plant$^{-1}$) of *C. maritima*. Each bar represents mean ± standard error of 6 determinations. Bars with the same letter are not significantly different at P ≥ 0.05.

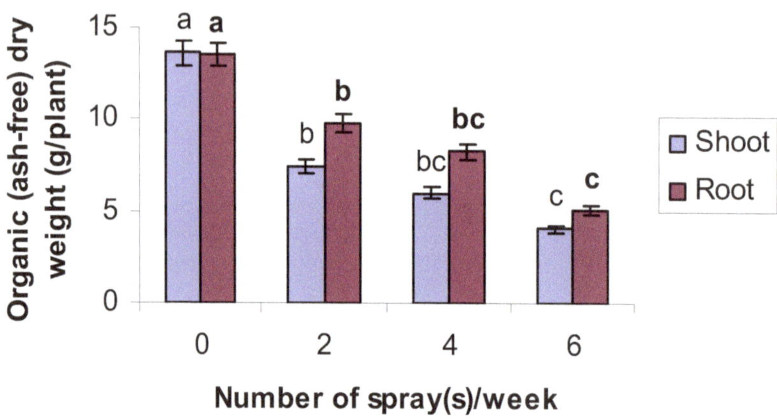


**Figure 5.** Effect of salt sprays on the organic mass (g plant$^{-1}$) of *C. maritima*. Each bar represents mean ± standard error of 6 determinations, taken at 12 weeks after initiation of treatments. Bars with the same letter are not significantly different at P ≥ 0.05.

water use efficiency in plants, which have been identified to be adaptive mechanisms under water stress (Morant-Manceau et al., 2004). Reduced photosynthetic leaf area was due primarily to chloride and sodium ions from the salt spray (Cartica and Quinn, 1980). However, leaf area reduction is in contrast with that of Touchette et al. (2009) on *Spartina alterniflora*. It has been reported that salt penetrates leaves through lesions or via stomata and is consequently translocated to other plant parts (Boyce, 1954; Barbour et al., 1985) which could affect root development.

Reduction in chlorophyll content was due to leaf damage resulting in increased necrotic spots. Necrotic damage by salt spray has been reported by many authors (Sykes and Wilson, 1988; Griffiths and Orians, 2004). This usually results in a decreased net photosynthesis that might be expressed in reduced growth and long-term survivorship (Griffiths and Orians, 2004). Chlorophyll reduction can also be due to ion deficiency or ion toxicity.

Factors such as soil salinity, salt spray, sand burial, water deficit, flooding and nutrient deficiency are the natural factors that determine the distribution of strand vegetation (Lee and Ignaciuk, 1985). Strand plants vary considerably in the factors that affect their survival and growth, with each considered as a micro-environmental factor among the various environmental conditions including soil salinity, sand burial, drought, waterlogging and nutrient deficiency that affect strand vegetation (Barbour, 1978). This experiment has shown that *C. maritima* growth was negatively affected by salt spray. Therefore, salt spray is a micro-environmental factor affecting *C. maritima* growth, which determines its presence and distribution in the strandline.

## Conclusions

Strand plants vary considerably in the specific factor(s) among the various abiotic factors that affect them in the strandline, hence the question: which of the factors determine the distribution of every plant species that grow naturally in the environment? This research has revealed that salt spray affects the growth of *C. maritima*, and it can be categorized as a salt spray-sensitive plant. Salt spray is therefore a micro-environmental factor (among the various abiotic factors) that contributes to the survival, growth and distribution of *C. maritima* in the strandline.

## Conflict of Interests

The author(s) have not declared any conflict of interests.

### REFERENCES


Arnon DI (1949). Copper enzymes in isolated chloroplast and polyperoxidase in *Phaseolus vulgaris*. Plant Physiol. 24:1-15.

Association of Official Analytical Chemists (AOAC) (1985). Official methods of analysis. 12th edition.

Barbour M, DeJong TM, Pavlik BM (1985). Marine beach and dune plant communities. In: Physiological ecology of North American plant communities, Chabot, B.F. and Mooney, H. A. (eds.), Chapman and Hall, New York, NY. pp. 296-322.

Barbour MG (1978). Salt spray as a micro-environmental factor in the distribution of beach plants at Point Reyes, California. *Oecologia*, 32:213-224.

Boyce SG (1954). The salt spray community. Ecol. Monogr. 24:29-67.

Cartica RJ, Quinn JA (1980). Responses of populations of *Solidago sempervirens* (Compositae) to salt spray across a barrier beach. Bull. Torrey Bot. Club. 104:29-34.

Cheplick GP, Demetri H (1999). Impact of saltwater spray and sand deposition on the coastal annual *Triplasis purpurea* (Poaceae). Am. J. Bot. 86(5):703-710.

Cheplick GP, White TP (2002). Saltwater spray as an agent of natural selection: no evidence of local adaptation within a coastal population of *Triplasis purpurea* (Poaceae). Am. J. Bot. 89:623-631.

De Vos AC, Broekman R, Groot MP Rozema J (2010). Ecophysiological response of *Crambe maritima* to airborne and soil-borne salinity. Ann. Bot. 105(6):925-937.

Gagne JM, Houle G (2002). Factors responsible for *Honckenya peploides* (Caryophyllaceae) and *Leymus mollis* (Poaceae) spatial segregation on subarctic coastal dunes. Am. J. Bot. 89:479-485.

Greaver TL, Sternberg LSL (2007). Fluctuating deposition of ocean water drives plant function on coastal sand dunes. Glob. Change Biol. 13:216-223.

Goldstein G, Alpha CG, Drake DR (1996). Morphological and physiological responses of *Scaevola sericea* (Goodeniaceae) seedlings to salt spray and substrate salinity. Am. J. Bot. 83:86-92.

Griffiths ME (2006). Salt spray and edaphic factors maintain dwarf stature and community composition in coastal sandplain heathlands. Plant Ecol. 186:69-86.

Griffiths ME, Keithac RP, Oriansa CM (2006). Direct and indirect effects of salt spray and fire on coastal heathland plant physiology and community composition. *Rhodora*, 108(933):32-42.

Griffiths ME, Orians CM (2003). Responses of common and successional heathland species to manipulated salt spray and water availability. Am. J. Bot. 90:1720-1728.

Griffiths ME, Orians CM (2004). Salt spray effects on forest succession in rare coastal sandplain heathlands: evidence from field surveys and *Pinus rigida* transplant experiments. J. Torrey Bot. Soc. 131:23-31.

Hesp PA (1991). Ecological processes and plant adaptations on coastal dunes. J. Arid Environ. 1:165-191.

Hutchinson J, Dalziel JM, Hepper FN (1968). Phylogenic sequence of orders and families. In: Flora of West Tropical Africa. Published by Crown Agents for Overseas Governments and Administrations, Volume III, Part 1, Millibank, London, S. W. 1. 574p.

Khan MA, Ungar IA, Showalter AM (2000a). Effects of salinity on growth, water relations and ion accumulation of the subtropical perennial halophyte, A*triplex griffithii* var. *stocksii*. Ann. Bot. 85:225-232.

Khan MA, Ungar IA, Showalter AM (2000b). The effect of salinity on the growth, water status, and ion content of a leaf succulent perennial halophyte, *Suaeda fruticosa* (L.) Forssk. J. Arid Environ. 45:73-84.

Kim S, Seo H, Kim J, Park MY, Kim S (2004). Leaf and bud responses to foliar spray of saline solutions in apple, pear, peach and grape. Korean J. Hortic. Sci. Technol, 45(6):340−344.

Lee JA, Ignaciuk R (1985). The physiological ecology of strandline plants. Plant Ecol. 62(1-3):15-19.

Morant-Manceau A, Pradier E, Tremblin G (2004). Osmotic adjustment, gas exchanges and chlorophyll fluorescence of a hexaploid triticale and its parental species under salt stress. J. Plant Physiol. 161(1):25-33.

Ogura A, Yura H (2007). Effects of sandblasting and salt spray on inland plants transplanted to coastal sand dunes. Ecol. Res. 23(1):107-112.

Owen N, Kent M, Dale P (2001). Spatial and temporal variability in seed dynamics of machair sand dune plant communities, the Outer Hebrides, Scotland. J. Biogeogr. 28:565-588.

Peter CI, Ripley BS, Robertson MP (2003). The distribution of *Scaevola plumieri* along the South African coast is limited by seasonal water balance and temperature. J. Veg. Sci. 14:89-98.

Pichler P, Oberhuber W (2007). Radial growth response of coniferous forest trees in an Inner alpine environment to heat-wave in 2003. For. Ecol. Manage. 242:688-699.

Sykes MT, Wilson JB (1988). An experimental investigation into the response of some New Zealand sand dune species to salt spray Ann. Bot. 62:159-166.

Tominaga TH, Ueki KK (1991). Clonal variation in salt tolerance of *Imperata cylindrica* (L.) Beauv. var. koenigii (Retz.) *et* Schinz. J. Jpn. Grassl. Sci. 37:69-75.

Touchette BW ,Rhodes KL, Smith GA, Poole M (2009). Salt spray induces osmotic adjustment and tissue rigidity in smooth cordgrass, *Spartina alterniflora* (Loisel). Estuar. Coast. 32(5):917-925.

Williamson K (2005). Action plan scope. In: Strandlines. Nature Gwynedd Bulletin, version 1, 2005. 6p.

# 7

# Genetic variability and divergence of seed traits and seed germination of five provenances of *Faidherbia albida (Delile)* A. Chev

**Grace Koech[1,2], Daniel Ofori[1], Anne W.T. Muigai[2], Martha Makobe[2], Muriuki, Jonathan[1], Mowo, G. Jeremias[1] and Jamnadass, Ramni[1]**

[1]World Agroforestry Centre, United Nations Avenue, Gigiri, P. O. Box 30677-00100, Nairobi, Kenya.
[2]Jomo Kenyatta University of Agriculture and Technology, P. O. BOX 62000-00100, Nairobi, Kenya.

**Establishment of *Faidherbia albida* trees on farm is often difficult despite the plant survival adaptive mechanisms such as drought and disease resistance. Adoption of the tree to agroforestry systems is also limited due to lack of knowledge on genetic variation of its provenances. Morphological characterization of *F. albida* provenances is therefore necessary to screen for natural genetic variation in seeds traits for selection of germplasm for long term agroforestry, timber production, fodder, soil fertility increment and environmental sustainability. In this study, seed traits of five provenances of *F. albida*: Taveta Wangingombe, Lupaso, Kuiseb and Manapools were examined. Divergent studies were analyzed based on seed morphology and geo-climatic conditions of the provenances. Seed length, width, thickness and weight were analyzed to determine the extent of phenotypic and genotypic variance and heritability. This study revealed significant differences among provenances (P≤0.05) for all the studied characters indicating substantial genetic variability. Genetic variance for all seed traits was higher than environmental variance suggesting that the expressions of these traits are under genetic control. This result was supported by high heritability values for seed length (0.92), width (0.99), thickness (0.99) and weight (0.99). Seed germination test involved 4 replicates of 25 randomly selected seeds per provenance. Mean germination percentage among provenances was 83.3% with the highest being 97% and the lowest 71%, P≤0.05. Relationships among these variables were analyzed using principal component analysis and cluster analysis resulting in separation of provenances into three distinct clusters. Manapools (760 mm), Lupaso (1165 mm) and Wangingombe (628 mm) with high rainfall were placed in cluster one. Taveta (545 mm) cluster two and Kuiseb (<50 mm) cluster three. Wangingombe (1700 m a.s.l.) clustered closely to Lupaso (500 m a.s.l.) than Taveta (760 m a.s.l.). High heritability ($h^2$>0.5) for all traits suggests that selection based on morphological traits can be made with a high degree of confidence.**

**Key words:** Provenances, selection, clinal distribution, geographical differentiation, genetic variation, heritability.

## INTRODUCTION

*Faidherbia albida* (Del.) A. Chev. is a leguminous trees species belonging to *Mimosoideacea* subfamily, tribe *Acaciacea*. It is locally referred to as Apple-Ring Acacia, Ana tree, Balanzan tree and Winter Thorn (Fagg and Barnes, 2003). Due to its phytochemical properties, pollen structure and phenology, it is placed in a monotypic genus

*Faidherbia* (Bernard, 2002; Hyde and Wursten, 2010). *F. albida* is an important agroforestry appreciated due to it compatibility with cropping systems (Roupsard et al., 1999; Payne, 2000; Ibrahim and Tibin, 2003). The tree is used in dry lands for soil conservation (Dangasuk et al., 2001). It can grow among field crops without over-shadowing them during the rainy season and provides shade during dry season (Orwa et al., 2009). Falling leaf mulch and the canopy shade creates a microclimate with better infiltration and reduced evapotranspiration which is crucial for plants (Gassama-Dia et al., 2003).

Despite the immense benefits of *F. albida*, broadening the utilization, breeding and conservation of the species still remain a challenge. *F. albida* is faced with threats to its gene pool due to drought and land use pressure; the situation is worsened by lack of natural regeneration and artificial propagation by seed or suckers among the people living in the Sahel (Weber and Hoskins, 1983; Bonkoungou, 1992; McGahuey, 1992). However, the *F. albida* trees currently growing in the Sahel regenerate naturally (Weber and Hoskins, 1983) and attempts of artificial regeneration have met failures due to poor survival and highly variable growth in early stages (McGahuey, 1985). It is therefore important to test seeds in order to screen for naturally available genetic variation for selection of germplasm for long term agroforestry, sustainability of timber production, fodder and soil fertility increment and environmental sustainability.

Seed morphology is an evolutionary trait contributing to genetic diversity (Aniszewski et al., 2001). Seed morphology influence water relation and seed dispersal, emergence, survival and seedling establishment (Milberg and Lamont, 1997); large and heavy seeds germinate rapidly with high survival and fast growth as compared to small seeds. Seed morphology is linked to fitness hence successful establishment (Zhang, 1998). Seeds traits that show high morphological variation are useful for selection of germplasm for conservation and propagation (Khurana and Singh, 2001). It has generally been reported that large seeds produce seedlings that have high survival rates (Moles and Westoby, 2004; Turnbull et al., 2008; Westoby, 1998) and trees with small seeds may produce more seeds per individual but with low seedling survival rates (Rees, 1995; Turnbull et al., 1999; Eriksson et al., 2000; Leishman and Murray, 2001). Knowledge of genetic variation among seed traits provides useful information for adaptation to heterogeneous environment. Seed traits that have high heritability values ($h^2 \geq 0.05$) are useful markers for selection (Akbar et al., 2003).

Variability studies are important in improvement of tree as it provides information relevant for selection of quality germplasm (Bhat and Chauhan, 2002). Germplasm quality affects the quality of trees propagated (IFSP, 2000) and should be considered during selection. Ibrahim (1996) and Dangasuk et al. (1997) examined the variation in seed and seedling traits of *F. albida* but did not provide any information on genetic and phenotypic variance, phenotypic and genotypic coefficient of variation and heritability of *F. albida* provenances. The current study is novel and aims at testing the reliability of selection of *F. albida* provenances based on seed morphological traits and to determine the genetic relationships among the provenances under study based on seed morphology and geo-climatic conditions.

## MATERIALS AND METHODS

### Study site

Morphological characterization of the seeds was conducted in April 2014 at the seed laboratory of the World Agroforestry Centre (ICRAF) in Nairobi, while the seeds were germinated in the ICRAF nursery greenhouse. ICRAF is located about 20 km south east of Nairobi, Kenya at latitude of 1° 33' S longitude 37° 14' E and an altitude of 1580 m above sea level. The mean annual rainfall ranges between 500 and 1370 mm and the mean temperature of 21°C.

### Seed sources

The seeds of five provenances, obtained from ICRAF seed bank were used for the study. The seeds were collected by Oxford Forest Institute (OFI) in 1990 for international provenance trials. The seeds were then processed and sent to ICRAF seed bank. For each provenance, OFI selected trees that had geographical discontinuity from other populations. The selected trees were obtained from sites with varying climate, soil, altitude and ecology. Undisturbed natural populations with distinct morphological and phenological traits were selected.

OFI collected 25 mother trees to represent a provenance. The mother trees were spaced 100 m apart to avoid collections from related individuals and to capture a large proportion of the natural genetic variation within the provenance (Ofori, 2001). The geographical range of the seeds varied from 3°24'S, 37°42'E to 23°34'S, 15°02'E longitude and altitude of 360 to 1700 m above the sea level. Taveta and Wangingombe represented eastern Africa provenances while Lupaso, Kuiseb and Manapools represented southern Africa provenances (Figure 1 and Table 1).

### Characterization of seed morphology

To investigate variation in seed length, width, weight and thickness, 100 seeds per provenance were randomly selected from the seed bulk. The seed were organized in a completely randomized experimental design with 25 seeds replicated four times per provenance (25 x 4 x 5) = 500 experimental units (Table 2). A Vernier caliper calibrated to two decimal places was used to measure seed length, width and thickness. Seed length was measured over the seed coat along the longest axis of the seed;

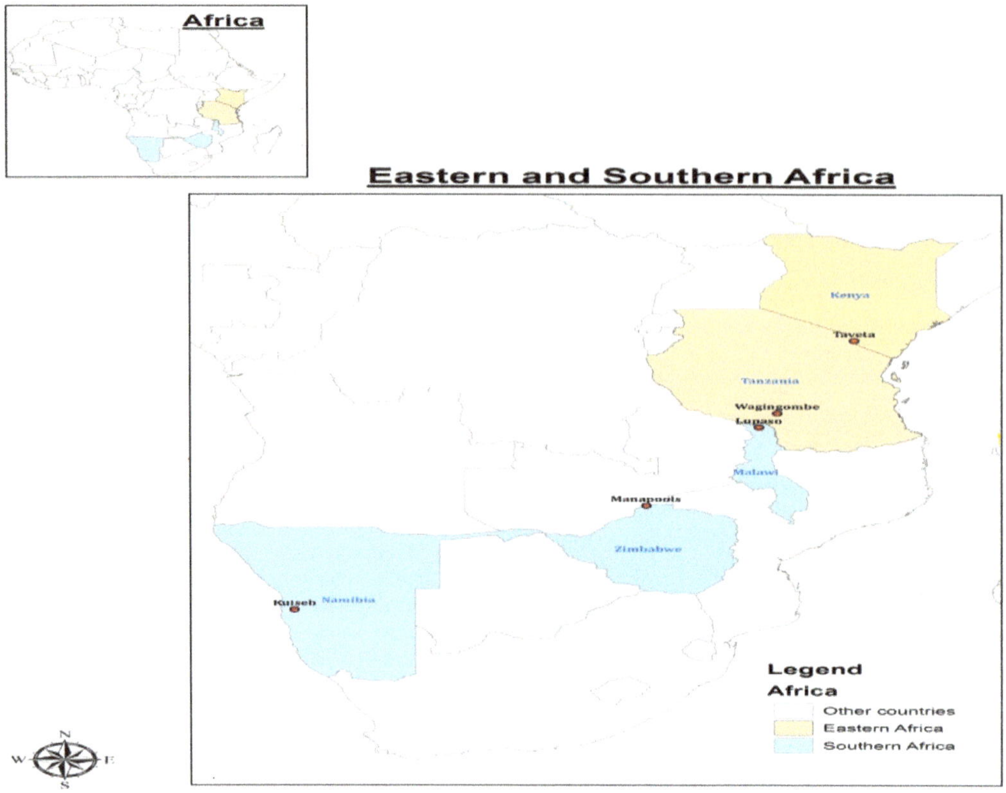


**Figure 1.** Map of the collection site for *Faidherbia albida* provenances under study.

**Table 1.** Geo-climatic (latitude, longitude, altitude, rainfall and temperature) of the different provenances of *F. albida* under study.

| Country | Region | Provenance | Location | Altitude (m asl) | Rainfall (mm) | Temperatures (°C) |
|---|---|---|---|---|---|---|
| Kenya | Eastern Africa | Taveta | 3°24'S, 37°42"E | 760 | 545 | 28.1 |
| Malawi | Southern Africa | Lupaso | 9°55'S, 33°53'E | 500 | 1165 | 24.8 |
| Namibia | Southern Africa | Kuiseb | 23°34'S, 15°02'E | 400 | <50 | 15.2 |
| Tanzania | Eastern Africa | Wangingombe | 8°51'S, 34°38'E | 1700 | 760 | 27.1 |
| Zimbabwe | Southern Africa | Manapools | 15°45'S, 29°20'E | 360 | 628 | 25.1 |

**Table 2.** Completely randomized experimental design for characterization of seed morphology and seed germination test

| Provenance | Number of seeds per replicate | | | | Total |
|---|---|---|---|---|---|
| | Replicate 1 | Replicate 2 | Replicate 3 | Replicate 4 | |
| Taveta | 25 | 25 | 25 | 25 | 100 |
| Wangingombe | 25 | 25 | 25 | 25 | 100 |
| Lupaso | 25 | 25 | 25 | 25 | 100 |
| Kuiseb | 25 | 25 | 25 | 25 | 100 |
| Manapools | 25 | 25 | 25 | 25 | 100 |
| Total | 100 | 100 | 100 | 100 | 500 |

seed width measurements were taken on one of the widest faces at the middle of the seed; seed coat thickness was measured without the removal of seed coat from the seed. An electronic top pan model was used to weigh each seed for all replicates.

**Table 3.** Morphological variation in seed length, width, thickness and seed weight among the five provenances of *F. albida*

| Provenance | Seed length(mm) | Seed width(mm) | Seed thickness (mm) | Seed weight (mg) |
|---|---|---|---|---|
| Taveta | 6.73c | 4.98d | 2.96c | 83.83d |
| Lupaso | 8.86b | 5.18c | 3.05a | 126.43c |
| Kuiseb | 10.21a | 5.89ab | 3.01ab | 160.62a |
| Manapools | 10.44a | 6.00a | 3.03ab | 167.15a |
| Wagingombe | 9.30b | 5.75b | 3.04ab | 143.20b |
| CV | 12.26% | 6.80% | 0.16% | 30.41% |
| SD | 1.74 | 0.55 | 0.014 | 50.9 |

CV = coefficient of variation, SD = standard deviation. Means followed by a different letter within a column are significant different according to LSD post-hoc test for all the traits.

### Seed germination test

After recording the variation in seed morphology, the seeds were germinated. The seeds were nicked (Bewley and Black, 1994) at the distal end near the microphyle (Manz et al., 2005; Wojtyla et al., 2006) using a nicking caliper. The seeds were then soaked in water for 24 h before sowing. Rapid influx of water was observed during the first twelve hours due to low water potential of dry seeds (Obroucheva and Antipora, 1997). Polythene tubes measuring 10 x 20 cm, filled with sterilized sand were used to germinate the seeds. The sand was sterilized using the oven method. Prior to sterilization, the sand was cleaned in tap water then placed in metal baking pans up to four inches in depth. The metal pans were then tightly covered with aluminum foil and placed in the oven at temperatures between 180 to 200°F; the temperatures were maintained for 30 min. After heating, the oven was cooled and metal pans removed. The aluminum foil covering the metal pans was left intact to prevent the sand from contamination. The sand was watered to field capacity before sowing the seeds. The polythene tubes containing the seeds were kept under a temperature range of 25-30°C and photoperiod of 12 h light and 12 h dark. Seed germination was monitored for thirty days and mean germination percentage was calculated following ISTA (1993).

### Statistical analysis

Seed morphological data and germination percentage were subjected to one-way analysis of variance after testing for homogeneity of variance and normality (Zar, 1996). The genotype means were further separated and compared using least significant difference test at a 0.05 significance level. The genetic, phenotypic and environmental variance, genotypic, environmental and phenotypic coefficient of variation and heritability for each trait were calculated from one way analysis of variance using formulas formulated by Zobel and Talbert (1991) as shown below:

$$\sigma^2_g = \frac{(MSG - MSE)}{r}$$

$$\sigma^2_p = \frac{MSG}{r}$$

$$\sigma^2_e = \left(\frac{MSE}{r}\right)$$

$$h^2 = \frac{\sigma^2_g}{\sigma^2_p}$$

$$PCV = \frac{\sqrt{\sigma^2_p}}{\bar{x}} \times 100$$

$$GCV = \frac{\sqrt{\sigma^2_g}}{\bar{x}} \times 100$$

Where

$\sigma^2_g$ = Genotypic variance, $\sigma^2_p$ = phenotypic variance, $\sigma^2_e$ = environmental variance $MSG$ = mean sum square for genotype, $MSE$ = mean sum square for error and $r$, is the number of replications, h² = narrow sense heritability, PCV = phenotypic coefficient of variation, GCV = genotypic coefficient of variation and $\bar{x}$ = grand mean.

Pearson correlation coefficient for seed traits and among seed traits with geo-climatic conditions of seed source was determined. In order to examine differences and relationship among seed traits and geo-climatic conditions, principal component analysis and cluster analysis were used in Multivariate Statistical Package (MVSP). PCA solutions were accepted when Eigen values were greater than one (Kaisers criterion) and compatible with Cat tells scree rule. Component scores and factor loading were calculated after varimax rotation. Factors equal or greater than 0.7 were considered as defining part of PCA. Hierarchical cluster analysis was used to group provenances. The nearest neighbour method was utilized for classification and Square Euclidean method used as dissimilarity. Discriminant analysis was used to determine the variables responsible for the cluster formation. Dendograms were plotted to determine phylogenetic relationships among the five provenance of *F. albida*.

## RESULTS

### Morphological variation in seed traits

There were significant differences ($p \leq 0.05$) among provenances in all seed traits measured except seed coat thickness (Table 3). The coefficient of variation (CV) was 30.41 % for seed weight and 12.26% for seed length. Seed thickness and seed width (CV 1.2 and 6.8% respectively) recorded a lower coefficient of variation suggesting minimal environmental effect on expression of these traits. On average, seed length among provenances ranged from 6.73 ± 0.01 (Taveta) to 1.04 ±

**Table 4.** Genetic variance components ($\sigma^2_g$, $\sigma^2_p$ and $\sigma^2_e$), genotypic (GCV) and phenotypic (PCV) coefficient of variation along with heritability ($h^2$) for seed traits of the five provenances of *F. albida*

| Seed traits | $\sigma^2_g$ | $\sigma^2_p$ | $\sigma^2_e$ | GCV | PCV | $h^2$ |
|---|---|---|---|---|---|---|
| Length | 0.53 | 0.57 | 0.04 | 75.7 | 79.1 | 0.92 |
| Width | 0.03 | 0.03 | 0 | 75.72 | 35.71 | 0.99 |
| Thickness | 0.0003 | 0.0002 | 0.0001 | 4.69 | 4.54 | 0.99 |
| Weight | 0.02 | 0.02 | 0 | 121.2 | 122.31 | 0.99 |

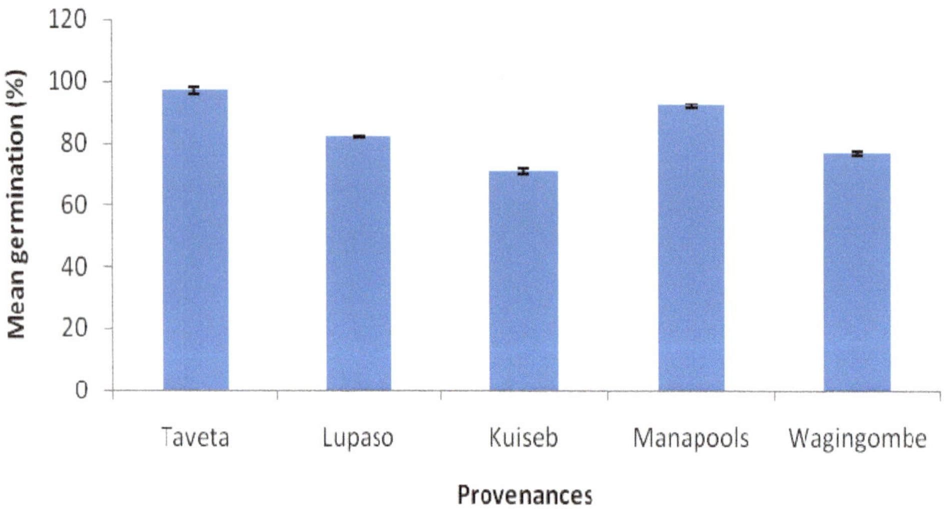


**Figure 2.** Mean germination percentage of five provenances of *F. albida* at four weeks after sowing.

0.008 mm (Manapools) with a grand mean of 9.11 ± 0.003 mm. The mean seed length for Wangingombe, and Eastern African provenance was higher than Lupaso (southern Africa provenance) though the difference was not statistically significant ($P \leq 0.05$). Southern African provenances recorded higher mean seed length as compared to East Africa provenances. Manapools recorded the maximum seed width 0.60 ± 0.003 mm followed by Kuiseb with 0.58 ± 0.004 mm. Taveta recorded lowest mean seed width as compared to the other provenances; seed width ranged from 0.49 ± 0.003 to 0.60 ± 0.003 mm. Seed weight was smaller in eastern Africa provenances than in southern African provenances. One way analysis of variance of seed thickness separated *F. albida* seeds into two thickness groups ranging from 0.29 to 0.305 mm. Seed thickness did not follow the same pattern as for seed length and seed width.

## Genetic variance components

Genetic variance components are represented in Table 4. Seed length recorded the highest variation but the lowest heritability index. Nonetheless, heritability indices ranging from 0.92 to 0.99 suggest that environmental influence on seed characteristics is minimal and hence genetic factors have a lot of influence on the seed traits analyzed. This suggests that selection based on morphological traits can be made with a high degree of confidence.

## Seed germination

There were significant differences ($p \leq 0.05$) among provenances in seed germination percentage (Figure 2). Seed germination percentage was highest for Taveta (97%) followed by Manapools (95%) and was lowest in Kuiseb (71%). Mean germination in nursery averaged 83.3% varying from 97 to 71% and was significantly different among provenances ($p \leq 0.05$). Taveta with the smallest seeds recorded the highest seed germination. No significant correlation ($r=0.15$ $p>0.05$) was observed between seed size or seed weight with mean germination.

## Correlation among seed traits

Correlation of seed traits showed that seed weight correlated with seed length ($r = 0.494$, $p \leq 0.05$) and seed width ($r = 0.433$, $p \leq 0.05$) (Table 5). Seed length and

**Table 5.** Mean correlation coefficient (r) between seed traits and germination percentage of *F. albida* provenances

| Seed traits | Length | Width | Thickness | Weight |
|---|---|---|---|---|
| Width | 0.597** | | | |
| Thickness | 0.133** | 0.104* | | |
| Weight | 0.494** | 0.433** | 0.094* | |
| Germination percentage | 0.12 | 0.09 | 0.04 | 0.15 |

*, **Significance at 5 and 1% probability level, respectively.

**Table 6.** Pearson correlation coefficient (r) for seed trait and geo-climatic condition of the seed origin of the five provenances of *F. albida.*

| Variables | Length | Width | Thickness | Weight | Altitude | Rainfall |
|---|---|---|---|---|---|---|
| Width | 0.562** | | | | | |
| thickness | -0.054 | -0.095 | | | | |
| weight | 0.545** | 0.527** | -0.053 | | | |
| altitude | -0.306** | -0.165 | 0.064 | -0.117 | | |
| Rainfall | -0.262** | -0.353** | 0.139 | -0.121 | 0.219* | |
| Temperature | -0.435** | -0.371** | 0.023 | -0.264** | 0.248* | 0.659** |

*, **Significance at 5 and 1% probability level, respectively.

**Table 7.** Factor loading for each variable on the component of PCA analysis.

| Variables | PC1 | PC2 |
|---|---|---|
| Length | **0.964** | -0.21 |
| Width | **0.93** | -0.247 |
| Thickness | **0.921** | 0.372 |
| Weight | **0.966** | -0.222 |
| Altitude | 0.077 | 0.618 |
| Rainfall | -0.127 | **0.864** |
| Temperatures | -0.37 | **0.756** |
| Eigen values | 3.74 | 1.99 |
| Percentage of variance | 53.35 | 28.48 |
| Cumulative variance (%) | 53.35 | 81.83 |

Values in bold (>0.7) are the principal components that explain most of the observed variation.

seed width also showed a significant positive correlation ($r=0.597$ $p \le 0.05$). Seed thickness showed a weak correlation with seed weight, width and length ($r$ = 0.094, 0.104, and 0.133 $p \le 0.05$). No significant correlation was recorded between seed size and seed weight with mean germination percentage.

## Correlation between seed traits and geo-climatic conditions of the seed origin

Correlation among seed data and geo-climatic data showed that seed length, width and weight were significantly negatively correlated to altitude, rainfall and temperature of the seed collection zone. Correlations among seed traits with geo-climatic condition of the provenances suggest the possibility of ecoclinal distribution of *F. albida* provenances (Table 6).

## Genetic divergence

Genetic divergence of seed traits were analyzed using principal component analysis and cluster analysis. An acceptable solution of component analysis was reached when two dimension models were found to be significant and explained 81.83% of the total variance observed. The total variance was partitioned into the two principal components as: 53.35% of the total variance/variation for the first component (PC1) which was dominated by seed weight, seed width and seed length and seed thickness, 28.48% for the second component (PC 2) defined by rainfall and temperature (Table 7). PCA dendrogram analysis based on squared Euclidean distance for dissimilarity from these three principal components revealed three clusters (Figure 3). Cluster 1 composes of Lupaso, Manapools and Wangingombe which receive high rainfall and temperatures; their seeds have large to moderate sizes and are heavy. Kuiseb and Taveta were placed in separate clusters. Taveta (cluster 2) on the other hand has the smallest and lightest seeds and receive higher rainfall and temperature as compared to Kuiseb (545 mm, 28.1°C annually). Kuiseb in cluster

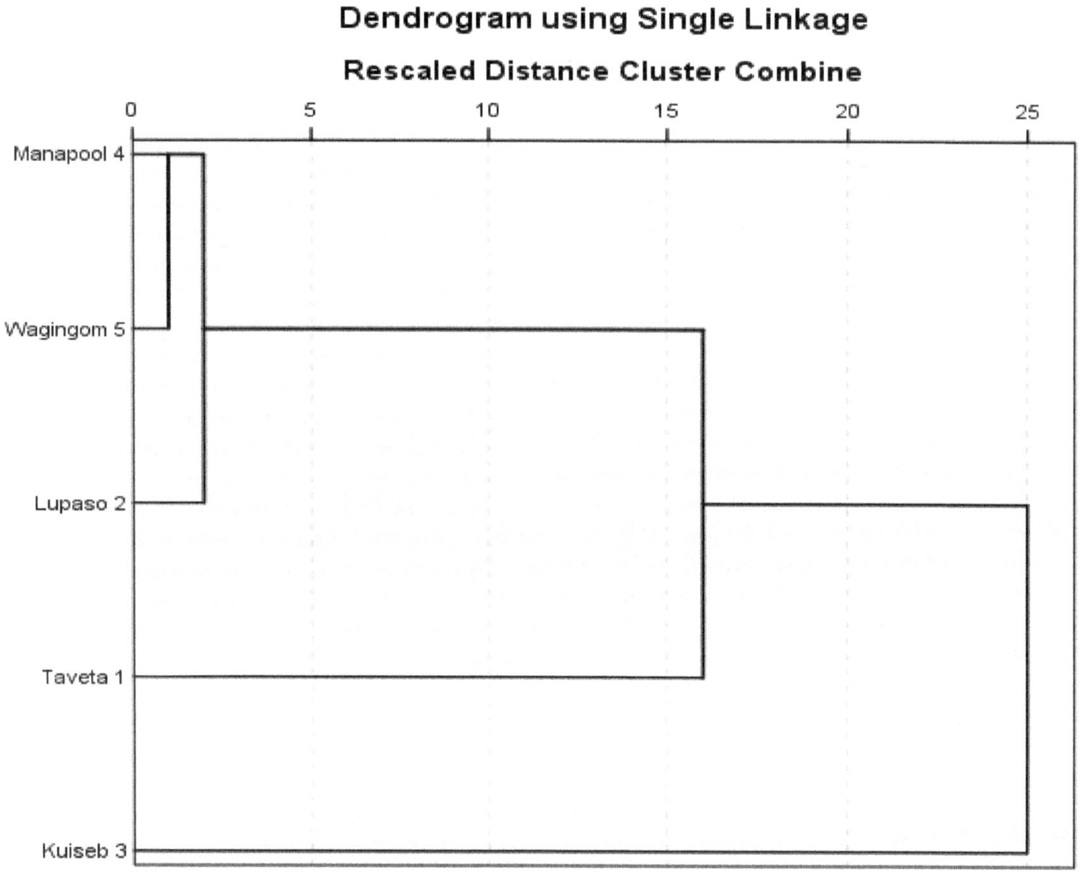


**Figure 3.** Dendogram for classification of five provenances of *F. albida* based on seed and geo-climatic conditions components from PCA.

three receives extremely low rainfall and temperatures as compared to other provenances under study (<50 mm, 15.2°C annually). Kuiseb has large and heavy seeds.

## DISCUSSION

### Variation in seed morphology

*F. albida* exhibited considerable amount of genetic variation in seed morphology, which could be due to its occurrence over a wide range of geo-climatic condition (Joly et al., 1992). Analysis of variance of seed traits showed significant differences, $p<0.05$ indicating high genetic variation among the five provenances of *F. albida* obtained from eastern and southern African provenances. Provenances were not regionally structured, and also showed a higher variability between eastern Africa provenances (Taveta and Wangingombe) than southern African provenances (Manapools, Kuiseb and Lupaso) as revealed by Euclidian distances of the cluster analysis. This could be due to the high variability in environmental conditions within Eastern and Southern Africa regions which is supported by PC2, pointing out that rainfall and temperature have great influence on the phenotypic variation. Ibrahim (1996), Harris et al. (1997) and Dangasuk and Gudu (2000) reported similar results in earlier studies. The variation in seed morphology could be due to interplay of both genetic and varying environmental conditions (interaction among rainfall, temperatures and altitude), exposing the trees to different selective pres-sures, resulting to different degrees of adaptation to climatic conditions (Mathur et al., 1984). The role of genetic and environment in seed morphology is also supported by the correlation between seed trait with geo-climatic conditions of the seed origin. Among all the traits studied, seed weight had the highest coefficient of variation indicating its sensitivity to environmental conditions. In the previous studies of tree species (Gera et al., 2000; Sivakumar et al., 2002; Mkonda et al., 2003), genetic control of seed morphology was observed which is consistent with our finding.

### Analysis of genetic components of seed traits

Genetic analysis of seed trait revealed the genetic control of seed traits. The phenotypic variance was greater than genotypic variance for seed length and seed width which agrees with earlier findings of Jonah et al. (2013), Tanimu and Aliyu (1997) and Tanimu et al. (1990). Genotypic variance was higher than environmental variance except for seed thickness indicating the role of the environment in determining the extent of the seed thickness. The genotypic coefficient of variation was either equal or greater than phenotypic coefficient of variation which shows the contribution of the environment in expression of seed traits. Jonah et al. (2013) and Agbo and Obi (2005) in their study of Bambara groundnut reported similar finding; indicating that the interaction of genetics and environment plays an important role in selection and tree improvement programs. The 100% heritability for seed width, weight and thickness indicates their high response to selection which is supported by the findings of Vanaja and Luckins (2006).

## Seed germination

Seed germination percentage varied significantly among provenances, $p<0.05$ and did not follow a regional pattern as was also shown by the morphological traits. Generally, large seeds are expected to have higher seed germination ability than small seeds (Isik, 1986). Seeds from Manapools provenance, having largest seeds had the second highest seed germination percentage after Taveta provenance which had the smallest sized seeds. This deviates from most findings (Vakshasya et al., 1992; Ginwal et al., 1994), except for Fenner (1991) who observed no significant correlation between seed size or seed weight with the seed germination hence caution should be taken in making such decisions based on seed morphology alone. Seed coat thickness was measured without separating it from the seed which may be another factor; however scarification and soaking of scarified seeds in water for 24 h before sowing might address the physical barriers imposed by seed coat thickness. Hence genotypic factors should not be overlooked and must always be taken into consideration (Ibrahim, 1996; Gera et al., 2000; Jayasankar et al., 1999; Mkonda et al., 2003; Sivakumar et al., 2002). This is supported by the high heritability for the various traits studied (low environmental variance), and thus suggest that most of these traits are under genetic control.

## Correlations of seed among seed traits and with geo-climatic conditions of seed sources

Seed characters were highly correlated. Correlated traits are of interest to tree breeding because change of one trait leads to simultaneous improvement of the correlated trait. Correlation among seed traits have been documented in tree species for example, Barracosa et al. (2007) and Dangasuk et al. (1997). The significant correlation among seed traits and geo-climatic conditions of the seed source shows that the environments affect the expression of seed traits and a possibility of clinal distribution of *F. albida* provenance under study.

## Seed divergence

High rate of environmental variability within Eastern and Southern Africa regions show that species distribution is discontinuous, with barriers to gene flow among populations, leading to isolation of populations. It is there-fore not expected to observe regional based population structure but rather habitat based structure for instance, Lupaso (Malawi), Manapools (Zimbabwe) and Wangingombe (Tanzania) that clustered together from regions receiving high rainfall and temperatures. Comparing the Eastern Africa provenances, Wangingombe has a higher rainfall (760 mm) and also located in a high altitude (1700 m above sea level) as compared to Taveta that is located in low rainfall area (545 mm) and in a lower altitude (760 m asl). Similarly, Kuiseb from the southern Africa region is located in extremely low rainfall area (<50 mm annually), very low temperature (15.2°C) as well as low altitude (400 m asl) as compared to the other two provenances hence its separation into a different cluster. The results of principal component analysis and cluster analysis indicated the role of interaction among different factors in genetic differen-tiation of *F. albida*.

## Conclusions

The high heritability for the traits under study show minimal environmental influence on seed characteristics hence genetic factors have a lot of influence on the seed traits analyzed. This suggests that selection based on morphological traits can be made with a high degree of confidence. The variation in the total germination percentage at the nursery level emphasizes the need for wide-range trials to enhance selection of the best provenances for breeding and conservation of *F. albida* genetic resources. The low germination percentage of Kuiseb could be due to absence of ripen pods during collection or poor handling and processing of the seeds, there is need therefore to recollect the seeds with proper handling for fairer assessments.

The current work is a baseline study aimed at identifying useful traits for selection of the best provenances of *F. albida* for tree improvement, breeding and conservation of its genetic resources. More research is needed to provide more information on the identified traits before a general conclusion is made.

## Conflict of Interests

The author(s) have not declared any conflict of interests.

## ACKNOWLEDGEMENTS

The authors wish to thank the World Agroforestry Centre (ICRAF, Nairobi) for providing the seeds for the study and International Fund for Agricultural Development (IFAD) for funding the Evergreen Agriculture Project in Kenya, which provided the budget for the study.

### REFERENCES

Agbo CU, Baiyeri KP, Obi IU (2005). Indigenous knowledge and utilization of Gongronema latifolia Benth: A case study of Women in University of Nigeria, Nsukka. Bio-Res. J. 3:66-69.

Akbar MT, Mahmood M, Yaqub M, Anwar M, Ali, Iqbal N (2003). Variability, correlation and path coefficient studies in summer mustard (Brassica juncea L.). Asian J. Plant Sci. 2:696-698.

Aniszewski TMH, Kupari AJ, Leinonen (2001). Seed number, seed size and seed diversity in Washington lupin (Lupinus poly-phyllus Lindl.). Ann. Bot. 87:77-82.

Barnes RD, Fagg CW (2003). *Faidherbia albida* monograph and annotated bibliography. Tropical Forestry Papers no. 41. Oxford Forestry Institute (OFI). Oxford, UK. pp. 267.

Bernard C (2002). *Faidherbia albida* (Delile) A. Chev. Record from Protabase. Oyen, L.P.A. and Lemmens, R.H.M.J. (Editors). PROTA (Plant Resources of Tropical Africa / Resources végétales de l'Afrique tropicale), Wageningen, Netherlands.

Bewley JD, Black M (1994). Seeds Physiology of Development and Germination, 2nd edition Plenum press, New York.

Bhat GH, Chauhan PS (2002). Provenance variation in seed and seedling traits of *Albizzia lebbek* Benth. J. Tree Sci. 21:52-57.

Bonkoungou EG (1992). Socio-cultural and Economic Functions of *Acacia albida* in West Africa. In: *Faidherbia albida* in the West African Semi-Arid Tropics. Vandenbeldt RJ, (Editor), Proceedings of a Workshop, International Center for Research in Agroforestry, Nairobi, Kenya. pp. 1-6.

Dangasuk OG, Gudu S (2000). Allozyme variation in 16 natural populations of *Faidherbia albida* (Del.) A. Chev. Hereditas 133:133-145.

Dangasuk OG, Wachira MR (2001). Interaction between soil properties and 16 *Faidherbia albida* provenances four years after planting in semi-arid Baringo district of Kenya. BOT: 421. Special Project Report. Department of Botany, Moi University, Eldoret, Kenya. pp. 29.

Dangasuk OG, Seurei P, Gudu S (1997). Genetic variation in seed and seedling traits in 12 African provenances of *Faidherbia albida* (Del.) A. Chev. at Lodwar, Kenya. Agrofor. Syst. 37:133-141.

Eriksson O, Friis EM, Pedersen KR, Crane PR (2000). Seed size and dispersal systems of Early Cretaceous angiosperms from Famalicão, Portugal. Int. J. Plant Sci. 161:39-329.

Fenner M (1991). The effects of the parent environment on seed germinability. Seed Sci. Res. pp. 75-84.

Gassama-Dia, Sané YK, N'Doye D (2003). Reproductive biology of *Faidherbia albida* (Del.) A. Chev. Silva Fennica 37:429-436.

Gera M, Gera N, Ginwal HS (2000). Seed trait variation in Dalbergia sissoo Roxb. Seed Sci. Technol. 28:467-475.

Ginwal HS, Gera M (2000) .Genetic variation in seed germination and growth performance of 12 Acacia nilotica provenances in India. J. Trop. For. Sci. 12:286-297.

Harris SA, Fagg CW, Barnes, RD, (1997). Isozyme variation in *Faidherbia albida* (Leguminosae, Mimosoideae). Plant Syst. Evol. 207:119-132.

Ibrahim A, Tibin IM (2003) .Feeding potential of *Faidherbia albida* ripe pods for Sudan desert goats. Sci. J. King Faisal Univ. 4:14-24.

Ibrahim AM (1996). Genetic variation in *Faidherbia albida*: implications for conservation of genetic resources and tree improvement. University Helsinki Tropical Forest. Doctoral thesis, pp. 86.

ISTA (1993). International Rules for Seed Testing. International Seed Testing Association. Seed Sci. Technol. 21:1-288.

Jayasankar S, Bondada BR, Li Z, Gray DJ (2002). A unique morphotype of grapevine somatic embryo exhibits accelerated germination and early plant development. Plant Cell Reports 20:907–911.

Joly HI, Zeh-Nlo M, Danthu P, Aygalent C (1992). Population genetics of an African acacia, Acacia albida I. Genetic diversity of populations from West Africa. Australian J. Bot. 40:59-73.

Jonah PM, Aliyu B, Jibung GG, Abimiku OE (2013). Phenotypic and Genotypic Variance and Heritability Estimates inBambara Groundnut (Vigna subterranea[L.] Verdc) in Mubi, Adamawa State, Nigeria. Int. J. IT Eng. Appl. Sci. Res.2:66-71.

Khurana E, Singh JS (2001). Ecology of seed and seedling growth for conservation and restoration of tropical dry forest: a review. Environ. Conserv. 28:39-52.

Leishman MR, Murray BR (2001). The relationship between seed size and abundance in plant communities: model predictions and observed patterns. Oikos 94:151-161.

Manz B, Muller K, Kucera B, Volke F, Leubner-Metzger G (2005).Water uptake and distribution in germinating tobacco seeds investigated invivo by nuclear magnetic resonance imaging.Plant Physiol. 138:1538-1551.

Mathur RS, Sharma KK, Rawat MMS (1984). Germination behavior of provenances of *Acacia nilotica* sp. indica. Indian Forester 110:435-449.

McGahuey M (1985). Assessment of the *Acacia albida* extension projects in Chad. Consultant's report prepared for USAJD Forestry Support Program. Chemonics International. Washington DC.

McGahuey M (1992). Extension of Acacia ulhrrla: Recapitahza- tion of the natural resource base. In: Vandenbeldt, R.J. (Ed.). *Faidherbia albida* in the West African Semi-Arid Tropics. Proceedings of a Workshop. International Center for Research in Agroforestry. Nairobi, Kenya, pp. 22-26.

Mkonda A, Lungu S, Maghembe JA, Mafongoya PL (2003). Fruit- and seed-germination characteristics of *Strychnos cocculoidesan* indigenous fruit tree from natural populations in Zambia. Agrofor. Syst. 58:25-31.

Moles AT, Westoby M (2004). Seedling survival and seed size: a synthesis of the literature. J. Ecol. 92:372-383.

Obroucheva NV, Antipova OV (1997). Physiology of the initiation of seed germination. Russian J. Plant Physiol. 44:250-264.

Ofori DA (2001). Genetic Diversity and its Implications for the Management and Conservation of *Milicia* species Ph.D. Thesis, University of Aberdeen, pp.158.

Orwa C, Mutua A, Kindt R, Jamnadass R, Anthony S (2009). Agro forest tree Database: a tree reference and selection guide version 4.0. World Agroforestry Centre, Kenya.

Payne WA (2000). Optimizing crop water use in sparse stands of Pearl Millet. Agron. J. 92:808-814.

Rees M (1995). Community structure in sand dune annuals: is seed weight a key quantity? J. Ecol. 83:857-886.

Roupsard O, Ferhi A, Granier A, Pallo F, Depommier D, Mallet B, Joly HI, Dreyer E (1999). Reverse phenology and dry-season water uptake by *Faidherbia albida* (Del.) A. Chev. in an agroforestry parkland of Sudanese West Africa. Funct. Ecol. 13:460-472.

Sivakumar P, Sharmila P, Saradhi PP (1998). Proline suppresses Rubisco activity in higher plants. Biochem. Biophys. Res. Commun. 252:428-432.

Tanimu B, Aliyu L (1997). The status of Bambara groundnut genetic Resources in Nigeria. Country reports. In Heller J, Begemann F, Mushonga J Editors.

Tanimu BS, Ado G, Aliyu L (1990). Genotypic variability in Bambara groundnut cultivars at samaru, Nigeria. In: Proceedings of the 17th Annual conference of the Genetics society of Nigeria (I.O. Obigbesban, Ed.). Institute for Agricultural Research and Training, Obafemi Awolowo University Nigeria pp. 54-60.

Turnbull LA, Paul-Victor C, Schmid B, Purves DW (2008). Growth rates, seed size, and physiology: do small-seeded species really grow faster? Ecol. 89:1352-1363.

Turnbull LA, Rees M, Crawley MJ (1999). Seed mass and the competition/colonization trade-off: a sowing experiment. J. Ecol. 87:899-912.

Vakshasya RK, Rajora OP, Rawat MS (1992). Seed seedling traits of

*Dalbergia sissoo* Roxb. Seed source variation studies in India. For. Ecol. Manag. 48:265-279.

Vanaja T, Luckins CB (2006). Variability in grain quality attributes of high yielding rice varieties (Oryza sativa L.) of diverse origin. J. Trop. Agric. 44:61-63.

Weber F, Hoskins MW (1983). Agroforestry in the Sahel. Department of Sociology, Virginia Polytechnic Institute and State University, Blacksburg, Virginia 2406. pp.102.

Westoby M (1998). A leaf-height-seed (LHS) plant ecology strategy scheme. Plant Soil 199:213-227.

Wojtyla L, Garnczarska M, Zalewski T, Bednarski W, Ratajczak L, Jurga S (2006). A comparative study of water distribution, free radical production and activation of antioxidative metabolism in germinating pea seeds. J. Plant Physiol. 163:1207-1220.

Zar JH (1996). Biostatistical Analysis, 3rd Edition. Prentice-Hall, Englewood Cliffs, NJ, pp. 662.

Zhang ZB, Shan L (1998). Comparison study on water use efficiency of wheat flag leaf. Chinese Sci. Bull. 43:1205-1210.

Zobel B, Talbert J (1991). Vegetative propagation. In: Applied forest tree improvement. Woveland Press, Inc., Prospect Heights, Illionois. ISBN 0 88133 604 1.

# Drought stress and its intensity, the factor of strategies selection for drought tolerance in *Haloxylon aphyllum*

**Naser Arabzadeh**

Agricultural and Natural Resources Researches Center, Sodoughi Avenue, P.O. Box 76175-538, Kerman, Iran.

**Osmotic parameters of *Haloxylon aphyllum* were studied after inducing dryness. Water relations parameters with improving resistance to dryness of this species through inducing severe dryness and clarification of physiologic mechanisms of this plant in response to a low water and dryness were among the objectives of this study. For this purpose, the method of pressure chamber was employed. By this method, the pressure-volume curve was drawn and the parameters of water relations of the plant were obtained from analyzing them. A relatively mild dryness was induced to plants through a lack of irrigation. After two weeks, Haloxylon water potential reached -16.5 bars. A severe dryness was also induced to that but after four weeks of no irrigation, it was reduced to -27.2 bars. Relatively mild and severe dryness were repeated for six and 11 periods respectively. In both series of experiment, the control water potential that were being watered every two days once, remained fixed at about -12.7 bars. Based on the results, although the relatively mild dryness increased the elasticity of plant textures, but it had not a meaningful impact on its osmotic potential. Although the use of a relatively severe dryness decreased both osmotic potential and osmotic adjustment, but at the same time, it increased the elasticity too.**

**Key words:** Haloxylon, water relations, water stress, drought resistance, osmotic potential, elasticity.

## INTRODUCTION

Plants use two mechanisms which are: 1) tolerating dryness and 2) escaping from it in confronting with dryness (Turner, 1979; May and Milthorpe, 1962). *Haloxylon aphyllum* is among species which with the help of endurance mechanism is able to spend dry periods. These mechanisms have been studied in some of the species like various kinds of pine (Emadian, 1988; Grime, 1979; Bilan et al., 1979:1978; Youngman, 1965).

Haloxylon species like many other species, by osmotic adjusting or increasing the elasticity of cellular wall in the condition of water tension maintain their turgescence better and consequently tolerate better drought periods (Emadian, 1988).

*H. aphyllum* like other multi-functional plants of haloxylon type is of great value and importance in protecting and supporting breakable ecosystems like deserts.

It is such that as a live windbreaker prevents soil

erosion (Tokasi et al., 2007; Safarnejad, 2005; Jafari et al., 2004), and as a correcting element increases the organic materials of soil and in a long term improves the structure of soil (Jafari et al., 2004), and increases the plant enrichment of the area under coverage (Bakhshi and Biroudian, 2008). In addition for those who live in deserts, it is an important source for the provision of fuel and fodder for cattle (Tokasi et al., 2007). As it has a broad emission spectrum in different soils from the viewpoint of texture (Javanshir et al., 1996), it can be noticed much more.

However, unfortunately, despite the high importance of haloxylon in biologic stabilization of sand lands, most of the planted plants in desert regions are facing the problem of dryness. Based on the confirmation of authorities and those in charge of desertification projects, after transferring the plants of *H. aphyllum* planted in vase to a natural plantation bed and despite observing principles related with post-plantation stages, after sometimes, a high percentage of transferred plants were afflicted with dryness and they had to be replanted. Taking action to replant instead of dried plants imposes relatively extravagant costs to the executive system of natural resources. According to the existing documents in the File Keeping Department of Agriculture Ministry of Iran, in average in each period, more than 10% of the vases with *H. aphyllum* were fading and became dried in the fields, and required replanting. In order to help reduce mortality percentage of planted plants in main beds of plantation, it seemed that placing one-year plants of *H. aphyllum* which were exposed to periodic dryness and possibility of their compatibility with unfavorable condition resulting reduction of rainfall and periodic droughts could be tested as an appropriate approach.

The main goal in conducting this research was firstly the study the changes of water relations parameters in order to improve resistance to dryness of *H. aphyllum* through induction of periodic dryness and also clarification of physiological mechanisms of this species in response to water shortage and dryness.

## MATERIALS AND METHODS

The identified seeds of *H. aphyllum* were planted in plastic vases and were taken into care for one year. Plastic vases with an approximate capacity of three liters were selected in order to pave way for a better and a greater growth of the plants roots. The soil consisted of wind sand, soil and leaf-soil in proportions of 2, 1 and 1 respectively. Supply of necessary nutrition for plants during the experiment was done by including leaf-soil in the mentioned combination. At the same time, in order to prevent unwanted accumulation of water in vases, some fine holes were made in their bottom. After one year, the plants were transferred to a greenhouse.

Upon completion of one-month period of plants compatibility in greenhouse, the treatments of dryness induction was applied. For this purpose, a sufficient number of good and healthy plants were selected for the experiment. Half of them were considered for induction of dryness and the rest were selected as the controls that were being watered every two days once. In this research, two series of experiments were conducted. The plants of the experiment of the first series received 6 periods of 7 to 14 days of dryness. During this period, the water potential (before sunrise) of their plant was measured every two days once by using the pressure chamber and according to Scholander method (Scholander et al., 1965). At this state, the water potential of haloxylon plants at the end of each period was reduced to -16.5 bars. The experiment second series plants, in addition to the mentioned dry periods, received five periods of dryness of 14 to 28 days too. In this series of tests, water potential of *H. aphyllum* plant was reduced at the end of each period to -27.2 bars. Induction of stress in mild (14-7 days) and severe (28-14 days) stresses were 6 and 6+5 (11) periods (replications) respectively. So every place (points) on the curve 3 is resultant 6 stress periods, and on the curve 4 is resultant 11 stress periods (repetitions).

It is worth mentioning that in order to measure the water potential of plants, separate plants were considered and in each measuring, 5 plants were cut and used. Plants were watered fully at the end of dryness period. Their before sunrise water potential in the day after irrigation was increased by -5.3 to -4.3 bars. Also, the control plants water potential was increased from -12.7 bars in both series of test to about -9.7 to -8.7 bars.

The impact of dryness induction on the elasticity and osmotic parameters of *H. aphyllum* plants became possible through an analysis of pressure-volume curves. These curves were prepared by using *Sholander method* (Figure 1).

The horizontal axis is the exit liquid volume ($W_e$) and its vertical axis is balancing pressure or $(\Psi_w)^1$. Point E, is the place of intersection of osmotic line with $W_e$ axis showing the rate of water which is exited under the infinite pressure from plant known as simplistic water ($W_s$). Point B is the place of conjunction of osmotic line with the axis of $(\Psi_w)^{-1}$ showing that $\Psi_s$ of plant is in the condition of full turgescence.

As Figure 1 shows each pressure-volume curve has two outstanding parts: 1) curve parts which encompasses about 5 to 8 points; 2) direct part which includes 8 to 11 points. The direct part of curve was used to estimate $\Psi_w$, $\Psi_p$, & $\Psi_s$. According to the recommendation of Cutler et al. (1979), this action was performed by drawing a regression curve on at least 7 to 8 points of the last spots of pressure-volume curves. The overall equation of the curve is as follows:

$$(\Psi_{st})^{-1} = (\Psi_{s0})^{-1} - m\sum W_i$$

In the relations, $(\Psi_{s0})^{-1}$ is the inverse of primary osmotic potential of plant at full turgescence condition; m is the slope of regression which is under the influence of plant size and osmotic feature of texture and rate of exchange of plant water, $(\Psi_{st})^{-1}$ is the opposite of $\Psi_s$ for t times of a pair from the data of pressure-volume and $\sum W_i$ is its corresponding accumulated quantity of exited liquid. With this assumption that the mentioned relation to be fully true in the considered range, by placing each pair of $P_i$ and $\Sigma W_i$, it is possible to get the $\Psi_s$. The continuation of this line crosses the vertical axis in point B which specifies the inverse of osmosis potential in full turgescence $(\Psi_{s0})^{-1}$ (Tyree and Jarvis, 1982). Subsequently, the turgor potential ($\Psi_p$) of the plant in each point of pressure-volume curve was obtained through difference of $\Psi_w$ and $\Psi_s$ related to the same point. On the other hand, the correlation line cut off the horizontal axis in point E, which determines the volume of exited liquid from plant in an infinite pressure $[(\Psi_{s0})^{-1} m^{-1}]$ (Figure 1). Active osmotic water of plant ($W\Psi_s$) was calculated in form of $W_s (W_0 - W_d)^{-1}$ and its inactive osmotic water in form of $1 - W_s (W_0 - W_d)^{-1}$ (Cutler et al., 1979).

The existing data in the part of curvature of pressure-volume curve was used to estimate the average of the absolute value of elasticity ($\bar{\varepsilon}$) of *H. aphyllum* (Tyree and Jarvis, 1982). To achieve

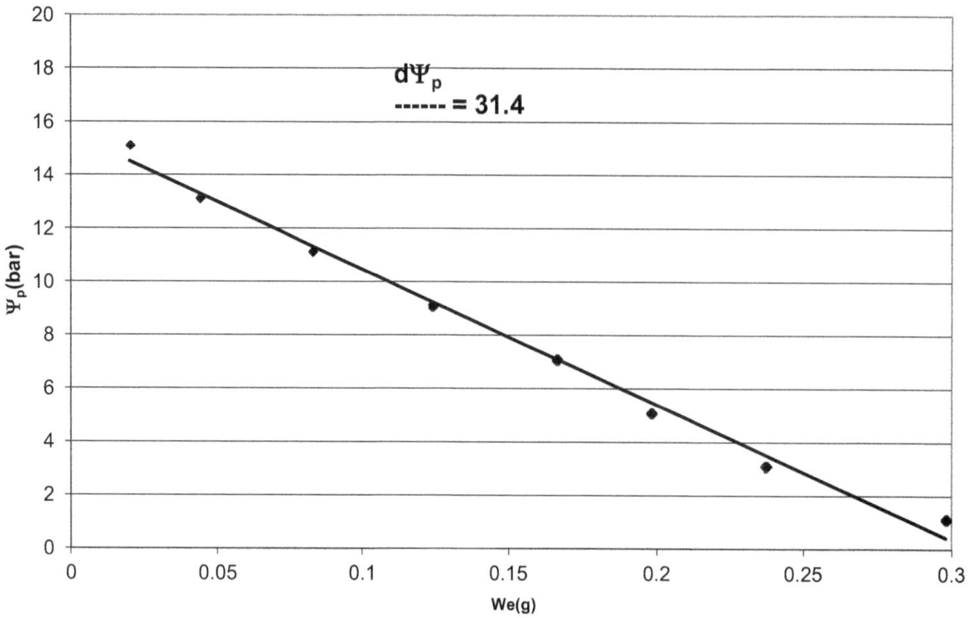


**Figure 1.** Pressure-volume curve of a plant of *Haloxylon aphyllum*.

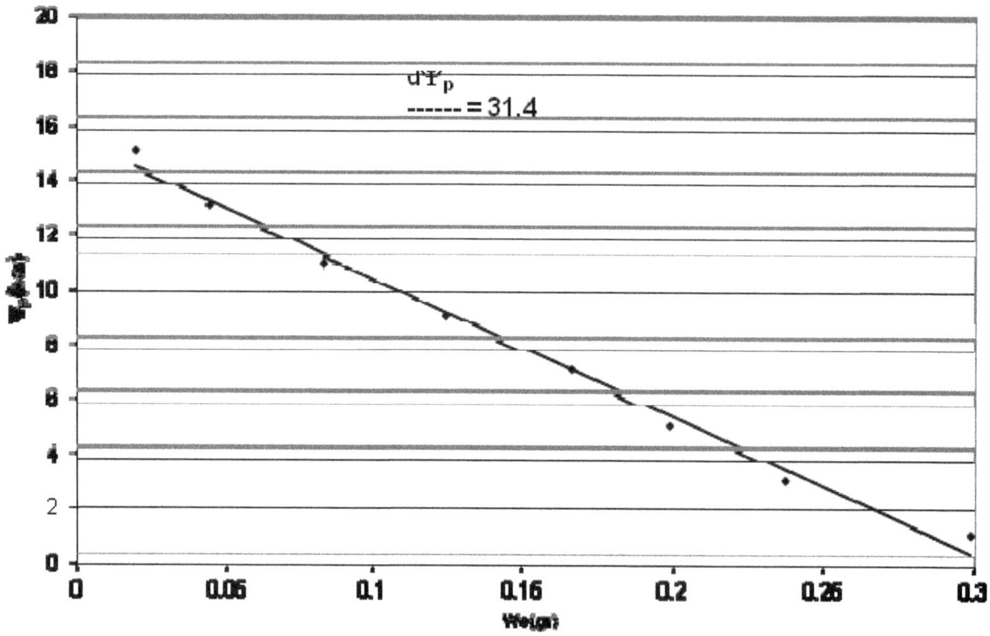


**Figure 2.** Calculating the rate of elasticity changes of cellular wall in *Haloxylon aphyllum* textures by using the curvature part of pressure-volume curve. Horizontal axis is exited volume ($W_e$), and vertical axis is swollen potential. In order to calculate the average of absolute value of the elasticity ($\bar{\mathcal{E}}$), the curve slope of $d\Psi_p\ (dW_e)^{-1}$ of plant was used.

this parameter, it was necessary that firstly $\Psi_p$ of the plant in the part of curvature of the pressure-volume curve be calculated. This action was performed by using the osmotic line and method of estimation of osmotic, turgor and water potentials. Then the obtained turgor potentials [($\Psi_p$)s] in the mentioned limit with the volume of corresponding condensed exited liquid was drawn in coordinates sheet and their regression equation was calculated (Figure 2). The ($\bar{\mathcal{E}}$) was also calculated from the product of the slope of line [($d\Psi_p$) $(dW_e)^{-1}$] in Ws.

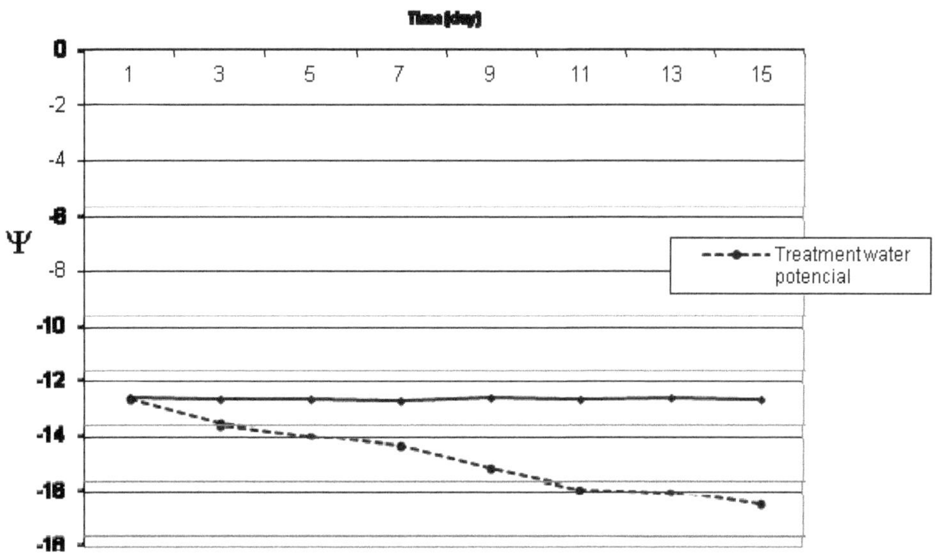


**Figure 3.** Relationship between plant water potential ($\Psi_w$) and time in *Haloxylon aphyllum* after watering. Pre-dawn plant water potential was measured every other day during two-week no-watering period.

**Table 1.** Comparing the result of water relations in *Haloxylon aphyllum* under mild dryness induction with control plants.

| Treatment | $\Psi_{wo}$ (Bar) | $\Psi_{po}$ (Bar) | $\Psi_{so}$ (Bar) | $\Psi_w$TLP (Bar) | $\Psi_s$TLP (Bar) | ε (Bar) | $W\Psi_s$ (%) |
|---|---|---|---|---|---|---|---|
| STRESSED[1] | -2.50** | 12.30** | -14.80 | -20.90 | -23.10 | 44.8** | 14.00** |
| CONTROL[2] | -5.90 | 07.30 | -13.20 | -22.50 | -21.40 | 30.50 | 28.80 |

[1]Seedlings, in addition to six early dry periods, five-day dry period 7 to 14 also received. The average water potential (before sunrise) at the end of their term limit had been reduced to -16.5 bars. Value of each parameter treated seedlings is related to seven seedlings. [2]Both seedlings were irrigated once a day. Amount of each parameter of control seedlings is related to five seedlings.

$\bar{\varepsilon} = W_s [(d\Psi_p) (dW_e)^{-1}]$

The used statistical method was t- student (Snedecor and Coohran, 1980).

## RESULTS

### Results of mild dryness induction

At the end of each period of mild dryness induction, water potential ($\Psi_w$) of *H. aphyllum* reduced about -16.5 bars; whereas $\Psi_w$ of control remained at a higher range, that is in average -12.7 bars (Figure 3).

Parameters of active osmotic water ($W\Psi_s$) and pressure potential at the condition of moisture full saturation ($\Psi_{p0}$) reduced and increased significantly and reached from 28.8 to 14% and was promoted from 7.3 to 12.3 bars, respectively. Their elasticity modulus ($\bar{\varepsilon}$) and water potential ($\Psi_{w0}$) had a highly significant increase (level 99%). Thus they accelerated from 30.5 to 44.8% and from -5.9 to -2.5 bars but the osmotic potential of tension treatments in full turgescence did not show a meaningful difference as compared with control (Table 1).

### Results of severe dryness induction

After applying severe dryness induction, the water potential of *H. aphyllum* diminished in average up to about -27.2 bars, whereas the water potential of controls more or less remained fixed at the level of mild dryness induction (-12.7 bars) (Figure 4).

Water potential and pressure potential in fully saturation moisture condition increased significantly (P<1%), so that with the promotion to a higher level, it reached to the level of -2 and 18.8 bars respectively. At the same time, like the mild dryness tension, very meaningful increase of elasticity ($\bar{\varepsilon}$) and very meaningful reduction of osmotic active water ($W\Psi_s$) were observed.

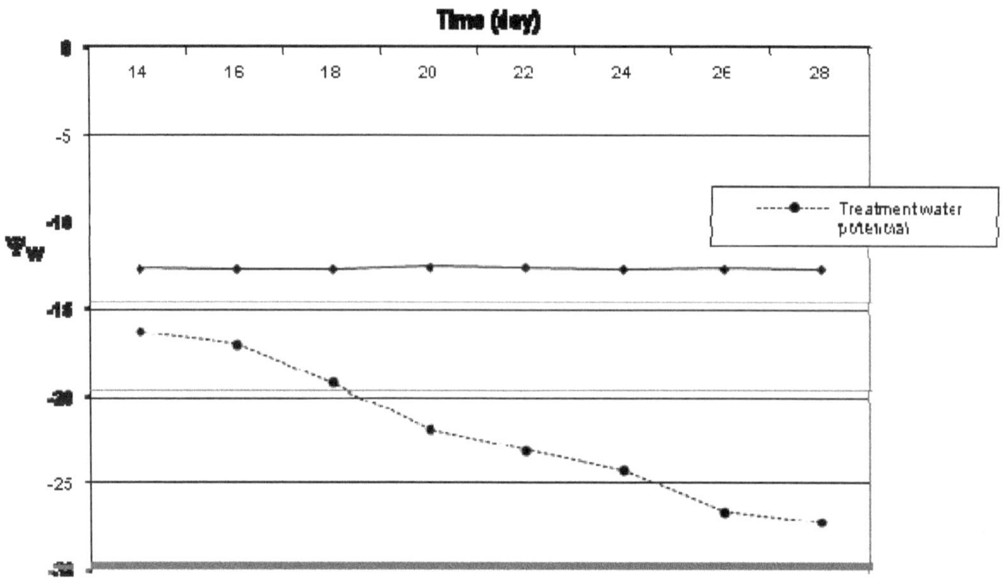


**Figure 4.** Relationship between plant water potential ($\Psi_w$) and time in *Haloxylon aphyllum* after watering. Pre-dawn plant water potential was measured every other day during a four-week period of no-watering.

However, concurrent with these changes, the osmotic potential in the condition of moisture full saturation ($\Psi_{s0}$) and dryness threshold ($\Psi_{stlp}$) reduced significantly (P<1%) and decreased from -13.5 to -20.8 bars and from -20.9 to -30.6 bars respectively.

## DISCUSSION

Primarily it seemed that the reaction of parameters of water relations (like $\Psi_{w0}$, $\Psi_{s0}$, $\Psi_{p0}$, $\Psi_{wtlp}$, $\bar{\varepsilon}$) of *Haloxylon aphyllum* in mild dryness condition was in conflict with their corresponding parameters in the control treatment (Table 1), because measured absolute value of elasticity ($\bar{\varepsilon}$) of under the treatment plants was 44.8 bars that was more elastic than $\bar{\varepsilon}$ of control plants (30.5 bars) significantly. On the other, the measured osmotic active water ($W\Psi_s$) in treated and controlled plants was 14 and 28.8% respectively, and this difference was meaningful in the level of 99%. Possibly, the concept of this with regard to the relation of ($\bar{\varepsilon}$) is that the more elastic plant needs a less water to maintain its turgescence. In addition, in the condition of full turgescence, the water potential of plants which were given dryness was higher than water potential of control plants.

Plant physiologists believe that many plants of dry regions are able to maintain their turgescence through mechanism of elasticity and consequently increase resistance to dryness. This feature is true for *Pinus taeda* (Emadian and Newton, 1989), *Pseudosuga menziesii* (Joly and Zaerr, 1987), *Dubautia ciliotala* (Robichaux and Canfield, 1985), and *Juglans nigra* (Parker and Pallardy, 1985). On the other, though the elasticity of treated plants increased, but their osmotic potential was without any meaningful change. In other words, in condition of mild dryness induction, no meaningful difference was observed at any statistical level between $\Psi s_0$ and $\Psi s_{tlp}$ in control and dryness tension treatments (Table 1). With regard to mentioned condition, it seems that in the mild dryness condition, *H. aphyllum* through increase of elasticity is able to maintain its turgescence and continues its growth. Apparently, this will be possible only in lieu of an exchange with reduction of osmotic adjustment.

Water relations parameters reaction in severe stress (28-14 days without irrigation) took place in continuing mild dryness condition, and osmotic adjustment was activated (Table 2) so that in addition to increase cell wall elasticity of shoot tissues (such as effect of periodic mild stress) osmotic potential of the tissue significantly decreased compared to control tissue, and continued the life of stressed-plants. After induction, severe periodic dryness on plants (*Haloxylon aphyllum*), the increase of water and turgor potentials (300 and 250%, respectively) and reduction of osmotic potential (more than 50%) in the condition of moisture saturation and at the dryness threshold was considerable.

The results showed that the induction of periodic mild dryness was not able to have an impact on the increase of resistance to dryness in *Haloxylon aphyllum.* It only changed the strategy of plant in confronting with stresses resulting from drought. Instead of that, periodic induction of severe dryness could make ideal changes in osmotic parameters, and not only caused the viable and

**Table 2.** Comparing the result of water relations in *Haloxylon aphyllum* under severe dryness induction with control plants.

| Treatment | $\Psi_{wo}$ (Bar) | $\Psi_{po}$ (Bar) | $\Psi_{so}$ (Bar) | $\Psi_{w}TLP$ (Bar) | $\Psi_{s}TLP$ (Bar) | ε (Bar) | $W\Psi_{s}$ (%) |
|---|---|---|---|---|---|---|---|
| MDI | -2.50 | 12.30 | -14.80 | -20.90 | -23.10 | 44.80 | 14.00 |
| SDI | -2.00 | 18.80 | -20.80 | -20.30 | -30.60 | 44.50 | 14.50 |

**Table 3.** Comparing the result of water relations in *Haloxylon aphyllum* under two water regimes: mild dryness induction and severe dryness induction.

| Treatment | $\Psi_{wo}$ (Bar) | $\Psi_{po}$ (Bar) | $\Psi_{so}$ (Bar) | $\Psi_{w}TLP$ (Bar) | $\Psi_{s}TLP$ (Bar) | ε (Bar) | $W\Psi_{s}$ (%) |
|---|---|---|---|---|---|---|---|
| Stressed | -2.00** | 18.80** | -20.80** | -20.30** | -30.60** | 44.50** | 14.50** |
| Control | -5.80 | 07.50 | -13.50 | -29.70 | -20.90 | 31.80 | 26.50 |

[1]Seedlings, in addition to six early dry periods, five-day dry period 14 to 28 also received. The average water potential (before sunrise) at the end of their term limit had been reduced to -27.2 bars. Value of each parameter treated seedlings is related to seven seedlings. [2]Both seedlings were irrigated once a day. Amount of each parameter of control seedlings is related to five seedlings.

freshness of plant but also enabled plants to bear the severe condition of dryness and could maintain their water potential at a very high level. Comparing the parameters of water relations of *H. aphyllum* in mild dryness with severe dryness condition (Table 3) confirm the importance of the mentioned subject matter.

## Mild dryness induction and severe dryness induction

Though under severe and longer dryness induction, cellular wall elasticity of *H. aphyllum* was maintained, however, osmotic adjustment became active for maintaining cellular turgescence. Therefore plants remained alive and continued their physiological and biochemical activities. Reduction of osmotic potential in the condition of saturation (by 6 bars) and reduce osmotic potential at threshold of wilt and dryness (by 7.5 bars) along with compatibility of plants to frequent dryness tension may enable them (compared to plants that were not exposed to any intensity of drought) tolerate drought periods resulting from factors affecting moisture shortage, and may be protected from the deleterious effects of drought periods in the same range of drought that seedlings similar to control wilt.

This idea can be put forth as the symbol of increase of resistance to dryness in *H. aphyllum* (which had received and accustomed to severe periodic dryness stresses) and to be used as a base for future research. Anyway, a decisive comment on this issue demands further studies and supplementary studies.

## Conclusion

The results of this research showed that *H. aphyllum* tolerates dryness well. If dryness be mild and its period be short, this species will be able to maintain its turgescence and continue its life by increasing elasticity mechanism. Apparently, this action is possible only in lieu of an exchange with reduction of osmotic adjustment. On the other hand, if dryness be severe and its period be long, for keeping its turgescence, in addition to increase elasticity, it uses the mechanism of osmotic adjustment too. In each of the two mentioned condition, maintaining turgescence in dryness condition cause the continuation of physiologic and biochemical activities of this species and make its growth and viability possible despite excessive and long dryness. Based on the results of this research, the executive officials of forestry and production of plants departments recommend that if they use *Haloxylon aphyllum* for biologic stability of sand lands in arid and semi-arid zones, prior to transfer of plants to main field, they should place them under five to eight periods of dryness induction for at least three to four weeks.

## Conflict of Interests

The author(s) have not declared any conflict of interests.

## ACKNOWLEDGEMENTS

I would like to express my sincere gratitude to Organization of Researches, Education and Expansion of Agriculture for supplying me with necessary credits and funds to implement this project. I also wish to thank Dr. Ramezanali Khavari-Nejad, Dr. Hossein Heidari Sharifabad and Dr. Fazlollah Emadian for their consultation in developing and implementing this research project.

## REFERENCES

Bilan MV, Hagan CT, Carter HB (1979). Stomata opening, transpiration, and needle moisture in loblolly pine seedlings from two Texas seed sources. For. Sci. 23:457-462.

Bilan MV, Leach H, Davies G (1978). Root development in loblolly pine (*pnus taeda*) from two Texas seed sources. In root from of planted trees (E. Van Eerden and J. M. Kinghron, eds.) Joint Rep., No*, P. 17-22. Brit Colum Ministry of Forests/ Can. For Serv. Victoria, Brit Colum.

Bakhshi J, Biroudian N (2008). A study on the effect of *Haloxylon aphyllum* plantation on richness and similarity understory species in Ardestan area. Pajouhesh & Sazandegi, 79:2-10. http://www.magiran.com/viewpdf.asp?no=1.

Cutler JM, Shahan KW, Steponkus PL (1979). Characterization of internal water relations of rice by a pressure-volum method. Crop Sci. 19:681-984.

Emadian SF (1988). Physiological responses of loblolly pine to silicon and water stress. Ph. D. Dissertation. Texas A&M University. Colleger Station, Texas, U.S. A.

Emadian SF, Newton RG (1989). Growth enhancement of Loblolly Pine, J. Plant Physiol. 134:98-103.

Grime JP (1979). Plant strategies and vegetation processes. Wiley, New York.

Jafari M, Azarnivand H, Tavakoli H, Zehtabian G, Esmailzadeh H (2004). Investigation on different vegetation effects on sand dunes stabilization and improvement in Kashan. Pajouhesh & Sazandegi, 64:16-21. http://www.sid.ir/fa/VEWSSID/J_pdf/560138364m03.pdf

Javanshir K, Dastmalchi H, Emarati A (1996). An ecologic surrvy of Hloxylon, Populus and Tamarix species in Iran deserts. The Second International Congress of Desertification, 1996, Iran.Jehade Sazandegi, 550pp, P. 1

Joly RJ, Zaerr JB (1987). Alternation of cell wall elasticity in Douglas-fir during periods of water deficit. J. Plant Physiol. 83:418-422.

May LH, Milthorpe FL (1962). Drought resistance of crop plants. Field Crop Abst., 15:171-179.

Parker WC, Pallardy SG (1985). Genotypic variation in tissue water relations of leaves and roots of black walnut (*Juglans nigra*) seedlings. Physiologia Plantarum J. 64:1:105-110. http://62.60.154.17/ebook2/16352.pdf

Robichaux RH, Canfield JE (1985). Tissue elastic properties of eight Hawaiian Dubautia species that differ in habitat and diploid chromosome number. Oecologia. 66:77-80.

Safarnejad A (2005). Comparison of saxaoul species (Haloxylon spp) for its improvement and expansion in desert areas. Pajouhesh & Sazandegi, 67:51-57. http://www.sid.ir/fa/VEWSSID/J_pdf/5601384MT07.pdf

Scholander PF, Hammel HT, Brad ED, Hemmingsen EA (1965). Sap pressure in the vascular plants. Crop Sci. 148:339-346.

Snedecor GW, Cochran WG (1980). Statistical Methods, 7th ed., The Lowa State University Press, Ames, Lowa.

Turner NC (1979). Drought resistance and adaptation to water deficit in crop plants. In Stress Physiology in crop plants (H. Mussell and R.C. Staples, eds.), pp. 87-103, John wiley and Sons, N.Y.

Tyree MT, Jarvis PG (1982). Water in tissues and cells. In Physiology and Ecology II. Water relations and carbon assimilation (O. L. Lange, P.S. Nobel, C.B. Osmond and H. Zigler, Eds). 12B:35-77. Springer-Verlag, Berlin.

Youngman AL (1965). An ecotypic differentiation approach to the study of isolated populations of pinus taeda in south central Texas, Ph. D. Thesis, Univ. Texas, Austin: Diss Abst. Int. B. 27, 3006.

# Effect of foliar application of humic acid, zinc and boron on biochemical changes related to productivity of pungent pepper (*Capsicum annuum* L.)

**Denre Manas[1], Bandopadhyay, P. K.[2], Chakravarty, A.[1], Pal, S.[1] and Bhattacharya, A.[1]**

[1]Department of Agricultural Biochemistry, Bidhan Chandra Krishi Viswavidyalaya, Mohanpur, Nadia-741252, West Bengal, India.
[2]Department of Plant Physiology, Bidhan Chandra Krishi Viswavidyalaya, Mohanpur, Nadia-741252, West Bengal, India.

**To study the effect of foliar application of humic acid at 0.05%, zinc at 0.05% and boron at 0.02% on biochemical changes related to productivity of pungent pepper cv. Bullet (*Capsicum annuum* L.), a pot experiment in randomized block design with three replications was conducted in the net house of Bidhan Chandra Krishi Viswavidyalaya, Mohanpur, Nadia-741252, West Bengal, India. In This experiment, we observed some physiological and biochemical characteristics of pungent pepper. Obtained results revealed that the highest value of height of plant, numbers of leaf plant$^{-1}$, leaf area, numbers of branch plant$^{-1}$, numbers of fruit plant$^{-1}$ and fruit length were exhibited by foliar application of HA + Zn + B. The highest value of fruit diameter, weight of 20 fresh fruits, numbers of seed fruit$^{-1}$ and weight of 100 seeds were exhibited by foliar application of Zn + B and HA + Zn + B, respectively. The total chlorophyll content significantly increased in HA + Zn + B followed by HA + B and Zn + B application. The highest value of reducing sugar, total sugar and starch content were obtained by application of HA + Zn + B. The highest content of protein, fat, energy and total phenol were exhibited by foliar application of Zn + B, Zn, HA + Zn + B and HA + Zn + B, respectively. The activity of polyphenol oxidase was significantly 'good performer' in HA + Zn + B followed by HA and HA + B application. There are significant and positive correlation of starch with some physiological characteristics such as leaf area, number of fruits plant$^{-1}$, fruit diameter and total chlorophyll. The correlation of energy value with carbohydrates, fat, protein, total chlorophyll and leaf area were positive and significant. Based on principle component analysis and considering average values of all variables, our results suggested that the good promising treatment in HA + Zn followed by Zn + B, HA + Zn + B and Zn alone application may bring about the proper value addition in quality as well as productivity of crop by enhancing the some physiological and biochemical characteristics in pungent pepper.**

**Key words:** *Capsicum annuum* L., humic acid, zinc, boron, quality, productivity.

## INTRODUCTION

Pungent pepper (*Capsicum annuum* L.) is one of the most important and popular crop grown in India and many countries all over the world. From time immemorial, it has been used as a common vegetable cum spice because of its color, test, pungency, flavor and aroma. Apart from basic quality, productivity of crop depends on their physical and biochemical properties. Concentrations of these properties in plant may be modified by various factors including foliar nutrition with organic component. Some studies estimates indicate that a large number of diverse materials can serve as sources of plant nutrients. The majority of nutrient input to agriculture comes from commercial mineral fertilizers. Organic manures are considered to play a significant but lesser role in nutrient contribution, leaving aside their beneficial effects on soil physicochemical and biological properties. Foliar feeding is a relatively new and controversial technique of feeding plants by applying liquid fertilizer directly to their leaves.

Most of the used organic-mineral fertilizers are humic substances, since, humic acid is one of the major components of humic substances. Nowadays, the use of humic acid has increased with increase in the agricultural production and the most economical humic acid almost applied directly to the soil and/or as a foliar application to the plants. The mode of action of humic acid on plant growth can be divided into direct and indirect effects as it affects the membranes resulting in improved transport of nutritional elements, enhanced protein synthesis, enhanced photosynthesis, solublization of micronutrients, reduction of active levels of toxic elements, enhancement of microbial population, enhanced soil structure improvement and increased both cation exchange capacity and water retention. MacCarthy et al. (1990). Singaroval et al. (1993) claimed that the increase in dry matter production with humic acid might be due to its direct action on plant growth auxin activity, contributing to increase in the dry matter.

Moreover, application of humic acid also increased the seed weight due to better mobilization of nutrients to seeds. Nardi et al. (1999) found that the biological activity of the humic acid was attributed to their chemical structure and their functional groups, which could interact with harmonic-binding proteins in the membrane system, evoking a hormone-like response.

Micronutrients have received a great deal of importance in crop production during recent years because of the widespread occurrences of their deficiencies from different parts of the country. Researchers from almost all the states in the country have also reported significant responses of many crops to micronutrient fertilization. Zinc (Zn) and boron (B) is an essential trace element for plants, being involved in many enzymatic reactions and is necessary for their good growth and development. The foliar application of micronutrients (Zn and B), increased total sugars (TS), reducing sugar (RS), non-reducing sugar (NRS), ascorbic acid (AA) and TSS/acid ratio of papaya fruit (Sing et al., 2002). The foliar application of micronutrients (Zn and B) also significantly enhanced fruit juice content, TSS, AA and NRS of sweet orange fruits.

Zn is also involved in regulating the protein and carbohydrate metabolism (Swietlik, 1999). Under Zn deficiency, the leaves shoot tips do not elongate fully, resulting in compressed internodes length and a tuft or rosette of leaves at the terminal (Tariq et al., 2007). However, in the light of better fruit quality development, Zn holds more significance besides imparting sustainability in production/productivity by reducing the fruit drop (Malik et al., 1999) and granulation (Kaur et al., 1990). External symptoms of Zn-deficiency in citrus as described by Wutscher (1979) are reduction in leaf size somewhat in proportion to the Zn-concentration in leaf. Zinc rates from 4 to 12 kg $ha^{-1}$ had beneficial effect on physical and chemical characters of fruits, but best results with regard to flavor and contents of juice, vitamin-C, and total sugars were obtained from the lowest Zn-rate (Mdwaradze, 1981). Application of Zn at 0.6% plus 2,4-D at 20 ppm as foliar spray gave the best results with regard to fruit weight, diameter, juice percentage, TSS and ascorbic acid content in Kagzi lime (Batru et al., 1984). Bahadur et al. (1998) claimed that the Zn uptake rate was faster in mango trees when zinc sulfate was foliar applied as compared with its soil application. While, the boron requirement is much higher for reproductive growth than for vegetative growth and increases flower production and retention, pollen tube elongation and germination, and seed and fruit development (Peres and Reyes, 1983). The application of boron as foliar spray also enhanced the fruit set in papaya (Jeyakumar et al., 2001). Furthermore, the supply of boron needed for reproductive growth in many crops is more needed than that needed for vegetative growth (Mengel and Kirkby, 1982; Marschner, 1986; Hanson, 1991) and same may be true in citrus. In citrus B deficiency leads to low sugar content, granulation and excessive fruit abortion (Reuther et al., 1968) as well as rind thickening. There is also concern that B is one of the micronutrient responsible for the changes in concentration and a number of metabolic pathways such as carbohydrate metabolism, nitrogen metabolism, phenol metabolism and ascorbate metabolism in plants (Marschner, 1995; Dordas and Brown, 2005; Lukaszewski and Blevins, 1996).

Therefore, in the present work, we have studied this response in pungent pepper grown at different levels of foliar application of HA and micronutrients (Zn and B), as well as the physiological and biochemical changes that take place during vegetative growth and fruit formation, in order to improve the quality, as well as productivity of crop.

**Table 1.** Detail doses of treatments which used as foliar spray for pungent pepper.

| Treatment | | Humic acid (HA) (%) | Zinc (Zn) (%) | Boron (B) (%) |
|---|---|---|---|---|
| Control : | $HA_0Zn_0B_0$ | 0.0 | 0.0 | 0.0 |
| HA: | $HA_1Zn_0B_0$ | 0.05 | 0.0 | 0.0 |
| Zn: | $HA_0Zn_1B_0$ | 0.0 | 0.05 | 0.0 |
| B: | $HA_0Zn_0B_1$ | 0.0 | 0.0 | 0.02 |
| HA+Zn: | $HA_1Zn_1B_0$ | 0.05 | 0.05 | 0.0 |
| HA+B: | $HA_1Zn_0B_1$ | 0.05 | 0.0 | 0.02 |
| Zn+B: | $HA_0Zn_1B_1$ | 0.0 | 0.05 | 0.02 |
| HA+ Zn+B: | $HA_1Zn_1B_1$ | 0.05 | 0.05 | 0.02 |

The source of humic acid was granular form, that of zinc (Zn) was $ZnSO_4.7H_2O$ and boron (B) from $Na_2B_4O_7.10H_2O$. Each spraying was done four times with sticker starting from 25 days after transplanting and subsequent ones at an interval of 10 days during vegetative stage.

## MATERIALS AND METHODS

### Field experiment

The seedling was grown in nursery beds prepared with a sandy loam soil and were 12 cm tall and 1.0 m wide. Weathered cow dung manure 4 kg/m, was mixed into the beds. Beds were drenched with formaldehyde (4.0%) and covered with polythene sheets for one week to avoid damping off disease. Seedling was treated with Dithane M-45 (2.5 $g.kg^{-1}$ of seed) (Hindustan Pulverizing Mills Industrial Growth Centre, Sumba, Jammu, India) prior to sowing. Fresh seeds of pungent pepper cv. bullet (*C. annuum* L.) collected from AICRP on Vegetable Crops were sown at 10 mm and 5 cm apart. After sowing, beds were covered with straw until seed germination and hand watering was done regularly. Seedlings were hardened by withholding water four days before transplanting.

### Pot experiment

A pot experiment was conducted in the net house of Bidhan Chandra Krishi Viswavidyalaya, Mohanpur, Nadia, West Bengal and India. The 40 days old seedlings were transplanted in earthen pot of 15 cm size (one $plant.pot^{-1}$) having a central drainage hole. Soil was prepared by mixing appropriate amount of well rotted cow dung and manures (soil 700 $g.pot^{-1}$; cow manure 100 $g.pot^{-1}$; urea 5 $g.pot^{-1}$; single superphosphate 20 $g.pot^{-1}$ and muriate of potash 6 $g.pot^{-1}$) (according to guidance of Department of Spices and Plantation Crops, Bidhan Chandra Krishi Viswavidyalaya, Mohanpur, Nadia, West Bengal, India). Additional fertilizer 5 $g.pot^{-1}$ was applied 21days after transplanting. The experiment consisting of eight treatments including control (only tap water sprayed) were arranged in a randomized block design with three replications. The detail treatments are summarized as in Table 1.

To prevent blossom and fruit drop, supplementary irrigation was required (hand water was applied at an interval of 1 day with the first being immediately after transplanting). No weeding was required. The insecticide Rogor at 2.5 $ml.L^{-1}$ (Rallis India Ltd., Mumbai, Maharashtra, India) was applied three times beginning just from flowering stage at 15 days intervals to control aphids. The following physiological and biochemical qualities were determined in the present experiment.

### Physiological qualities

All the recommended cultural practices for pungent pepper production were followed according to advice of Horticulture Faculty, Department of Spices and Plantation Crops, Bidhan Chandra Krishi Viswavidyalaya, Mohanpur, Nadia- 741252, West Bengal, India. Data were recorded as follows:

#### *Plant growth characteristics*

At the end of the growing season, (~90 days after transplanting) samples of ten selected plants were taken at random from each replication to determine the following characteristics:

1. Average height of plant (cm)
2. Average number of leaves $plant^{-1}$
3. Average leaf area (sq.cm) [Leaf area was determined followed the procedure of McKee (1964)]
4. Average number of $branches.plant^{-1}$

#### *Fresh fruit yield and its components*

Fresh fruit yield was started andharvested at 25 days after initiation of flowering. Samples of ten selected plants of each replication were counted until final harvesting to determine the following characteristics:

1. Average number of fresh fruits $plant^{-1}$
2. Average length of fresh fruit (cm)
3. Average diameter of fresh fruit (cm)
4. Weight of 20 fresh fruits (g)

#### *Seed yield and its compositions*

Twenty pods were taken at random from each replication to determine the following characteristics:

1. Average number of fresh seeds $fruit^{-1}$
2. Weight of 100 fresh seeds (g)

### Biochemical qualities

About 300 g weighed edible pungent pepper from each replication was collected. Fresh materials were chopped with a sharp knife into small pieces before analysis of total phenol and one oxidative enzyme [polyphenol oxidase (PPO)]. Samples were shredded and dried at 40°C for 96 h. This material was prepared for reducing sugar, total sugar, starch, protein and fat analyses by grinding to a fine powder using an electric grinder. A subsample of the dried shredded material was further dried at 100°C to constant dry weight to determine percent moisture.

#### *Analysis of chlorophyll a, chlorophyll b and total chlorophyll*

One gram fresh leaf mass of pungent pepper (*C. annuum* L.) was extracted by macerating with 10 ml of 80% solution of acetone. The procedure was repeated until the leaf mass was completely decolorized. The resulting extracts are kept in the dark in order to prevent the destruction of the chlorophyll molecules. The extracts are spectrophotometrically measured at the wave length of 663 and 645 nm (Arnon, 1956). The concentration (content) of chlorophylls in the solution was calculated with the following formulas:

$$\text{Chlorophyll a } (mg.g^{-1}) = [12.7(A_{663}) - 2.69(A_{645})] \times V/1000 \times W$$

$$\text{Chlorophyll b } (mg.g^{-1}) = [22.9(A_{645}) - 4.68(A_{663})] \times V/1000 \times W$$

$$\text{Total chlorophyll } (mg.g^{-1}) = [20.2(A_{645}) - 8.02(A_{663})] \times V/1000 \times W$$

Where, A = Absorbance of specific wavelength, V = final volume of chlorophyll extract in 80% acetone, W = weight of sample.

#### *Analysis of reducing sugar*

Reducing sugar were extracted in 10 ml of 80% alcohol by boiling 0.1 g dry powdered sample for 30 min at 80-90°C followed by centrifugation at 5,000 *g* for 10 min and subsequent analysis was followed using the Nelson-Somogyi's method (Sadasivam and Manickam, 1992). The amount of reducing sugar was determined against a glucose solution (0-500 µg) as standard curve.

#### *Analysis of total sugar and starch*

Sugar and starch were extracted in 10 ml of 80% anhydrous alcohol by boiling 0.1 g dry powdered sample for 30 min at 80°C followed by centrifugation at 5,000 *g* for 10 min and subsequent procedures was followed using the Anthrone reagent method (Sen et al., 2005). The amount of concentration was determined against a glucose solution (0-500 µg) as standard curve.

#### *Analysis of total protein*

Total nitrogen (N) was analyzed by the Kjeldahl method (Casanas et al., 2002) and used for the calculation of the protein concentration by multiplying with a conversion factor of 6.25.

#### *Analysis of fat*

Fat content of dry fruit of pungent pepper was analyzed following extraction with petrol ether in a Soxhlet apparatus and subsequent proceeding was followed by method of AOAC (1984).

#### *Analysis of energy*

The energy value was calculated using Atwater factors of 17 kJ/g of protein, 38 kJ/g of fat, 17 kJ/g of starch and 16 kJ/g of sugar (FAO/WHO/UNU, 1985).

#### *Analysis of total phenol*

Total phenol was extracted in 50% methanolic 1.2 N HCl by boiling 1 g of finely chopped tissue for 1.5 h at 80-90°C following the method of Vinson et al. (1995) and subsequent analysis was with the Folin-Ciocalteau reagent using gallic acid as standard.

#### *Analysis PPO Activity*

One gram of fresh tissue from each treatment was macerated in a pre-chilled pestle and mortar and extracted with 10 ml phosphate buffer (pH 6.0) to determine PPO activity. Triturated samples were centrifuged at 10,000 *g* for 30 min at 0°C and supernatants were assessed for enzyme activity. Estimation of polyphenol oxidase was made by adding 2 ml phosphate buffer and 0.5 ml (0.01 M) catechol solution with 0.5 ml extract. Changes in activities were measured following the procedure of Matto and Diamond (1963).

#### Statistical analysis

Data were subjected to ANOVA of a randomized block design, simple correlation were calculated and tested for significance and means. Principal component analysis (PCA), as the method of identifying the effect dimension of the data, was used to summarize the treatment information in a reduced number of effects for selection of the best performing treatment. Statistical analyses were done using SPSS Professional Statistics ver. 7.5 (SPSS Inc., Irvine, California).

## RESULTS AND DISCUSSION

### Physiological qualities

The changes in physiological qualities in response to foliar applications of HA, Zn and B in plants and fresh fruits of pungent pepper (*C. annum* L.), are shown in Tables 2, 3 and 4.

#### *Plant growth qualities*

The plant growth quality aspects were determined with regard to average height of plant, average number of leaves $plant^{-1}$, average leaf area and average number of branches $plant^{-1}$ (Table 2). In our present experiment, results revealed that with respect to plant height, there was significantly positive influence in all treatments except Zn and B individual application as compared to that of their corresponding control (only tap water spray) treatment, in which (Zn and B) treatments also positive influenced the height of plant, apparently, though not significantly. The higher performer was observed, when plants are treated with HA + Zn + B followed by HA + Zn and Zn + B.

On the other hand, in respect to number of leaves $plant^{-1}$, there was significantly positive effect in all treatments except B application alone over the control treatment, in which B treatment also positively enhanced, apparently, but not significantly. The maximum value of number of leaves was observed in HA + Zn + B treatment.

Regarding leaf area, significantly positive influence was

**Table 2.** Effect of foliar application of HA, Zn and B on plant growth quality in pungent pepper.

| Treatment | Average height of plant (cm) | Average number of leaves $plant^{-1}$ | Average leaf area (sq.cm) | Average number of branches $plant^{-1}$ |
|---|---|---|---|---|
| Control | 45.60 | 30.46 | 1.19 | 27.31 |
| HA | 53.02* | 40.85* | $1.40^{NS}$ | 45.87* |
| Zn | $51.09^{NS}$ | 46.34** | $1.28^{NS}$ | 49.02* |
| B | $48.27^{NS}$ | $39.39^{NS}$ | $1.33^{NS}$ | $35.95^{NS}$ |
| HA+Zn | 58.00** | 56.34** | 1.46* | 63.04** |
| HA+B | 51.62* | 42.63* | 1.69** | 56.26** |
| Zn+B | 57.00** | 53.91** | 1.71** | 71.17** |
| HA+Zn+B | 64.03** | 58.57** | 1.80** | 75.11** |
| SE(m)± | 1.96 | 3.21 | 0.07 | 5.56 |
| LSD (0.05) | 5.89 | 9.64 | 0.22 | 16.68 |
| LSD (0.01) | 8.47 | 13.87 | 0.30 | 24.02 |

*Significant at 5%, **Significant at 1% and $^{NS}$Non significant

seen in all combination treatments of HA, Zn and B (HA + Zn, HA + B, Zn + B and HA + Zn + B) as compared to that of their corresponding control treatment. While, individual treatments of HA, Zn and B also exerted positive influence, through not significantly. The best performer was observed, when plants are treated with HA + Zn + B followed by Zn + B and HA + Zn application.

In the case of number of branches $plant^{-1}$, it was significantly induced in all treatments except B application alone than the control treatment. While, B application alone also induced the number of branches but not significantly. The highest number was recorded from HA + Zn + B followed by Zn + B treatment.

Our results also indicated that, in respect to growth quality, Zn application gave the better flourishing than the HA and B application over the control treatment. Moreover, under HA treatment, the plant height and number of branches $plant^{-1}$ were increased significantly in pea plant, that reported by El-Hak et al. (2012). Azarpour et al. (2011) also reported that the foliar application of HA gave the best and highest plant height values of cowpea. Kaya et al. (2005) reported that, HA and Zn both alone and/or combination treatments significantly enhanced plant height as compared to control treatment. While, Hatwar et al. (2003) reported that, Zn (0.1%) and B (0.1%) both alone and/or combination treatments exerted significantly positive influence on the plant height and number of branches $plant^{-1}$ as compared to the control. Improvement of plant growth might be due to enhancement in photosynthetic and other metabolic activities which led to an increase in various plant metabolites responsible for cell division and cell elongation. Enhanced photosynthetic reactions in the presence of Zn and B might have increased the plant growth (Rawat and Mathpal, 1984). The further height of plant due to application of Zn explained by Mallick and Muthukrishnan (1979) was due to an active synthesis of tryptophan, an amino acid in the presence of Zn and it is the precursor of IAA which stimulates the growth of plant tissues. Besides, the Zn and B also plays an essential role in the development and growth of new cells in the plat. Plant requires B for synthesis of amino acids and proteins and regulation of carbohydrates metabolism (Dyar and Webb, 1961).

### *Fresh fruit yield and it's components*

The fruit yield aspects were determined with regard to average number of fresh fruits $plant^{-1}$, average length of fresh fruit, average diameter of fresh fruit and average weight of 20 fresh fruits (Table 3).

Our present experiment indicated that, with respect to number of fresh fruits $plant^{-1}$, all the treatments except HA application alone exerted significantly positive influence as compared to control. The HA application also exerted positive influence, through not significantly. The maximum number of fruit $plant^{-1}$ was observed by application of HA + Zn + B.

In the case of length of fruit, Zn, B, HA + B, Zn + B and HA + Zn + B application exerted significantly positive influence over the control. The remaining treatments, HA and HA + Zn application also influenced, apparently, though not significantly. The highest values of length of fruit were exhibited by HA + Zn + B followed by Zn + B and HA + B application.

All the treatments except HA alone exerted significantly positive influence on the diameter of fresh fruit as compared to control treatment, in which HA had positive influence but not significantly. The maximum value was recorded by Zn + B application.

Regarding, weight of 20 fresh fruits, HA + Zn + B, Zn + B and HA + Zn application exerted significantly positive effect as compared to the control. The remaining treatments also had positive effect, though not significantly.

The best performer was observed by HA + Zn + B

**Table 3.** Effect of foliar application of HA, Zn and B on fresh fruit yield and its components in pungent pepper.

| Treatment | Average number of fresh fruits plant$^{-1}$ | Average length of fresh fruit (cm) | Average diameter of fresh fruit (cm) | Average weight of 20 fresh fruits (g) |
|---|---|---|---|---|
| Control | 3.37 | 4.47 | 3.00 | 45.67 |
| HA | 5.50$^{NS}$ | 5.27$^{NS}$ | 3.50$^{NS}$ | 48.33$^{NS}$ |
| Zn | 6.58* | 5.90** | 4.10** | 52.73$^{NS}$ |
| B | 5.78* | 5.57* | 4.00** | 49.60$^{NS}$ |
| HA+Zn | 8.66** | 5.06$^{NS}$ | 4.70** | 60.33* |
| HA+B | 8.33** | 6.20** | 4.60** | 57.60NS |
| Zn+B | 9.53** | 6.90** | 4.80** | 69.33** |
| HA+Zn+B | 9.80** | 7.30** | 4.50** | 81.60** |
| SE(m)± | 0.75 | 0.31 | 0.21 | 4.05 |
| LSD (0.05) | 2.25 | 0.95 | 0.63 | 12.16 |
| LSD (0.01) | 3.24 | 1.34 | 0.91 | 17.50 |

*Significant at 5%, **Significant at 1% and $^{NS}$Non significant.

followed by Zn + B application.

Whatsoever, Halwar et al. (2003) reported that, Zn and B both alone and/or combination treatments exerted significantly positive influence on the number of fruit plant$^{-1}$ as compared to control. While Dongre et al. (2000) reported that the application of $ZnSO_4$ (0.10, 0.25 and 0.50%) exerted significantly positive influence on the average length of fruit and average diameter of fruit as compared to control. Concerning the effect of foliar application with HA, number of fruits plant$^{-1}$, fruit length, fruit diameter and 20 fruits weight were significantly increased during first season, 2009/2010 in pea (El-Hak et al., 2012). Azarpour et al. (2011) reported that HA foliar application gave significant increasing results on number of pods plant$^{-1}$ and pod length of cowpea. Kaya et al. (2005) reported that HA + Zn treatments significantly enhanced number of pods plant$^{-1}$ over the control treatment. Many researchers e.g., Zaky et al. (2006) on beans, Neri et al. (2002) on strawberry and Forgac and Czimbalmos (2011) on pea plants used the HA foliar application in green house and/or open field cultivations and found that number of pods plant$^{-1}$, total yield plant$^{-1}$ and average pod fresh weight were markedly increased by the treatment at the rate of 1 g.L$^{-1}$ combined with irrigation water.

### *Fresh seed yield and its compositions*

Data regarding number of fresh seeds fruit$^{-1}$ and weight of 100 fresh seeds are presented in Table 4. Analysis of variance revealed that the number of seeds fruit$^{-1}$, was significantly positive as observed by foliar application of Zn + B, HA + Zn + B, HA + B, HA + Zn and B as compared to control. HA and Zn application also occurred the positive effect, though not significantly. The good promising result was recorded by Zn + B followed by HA + Zn + B, HA + B and HA + Zn application.

On the other hand, in respect to average weight of 100 fresh seeds, all the combination treatments (HA + Zn, HA + B, Zn + B and HA + Zn + B) exerted significantly positive influence as compared to control treatment. While, individual application of HA, Zn and B also had positive influence, though not significantly. The maximum weight of 100 fresh seeds were recorded by application of HA + Zn + B. Moreover, Dongre et al. (2000) reported that the application of $ZnSO_4$ (0.10, 0.25 and 0.50%) exerted significantly positive influence on the average number of seeds fruit$^{-1}$ and weight of 500 seeds as compared to the control. While, number of fresh seeds pod$^{-1}$, seeds weight and green pods yield significantly increased in pea (El-Hak et al., 2012). Similar results were obtained by Malik and Azam (1986) on wheat; Putintsev and Platonova (1991) on pea; Salib (2002) on peanut; Habashy et al. (2005) on peanut and faba bean and Azarpour et al. (2011) on cowpea plants as they reported that foliar spray with humic acid increased the dry seed yield and its parameters.

## Biochemical qualities

### *Green pigments content*

Beneficial effect of Zn on photosynthetic pigments may be due to its role in increasing the rates of photochemical reduction (Kumar et al., 1988), chloroplast structure, photosynthetic electron transfer as well as photosynthesis (Romheld and Marschner, 1991). In our present experiment, results revealed that, all the treatments exerted significantly positive influence on content of chlorophyll a, chlorophyll b and total chlorophyll as compared to control treatment (Table 5). The maximum chlorophylls (chl. a, chl. b and total chl.) content were recorded from HA + Zn + B application. Increased chlorophyll concentration by foliar application of B and Zn were reported in Irish plant over the control treatment (Khalifa et al., 2011). However

**Table 4.** Effect of foliar application of HA, Zn and B on fresh seed yield and its compositions in pungent pepper.

| Treatment | Average number of fresh seeds.fruit$^{-1}$ | Average weight of 100 fresh seeds (g) |
|---|---|---|
| Control | 56.06 | 0.51 |
| HA | 61.46$^{NS}$ | 0.54$^{NS}$ |
| Zn | 59.26$^{NS}$ | 0.54$^{NS}$ |
| B | 66.37* | 0.53NS |
| HA+Zn | 71.52** | 0.57* |
| HA+B | 72.41** | 0.58* |
| Zn+B | 78.43** | 0.62** |
| HA+Zn+B | 76.54** | 0.65** |
| SE(m)± | 2.75 | 0.02 |
| LSD (0.05) | 8.26 | 0.05 |
| LSD (0.01) | 11.88 | 0.08 |

*Significant at 5%, **Significant at 1% and $^{NS}$Non significant.

**Table 5.** Effect of foliar application of HA, Zn and B on green pigments content in pungent pepper.

| Treatment | Chlorophyll a [mg.g$^{-1}$ (FW)] | Chlorophyll b [mg.g$^{-1}$ (FW)] | Ratio of chlorophyll a:b | Total Chlorophyll [mg.g$^{-1}$ (FW)] |
|---|---|---|---|---|
| Control | 0.91 | 0.44 | 2.07 | 1.32 |
| HA | 1.26* | 0.71* | 1.78* | 1.97* |
| Zn | 1.23* | 0.70* | 1.75** | 1.93* |
| B | 1.38** | 0.89** | 1.56** | 2.27** |
| HA+Zn | 1.45** | 1.05** | 1.46** | 2.25** |
| HA+B | 1.56** | 1.05** | 1.51** | 2.72** |
| Zn+B | 1.55** | 1.04** | 1.49** | 2.60** |
| HA+Zn+B | 1.72** | 1.15** | 1.50** | 2.83** |
| SE(m)± | 0.08 | 0.08 | 0.07 | 0.17 |
| LSD (0.05) | 0.25 | 0.24 | 0.21 | 0.50 |
| LSD (0.01) | 0.36 | 0.34 | 0.30 | 0.73 |

*Significant at 5%, **Significant at 1% and $^{NS}$Non significant.

Farouk et al. (2011) reported that the application of HA also enhanced the chlorophyll concentration in radish plant. Whereas, in the case of values of chlorophyll a:b ratio opposite trend result of total chlorophyll were given, which exerted significantly negative influence in all treatments as compared to that of control (only water sprayed) treatment.

### *Moisture content*

There are differences in moisture content and nutrient contents among the several treatments are presented in Table 6. In our present experiment, foliar applications of HA, Zn and B through different treatments could produce negative significant differences in the content of moisture. Therefore, no positive significant effects could be observed by applications of HA, Zn and B on the content of moisture.

### *Carbohydrates constituents*

The beneficial effect of B may be due to its role in facilitating transport of carbohydrates, that is, starch and sugar (Donald et al., 1998). The obtained results are in conformity with those of Farahat et al. (2007) on *Cupressus sempervirens* and Nahed and Laila (2007) on *Salvia farinacea*. In our present study, results showed that in respect to reducing sugar, significantly enhanced effects were seen in all treatments except HA alone as compared to the control, in which (HA) also positive enhanced, apparently, though not significantly (Table 6). The maximum content of reducing sugar was recorded by HA + Zn + B and HA + B application.

Regarding, total sugar content (Table 6), exerted significantly positive influenced results were observed in HA + Zn + B, Zn + B, HA + B and B applications as compared to control. HA, Zn and HA + Zn also had positive effect but not significantly. The highest concentration of total

**Table 6.** Effects of HA, Zn and B on biochemical qualities in pungent pepper cv. Bullet (*C. annuum* L.)

| Treatment | Moisture content (%) | Reducing sugar [mg.g$^{-1}$ (DW)] | Total sugar [mg.g$^{-1}$ (DW)] | Starch [mg.g$^{-1}$ (DW)] | Protein [mg.g$^{-1}$ (DW)] | Fat [mg.g$^{-1}$ (DW)] | Energy [kJ.g$^{-1}$ (DW)] | Total phenol [mg.g$^{-1}$ (QE) (FW)] | Polyphenol oxidase (ΔA/min/g) |
|---|---|---|---|---|---|---|---|---|---|
| Control | 86.02 | 7.49 | 26.65 | 213.80 | 111.30 | 71.00 | 8.65 | 0.97 | 0.56 |
| HA | 82.30* | 8.90$^{NS}$ | 33.40$^{NS}$ | 270.60** | 177.67** | 93.00$^{NS}$ | 11.69** | 1.14$^{NS}$ | 1.7** |
| Zn | 83.22* | 9.99* | 29.21$^{NS}$ | 251.60* | 200.67** | 122.00** | 12.79** | 1.15$^{NS}$ | 0.67$^{NS}$ |
| B | 82.37* | 11.61** | 39.27* | 301.40** | 153.67* | 60.00$^{NS}$ | 10.65* | 1.58* | 0.66$^{NS}$ |
| HA+Zn | 79.66** | 12.21** | 35.67$^{NS}$ | 291.70** | 211.00** | 111.00** | 13.33** | 1.40$^{NS}$ | 0.95$^{NS}$ |
| HA+B | 81.01** | 13.05** | 44.80** | 308.23** | 203.74** | 91.00$^{NS}$ | 12.88** | 1.86** | 1.61** |
| Zn+B | 80.59** | 12.21** | 47.85** | 314.76** | 221.11** | 115.83** | 14.28** | 1.99** | 1.12* |
| HA+Zn+B | 79.03** | 13.05** | 52.00** | 322.67** | 215.32** | 121.20** | 14.58** | 2.25** | 1.72** |
| SE(m)± | 0.76 | 0.68 | 3.01 | 12.33 | 12.56 | 7.82 | 0.66 | 0.15 | 0.16 |
| LSD (0.05) | 2.23 | 2.05 | 9.03 | 36.93 | 37.68 | 23.45 | 1.97 | 0.46 | 0.49 |
| LSD (0.01) | 3.28 | 2.94 | 13.00 | 53.26 | 54.26 | 33.78 | 2.85 | 0.65 | 0.69 |

*Significant at 5%, **Significant at 1% and $^{NS}$Non significant.

sugar was observed from application of HA + Zn + B followed by Zn + B and HA + B.

On the other hand, the starch contents (Table 6), exerted significant positive influence in application of HA, Zn and B alone and/or combination treatments. The highest concentration of starch was observed from HA + Zn + B followed by Zn + B and HA + B application. However Khan et al. (2012) observed a decreased trend in reducing sugar concentration in relation to foliar application of Zn and B in citrus fruit. While, Kalifa et al. (2011) also reported that application of Zn and B increased the carbohydrates percentage in Irish plant. A similar trend of results was found by El-Khayat (1999), Gomaa (2001) and Samia and Mahmoud (2009) showing that Zn increased total carbohydrate in *Antholyza aethiopica* and *Tritonia crocata* plants, respectively. However, under Zn treatment, the values of total and non-reducing sugar increase were observed in citrus fruit (Khan et al., 2012). Increased total sugar concentration by application of HA was reported in radish plant (Farouk et al., 2011).

### *Protein content*

Protein content of groundnut pod was significantly influenced by B application as reported by many workers (Bhuiyan et al., 1997; Murthy, 2006). Application of B stimulated nitrogen content of potato tubers and might have increased protein synthesis and subsequent storage of protein as suggested by Yadav and Manchanda (1979). In our present study, results observed that with respect to protein content, significant increasing trend were observed in all treatments over the control treatment (Table 6). The highest concentration of protein was recorded from Zn + B followed by HA + Zn + B and HA + Zn application. Moreover, increasing the concentration of protein by application of HA was reported in radish plant (Farouk et al., 2011) and in common vetch (*Vicia sativa* L.) (Saruham et al., 2011).

### *Lipid content*

In the case of fat content (Table 6), HA, B and HA+B foliar spray failed to produce any significant effect over the control. The highest value of fat was exhibited by Zn followed by HA + Zn + B and Zn + B application. However, foliar application of Zn and B significantly increased the oil content in flowers of Irish plant over the control as observed by Khalifa et al. (2011).

### *Energy content*

Regarding energy values in pungent pepper fruits (Table 6), all the treatments produced significantly increasing effects over the control treatment. Whereas, plants treated with foliar application of combination treatments gave best results than individual treatments. The highest value of energy was observed from application of HA + Zn + B followed by Zn + B and HA + Zn.

### *Total phenol*

Phenols are ubiquitous phytochemicals that contributed largely to antioxidant potential of any plant. The B is presumably responsible for the metabolism changes and cell damages in boron deficient tissue (Marschner, 1986) and it is thought that B complexes the phenolic compounds in plant cells, reducing their potential toxicity (Lee and Arnoff, 1967). However, the increase in total phenol effectuated by Zn application in wheat leaves was reported by Vinod et al. (2012). Hamid et al. (2010) found that phenolic content of plant were increasing with increasing level of heavy metal. In our present experiment, results showed that the total phenol content significantly increased in all treatments exception HA, Zn and HA + Zn application as compared to the control, which also followed an increasing trend over the control, apparently, though not significantly (Table 6). The highest concentration of total phenol was observed from HA + Zn + B followed by Zn + B and HA + B application.

### *Polyphenol oxidase*

Polyphenol oxidases (PPO) proteins containing copper catalyze oxidation of hydroxyphenols to their quinine derivatives, which then spontaneously polymerize. Enzymatic browning in fruits and vegetables are due to presence of PPO and is often responsible for unpleasant sensory qualities and losses in nutrient quality (Sanchez-Amat and Solano, 1997). Cakmak et al. (1995) reported that under B-deficiency conditions, sunflower leaves are characterized by a rapid loss of $K^+$ and a strong discoloration (browning), as occurs with high PPO activity. Our present study showed that the polyphenol oxidase (PPO) activity was significantly increased in all treatments except Zn, B and HA + Zn application as compared to control, where Zn, B and HA+Zn also increased the activity of PPO but not significantly (Table 6). The highest activity of PPO was observed in foliar spray with HA+Zn+B followed by HA and HA + B. While, the PPO activity declined significantly in plant with B deficiency as reported by Cara et al. (2002). However, recently Pfeffer et al. (1998), working with B-deficient sunflower leaves, observed that B does not maintain plasma-membrane integrity by complexing phenols and inhibiting PPO activity to prevent damage by oxygen-free radicals or by regulating ascorbate metabolism. The authors concluded that the resupply of B to deficient leaves might be the result of the inactivation (detoxification) of phenolics by the formation of stable phenol-borate complexes and, thus, repression of phenol oxidation.

### Correlation among variables

Correlation coefficients (Table 7) showed that the number of branches plant$^{-1}$ had positive and significant correlation with length of plant ($r^2$=0.932), number of leaves plant$^{-1}$ ($r^2$=0.943) and leaf area ($r^2$=0.882). While number of fruits plant$^{-1}$ had positive and significant correlation with length of fruit, number of leaves plant$^{-1}$, leaf area and number of branches plant$^{-1}$. The correlation of number of seeds fruit$^{-1}$ with fruit length ($r^2$=0.774) and fruit diameter ($r^2$=0.882) were positive and significant. There are significant and positive correlation between fruit length and weight of 100 seeds ($r^2$=0.890), plant height and weight of 100 seeds, total chlorophyll and leaf area ($r^2$=0.921). The correlation of energy with carbohydrates, fat, protein, total chlorophyll and leaf area were positive and significant. Another, significant and positive correlation was between polyphenol oxidase and leaf area ($r^2$=0.742). Moreover, there was significant and positive correlation of starch with some physiological characteristics such as leaf area, number of fruits.plant$^{-1}$, fruit diameter and total chlorophyll. The correlation of total phenol with carbohydrates and chlorophyll were positive and significant. Our results revealed that the positive and significant correlation between number of fruits plant$^{-1}$ and 100 seeds weight ($r^2$=0.919), agree with that of result by Kaya et al. (2005), they observed positive and non significant correlation. The significant inverse correlation between moisture and starch ($r^2$=0.901) obtained supports results of Casanas et al. (2002) and Guchhait et al. (2008). Above relationships indicate that improving number of leaves and leaf area as well as increasing number of fruit and chlorophyll contents could accompany improvement in energy value as well as quality of fruits.

### Principal component analysis (PCA)

In our present experiment, PCA was used to summarize the treatment information in a reduced number of components, where a total of three components were chosen (PC1, PC2 and PC3) due to their Eigen value being greater than 1.0 and they together explained 93.07% of total variance (Table 8).

The PC1 explained 80.99% of total variance where all the variables other than ratio of chlorophyll a:b and moisture content are positively loaded meaning that all other variables limit ratio of chlorophyll a:b and moisture content, in which the former is not desirable (Table 8). Therefore, on the basis of the first component, the treatments such as HA + Zn + B, Zn + B, HA + Zn and HA + B can be selected as having all desirable traits.

PC2, on the other hand, explained another 7.53% of total variance in which an increase in length of plant, number

**Table 7.** Pearson's correlation matrix of all variables.

| Variable | Plant height | Number of leaves | Leaf area | Number of branches | Number of fruits | Fruit length | fruit diameter | Weight of 20 fruits | Number of seeds |
|---|---|---|---|---|---|---|---|---|---|
| Plant height | | | | | | | | | |
| Number of leaves | 0.936** | | | | | | | | |
| Leaf area | 0.787* | 0.711* | | | | | | | |
| Number of branches | 0.932** | 0.943** | 0.882** | | | | | | |
| Number of fruits | 0.861** | 0.919** | 0.892** | 0.970** | | | | | |
| Fruit length | 0.699 | 0.681 | 0.856** | 0.801* | 0.810* | | | | |
| Fruit diameter | 0.684 | 0.838** | 0.765** | 0.856** | 0.948** | 0.694 | | | |
| Weight of 20 fruits | 0.914** | 0.861** | 0.874** | 0.916** | 0.884** | 0.855** | 0.717* | | |
| Number of seeds | 0.767* | 0.787* | 0.922** | 0.875** | 0.930** | 0.774* | 0.882** | 0.849** | |
| Weight of100 seeds | 0.900** | 0.841** | 0.947** | 0.945** | 0.919** | 0.890** | 0.759* | 0.979** | 0.900** |
| Chlorophyll a | 0.803* | 0.804* | 0.915** | 0.869** | 0.931** | 0.837** | 0.874** | 0.834* | 0.927** |
| Chlorophyll b | 0.780* | 0.821* | 0.870** | 0.855** | 0.938** | 0.734* | 0.922** | 0.797* | 0.945** |
| Ratio of chlorophyll a:b | -0.673 | -0.784* | -0.747* | -0.775** | -0.886** | -0.648 | -0.938** | -0.662 | -0.885** |
| Total chlorophyll | 0.714* | 0.721* | 0.921** | 0.820* | 0.902** | 0.850** | 0.865** | 0.785* | 0.918** |
| moisture | -0.888** | -0.900** | -0.834** | -0.902** | -0.932** | -0.683 | -0.871** | -0.806* | -0.888** |
| Reducing sugar | 0.655 | 0.736* | 0.818* | 0.772* | 0.897** | 0.725* | 0.928** | 0.732* | 0.903** |
| Total sugar | 0.711* | 0.649 | 0.947** | 0.783* | 0.833* | 0.877** | 0.732* | 0.850** | 0.930** |
| Starch | 0.691 | 0.707 | 0.856** | 0.766* | 0.855** | 0.772* | 0.836** | 0.728* | 0.921** |
| Protein | 0.797* | 0.892** | 0.759* | 0.924** | 0.928** | 0.736* | 0.898** | 0.728* | 0.760* |
| Fat | 0.760* | 0.819* | 0.514 | 0.813* | 0.714* | 0.590 | 0.600 | 0.680 | 0.447 |
| Energy | 0.875** | 0.932** | 0.821* | 0.969** | 0.958** | 0.817* | 0.883** | 0.838** | 0.812* |
| Total phenol | 0.702 | 0.663 | 0.931** | 0.783* | 0.845** | 0.894** | 0.756* | 0.873** | 0.924** |
| Polyphenol oxidase | 0.605 | 0.384 | 0.742* | 0.575 | 0.510 | 0.528 | 0.318 | 0.504 | 0.502 |

*Significant at 5%, **Significant at 1%, Student's t-test.

**Table 7.** Contd.

| Variable | Weight of 100 seeds | Chlorophyll a | Chlorophyll b | Ratio of chlorophyll a:b | Total Chlorophyll | Moisture | Reducing sugar | Total sugar | Starch | Protein | Fat | Energy | Total phenol |
|---|---|---|---|---|---|---|---|---|---|---|---|---|---|
| Plant height | | | | | | | | | | | | | |
| Number of leaves | | | | | | | | | | | | | |
| Leaf area | | | | | | | | | | | | | |
| Number of branches | | | | | | | | | | | | | |
| Number of fruits | | | | | | | | | | | | | |
| Fruit length | | | | | | | | | | | | | |
| Fruit diameter | | | | | | | | | | | | | |
| Weight of 20 fruits | | | | | | | | | | | | | |

**Table 7.** Contd.

| Variable | Weight of 100 seeds | Chlorophyll a | Chlorophyll b | Ratio of chlorophyll a:b | Total chlorophyll | Moisture | Reducing sugar | Total sugar | Starch | Protein | Fat | Energy | Total phenol |
|---|---|---|---|---|---|---|---|---|---|---|---|---|---|
| Number of seeds | | | | | | | | | | | | | |
| Weight of 100 seeds | | | | | | | | | | | | | |
| Chlorophyll a | 0.883** | | | | | | | | | | | | |
| Chlorophyll b | 0.837** | 0.977** | | | | | | | | | | | |
| Ratio of chlorophyll a:b | -0.707* | -0.926** | -0.968** | | | | | | | | | | |
| Total chlorophyll | 0.852** | 0.988** | 0.958** | -0.911** | | | | | | | | | |
| moisture | -0.838** | -0.946** | -0.961** | 0.924** | -0.893** | | | | | | | | |
| Reducing sugar | 0.767* | 0.947** | 0.975** | -0.959** | 0.951** | -0.885** | | | | | | | |
| Total sugar | 0.903** | 0.928** | 0.881** | -0.784* | 0.941** | -0.802* | 0.853** | | | | | | |
| Starch | 0.792* | 0.968** | 0.957** | -0.941** | 0.969** | -0.901** | 0.931** | 0.923** | | | | | |
| Protein | 0.785* | 0.828* | 0.815* | -0.810* | 0.796* | -0.865** | 0.766* | 0.643 | 0.734* | | | | |
| Fat | 0.670 | 0.492 | 0.457 | -0.402 | 0.423 | -0.580 | 0.381 | 0.332 | 0.313 | 0.832* | | | |
| Energy | 0.877** | 0.869** | 0.841** | -0.803* | 0.829* | -0.890** | 0.781* | 0.727* | 0.767* | 0.983** | 0.846** | | |
| Total phenol | 0.908** | 0.922** | 0.880** | -0.781* | 0.935** | -0.784* | 0.874** | 0.990** | 0.901** | 0.643 | 0.359 | 0.732* | |
| Polyphenol oxidase | 0.615 | 0.630 | 0.528 | -0.397 | 0.627 | -0.608 | 0.418 | 0.617 | 0.565 | 0.519 | 0.337 | 0.546 | 0.540 |

*Significant at 5%, **Significant at 1%, Student's t-test.

**Table 3.** Results of PCA for effect of foliar application of HA, Zn and B on biochemical changes related to productivity of pungent pepper (*C. annuum* L.).

| Principle component | Eigen value | Variance | Cumulative variance (%) |
|---|---|---|---|
| Eigen value and variance accounted for (%) by PCA based on correlation matrix | | | |
| 1 | 18.63 | 80.99 | 80.99 |
| 2 | 1.73 | 7.53 | 88.53 |
| 3 | 1.04 | 4.53 | 93.07 |
| **Variables** | **PC1** | **PC2** | **PC3** |
| Factor loadings due to PCs with Eigen values greater than 1 | | | |
| Length of plant | 0.872 | 0.324 | -0.143 |
| Number of leaves | 0.889 | 0.369 | 0.139 |
| Leaf area | 0.933 | -0.118 | -0.261 |
| Number of branches | 0.953 | 0.281 | -0.049 |
| Number of fruits | 0.984 | 0.111 | 0.092 |
| Length of fruit | 0.857 | -0.019 | -0.309 |
| Diameter of fruit | 0.909 | 0.012 | 0.361 |
| Weight of 20 fruits | 0.904 | 0.147 | -0.304 |
| Number of seeds | 0.950 | -0.185 | -0.012 |
| Weight of 100 seeds | 0.944 | 0.092 | -0.301 |

**Table 8. Contd.**

| Principle component | Eigen value | Variance | Cumulative variance (%) |
|---|---|---|---|
| Chlorophyll a | 0.975 | -0.177 | 0.045 |
| Chlorophyll b | 0.960 | -0.188 | 0.186 |
| Ratio of Chl. a/Chl. b | -0.896 | 0.209 | -0.389 |
| Total chlorophyll | 0.950 | -0.267 | 0.031 |
| Moisture | -0.948 | -0.010 | -0.177 |
| Reducing sugar | 0.913 | -0.279 | 0.251 |
| Total sugar | 0.906 | -0.333 | -0.252 |
| Starch | 0.915 | -0.350 | 0.101 |
| Protein | 0.889 | 0.322 | 0.225 |
| Fat | 0.638 | 0.754 | -0.011 |
| Energy | 0.936 | 0.309 | 0.082 |
| Total phenol | 0.907 | -0.311 | -0.233 |
| Polyphenol oxidase | 0.590 | -0.049 | -0.358 |

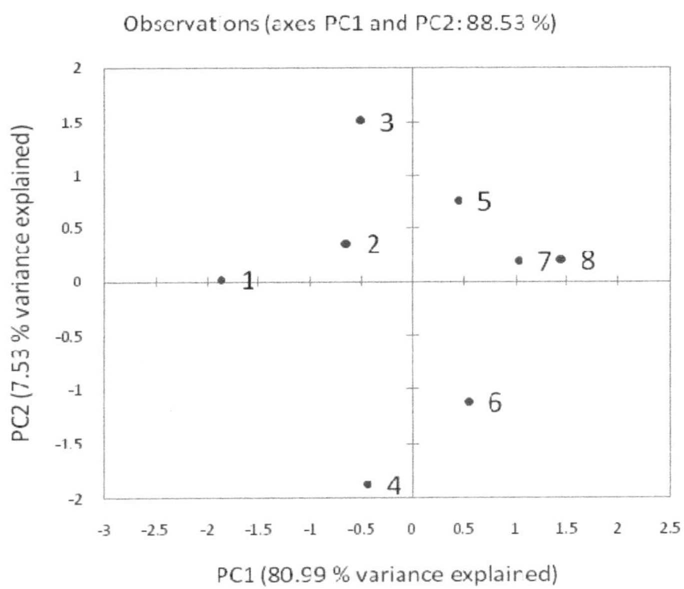


**Figure 1.** Biplot of regression factor scores for the first and second components produced by PCA: Legend: 1 (Control); 2 (HA); 3 (Zn); 4 (B); 5 (HA+Zn); 6 (HA+B); 7 (Zn+B); 8 (HA+Zn+B).

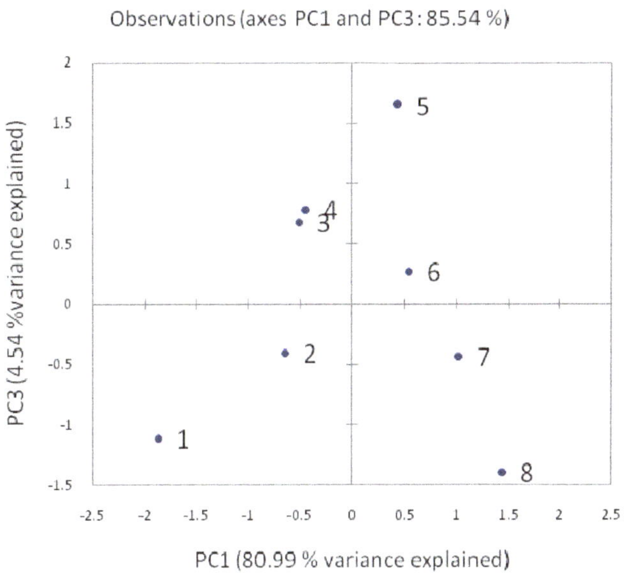


**Figure 2.** Biplot of regression factor scores for the first and third components produced by PCA: Legend: 1 (Control); 2 (HA); 3 (Zn); 4 (B); 5 (HA+Zn); 6 (HA+B); 7 (Zn+B); 8 (HA+Zn+B)

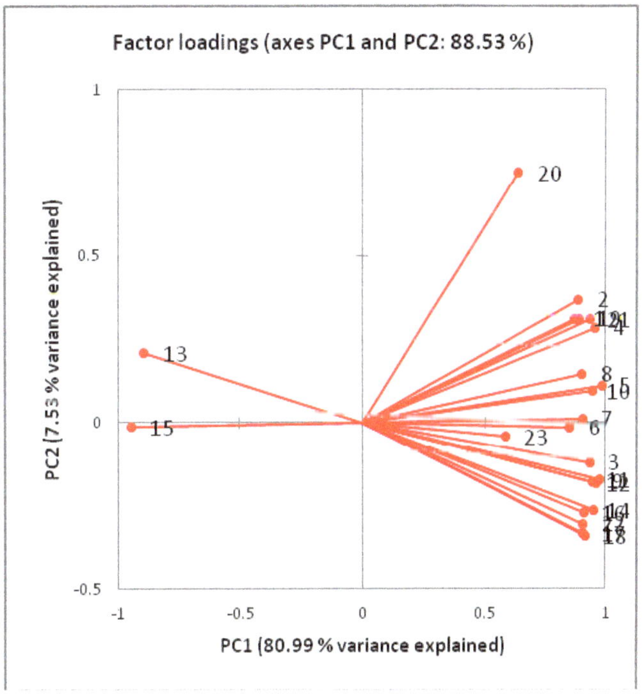


**Figure 3.** Scatter diagram of correlations between variables and factors loading for the first and second components produced by PCA: Legend: 1 (length of plant); 2 (number of leaves); 3 (Leaf area); 4 (number of branches); 5 (number of fruits); 6 (Length of fruit); 7 (diameter of fruit); 8 (weight of 20 fruits); 9 (Number of seeds); 10 (weight of 100 seeds); 11 (chlorophyll a); 12 (Chlorophyll b); 13 (ratio of Chl. a/Chl. b); 14 (total chlorophyll); 15 (moisture); 16 (reducing sugar); 17 (total sugar); 18 (starch); 19 (protein); 20 (fat); 21 (energy); 22 (total phenol); 23 (polyphenol oxidase).

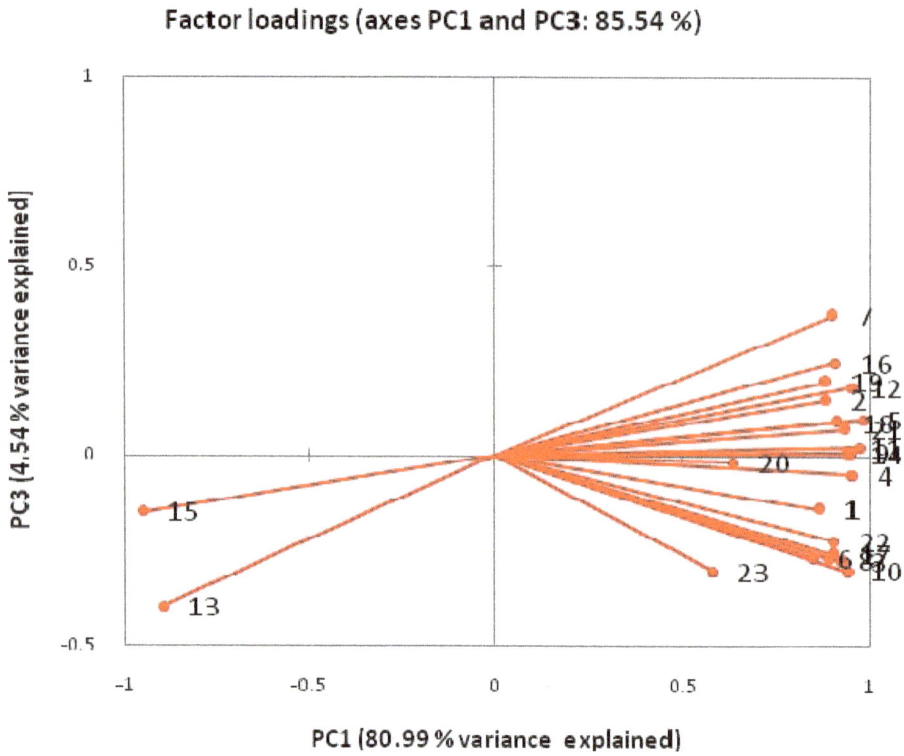


**Figure 4.** Scatter diagram of correlations between variables and factors loading for the first and third components produced by PCA: Legend: 1 (Length of plant); 2 (Number of leaves); 3 (leaf area); 4 (number of branches); 5 (number of fruits); 6 (Length of fruit); 7 (Diameter of fruit); 8 (weight of 20 fruits); 9 (number of seeds); 10 (weight of 100 seeds); 11 (chlorophyll a); 12 (chlorophyll b); 13 (ratio of Chl. a/Chl. b); 14 (total chlorophyll); 15 (moisture); 16 (reducing sugar); 17 (total sugar); 18 (starch); 19 (protein); 20 (fat); 21 (energy); 22 (total phenol); 23 (polyphenol oxidase).

of leaves, number of branches, number of fruits, diameter of fruit, weight of 20 fruits, weight of 100 seeds, ratio of chlorophyll a:b, protein, fat and energy were associated with a decrease in leaf area, length of fruit, number of seeds, chlorophyll a, chlorophyll b, total chlorophyll, moisture, reducing sugar, total sugar, starch, total phenol and polyphenol oxidase (Table 8 and Figure 3). Based on PC1 and PC2, the treatments such as HA + Zn + B followed by Zn + B and HA + Zn can be selected as performers (Figure 1).

The PC3 explained an additional 4.53% of total variance, in which number of leaves, number of fruits, diameter of fruit, chlorophyll a, chlorophyll b, total chlorophyll, reducing sugar, starch, protein, and energy were positively loaded in contrast to length of plant, leaf area, number of branches, length of fruit, weight of 20 fruits, number of seeds, weight of 100 seeds, ratio of chlorophyll a:b, moisture, total sugar, fat, total phenol and polyphenol oxidase, which were negatively loaded (Table 8 and Figure 4). According to the plot of regression factor scores due to PC1 and PC3, the performing treatment such as HA + Zn followed by Zn can be selected as having all desirable trails (Figure 2).

## Conclusions

Based on principle component analysis and considering average values of all variables, our results suggested that the good promising treatment in HA + Zn followed by Zn + B, HA + Zn + B and Zn alone application may bring about the proper value addition in quality as well as productivity of crop by enhancing some physiological and biochemical characteristics in pungent pepper.

## Conflict of Interests

The author(s) have not declared any conflict of interests.

## ACKNOWLEDGEMENTS

We would like to thank the Bidhan Chandra Krishi Viswavidyalaya for supporting this work and also wish to express our grateful thanks to Mr. Soumya Ghanti, Department of Spices and Plantation Crops, Faculty of Horticulture, BCKV, Mohanpur, Nadia-741252, West Bengal,

India, who have supported the cultural practices for pungent pepper cultivation in earthen pot.

## REFERENCES


Arnon DI (1956). Photosynthesis by isolated chloroplast. Biochem. Biphys. 20:440-461.

Association of Official Analytical Chemists (1984). *Official method of analisis*. 14th ed. Association of Official Analytical Chemists, Washington, D.C.

Azarpour E, Danesh RK, Mohammadi S, Bozorgi HR, Moraditochaee M (2011). Effects of nitrogen fertilizer under foliar spraying of humic acid on yield and yield components of cowpea (*Vigna unguiculata*). World Appl. Sci. J. 13:1445-1449.

Bahadur L, Malhi CS, Singh Z (1998). Effect of foliar and soil applications of zinc sulphate on zinc uptake, tree size, yield, and fruit quality of mango. J. Plant Nutr. 21(3):589-600.

Batru RSH, Rajpal CBS, Rath S (1984). Effect of zinc, 2,4-D andGA3 in Kagzi lime (*Citrus aurantifolia* Swingle) in fruit quality. Haryana J. Hort. Sci. 11(1,2):59-65.

Bhuiyan Md. M, Nishimura M, Matsumura S, Shimono T (1997). Antibacterial effects of the crude *Azadirachta indica* Neem bark extract on *Streptococcus* sobrinus. Pediatr. Dental J. 7(1):61-64.

Cakmak I, Kurz H, Marschner H (1995). Short-term effects of boron, germanium and high light intensity on membrane permeability in boron deficient leaves of sunflower. Physiol. Plant. 95(1):11-18.

Cara FA, Sánchez E, Ruiz JM, Romero L (2002). Is phenol oxidation responsible for the short-term effects of boron deficiency on plasma-membrane permeability and function in squash roots? Plant Physiol. Biochem. 40:853-858.

Casanas R, Gonzalez M, Rodriguez E, Marrero A, Diaz C (2002). Chemometric studies of chemical compounds in five cultivars of potatoes from Tenerife. J. Agric. Food Chem. 50(7):2076-2082.

Donald DH, Gwathmey CO, Sams CE (1998). Foliar feeding on cotton: Evaluation potassium sources, potassium solution buffering and boron. Agron. J. 90:740-746.

Dongre SM, Mahorkar VK, Joshi PS, Deo DD (2000). Effect of micronutrients spray on yield and quality of chilli (*Capsicum annuum* L.) var Jayanti. Agric. Sci. Digest. 20(2):106-107.

Dordas C, Brown PH (2005). Boron deficiency affects cell viability, phenolics leakage and oxidative burst in rose cell cultures. Plant Soil 268:293-301.

Dyar JJ, Webb KL (1961). A relation between boron and auxin in $C^{14}$ translocation in bean plants. Pl. Physiol. 36:672-676.

El-Hak SHG, Ahmed AM, Moustafa YMM (2012). Effect of foliar application with two antioxidants and humic acid on growth, yield and yield components of peas (*Pisum sativum L.*). J. Hort. Sci. Ornam. Plants 4(3):318-328.

Farahat MM, Ibrahim MMS, Taha SL, El-Quesni EMF (2007). Response of vegetative growth and some chemical constituents of *Cupressus sempervirens* L. to foliar application of ascorbic acid and zinc at Nubaria. World J. Agric Sci. 3(4):496-502.

Farouk S, Mosa AA, Taha AA, Ibrahim HM, EL-Gahmery AM (2011). Protective effect of humic acid and chitosan on radish (*Raphanus sativus*, L. var. sativus) plants subjected to cadmium stress. J. Stress Physiol. Biochem. 7(2):99-116.

Forgac L, Czimbalmos R (2011). The applied soil protective cultivation system-a method to reduce and prevent the soil degradation processes. Novenytermeles 60:279-282.

Gomaa AO (2001). Effect of foliar spray with some amino acids and nutrient elements on *Antholyza aethiopica*, L. plants. Proc. The Fifth Arabian Horticulture Conference, Ismailia, Egypt, March 24-28, 2:63-73.

Guchhait S, Bhattacharya A, Pal S, Mazumdar D, Chattapadhyay A, Das AK (2008). Quality evalution of cormels of new germplasm of Taro. Int. J. Veg. Sci. 14(4):304-321.

Habashy NR, Ragab AAM, El-Rasoul Sh. MA (2005). Effect of some bio-organic amendments on a sequence cropping pattern of peanut-faba bean grown on a sandy soil. Egypt J. Appl. Sci. 20:752-763.

Hamid N, Bukhari N, Jawaid F (2010). Physiological responses of *phaseolus vulgaris* to different lead concentrations. Pak. J. Bot. 42(1):239-246.

Hanson EJ (1991). Movement of boron out of tree fruit leaves. Hort. Sci. 26:271-273.

Hatwar GP, Gondane SU, Urkude SM, Gahukar OV (2003). Effect of micronutrients on growth and yield of chilli. J. Soil Crops 13(1):123-126.

Jeyakumar P, Durgadevi D, Kumar N (2001). Effect of zinc and boron fertilization on improving fruit yields in papaya (*Carica papaya* L.) cv. Co5. J. Plant Nutr. Food Security and Sustainability of agroecosystems. Kluwer Academic, Netherland, pp. 356-357.

Kaur H, Aulakh PS, Kapur SP, Singh SN (1990). Effect of growth regulators and micronutrients on granulation and fruit quality of sweet orange cv. Jaffa. Punjab Hort. J., 30(1-4):13-19.

Kaya M, Atak M, Khawar KM, Çiftci CY, Özcan S (2005). Effect of Pre-Sowing Seed Treatment with Zinc and Foliar Spray of Humic Acids on Yield of Common Bean (*Phaseolus vulgaris* L.). Int. J. Agri. Biol. 7(6):875-878.

Khalifa RKHM, Shaaban SHA, Rawia A (2011). Effect of foliar application of zinc sulfate and boric acid on growth, yield and chemical constituents of Iris plants, Ozean J. Appl. Sci. 4(2):129-144.

Khan AS, Ullah W, Malik AU, Ahmad R, Saleem BA, Rajwana IA (2012). Exogenous applications of boron and zinc influence leaf nutrient status, tree growth and fruit quality of Feutrell;s early (*citrus reticulate* Blanco). Pak. J. Agric. Sci. 49(2):113-119.

Kumar K, Arvind K, Vidyasagar R, Rao K (1988). Studies on growth and activity of photosynthetic enzymes on Sorghum bicolor L. as influenced by micronutrients. *Procendian Natl. Sci.* Acard Part B Biol. Sci. 54:75-80.

Lee S, Aronoff S (1967). Boron in plants: A biochemical role. Sci. 158:798-799.

Lukaszewski KM, Blevins DG (1996). Root growth inhibition in boron deficient and Aluminum stressed squash may be a result of ascorbate metabolism. Plant Physiol. 112:1135-1140.

MacCarthy P, Clapp CE, Malcolm RL, Bloom PR (1990). Humic substances in soil and crop sciences: selected readings. Madison: Am. Soc. Agron. p. 281.

Malik KA, Azam F (1986). Effect of humic acid on wheat (*Triticum aestivum* L.) seedling growth. Environ. Exp. Bot. 25:245-252.

Malik RP, Ahlawat VP, Nain AS (1999). Effect of foliar spray of urea and zinc sulphate on growth and fruiting of kinnow - A mandarin hybrid. Haryana J. Hort. Sci. 28(12):33-34.

Mallick MFR, Muthukrishnan CR (1979). Effect of micronutrients on tomato (*Lycopersicon esculentum* Mill) I- Effect on growth and development. South Indian Hort. 21(1-2):121-123.

Marschner H (1986). Mineral nutrition of higher plants. Academic Press, San Diego, CA.

Marschner H (1995). Mineral Nutrition of Higher Plants. 2nd eds. Academic Press, San Diego, pp. 379-396.

Matto A, Diamond AE (1963). Symptoms of *Fusarium* wilt in relation to quality of fungus and enzyme activity in tomato stem. Phytopathol. 53:574-575.

McKee GW (1964). A coefficient for computing leaf area in hybrid corm. J. Agron. 56:240-241.

Mdwaradze TD (1981). Effect of different rates of zinc fertilizers on qualitative indices of mandarin fruit. *Subtrophicheskie Kul'tury,* 5:49-51.

Mengel K, Kirkby EA (1982). Principles of plant nutrition. Int'l Potash Instt. Bern, Switzerland. Peres, L.A. and R.D. Reyes. 1983. J. Agric. Univ. Puerto Rico. 67:181-187.

Murthy IYLN (2006). Boron studies in major oilseed crops. Indian J. Fertilizers 1(2):11-20.

Nahed GAE-A, Laila KB (2007). Influence of tyrosine and zinc on growth, flowering and chemical constituents of *Salvia farinacea* plants. J. Appl. Sci. Res. 3(11):1479-1489.

Nardi S, Pizzeghello D, Renero F, Muscobo A (1999). Biological activity of humic substances extracted from soil under different vegetation cover. Commun. Soil Sci. Plant Anal. 30:621-634.

Neri D, Lodolini EM, Savini G, Sabbatini P, Bonanomi G, Zucconi F

(2002). Foliar application of humic acids on strawberry (cv. Onda). Acta Hort. 594:297-302.

Pfeffer H, Dannel F, Römheld V (1998). Are there connections between phenol metabolism, ascorbate metabolism and membrane integrity in leaves of boron deficient sunflower plants? Physiol. Plant. 104(3):479-485.

Putintsev AF, Platonova NA (1991). The effectiveness of treating pea seeds with humates before sowing. *Selektsiyai Semonovodstvo* (Moskva), 6:50-52.

Rawat PS, Mathpal KN (1984). Effect of micronutrients yield and sugar metabolism of some of the vegetables under kumaud hill condition. Sci. Cilt. 50(8):243-244.

Reuther W, Batchelor LD, Webber HJ (1968). The citrus industry. Vol. 2. Univ. of California. Div. Agric. Oakland, CA.

Romheld V, Marschner H (1991). Function of micronutrients in plants. In "Micronutrients in Agriculture" Published by *Soil Sci. Soc. Amer. Inc. Madison Wisconsin,* USA, pp. 297-299.

Sadasivam S, Manickam A (1992). *Biochemical methods.* New Delhi: Wiley Eastern Limited.

Salib MM (2002). The integrated effect of humic acid and micro nutrients in combination with effective microorganisms on wheat and peanut grown on sandy soil. Zagazig J. Agric. Res. 29:2033-2050.

Samia MZE-B, Mohmoud AMA (20090. Effects of corms storage, zinc application and their interaction on vegetative growth, flowering, corms productivity and chemical constituents of *Tritonia crocata* Ker Gawl Plant. J. Agric. Res. Kafr El-Sheikh Univ. 35(1):230-255.

Sanchez-Amat A, Solano F (1997). A pluripotent polyphenol oxidase from the melanogenic marine Alteromonas sp shares catalytic capabilities of tyrosinases and laccases. Biochem. Biophys. Res. Commun. 26;240(3):787-92.

Sen S, Bhattacharya A, Mazumder D, Sen H, Das AK, Pal S (2005). Nutrient and antinutrient composition of cormels of *colocasia esculenta* var *antiquorum.* Int. J. Veg. Sci. 11:17-34.

Sing DK, Paul PK, Ghosh SK (2002). Response of papaya to foliar application of boron, zinc and their combinations. Department of Pomology and Post Harvest Technol. Uttar Banga Krishi Viswavidyalaya, Pundibari-736 165, Cooch Behar (West Bengal), India.

Singaroval R, Balasubramanian TN, Govindasamy R (1993). Effect of humic acid on sesam (*Sesamum indicum*). Indian J. Agron. 38:147-149.

Swietlik D (1999). Zinc nutrition in horticultural crops. Horticultural Reviews. John Wiley & Sons, Inc. New York. 23:109-180.

Tariq M, Sharif M, Shah Z, Khan R (2007). Effect of foliar application of micronutrients on he yield and quality of sweet orange (*Citrus sinensis* L.). Pak. J. Biol. Sci. 10(11):1823-1828.

Vinod K, Awasthi G, Chauchan PK (2012). Cu and Zn tolerance and responses of the Biochemical and Physiochemical system of Wheat. J. Stress Physiol. Biochem. 8:3.

Vinson JA, Hao Y, Su X, Zubik L (1995). Phenol antioxidant quantity and quality in foods: vegetables. J. Agric. Food Chem. 46:3630-3634.

Wutscher HK, Hardesty C (1979). Concentration of 14 elements in tissues of blight affected and healthy Valencia orange trees. J. Amer. Soc. Hort. Sci. 104:9-11.

Yadav OP, Manchanda HR (1979). Boron tolerance studies in gram and wheat grown on Sierozem sandy soil. J. Indian Soc. Soil Sci. 27:174-180.

Zaky MH, El-Zeiny OAH, Ahmed ME (2006). Effects of humic acid on growth and productivity of bean plants grown under plastic low tunnels and open field. Egypt. J. Appl. Sci. 21:582-596.

# Effect of ethephon and planting density on lodged plant percentage and crop yield in maize (*Zea mays* L.)

**Misheck Chandiposha and Fanuel Chivende**

Department of Agronomy, Midlands State University, P Bag 9055, Gweru, Zimbabwe.

**Increasing maize production is vital in many developing countries. One way of increasing production is through adoption of high planting densities. However, high planting densities are associated with intra-specific competition between plants for resources like nutrients, water and sunlight, as a result weak stalks develops which are susceptible to lodging. The challenge of lodging in high planting densities can be reduced by ethephon (2-chloroethylphosphonic acid). A 3 x 2 factorial experiment in randomised complete block design was carried out at New Donnington Farm, Zimbabwe in the 2011/12 rainy season to study the effects of ethephon rates and planting densities on maize. Ethephon rate had the following levels 0 l/ha (control), 0.56 l/ha and 0.84 l/ha while the density consisted of 53,333 and 80,000 plants/ha. Ethephon rate significantly (p<0.05) reduced the plant height of maize as compared to the control plants across different levels of planting densities. Also a significant increase (p<0.05) in internode diameter in the treated stands over the control plants across different levels of planting densities was recorded. Ethephon treatment significantly reduced (p<0.05) the percentage of lodged plants from 6.54 (control) to 4.34 and 2% for medium and high rate, respectively. Grain yield increased significantly in response to increased rates of ethephon at high planting density by 28.5 and 29.1 tons/ha for medium and high rate, respectively.**

**Key words:** Maize, ethephon, planting, density, lodging.

## INTRODUCTION

Maize is an important crop in the world and is ranked third after wheat and rice (Hoshang, 2012). In Zimbabwe and other African countries, maize is the staple crop and is grown by majority of smallholder farmers. It is estimated that more than half of the world area under maize production is in developing countries including Africa (Mejía, 2003). In Zimbabwe, despite the extensive area under production of maize in developing countries, the yields are still generally low. The poor yields can be attributed to poor agronomic practices, frequency of drought, biotic stresses among other factors.

Since maize is economically important, increasing its production is vital in African countries. One way of improving production is through adopting high planting densities. High planting densities increase the number of ears per unit area and thereby improving the grain yield (Abuzar et al., 2011). However, too high populations may cause intra-specific competition between plants for resources like nutrients, water and sunlight (Zamir, 2011). In addition, high planting densities are associated with lodging of stalks (Shekoofa and Emam, 2006). Lodging can affect photosynthesis, pollination and yield of maize crop due to poor solar radiation interception.

The challenge of lodging in high planting densities could be reduced by using ethephon (2-chloroethyl-phosphonic acid) (Shekoofa and Emam, 2008). Ethephon

**Table 1.** The interaction between planting density and ethephon rates on final plant height (cm).

| Ethephon | 53, 3333 plants/ha (low density) | 80, 000 plants/ha (high density) |
|---|---|---|
| 0 | 277.25[a] | 285.15[b] |
| 0.56 l/ha | 265.29[c] | 267.6[d] |
| 0.84 l/ha | 254.15[e] | 257.55[f] |
| p value | < 0.001 | |
| Lsd | 1.38 | |
| cv (%) | 5.3 | |

is an ethylene releasing compound whose effect depends on crop species, amount applied and timing of application (Khuankaew et al., 2009). It is a growth regulator which upon application, increases resistance to lodging through thickening of lower internodes, termination of apical dominance (Shekoofa and Emam, 2006). Besides reducing lodging in crop plants, ethephon has also been found to improve the water use efficiency (WUE) of plants by making water available during critical reproductive growth stages (Campos et al., 2004). Therefore, ethephon can be used to improve crop yield in dry areas which receive very low rainfall (Langan et al., 1987).

Elsewhere, ethephon has been extensively studied as a plant growth regulator used to promote fruit ripening, abscission, flower induction, breaking of apical dominance but there is limited information to show that it has been evaluated as an anti-lodging agent in maize under high planting densities.

## MATERIALS AND METHODS

A field experiment was conducted at New Donnington Farm in Zimbabwe, which lies between 30°41' E and 17° 52' S in Norton during the 2011-2012 growing season. The farm is located in Agro-ecological region 11a about 38 km from Harare along Bulawayo road. The altitude of the area is 1360 m above sea level. The mean annual rainfall is between 950 and 1000 mm per annum whilst the mean annual temperature is 20.6°C. The maximum and minimum temperatures of the area are 26 and 17°C, respectively. The study site was dominated by light, well drained sandy loam soil textures (Nyamapfene, 1991). The design used was a 3 × 2 factorial experiment in randomized complete block design with four replications and the blocking factor was the slope. The first factor was ethephon with three levels, 0, 0.56 and 0.84 l/ha and the second factor was planting density with two levels, 53,333 and 80,000 plants/ha.

Land was conventionally prepared using a tractor drawn disk harrow and a roller to obtain a fine tilth from which seedbeds were raised. A marked wire cable was used to mark planting stations and the seed was hand sown in plots of 4.5 m wide and 5.0 m long. In row spacing was 16.6 (high density) and 25 cm (low density). Inter-row spacing was 75 cm in all plots and uniformity of sowing depth was achieved by using a hand dibbler to make holes of 5 cm deep. A basal fertiliser of Compound D (7% N, 14% $P_2O_5$ and 7% $K_2O$) was applied at a rate of 300 and 450 kg per hectare under the low and high planting densities, respectively. Top dressing was done using urea (46% nitrogen), the fertilizer was split applied at the rate of 250 and 375 kg per hectare at the low and high planting densities, respectively at four and eight weeks after planting. The field was kept weed free by the use of Atrazine and Alachlor at 31/ha and 21/ha, respectively. Late and persistent weeds were controlled mechanically by the use of hoes at eight weeks after crop emergence.

Stalk borer (*Busceola fusca*) was controlled by the use of dipterex (dimethyl 2.2.2-trichloro-1-hydroxyethyl-phosphate) which was applied at four weeks and six weeks after crop emergence at a rate of 4 kg per ha. Ethephon mixed with blood buff (surfactant) was applied in the morning at 8 to 9 leaf stage using a knapsack sprayer. Each plot was surrounded by plastic walls to avoid the drift of the solution to the adjacent plots.Confider was sprayed at physiological maturity to control termites using a knapsack sprayer.

The following measurements were recorded, plant height, internode length and diameter, percentage of lodged plants and grain yield. Data was subjected to analysis of variance using GENSTAT version 14. Number of lodged plants was square root transformed prior to analysis of variance. Treatments means were separated using least significant difference test at 5% level.

## RESULTS AND DISCUSSION

### Ethephon rate and planting density on final plant height and internode length

There was a significant interaction ($p< 0.05$) in the final plant height between planting densities and ethephon rates. Ethephon treatment resulted in a significant ($p<0.05$) reduction of plant height across the two plant densities (Table 1). The reduction in plant height can be attributed to the observed decrease in internode lengths (Figure 1).

This finding was also in agreement with the results of Shekoofa and Emam (2006) and Khuankaew et al. (2009). However, Poovaiah and Leopold (1973) reported that stem elongation was observed after ethephon was applied to blue grass (*Poa pratensis* L) and wheat (*Triticum aestivum* L). This confirms that plant species respond differently to ethephon application.

### The effect of ethephon on percentage of lodged plants and internode diameter (mm) of maize

There was significant ($p< 0.05$) reduction on the percentage of lodged plants following ethephon treatment. The minimum percentage of lodged plants were recorded where high ethephon rate was used (Figure 2). These results may be attributed to an increase in internode diameter following ethephon treatment as shown in Table 2. Shekoofa and Emam (2006) also found that ethephon increases internode diameter and is important for reducing lodging.

### Effect of planting density on the percentage of lodged plants

Planting densities had a significant effect ($p< 0.05$) on

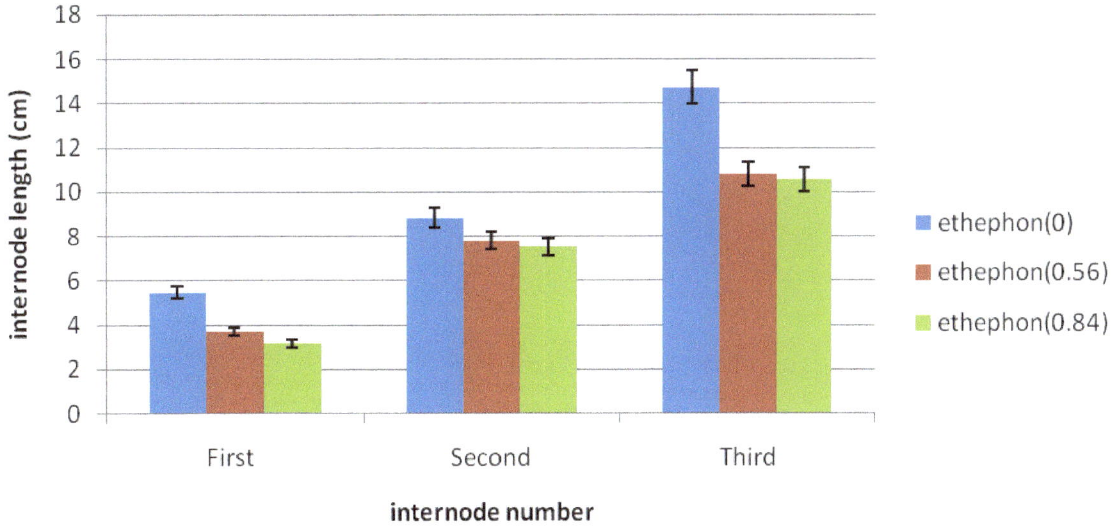


**Figure 1.** Effect of ethephon rates on final internode length of maize plant (cm).

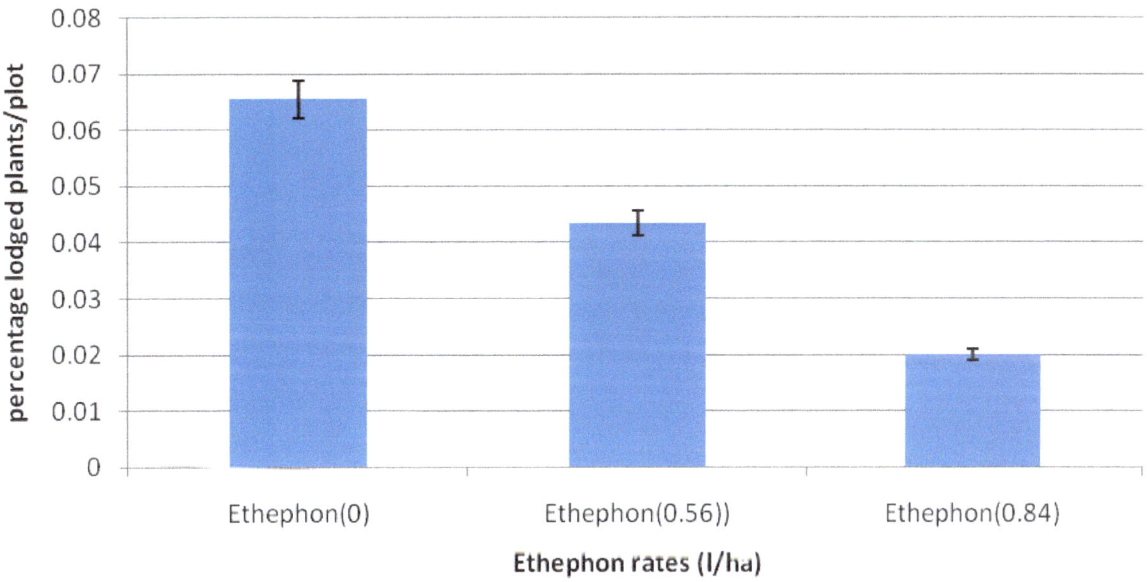


**Figure 2.** Effect of ethephon levels on percentage of lodged plants.

**Table 2.** The interaction between plant density and ethephon rates on first internode diameter of the maize plant.

| Ethephon | 53, 3333 plants/ha (low density) | 80, 000 plants/ha (high density) |
|---|---|---|
| 0 | 28.9070$^{a}$ | 28.8555$^{b}$ |
| 0.56 l/ha | 32.4180$^{c}$ | 31.2835$^{d}$ |
| 0.84 l/ha | 29.91$^{c}$ | 29.7995$^{e}$ |
| p. value | < 0.001 | |
| LSD | 0.06025 | |
| CV (%) | 5.1 | |

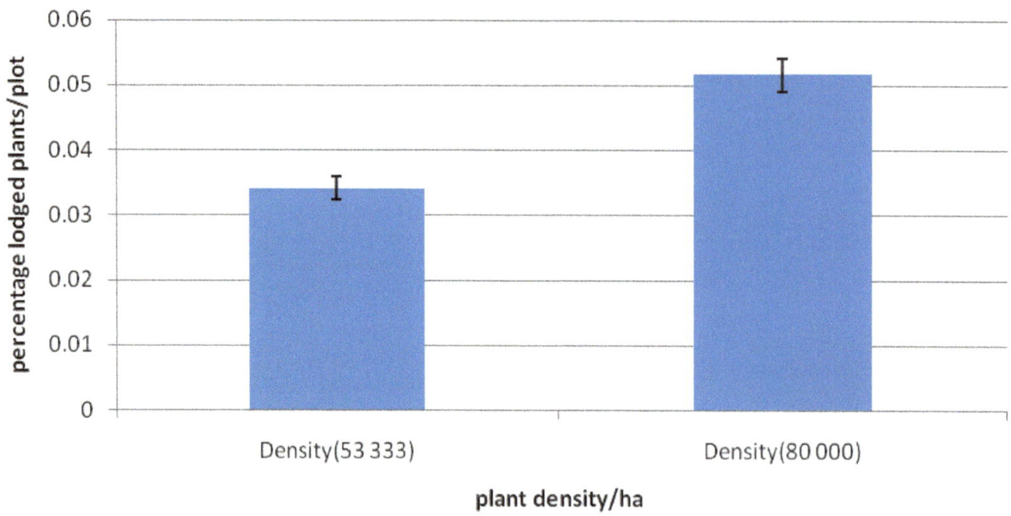


**Figure 3.** Effect of planting density levels on percentage of lodged plants.

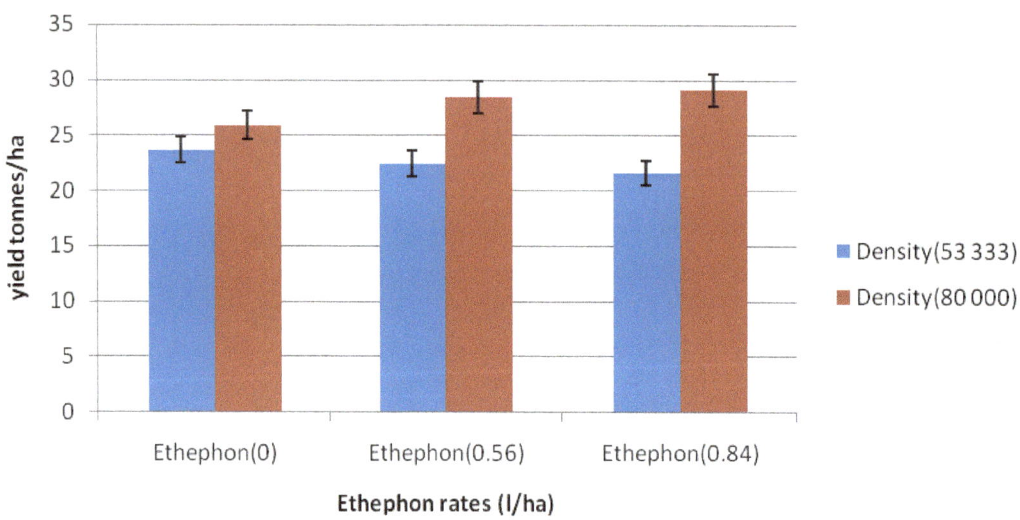


**Figure 4.** The interaction between different levels of planting density and ethephon rates on grain yield (t/ha).

percentage of lodged plants. There was a significant ($p<0.05$) increase in percentage of lodged plants following an increase in planting densities (Figure 3). This is in agreement with the findings by Liu et al. (2012), who reported that, an increase in planting densities result in risk of root lodging. The intraspecific competition between plants for resources like light, water and nutrients may result in etiolating of the crop thereby weakening the root system and anchoring ability.

### Ethephon and planting density on grain yield (tons per hectare)

A significant interaction ($p< 0.05$) between planting densities and ethephon rates was observed for the grain yield (Figure 4). Addition of ethephon resulted in significant differences ($p< 0.05$) in grain yield between planting densities. Ethephon has been reported to increase maize yield under drought conditions. Since this experiment was set under rain-fed conditions, there are certain times mid dry spells were felt and as such ethephon improved the maize yield. The plant growth regulator increases the grain yield under water stress conditions by extending water availability during critical stages like grain filling (Kasele et al., 1994). The results also showed a significant increase in grain yield under high planting density when ethephon was applied. Ethephon might have reduced lodging in treatments where high planting density was used resulting in more cobs being harvested.

Ethephon had no significant effect ($p < 0.05$) on grain yield under low planting density. The plant growth regulator is not beneficial under low planting densities, as lodging is low as shown in Figure 4.

## Conclusions

Ethephon reduces the plant height due to shortening of internodes. The plant growth regulator also reduces the percentage of lodged plants across all plant densities due to increase in internode diameter. Increase in planting densities tends to exacerbate lodging in maize crop. Ethephon increases grain yield under high planting densities and have no effect under low plant population.

## ACKNOWLEDGEMENTS

The authors would like to express gratitude to Pioneer Seeds for financial support. Thanks also go to New Donnington Farm for providing the experimental area. The authors would also want to thank the following people, Dr D. Myers, Mr. Z. Sigwadi, Mr. P. Chivaura and Mr. G. Nyamutowera for guidance and support.

### REFFERENCES

Abuzar MR, Sadozai GU, Baloch MS, Baloch AA , Shah IH, Javaid T, Hussain N (2011). Effect of plant population densities on yield of maize. J. Anim. Plant Sci. 2(4):692-695.

Campos H, Cooper M, Habben JE, Edmeades GO, Schussler JR (2004). Improving drought tolerance in maize: a view from industry. Field Crops Res. 90(1):19-34.

Hoshang R (2011). Effect of Plant Density and Nitrogen Rates on Morphological Characteristics Grain Maize. J. Basic Appl. Sci. Res. 2(5):4680-4683.

Kasele IN, Nyirenda F, Shanahan JF, Nielsen DC, d'Andria R (1994). Ethephon Alters Corn Growth, Water Use, and Grain Yield Under Drought Stress. Agron. J. 86 (2):283-288.

Khuankaew T, Ohyama T and Ruamrungsr S (2009). Effects of ethephon application on growth and development of *Curcuma alismatifolia gagnep*. Bulletin Faculty of Agriculture Niigata University 62(1):9-15.

Langan TD, Oplinger ES (1987). Growth and yield of ethephon treated maize. Agron. J. 79:130-134.

Liu S, Song F, Liu F, Zhu X, Xu H (2012). Effect of Planting Density on Root Lodging Resistance and Its Relationship to Nodal Root Growth Characteristics in Maize (*Zea mays* L.). J. Agric. Sci. 4(12):182-189.

Mejía D (2003). MAIZE: Post-Harvest Operation. Food and Agriculture Organization of the United Nations (FAO), AGST.

Nyamapfene K (1991). The soils of Zimbabwe. Nehanda Publishers. Harare, Zimbabwe.

Poovaiah BW, Leopold AC (1973). Effects of ethephon on growth of grasses. Crop Sci. 13:755-758.

Shekoofa A, Emam Y (2006). Maize (*Zea mays*) growth and yield response to ethephon application under water stress conditions. Iran Agric. Res. 24(2):39-52.

Shekoofa A, Emam Y (2008). Plant growth regulator (ethephon), alters maize (*Zea mays*) growth, water use and grain yield under water stress. Agron. J. 10:41-48.

Zamir MSI, Ahmad AH, Javeed HMR, Latif T (2011). Growth and yield behaviour of two maize hybrids (*Zea mays* L.) towards different plant spacing. Cercetări Agronomice în Moldova. 14(2):33-40.

# 11

# Variability, heritability and genetic advance in upland cotton (*Gossypium hirsutum* L.)

**R. Dhivya, P. Amalabalu, R. Pushpa and D. Kavithamani**

Centre for Plant Breeding and Genetics, Tamil Nadu Agricultural University, Coimbatore-641 003, India.

**The analysis of variance study indicated the presence of significant difference among all the traits in *Gossypium hirsutum* accessions. The highest phenotypic coefficient of variation (GCV) and genotypic coefficient of variation (GCV) were recorded by seed index, plant height, lint index and boll weight. Genotypic co-efficient of variation had a similar trend as PCV. High heritability along with high genetic advance was observed in traits viz., number of sympodia per plant, single plant yields, seed index and micronaire value. The combinations of high heritability with high genetic advance will provide a clear base on the reliability of that particular character in selection of variable entries. Based on *per se* performance, the accessions MCU5, TCH1715, TCH1716 and G cot 16 were identified as potential donors for single plant yield (g), number of bolls per plant, 2.5% span length (mm) and bundle strength (g/tex). So these accessions may be utilized for crossing programme to improve a particular character in crop improvement.**

**Key words:** Genetic variability, heritability, genetic advance, upland cotton.

## INTRODUCTION

Cotton is an important fibre crop of global importance which is grown in tropical and subtropical regions of more than 60 countries of the world. Despite threat from synthetic fibre or manmade fibre, cotton retains its reputation as "queen of the fibre plants"". For multiple uses of lint and by-products, cotton is also referred to as "white gold". In any crop improvement programme, knowledge on nature of gene action and inheritance of traits is essential so as to choose a suitable breeding methodology in crop improvement (Vineela et al., 2013). Development of an effective breeding programme depends on the existence of genetic variability for various economic characters in the gene pool. Selection is effective only when there is enough magnitude of variability in the breeding population. An understanding of precise magnitude of variability present in a population is important in formulating the most appropriate breeding technique for improvement of various characters. The present investigation was carried out with 54 *Gossypium hirsutum* lines of diverse origin to estimate their *per se* performance, variability, heritability and genetic advance on the genetic architecture of 16 yield, yield components and fibre quality traits.

## MATERIALS AND METHODS

The study was conducted in the Department of Cotton, Tamil Nadu Agricultural University, Coimbatore during winter 2010. Fifty four *G. hirsutum* cotton genotypes were planted in randomized block design with two replications. Uniform spacing of 90 x 45 cm and all the recommended field operations were carried out. In each replication five competitive plants were randomly selected and observations were recorded for 16 characters viz., days to 50% flowering, plant height (cm), internode length (cm), number of sympodia/plant, number of ovules/plant, number of bolls/plant, boll weight (g), number of seeds/plant, seed setting percentage, seed cotton yield/plant, lint index, seed index, ginning outturn (%), 2.5% span length (mm), bundle strength (g/tex) and fibre fineness.

Analysis of variance was carried out statistically utilizing the mean values (Panse and Sukhtame, 1995). The phenotypic and genotypic coefficient of variation was estimated using the formula suggested by Burton (1952) and expressed in percentage. The phenotypic and genotypic variances were calculated by utilizing the

**Table 1.** Mean and range performance for different characters among G. *hirsutum* accessions.

| Characters | Mean | Minimum | Maximum |
|---|---|---|---|
| Days to 50% flowering | 56.6 | 50.0 | 65.0 |
| Plant height (cm) | 108.8 | 78.0 | 147.0 |
| Inter node length(cm) | 5.4 | 4.0 | 6.5 |
| No of Sympodia per plant | 19.6 | 14.0 | 27.0 |
| No of bolls per plant | 23.6 | 19.5 | 38.5 |
| Boll weight (g) | 4.3 | 12.4 | 5.5 |
| No of seeds per boll | 27.4 | 23.0 | 35.5 |
| No of ovules per flower | 30.7 | 24.0 | 38.0 |
| Seed setting percentage | 89.6 | 71.5 | 100.0 |
| Single plant yield (g / plant) | 57.6 | 31.7 | 91.6 |
| Lint index | 5.7 | 4.0 | 7.9 |
| Seed index | 9.6 | 6.4 | 16.3 |
| Ginning outturn (%) | 37.3 | 30.8 | 46.1 |
| 2.5 % span length (mm) | 28.4 | 24.3 | 34.0 |
| Bundle strength (g / tex) | 20.4 | 16.2 | 23.7 |
| Micronaire value | 4.3 | 3.5 | 5.2 |

respective mean sum of square from variance table (Lush, 1940). Heritability, expected genetic advance and genetic gain in the broad sense was calculated according to the formula suggested by Johanson et al. (1955).

## RESULTS AND DISCUSSION

The *per se* performance of yield and fibre quality characters were recorded on 54 germplasm accessions and the range of variations observed in respect of all the 16 traits studied are presented in the Table 1. Based on *per se* performance, the accessions such as MCU 5, TCH1715, TCH1716 and G cot 16 in *G. hirsutum* were identified as potential donors which recorded highest mean values for single plant yield, number of bolls per plant, 2.5% span length (mm) and bundle strength (g/tex). These accessions may be exploited for further improvement of the above traits by breeding programmes. Regarding the identification of donor for specific trait, the highest performer of that particular trait can be considered. Among the accessions, TCH 1715 (5.5 g of boll weight), RB 488 (46.1% of GOT) were expressed the highest *per se* for that particular traits. The culture TCH 1710 (34 mm, 23.1g/tex) recorded the highest 2.5% span length. So these accessions may be utilized for crossing programme to improve that particular character in crop improvement. Anjali was found to be compact type which can be utilized in development of genotype suitable for high density planting.

The analysis of variance showed highly significant differences among genotypes for all the characters studied (Table 2) and infers existence of considerable genetic diversity among genotypes. Phenotypic variance, genotypic variance, phenotypic coefficient of variation, genotypic coefficient of variation, heritability in broad sense and genetic advance as percentage of mean which were estimated for 16 characters are shown in Table 3. The knowledge of nature and magnitude of variability available in the genotypes for different characters is an important prerequisite for making simultaneous selection over more number of characters to bring remarkable improvement in cotton. The analysis of variance study indicated the presence of significant difference among all the traits in the accessions. The heritable (genotypic) variation is usually masked by non-heritable variation creating

**Table 2.** Analysis of variance for the different characters among the germplasm accessions of *G. hirsutum*.

| Source of variation | Degrees of freedom | Days to 50% flowering | Plant height (cm) | nter node length (cm) | Number of sympodia / plant | Number of bolls / plant | Boll weight (g) | Number of sees / boll | Number of ovule / flower | Seed setting percentage | Single plant yield (g) | Lint index | Seed index | Ginning out turn (%) | 2.5 % span length (mm) | Bundle strength(g /tex) | Micronaire value |
|---|---|---|---|---|---|---|---|---|---|---|---|---|---|---|---|---|---|
| Genotypes | 53.0 | 10.4** | 466.6** | 0.7** | 14.1** | 18.3** | 0.6* | 10.8** | 17.1** | 90.4** | 441.9** | 1.3** | 4.6** | 21.1** | 10.2** | 6.1** | 0.3** |
| Error | 53.0 | 2.7 | 218.6 | 0.4 | 2.8 | 5.0 | 0.3 | 5.50 | 6.5 | 40.1 | 2.1 | 0.5 | 1.2 | 10.9 | 0.2 | 0.3 | 0.1 |

*,** Significant at 5 and 1% levels, respectively.

difficulty in exercising selection. Hence it becomes necessary to partition overall variability into heritable and non-heritable components to enable the breeders to plan for proper breeding programme. The plant height recorded the highest value for phenotypic variance (342.6) and the single plant yield recorded highest genotypic variance (219.9). The micronaire value exhibited the lowest phenol-typic variance (0.2) and in case of genotypic variance, the traits *viz*., inter node length, boll weight and micronaire value had recorded the lowest value (0.1). The co-efficient of phenotypic and genotypic variance were calculated for all the characters under study. The PCV ranged from 4.5 (days to 50% flowering) to 25.9% (single plant yield). The highest PCV was followed by seed index (17.7%), plant height (17.0%), lint index (17.0%) and boll weight (15.4%). Genotypic co-efficient of variation had a similar trend as PCV. The range varied from 3.5 (Days to fifty percent flowering) to 25.8 (Single plant yield). The maximum genotypic coefficient of variation (GCV) was observed for single plant yield (25.8) followed by seed index (13.5), number of sympodia per plant (12.1), lint index (11.2) and number of bolls per plant (10.9), plant height (10.2). The lowest PCV (4.5%) and GCV (3.5) values was observed in days to 50% flowering. In the present study, there was a close correspondence between phenotypic and genotypic variance for days to 50% flowering, inter node length, boll weight, single plant yield, lint index, micronaire value and 2.5% span length indicating less environmental influence. But plant height, number of sympodia per plant, number of bolls per plant, number of seeds per boll, seed setting percentage and ginning outturn showed higher variation indicating the influence of environment on these characters. Since the variations are influenced by the magnitude of the units of measurements of different traits, a measure of coefficient of variation which is independent of the unit of measurement is more useful in comparing between populations. In *G. hirsutum* accessions PCV was higher than the GCV for all characters. From this, we can understand that the environment plays a major role on expression of all these traits leading to increase in the PCV more than GCV. The highest PCV and GCV estimates were recorded for single plant yield indicating the presence of significant genetic variability in this character. Selection pressure can be applied on this character to isolate promising genotypes. Similar observations in cotton was reported by Dheva and Potdukhe (2002) and Preetha and Raveendran (2007).

Moderate PCV and GCV estimates were noticed for the characters such as plant height, number of sympodia per plant, number of bolls per plant, lint index and seed index. Girase and Mehetne (2002) and Harshal (2010) also reported the moderate PCV and GCV in various traits with the suggestion that these characters can be improved though rigorous selection. The characters such as days to 50% flowering, 2.5% span length and micronaire value exhibited low PCV and GCV which indicated that the breeds should go for source of high variability for these characters to make improvement. Similar suggestion were given by Kowsalya and Raveendran (1996), Do Thi Ha An et al. ((2006) in their conclusion. In a population, the observed variability is a

**Table 3.** Components of variance for yield and fibre quality characters of *G. hirsutum* accessions.

| Characters | Phenotypic variance | Genotypic variance | PCV (%) | GCV (%) | $h^2$ (%) | A | GA as percentage of mean |
|---|---|---|---|---|---|---|---|
| Days to 50% flowering | 6.6 | 3.9 | 4.5 | 3.5 | 58.5 | 4.0 | 7.0 |
| Plant height (cm) | 342.6 | 124.0 | 17.0 | 10.2 | 36.2 | 17.7 | 16.3 |
| Inter node length(cm) | 0.5 | 0.1 | 13.5 | 6.6 | 23.5 | 0.5 | 8.4 |
| No of sympodia per plant | 8.5 | 5.7 | 14.8 | 12.1 | 66.9 | 5.1 | 26.2 |
| No of bolls per plant | 11.7 | 6.7 | 14.4 | 10.9 | 57.2 | 5.2 | 21.8 |
| Days to 50% flowering | 6.6 | 3.9 | 4.5 | 3.5 | 58.5 | 4.0 | 7.0 |
| Plant height (cm) | 342.6 | 124.0 | 17.0 | 10.2 | 36.2 | 17.7 | 16.3 |
| Boll weight (g) | 0.4 | 0.1 | 15.4 | 8.4 | 29.8 | 0.5 | 12.1 |
| No of seeds per boll | 8.2 | 2.6 | 10.4 | 6.0 | 32.5 | 2.5 | 9.0 |
| No of ovules per flower | 11.8 | 5.3 | 11.2 | 7.5 | 44.5 | 4.0 | 13.1 |
| Seed setting percentage | 65.3 | 25.1 | 9.0 | 5.6 | 38.5 | 8.2 | 9.2 |
| Single plant yield (g / plant) | 222.0 | 219.9 | 25.9 | 25.8 | 89.1 | 39.0 | 27.3 |
| Lint index | 0.9 | 0.4 | 17.0 | 11.2 | 43.9 | 1.1 | 19.6 |
| Seed index | 2.9 | 1.7 | 17.7 | 13.5 | 58.0 | 2.6 | 27.1 |
| Ginning outturn (%) | 16.0 | 5.1 | 10.7 | 6.0 | 31.6 | 3.3 | 9.0 |
| 2.5 % span length (mm) | 5.2 | 5.0 | 8.0 | 7.9 | 96.2 | 5.8 | 20.4 |

combined measure of genetic and environment causes, where as the genetic variability is the only estimate heritable from generation to generation. However, a measure of heritability alone does not give an idea about the expected gain in the next generation but it has to be considered in conjunction with genetic advance. The traits which expressed high heritability and high genetic advance as percentage of mean could be used as a powerful tool in selection process. According to Panes and Sukhatme (1995) such characters were found to be governed by additive genes and had minimum environment influence. The heritabi-lity estimates ranged from 23.5 (internode length) to 89.1% (single plant yield). The high heritability estimates of 89.1% were recorded by single plant yield followed by 2.5% span length (96.2 %) and bundle strength (91.8%) where as the low heritability was observed in inter node length (23.5%).

The high heritability was registered in the traits *viz.,* number of sympodia per plant, single plant yield, seed index, 2.5% span length and bundle strength. The inter node length exhibited low heritability in this investigation. For efficient selection, we cannot completely depend on heritability alone. The combinations of high heritability with high genetic advance will provide a clear base on the reliability of that particular character in selection of variable entries. The genotypic advance as percent of mean for 16 traits ranged from 7.0 to 27.3%. The higher genetic advance as percent of mean was recorded by single plant yield (27.3%) followed by seed index (27.1 %), number of sympodia per plant (26.2 %), number of bolls per plant (21.8 %). The lowest value of 7.0% was observed in the trait days to 50% flowering. High heritability along with high genetic advance was observed in traits *viz.,* number of sympodia per plant, single plant yields, seed index and micronaire value in *G. hirsutum* (Do Thi Ha An et al., 2008). These traits are highly reliable during selection. High heritability combined with moderate genetic advance was found in the 2.5% span length. It was in accordance with of Muhammad et al. (2004). Among the study materials some of the accessions were identified as potential donors for the improvement of different characters (Table 4). The accessions with high mean performance are generally preferred for all the traits except days to 50% flowering, since earliness is the preferred attribute and early flowering was taken into consideration. From the results of the present study, it can be concluded that direct selection can be done for most of the yield attributing traits since it exhibited high genetic variability and high range of variation. A high PCV over GCV for the characters studied indicated that environment influences the expression of these characters under study. High genetic advance, genetic gain and heritability were recorded for number of sympodia per plant, single plant yield, seed index, micronaire value and 2.5% span length indicated that selection can be resorted for the improvement of these characters in the future crop improvement programmes.

**Table 4.** Potential donors for yield and fibre quality traits.

| S/N | Characters | Potential accessions |
|---|---|---|
| 1 | Days to 50% flowering | MCU-7, SVPR-3, TCH-1716, SVPR-2 |
| 2 | Plant Height | SCS 102, T CH 1715, Okra narrow |
| 3 | Inter node length | MCU 5, RAC9740, SCS 102, F-1946 |
| 4 | No. of Sympodia | PSCL VII, Okra narrow, RHC 1694, SCS 102 |
| 5 | No. of ovules | NDLH 1588, TCH 1608 |
| 6 | No. of Bolls | Okra Narrow, G cot-16, F-1946 |
| 7 | Boll weight(g) | TCH 1715, MCU 13, LRA 5166 |
| 8 | No. of seeds per boll | NDLH 1588, SVPR 4 |
| 9 | Seed setting % | MCU 12, MCU 13, KC 3, TCH 1710 |
| 10 | Single plant yield | MCU 5, MCU 12, Surabhi, SVPR 2, SVPR 4, KC 2, TCH 1715, TCH 1716, TCH 1732, TCH 1734, TCH 1744, G. COT-16, CCH 2117, Pusa 953, ARB 2001, RHC 1694, LH 1961, RB 488, GISV201, MHIS-5, MHIS-7, Sara 2,Giza 1461, Sumangala |
| 11 | Lint index | GISV 201, RAC 9544 |
| 12 | Seed index | SVPR 4, L-752, TCH 1715 |
| 13 | Ginning Out Turn | RB 488, TCH 1705 |
| 14 | 2.5 % span length | TCH 1608, CNH 152, TCH 1705, RAC 9740, TCH 1732, MCU 5, MCU 13, TCH 1710, TCH 1715, TCH 1716 |
| 15 | Bundle strength | TCH 1710, TCH 1715, MCU 5, TCH 1732, TCH 1734, TCH 1744 |
| 16 | Micronaire value | MHIS-7, MCU 7, CCH 2117, NDLH 1588 |

## REFERENCES


Burton GW (1952). Quantitative inheritance in grasses. Proceedings of the Sixth international Congress, pp.277-283.

Dheva NG, Potdukhe NR (2002). Studies on variability and correlations in upland cotton for yield and its components. J. Indian Soc. Cotton Improv, pp.148-152.

Do Thi Ha An, Ravikesavan R (2006). Genetic diversity in cotton (*Gossypium sp*): In: national conference on plant sciences research and development, APSI scientist meet, PSG CAS, Coimbatore, pp. 11-13.

Do Thi Ha An, Ravikesavan R, Iyanar K (2008). Genetic advance and heritability as a selection index for improvement of yield and quality in cotton. J. Cotton Res. Dev. 22(1):14-18.

Girase VS, Mehetre SS (2002). Correlation and Path analysis in cotton. J. Cotton Res. Dev. 16(1):1-7.

Harshal EP (2010). Variability and correlation analysis by using various quantitative traits in released Bt cotton hybrids. J. Cotton Res. Dev. 24(2):141-144.

Johanson AW, Robinson HF, Comstock RF (1955). Estimates of genetic and environmental variability in soybean. Agric. J. 47:314-318.

Kowsalya R, Raveendran TS (1996). Genetic variability and $D^2$ analysis in upland cotton. Crop Res. 12(1):36-42.

Lush JL (1940). Intra-sire correlation and regression of offspring on dams as a method of estimating heritability of characters in Proc. Am. Soc. Anim. Prod. 33:293-301.

Muhammad I, Muhammad AC, Abdul J, Muhammad ZI, Muhammad-ul-Hassan and Noor-ul-Islam (2004). Inheritance of Earliness and other Characters in Upland Cotton. J. Biol. Sci. 3(6):585-590.

Panes VG, Sukhatme PV (1995). Statistical methods for agricultural workers, 3rd Ed., ICAR, New Delhi, p.58.

Preetha S, Raveendran TS (2007). Genetic variability and association studies in three different morphological groups of cotton (*Gossypium hirsutum* L.) Asian J. Plant Sci. 6(1):122-128.

Vineela N, Samba Murthy JSV, Ramakumar PV, Ratna KS (2013). Variability Studies for Physio Morphological and Yield Components Traits in American Cotton (*Gossypium hirsutum* L). J. Agric. Vet. Sci. 4(3):07-10.

# 12

# New technique for adventitious rooting and clonal propagation of *Piper longum* L. (pippali) through leaf cuttings

**Uday Chand Basak, Dipika Dash, Gourisankar J. P. Jena and Ajay Kumar Mahapatra**

Regional Plant Resource Centre, Nayapalli Bhubaneswar-751015, Odisha, India.

**A simple but unique protocol was developed for root production and clonal propagation of *Piper longum* L. (pippali), an important medicinal plant of India. Roots and shoots were induced in apical and basal petiolar halves of leaf using auxins. Average root number, root length and survival of rooted leaf cuttings were not significantly affected by type of auxin and leaf cuttings. Highest number of roots (13.40), root length (9.65 cm), rooting behaviour (91.69%) and survival of rooted cuttings (83.33%) were recorded in petiolar leaf cuttings treated with indole 3-butyric acid (IBA)/naphthalene acetic acid (NAA) (1000 ppm each). The petiolar leaf cuttings treated with IBA/NAA showed significantly higher percentage of shooting (83.33). The shoot number (2.0 per cutting) was also significantly highest in IBA/NAA treated leaf cuttings. *Pippali* can be regenerated via leaf-cuttings, either without hormone treatment, or for better results by using low concentrations of IBA or combination of IBA and NAA hormones. Production of planting material using leaf can substitute traditional propagules such as sucker, vine cutting, etc.**

**Key words:** Auxins, indole 3-butyric acid (IBA), medicinal plants, naphthalene acetic acid (NAA), vegetative propagation.

## INTRODUCTION

*Piper longum* L. (long pepper, pippali), an unisexual perennial climber with woody roots belonging to the family piperaceae is distributed throughout India. Almost all parts of it, namely roots, stems and fruits are medicinally important and used especially in the treatment of diseases of respiratory tract like bronchitis, asthma, cough, etc (Sivarajan and Balachandran, 1994). The principal pharmacological constituents are piperine and piplartine. The crude extract of *P. longum* contains 3-8% of piperine (James, 1999).

Collection of roots from wild habitats and deforestation has made this plant species a threatened taxon. As the plants are extracted from its natural habitat for use in drug formulation, the species has become very rare in the forests of Kerala (Nair, 2000). Conventionally, *P. longum* is propagated through seeds, suckers or cuttings or by layering of mature branches at the beginning of rainy season. Although, conventional propagation is beset with problems of poor seed viability, low percentage of germination and scanty or delayed rooting of vegetative cuttings. Therefore, there is a need to explore alternate propagation methods (Sarasan et al., 1993).

Vegetative propagation through leaf cutting can be a suitable way to develop plants economically and in a controlled manner. The advantage of this type of propagation is that, with this technique plants can be raised throughout the year and the mother plant is less disturbed unlike the stem cutting method. Several reports are available regarding vegetative propagation of *P. longum*. Vegetative propagation by the application of root

**Table 1.** Experimental layout and allocation of treatments for rooting in leaf-cuttings of *P. longum.*

| Leaf cutting | Treatment | | | | |
|---|---|---|---|---|---|
| | $T_0$: Control (without auxins) | T1: IBA (1000 ppm) | T2: NAA (1000 ppm) | T3: IBA (1000 ppm) + NAA (1000 ppm) | Total |
| Apical halves | 30 | 30 | 30 | 30 | 120 |
| Petiolar halves | 30 | 30 | 30 | 30 | 120 |
| Total | 60 | 60 | 60 | 60 | 240 |

promoting substances in stem cutting has been studied (Bhat et al., 1995). Soniya et al. (2002) has also reported micro propagation of *P. longum*. However, regeneration of adventitious roots and shoots from the leaf cuttings of *P. longum* at *in vivo* condition has not been attempted. As root of *P. longum* have great medicinal value this study aimed at inducing growth of root. This could be an effective and less expensive way for production of root from leaves of *P. longum*.

## MATERIALS AND METHODS

This experiment was carried out during July-September, 2009 at the institutional nursery of Regional Plant Resource Centre, Bhubaneswar, Odisha, India. Leaves were obtained from stock plants grown under shade-net house. Semi-mature and fresh leaves having 5 prominent veins were used for rooting and shooting experiment. A total of 240 leaf-cuttings (both apical and petiolar halves) were prepared out of 120 mature leaves by transverse cutting in the middle of the leaf blade. Two types of common rooting hormones viz. indole 3-butyric acid (IBA) and naphthalene acetic acid (NAA), all at four levels were used as per details given:$T_0$- Control (without auxins); $T_1$ - IBA (1000 ppm); $T_2$- NAA (1000 ppm); $T_3$- IBA (1000 ppm) + NAA (1000 ppm).

A total of 60 leaf-cuttings were treated with each type of auxin treatment in 3 replications (10 cuttings per replication) (Table 1). The rooting hormones were applied following quick dip method (Basak et al., 2000). The treated cuttings were put (at 45 deg angle) in rooting bed filled with sand:soil mixture (1:1) at the rate of 10 cuttings per block (representing one replication).

Rooting/shooting was recorded after 60 days in respect of the following parameters:

i. Percentage of rooted leaf-cuttings (rooting %)
ii. Root number per cutting
iii. Root length per cutting
iv. Percentage of shoot leaf-cuttings (shooting %)
v. Shoot number per cutting
vi. Survivability of shoot cuttings (survival %)

The data were subjected to analysis of variance followed by Student-Newman-Keuls test using GraphPad Prism (Ver,5,0). All the percentile values were converted into angular transformation for analysis.

## RESULTS

### Percentage of rooting

The type of auxins applied and leaf cuttings used had no relationship with number of roots developing out of plant ($p>0.05$) (Tables 2 and 3). The percentage of leaf-cuttings producing root was impacted by all four treatments ($T_0$- Control, $T_1$ - IBA (1000 ppm), $T_2$- NAA (1000 ppm), $T_3$- IBA (1000 ppm) + NAA (1000 ppm). Petiolar leaf-cuttings treated with T3 developed maximum roots (91.69%). Although there was variation in the rooting proficiency in relation to treatments, all leaf cuttings nevertheless developed roots in sand : soil mixture (1:1). But the rooting percentage varied from 70% (in apical leaf cuttings with control) to 91.69% (in petiolar leaf-cuttings with T3).

### Root number per leaf-cutting

Average root number was not significantly ($p>0.05$) affected by auxin treatments in leaf-cuttings (both apical and petiolar halves) (Tables 2 and 3). The highest number of roots (13.40) was obtained in leaf cuttings treated with T3 (IBA 1000 ppm). Root number varied from 8.57 (in apical leaf cuttings with control) to 13.40 (in petiolar leaf cuttings treated with T3) (Figure 1).

### Effect on length of root

The effect of auxin and leaf parts used (apical and petiolar halves) had no significant ($p>0.05$) relationship on the root length per cutting. Longest roots (9.65 cm) were obtained from leaf-cuttings treated with T3. The average length of roots ranged from 6.18 cm (in apical leaf cuttings with control) to 9.65 cm (in petiolar leaf-cuttings with T3 (Figure 1).

### Shooting percentage

The variation in shooting out of leaf cutting was significantly influenced ($p<0.05$) by the type and dosage of rooting hormones (Tables 2 and 3). Petiolar leaf-cuttings treated with T3 exhibited highest percentage of shooted cuttings (83.33%). As seen in the case of rooting, shoots are also formed in leaf cutting treated with auxins irrespective of concentrations. Percentage of shoots in the treated leaf-cuttings ranged from 60% (in apical leaf cuttings with control) to 83.33% (in petiolar leaf-cuttings with T3) (Figure 1).

**Table 2.** Performance of adventitious rooting and multiple shoots regeneration from the leaf cutting of *Piper longum* as influenced by plant growth regulators.

| Parameter/Type of leaf-cuttings | Treatment | | | |
|---|---|---|---|---|
| | $T_0$: (Control) | T1: (IBA 1000 ppm) | T2: (NAA 1000 ppm) | T3: (IBA 1000 ppm + NAA 1000 ppm) |
| **Rooting (%)** | | | | |
| Apical leaf cutting | **70 ± 1.93** | 80.24 ± 1.92 | 78.29 ± 2.52 | 87.65 ± 2.6 |
| Petiolar leaf cutting | 73.33 ± 1.92 | 81.47 ± 0.98 | 75.55 ± 4.05 | **91.69 ± 2.25** |
| **Root number** | | | | |
| Apical leaf cutting | **8.57 ± 0.43** | 11 ± 0.12 | 11.03 ± 0.12 | 12.11 ± 0.56 |
| Petiolar leaf cutting | 8.73 ± 0.36 | 11.67 ± 0.52 | 11.21 ± 0.56 | **13.4 ± 0.47** |
| **Root length** | | | | |
| Apical leaf cutting | **6.18 ± 0.48** | 8.79 ± 0.53 | 8.78 ± 0.56 | 9.1 ± 0.25 |
| Petiolar leaf cutting | 6.87 ± 0.43 | 8.6 ± 0.49 | 8.23 ± 0.69 | **9.65 ± 0.51** |
| **Shooting (%)** | | | | |
| Apical leaf cutting | **60 ± 1.92** | 69.99 ± 1.93 | 63.33 ± 1.92 | 73.33 ± 1.92 |
| Petiolar leaf cutting | 63.33 ± 1.92 | 76.67 ± 3.85 | 70 ± 1.93 | **83.3 ± 1.93** |
| **shoot number** | | | | |
| apical leaf cutting | 1.4 ± 0.06 | 1.4 ± 0.1 | 1.5 ± 0.12 | 1.7 ± 0.06 |
| petiolar leaf cutting | 1.6 ± 0.24 | 1.5 ± 0.15 | 1.8 ± 0.17 | 2 ± 0.12 |
| **Survival (%)** | | | | |
| Apical leaf cutting | **50.01 ± 3.86** | 73.33 ± 1.92 | 73.33 ± 3.85 | 80 ± 1.92 |
| Petiolar leaf cutting | 53.33 ± 6.94 | 80 ± 5.17 | 76.66 ± 1.92 | **83.33 ± 2.59** |

**Table 3.** Statistical analysis of variance for different parameters, that is, rooting (%), shooting (%), root number and length, shoot number per leaf cutting of *P. longum.*

| Parameter | Source of variation | Df | F | P |
|---|---|---|---|---|
| | Treatment | 2 | 0.2565 | 0.7765 |
| Rooting (%) | Type of leaf-cuttings | 1 | 0.2466 | 0.6255 |
| | No. of replications | 2 | 0.8007 | 0.8997 |
| | Treatment | 2 | 0.04909 | 0.9522 |
| Root Number | Type of leaf-cuttings | 1 | 0.001259 | 0.9721 |
| | No. of replications | 2 | 0.1621 | 0.8516 |
| | Treatment | 2 | 0.04321 | 0.9578 |
| Root Length | Type of leaf-cuttings | 1 | 0.04364 | 0.8369 |
| | No. of replications | 2 | 0.3221 | 0.7287 |
| | Treatment | 2 | **1.155*** | **0.3373** |
| Shooting (%) | Type of leaf-cuttings | 1 | **4.575*** | **0.0464** |
| | No. of replications | 2 | 0.2178 | 0.8064 |
| | Treatment | 2 | **3.481*** | **0.0527** |
| Shoot number | Type of leaf-cuttings | 1 | **5.005*** | **0.0382** |
| | No. of replications | 2 | 0.3913 | 0.6818 |

**Table 3.** Contd.

| | | | | |
|---|---|---|---|---|
| **Survival (%)** | Treatment | 2 | 0.4505 | 0.6443 |
| | Type of leaf-cuttings | 1 | 0.05247 | 0.8214 |
| | No. of replications | 2 | 0.2512 | 0.7806 |

*Significant at 0.05% level of probability.

**Figure 1.** Induction of adventitious roots and shoots in apical (upper row) and petiolar halves of leaves (lower row) of *P. longum* treated with auxin (T3, IBA/NAA at 1000 ppm each).

### Shoot number per leaf-cutting

Average shoot number was significantly (p<0.05) affected due to various auxin treatments and varied in different leaf-cuttings (Tables 2 and 3). The highest shoot number (2.0) was obtained in leaf cuttings under the T3 treatment. Shoot number varied from 1.40 (in apical leaf cuttings with control and IBA) to 2.00 (in petiolar leaf cuttings with IBA plus NAA, that is, T3) (Figure 1).

### Survival percentage in leaf cuttings

Survival percentage of rooted (and shooted) leaf-cuttings under different auxin treatments and varying doses is presented in Tables 2 and 3. Though survival percentage differed in leaf cuttings treated with different type of auxins and their combined treatments, the differential response was not significant (p>0.05). Petiolar leaf cuttings with T3 (IBA 1000 + NAA 1000 ppm) had the highest survival of 83.33%. Whereas the survival percentage of rooted and shooted leaf-cuttings varied from 50.01 (in apical leaf-cuttings under control) to 83.33% (petiolar leaf-cuttings with T3).

## DISCUSSION

In this study, adventitious rooting and shooting were induced in leaf-cuttings of *P. longum*. Effect of growth hormones on leaf/petiolar rooting and shooting behavior was assessed under individual and combined treatment of hormones. IBA in combination with NAA was found most effective than their individual effect on rooting. These findings supported the result on adventitious rooting in stem cuttings of some mangrove species (Basak et al., 2000). In leaf cuttings, new growing points usually originate in the parenchymatous tissue closely associated with vascular tissues in the leaves. When vascular bundles of the leaf are severed but suitable growth conditions are provided new roots can be initiated but there was difficulty, in shoot regeneration. Leaf cuttings of most plants do not generate a new plant and produce only a few roots or just decay. Due to the fact that leaf cuttings do not include an axillary bud, they can be used only for plants that are capable of forming adventitious buds (Hartmann et al., 1996). Probably, *P. longum* has the capacity to form adventitious buds in leaf-cuttings with or without been triggered by exogenous application of auxins.

In conclusion, *P. longum* can be regenerated via leaf-cuttings, either without hormone treatment, or with even better results using low concentrations of IBA or combination of IBA and NAA hormones. The non-significant differences for almost all the studied parameters in treated and control implies that *P. longum* can be propagated vegetatively at reduced cost through leaf cutting. This method, therefore, can be adopted with minimum capital to produce quality planting material.

## ACKNOWLEDGEMENTS

We are thankful to the National Medicinal Plants Board (NMPB), Government of India for financial support vide Project Grant No.GO/OR-1/2008-09.

## REFERENCES

Basak UC, Das AB, Das P (2000). Rooting response in stem cuttings from five species of mangrove trees: Effect of auxins and enzyme activities, Mar. Biol. 36:185-189.

Bhat SR, Chandel KPS, Malik SK(1995). Plant regeneration from various explants of cultivated *Piper* species. Plant Cell Rep. 14:398-402.

Hartmann HT, Kester DE, Davies FT, Geneve RL (1996). Plant Propagation, Principles and Practices. 6th ed. Prentice Hall: Upper Saddle River, New Jersey.

James V (1999). Piperine: the treasured alkaloid in *Piper nigrum*. Spice India 12:9-12.

Nair KKN (2000). Manual of Nonwood Forest Produce Plants of Kerala (pp. 268-270). Kerala Forest Research Institute, Kerala, India.

Sarasan V, Thomas E, Lawrence B, Nair GM (1993). Plant regeneration in *Piper longum* L. (Piperaceae) through direct and indirect shoot development. J. Spices Arom. Crops 2:34-40.

Sivarajan VV, Balachandran I (1994) Ayurvedic Drugs and their Plant Sources Oxford and IBH Publishing Co. Pvt. Ltd. pp. 374-376.

Soniya EV, Das MR (2002). *In vitro* micropropagation of *Piper longum* - an important medicinal plant Plant. Cell, Tissue Org. Cult. 70:325-327.

# 13

# Skip-row planting of maize and sorghum in semi-arid Ethiopia

**Tewodros Mesfin[1], Jibril Mohammed[1], Abuhay Taklete[1], Feyera Merga[1] and Charles Wortmann[1,2]**

[1]Melkassa Agriculture Research Center-EARI, P. O. Box 436, Adama Ethiopia
[2]University of Nebraska-Lincoln, 369 Keim Hall, Lincoln EN 68583-0915, Ethiopia.

**Soil water deficits during grain fill constrain crop production in semi-arid areas of Ethiopia. Skip-row planting is a means of saving soil water for grain fill while tie-ridging can improve soil water availability throughout the season by reducing runoff. The hypotheses were that where rainfall ceases before or during early grain fill 1) maize (*Zea mays* L.) and sorghum (*Sorghum bicolor* L. Moench) yield can be increased by skip-row planting, 2) skip-row planting and tie-ridging interact positively, and 3) productivity can be further increased by planting an early maturity crop in the skip-row area. Skipping a row after planting one or two rows resulted in similar yields compared with planting all rows. Maize yield was 43% greater with tie-ridge compared with flat tillage. There was no tillage by skip-row interaction. Productivity was increased by 20% when a relatively short season bean (*Phaseolus vulgaris* L.) was planted in the skipped rows of both maize and sorghum as cereal yield was not affected but bean added to productivity. Tie-ridging presents an opportunity for increasing maize yield. Skip-row planting for similar conditions is unlikely to increase productivity unless bean or another crop is planted in the skip-row area.**

**Key words:** Dry bean, intercrop, Sahel, soil water deficits, late season stress, tie-ridging.

## INTRODUCTION

Rainfall often ceases before or during early grain fill of maize and sorghum in many semi-arid areas of Ethiopia. Severe stress due to soil water deficits during grain fill is common resulting in low grain weight, grain yield, and harvest index. Such stress has been estimated to account for more than 300,000 Mg year$^{-1}$ grain yield loss for grain sorghum in Ethiopia (Wortmann et al., 2009).

Skip-row planting is a means of delaying root access to available soil water until later growth stages as the root system extends (Milroy et al., 2004). Within-row plant density is commonly increased with skip-row planting to compensate for fewer planted rows and the crop is more likely to experience stress during the vegetative stage while having greater soil water availability during grain fill (Nielsen et al., 2007). Skip-row planting may therefore result in increased kernel weight, improved harvest index, and increased grain yield where soil water deficit stress is common during grain fill. Skip-row planting is expected to result in reduced yield where terminal stress due to soil water deficit is not severe or not common but may increase yield and safeguard against crop failure when such stress is severe (Whish et al., 2005). Potential for increased yield with skip-row planting increases as frequency of severe soil water deficits during grain fill increases, accompanied by potential for deep rooting depth and good soil available water holding capacity. Increased evaporative loss of soil water with skip-row planting is a concern, although saving more deep soil water for later in the season can compensate for these losses (Myers et al., 1986; Spackman et al., 2000). Maintenance of good crop residue ground cover to reduce soil water evaporation enhances the potential of skip-row planting (Klein et al., 2007).

In Australia, skip-row planting was advantageous when mean sorghum grain yield was less than 2.5 Mg ha$^{-1}$ because of soil water deficits (Spackman et al., 2000). Clark and Knight (1996) observed increased sorghum grain yield with skip-row planting when mean grain yield was less than 2 Mg ha$^{-1}$, but decreased grain yield when mean yield was more than 3 Mg ha$^{-1}$. In Ethiopia, planting two rows alternated with a skipped row resulted in increased sorghum grain yield in northern Ethiopia where mean yields were less than 2.5 Mg ha$^{-1}$ but not in the Central Rift Valley where mean yields were more than 2.5 Mg ha$^{-1}$ (Mesfin et al., 2010).

On the US Great Plains, skip-row planting resulted in a corn and sorghum grain yield advantage compared with conventional planting if mean grain yield was less than 4 Mg ha$^{-1}$ but a disadvantage when yield was more than 5 Mg ha$^{-1}$ (Vigil et al., 2008). Planting maize skipping alternate pairs of rows had more yield compared with planting all rows if soil water deficits limited yield to 4.7 Mg ha$^{-1}$ and skipping alternate rows was advantageous to 6.9 Mg ha$^{-1}$ (Lyon et al., 2009). In Nebraska, skipping alternate rows was advantageous for grain sorghum when soil water deficits limited grain yield to less than 4.5 Mg ha$^{-1}$ (Abunyewa et al. 2010). Water use of sorghum was more and less efficient with skipping alternate rows compared with planting all rows in Nebraska when mean in-season precipitation was less than 2 and more than 2.5 mm day$^{-1}$, respectively (Abunyewa et al., 2011).

Runoff loss of water with intense rainfall events can be reduced by tie-ridging with inter-row furrows of 20 to 30-cm depth blocked with earthen ties spaced according to the slope of the land, water infiltration rate, and expected intensity of rainfall (Lal, 1977; Gusha, 2002; Gebrekidan, 2003; Pendke et al., 2004). Sorghum grain yield in northern Ethiopia was 62% more with tie-ridging compared with flat planting (Brhane et al., 2006). Mesfin et al. (2009) reported 33 and 18% more sorghum grain yield in northern and Central Rift Valley locations of Ethiopia, respectively, with tie-ridging compared with flat planting. Highland pulse grain yield was increased with tie-ridging by 31 to 96% in northern Ethiopia (Brhane and Wortmann, 2008). In eastern Kenya, maize and cowpea (*Vigna unguiculata* [(L.)] yield was inconsistently increased with tie-ridging (Miriti et al., 2007). The tie-ridging effect was enhanced with skip-row planting of sorghum in northern Ethiopia but this interaction did not occur in the Central Rift Valley (Mesfin et al., 2009). The tie-ridging by skip-row planting interaction has not been previously investigated for maize.

Smallholder farmers of Ethiopia are reluctant to leave land unplanted as required with skip-row planting. Planting an early maturing crop with a relatively shallow root system, such as bean, in the skip-row area of maize or sorghum may preserve benefits of skip-row planting while giving some bean yield. Workayehu and Wortmann (2011) found that maize planted skipping alternate rows with bean in the skipped row gave a mean land equivalent ratio of 1.6. In another study, a maize intercrop was 55% more productive compared with the sole crops (Bhatnagar and Chaplot, 1991).

The objectives of this research were to determine the impact of skip-row planting and tie-ridge tillage on maize yield, and of skip-row planting of maize and sorghum with an intercrop of bean on yields in an important semi-arid agricultural production area of the Central Rift Valley in Ethiopia. The hypotheses addressed were that where water deficit stress is severe during grain fill 1) maize and sorghum yield can be increased by skip-row planting to save soil water for the reproductive stage, 2) there is a positive interaction of skip-row planting with tie-ridging, and 3) productivity can be further increased by planting an early maturity crop in the skip-row area.

## MATERIALS AND METHODS

Trials were conducted in 2006 to 2010 for study I and in 2010 to 2012 for study II at Melkassa Agricultural Research Center of the Ethiopia Agricultural Research Institute located at 8°24'N, 39°12'E, and 1500 m elevation. The study I trial of 2009 was not completed because of severe drought stress. The soil was a calcareous clay loam of volcanic parent material classified as a Typic Haplustand with low wet aggregate stability, a propensity for crusting, more than 1-m rooting depth, and slopes ranging from 0.02 to 0.04 m m$^{-1}$. Soil pH was 7.4 and soil organic matter was 12 g kg$^{-1}$ for the 0 to 20-cm depth. Cumulative rainfall for the growing season of June to October was more than the long term average and ranged from 635 mm in 2011 to 818 mm in 2012 (Figure 1) with a mean of 65% falling in July and August. The trial site was moved each year. The previous crops were onion (*Allium cepa* L.), maize, and dry bean in 2009, 2010, and 2011, respectively.

The complete factorial combination of treatments for study I consisted of three or four planting treatments and two tillage treatments. The maize planting treatments, based on 0.75 m inter-row spacing, were: planting all rows; skipping alternate rows; skipping alternate pairs of rows; and planting two rows alternated with a skipped row. The treatment of planting two rows alternated with a skipped row was included in 2008 and 2010 only. The tillage treatments were flat and tie-ridge tillage. The complete factorial combination of treatments for study II was in a split plot design with corn or sorghum as the main plots and three planting practices as the sub-plots including planting all rows, planting two rows alternated with a skipped row, and planting two rows alternated with bean planted is the skipped row. Treatments included a bean sole

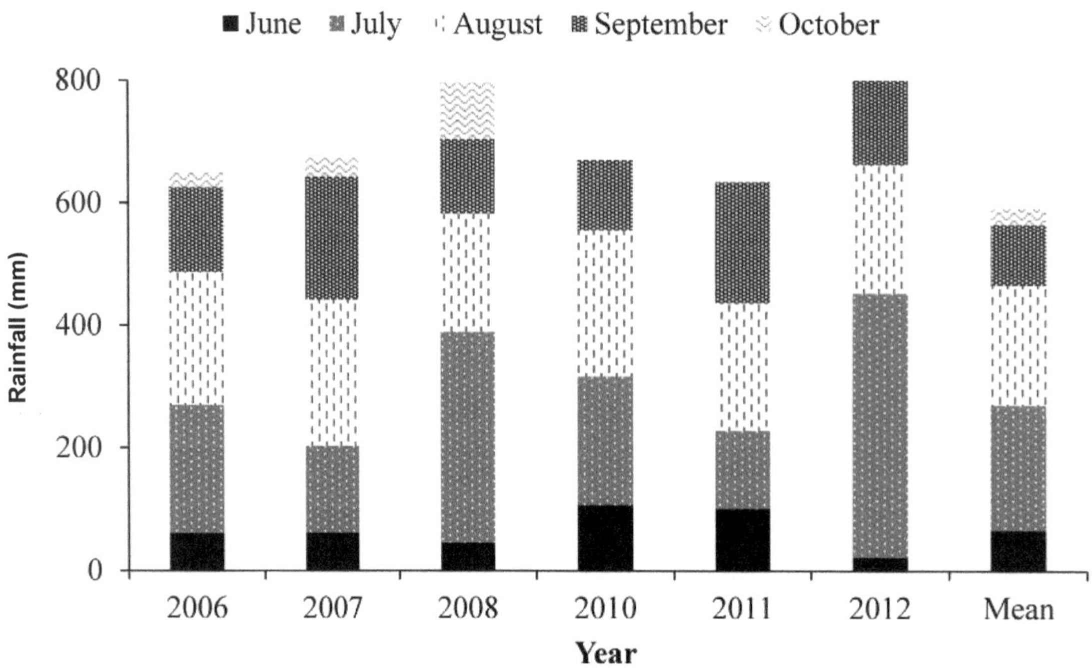


**Figure 1.** Growing season (May to October) monthly rainfall for trials conducted to evaluate skip-row planting and tie-ridging effects on maize in central Ethiopia.

crop. Study II was planted with tie-ridging. Individual plot size was 10 × 6 m for each study and each had three replications.

The entire experimental area was tilled with a tractor drawn plow and then ridged with a tractor drawn 4-furrow ridger before planting and application of the treatments. Ridges were of 0.30 m height with 0.75 m spacing that were tied at the ends of the plots to create the tie-ridges. Planting followed the on-set of the rains and occurred during the first two weeks of July. Harvest was in November. Maize and sorghum were planted in the furrow. Ridges were reshaped by ridging the soil around the lower stems of the plants and retied at weeding time.

Plant stands were thinned to desired density by manual uprooting at about two weeks after emergence. The targeted maize and sorghum plant densities were 6.7 plants $m^{-2}$ with greater within-row densities with skip-row planting. Bean was also planted in the furrow at 25 and 6.3 plants $m^{-2}$ for sole crop and intercrop, respectively. The maize, sorghum, and bean cultivars were Melkassa II, Meko, and Awash Melka, respectively. Melkassa II, Meko, and Awash Melka require 125, 115, 90 to 95 days to maturity, respectively, at 1500 m elevation. Fertilizer was band-applied at the rate of 20 kg $ha^{-1}$ each of N and P at planting time as di-ammonium phosphate and 23 kg $ha^{-1}$ of N was side-dress applied as urea for all plots. The sites were manually weeded twice. No pesticides were used for pest and weed control.

Five randomly selected plants per row were measured to determine maize and sorghum plant height at maturity. The maize and sorghum plants from 7 m of row from the center three rows of each plot (15.75 $m^2$) were cut at the soil surface and air-dried after counting the plants and ears. Grain was shelled from the ears and the cobs were combined with the other stover to obtain stover dry weight. Air-dried grain water content was not affected by planting treatments and was found to average 123 g $kg^{-1}$ for maize and bean and 118 g $kg^{-1}$ for sorghum. Grain yields were expressed at 150 g $kg^{-1}$ water content.

The air-dried weight of 100 kernels was determined. Kernel $ear^{-1}$ and $panicle^{-1}$ were calculated. The harvest index was calculated as the ratio of grain yield to the above-ground biomass yield with grain and stover water content adjusted to a water content of 80 g $kg^{-1}$.

For P2S1 with bean treatments, the one bean row in the area for yield determination was harvested by uprooting the plants and threshing after air-drying. Bean grain yield was also expressed at 150 g $kg^{-1}$ water content.

The analyses of variance for study I were conducted combining 2006 with 2007 data and 2008 with 2010 data using Statistix 9 (Analytical Software, Tallahassee, FL). The ANOVA for study II was combined for 2010-2012. Grain yield was related to yield components using regression and correlation analyses. Means were compared with the ANOVA protected LSD means separation test. Treatment effects and relationships were considered significant at $P$ less than 0.05.

## RESULTS

### Study I

There were no treatment effect on days to anthesis ($\overline{X}$ = 81 in 2006-7 and $\overline{X}$ = 69 in 2008 and 2010) and days to physiological maturity ($\overline{X}$ = 135 in 2006-7 and $\overline{X}$ = 120 in 2008 and 2010). Mean plant height was 11 cm taller with tie-ridging compared with flat tillage but not affected by row configuration (Table 1). In 2006 and 2007, grain yield and grain yield components were not affected by interactions. Grain yield and kernel weight were increased by 29 and 19% with tie ridge compared with flat tillage, respectively. Tillage effects were not significant for the other yield components in these years. Grain yield was 24% less with skipping alternate pairs of rows compared with planting all rows. Grain yield and its components were not different with skipping alternate rows and planting two rows alternated with a skipped

**Table 1.** Maize performance as affected by tie-ridging and skip-row planting in the Central Rift Valley of Ethiopia. PF, P1:S1, P2:S1B and P2:S1, and P2:S2 are planting configurations with all rows planted, single planted and skipped rows alternated, two planted alternated with one skipped row with and without bean planted in the skipped row, and planted with skipped pairs of rows alternated, respectively.

| Factor level | Plant height (cm) | Ear plant$^{-1}$ | Kernel ear$^{-1}$ | 100 kernel weight (g) | Grain yield (Mg ha$^{-1}$) | Stover yield (Mg ha$^{-1}$) | HI |
|---|---|---|---|---|---|---|---|
| **2006 and 2007; three row configurations‡** | | | | | | | |
| **Year** | | | | | | | |
| 2006 | 242 | | 363 | 23.6 | 3.96 | 4.89 | 0.42 |
| 2007 | 241 | 0.90 | 347 | 23.6 | 4.78 | 3.72 | 0.53 |
| **Tillage** | | | | | | | |
| Flat tillage | 236 | 0.88 | 359 | 21.6 | 3.81 | 4.27 | 0.44 |
| Tie-ridge | 246 | 0.92 | 350 | 25.6 | 4.93 | 4.34 | 0.50 |
| | ** | ns | ns | * | * | Ns | ** |
| **Row configuration** | | | | | | | |
| PF | 246 | 0.92 | 339 | 25.2 | 4.60a | 4.78a‡ | 0.47 |
| P1:S1 | 238 | 0.91 | 374 | 23.9 | 4.39ab | 4.29ab | 0.49 |
| P2:S2 | 240 | 0.88 | 352 | 21.7 | 3.50b | 3.84b | 0.46 |
| | Ns | ns | ns | ns | * | ** | Ns |
| **2008 and 2010; four row configurations§** | | | | | | | |
| **Year** | | | | | | | |
| 2008 | 286 | 0.87 | 419 | 25.9 | 3.87 | 6.22 | 0.37 |
| 2010 | 207 | 0.87 | 200 | 25.4 | 2.74 | 6.77 | 0.27 |
| **Tillage** | | | | | | | |
| Flat tillage | 240 | 0.84 | 304 | 24.2 | 2.65 | 5.83 | 0.30 |
| Tie-ridge | 252 | 0.90 | 315 | 27.3 | 3.96 | 7.16 | 0.34 |
| | * | ** | ns | ** | *** | *** | * |
| **Row configuration** | | | | | | | |
| PF†† | 252 | 0.89a | 332 | 26.0 | 3.63a | 6.91a | 0.33 |
| P1:S1 | 252 | 0.90a | 302 | 26.8 | 3.48a | 6.36ab | 0.33 |
| P2:S2 | 244 | 0.83b | 284 | 23.6 | 2.62b | 5.58b | 0.31 |
| P2:S1 | 240 | 0.86ab | 321 | 26.2 | 3.51a | 7.13a | 0.31 |
| | ns | * | ns | ns | *** | ** | ns |

Ns,*, *, **, ***: Not significant and significant at $P$ = 0.05, 0.01, and 0.001, respectively. Letters denote differences within columns and sets of years using the ANOVA-protected LSD 0.05 means comparison test.

row compared with planting all rows.

Grain yield was affected by the year x tillage interaction because yield was 43 and 17% less in 2008 and 2010, respectively, with flat tillage compared with tie-ridging (Table 1). Grain yield was not affected by other interactions but was consistently more with tie-ridge compared with flat tillage with a mean increase of 43%. Grain yield was consistently less with skipping alternate pairs of rows compared with other row configurations but planting all rows, skipping alternate rows, and planting two rows alternated with a skipped row had similar mean yields.

The increase in ear plant$^{-1}$ was greater with planting all rows and planting two rows alternated with a skipped row compared with skipping alternate rows and skipping alternate pairs of rows in 2008 than in 2010 (Table 1). The main effect of tie-ridging compared with flat tillage was a 49% increase in grain yield, a 13% increase in kernel weight, and a 7% increase in ears plant$^{-1}$. Mean grain yield and ear plant$^{-1}$ were 28 and 7% less with skipping alternate pairs of rows compared with planting all rows. Grain yield and its components were not different with skipping alternate rows and planting two rows alternated with a skipped row compared with planting all rows.

Stover yield was more in 2006 and 2008 and less in 2007 with tie-ridging compared with flat tillage with no significant tillage effect in 2010 (Table 1). The reduction in stover yield with skipping alternate pairs of rows compared with planting all rows was greater in 2007 than in 2006, with a mean reduction over all four years of 20% with skipping alternate pairs of rows compared with

**Table 2.** Analysis of variance results for the effect of planting pattern on maize, sorghum, and bean in the Central Rift Valley of Ethiopia.

| Factor level | Plant height (cm) | Ear ($m^{-2}$) | Kernel ($ear^{-1}$) | Kernel weight (mg) | Grain yield (Mg $ha^{-1}$) | Stover yield (Mg $ha^{-1}$) | HI |
|---|---|---|---|---|---|---|---|
| **Maize** | | | | | | | |
| 2010 | | 5.53 | 219 | 247 | 3.09 | 7.84 | 0.26 |
| 2011 | 202 | 6.38 | 273 | 218 | 3.87 | 7.05 | 0.33 |
| 2012 | 181 | 6.36 | 236 | 213 | 3.37 | 6.81 | 0.29 |
| Planting (P) | ns | ns | ns | ns | ns | ns | ns |
| Year x P | ns | ns | * | ns | * | ns | *** |
| **Sorghum** | **Plant height (cm)** | **Panicle $plant^{-1}$** | **Kernel $panicle^{-1}$** | **Kernel weight (mg)** | **Grain yield (Mg $ha^{-1}$)** | **Stover yield (Mg $ha^{-1}$)** | **HI** |
| 2010 | | 6.10 | 960 | 38.3 | 2.25 | 5.71 | 0.28 |
| 2011 | 161 | 5.76 | 1229 | 36.1 | 2.61 | 8.20 | 0.23 |
| 2012 | 150 | 6.39 | 1684 | 29.0 | 3.07 | | |
| Planting | ns | ns | ns | ns | ns | ns | ns |
| Year x P | ns | ns | * | ns | ns | ns | ns |
| **Bean** | | **Pod $plant^{-1}$** | **Kernel $pod^{-1}$** | **Kernel weight (mg)** | **Grain yield (Mg $ha^{-1}$)** | **Stover yield (Mg $ha^{-1}$)** | **HI** |
| 2010 | | 23.3 | 4.93 | 188 | 1.65 | 2.33 | 0.39 |
| 2011 | | 14.9 | 6.59 | 175 | 0.93 | 0.56 | 0.62 |
| 2012 | | | | 158 | 1.68 | | |
| Planting | | ns | ns | ns | *** | *** | ns |
| Year x P | | ns | ns | * | ** | ns | ns |

ns,*, *, **, ***: Not significant and significant at P = 0.05, 0.01, and 0.001, respectively.

planting all rows. Mean harvest index for the four years was 13% greater with tie-ridge compared with flat tillage but was not affected by interactions or row configuration.

Variation in kernel $ear^{-1}$ accounted for more variation in grain yield compared with other yield components. The significant Pearson correlation coefficients of yield components with grain yield were 0.18, 0.68, and 0.49 for ear $m^{-2}$, kernel $ear^{-1}$, and 100-kernel weight, respectively. In total, variation in yield components accounted for 99% of the variation in grain yield: Grain yield = -9.64 + 0.206 * 100-kernel weight + 0.014 kernel $ear^{-1}$ + 0.801 ear $m^{-2}$.

## Study II

Maize and sorghum plant height, ear or panicle $m^{-2}$, kernel wt., and stover yield were not affected by treatments (Table 2). Grain yield and harvest index of sorghum were also not affected by treatments. Maize grain yield was 30% more for planting all rows compared with planting two rows alternated with a skipped row in 2011 and kernel $ear^{-1}$ was 40% more in 2012 with planting all rows compared with planting two rows with bean planted in the skipped row but these variables were not affected in other years (Table 3). Maize harvest index was low and high, respectively, in 2010 and 2011 with planting all rows compared with planting two rows alternated with a skipped row. Sorghum panicle $m^{-2}$ was 12% more with planting two rows alternated with a skipped row compared with the other planting arrangements in 2010 but otherwise not affected by treatments.

Maize kernel $ear^{-1}$ and kernel $m^{-2}$ were highly correlated (r = 0.97) and both were related to maize grain yield (r = 0.86). Kernel weight was not related to maize yield but ear $m^{-2}$ was related to yield (r = 0.50). Kernel $m^{-2}$ accounted for 73% of the variation in maize yield. Sorghum kernel $panicle^{-1}$ and kernel $m^{-2}$ were highly correlated (r = 0.96), both were related to sorghum grain yield (r = 0.83 and 0.85, respectively), and together accounted for 96% of the variation in yield. Kernel weight and panicle $m^{-2}$ were not related to sorghum yield.

Bean pod $plant^{-1}$, kernel $pod^{-1}$, stover yield, and harvest index were not affected by treatments (Table 3). Intercrop bean yield was 31% of sole crop bean yield. Bean yield was less suppressed by intercropping with sorghum compared with maize in 2010 but yields were similar for the two intercrops in 2011 and 2012. Bean kernel weight was 11% less with intercropping compared with sole cropping in 2011. The maize: bean land equivalent ratio

**Table 3.** Planting configuration by year interaction effects on maize, sorghum, and bean in the Central Rift Valley of Ethiopia. PF and P2:S1 are planting configurations with all rows planted and two planted alternated with one skipped row, respectively.

| Factor level | 2010 | 2011 | 2012 | 2010 | 2011 | 2012 |
|---|---|---|---|---|---|---|
| | Maize yield (Mg $ha^{-1}$) | | | Maize kernel ($ear^{-1}$) | | |
| PF | $2.77^{c}$ | $4.38^{a}$ | $3.84^{ab}$ | $193^{c}$ | $299^{a}$ | $286^{ab}$ |
| P2:S1 | $3.46^{bc}$ | $3.37^{bc}$ | $2.90^{bc}$ | $251^{abc}$ | $226^{abc}$ | $220^{bc}$ |
| P2:S1 with bean | $3.03^{bc}$ | $3.85^{ab}$ | $3.32^{bc}$ | $213^{bc}$ | $296^{a}$ | $203^{c}$ |
| | Maize HI | | | Sorghum panicle ($m^{-2}$) | | |
| PF | $0.23^{c}$ | $0.42^{a}$ | $0.31^{b}$ | $5.76^{bc}$ | $5.74^{bc}$ | $6.38^{ab}$ |
| P2:S1 | $0.31^{b}$ | $0.31^{b}$ | $0.31^{b}$ | $6.55^{a}$ | $5.53^{c}$ | $6.38^{ab}$ |
| P2:S1 with bean | $0.30^{b}$ | $0.32^{b}$ | $0.29^{bc}$ | $6.00^{bc}$ | $6.00^{bc}$ | $6.42^{ab}$ |
| | Bean grain yield (Mg $ha^{-1}$) | | | Bean kernel weight (mg) | | |
| Bean, sole crop | $3.11^{a}$ | $1.49^{b}$ | $2.83^{a}$ | $180^{ab}$ | $189^{a}$ | $157^{c}$ |
| P2:S1, maize | $0.63^{d}$ | $0.74^{cd}$ | $0.71^{cd}$ | $191^{a}$ | $167^{bc}$ | $163^{bc}$ |
| P2:S1, sorghum | $1.23^{bc}$ | $0.57^{d}$ | $0.72^{cd}$ | $193^{a}$ | $170^{bc}$ | $156^{c}$ |

Ns,*, *, **, ***: Not significant and significant at P = 0.05, 0.01, and 0.001, respectively. Letters denote differences within columns using the ANOVA-protected LSD 0.05 means comparison test.

was 1.42, 1.43, and 1.01 in 2010, 2011, and 2012, respectively. The sorghum:bean land equivalent ratio was 1.19, 153, and 1.19 in 2010, 2011, and 2012, respectively. Overall mean land equivalent ratio for planting two rows with bean planted in the skipped row was 1.30.

## DISCUSSION

In all trials, there was little rainfall after flowering. Tie-ridge compared with flat tillage resulted in more maize grain yield in all years with planting all rows which is generally consistent with results for sorghum and pulses in northern Ethiopia (Brhane et al., 2006; Brhane and Wortmann 2008; Mesfin et al., 2010). Stover yield was increased with tie-ridging in two of four years. Tie-ridging is potentially widely beneficial to maize production on moderately sloping land with soil of medium texture and weak aggregate stability.

Results from the above cited studies indicate the benefits of tie-ridging for vertic soils that have slow infiltration when wet. All of these studies were done on moderately sloping land. The risk of ridges breaking during heavy rainfall when used on steeply sloping land is of concern and the practice should be used in such cases only if other practices are in place, such as terraces or vegetative barriers, that reduce runoff or otherwise stabilize the slope against erosion. Tie- ridging requires tillage but tillage is already the usual practice to break soil crusts and for weed control.

Skip-row planting did not result in increased maize yield. Grain yield was generally well above 2.0 to 2.5 Mg $ha^{-1}$, the upper grain yield level at which skip-row planting was found to be advantageous compared with planting all rows for grain sorghum when practiced without good crop residue cover (Spackman et al. 2000; Clark and Knight 1996). Mean grain yields were less but near the 4.7 Mg $ha^{-1}$ yield level identified by Lyon et al. (2009) for skip-row planting to be advantageous to maize yield under no-till conditions with good ground cover on the US Great Plains.

However, even with flat tillage where mean maize grain yield was 3.1 Mg $ha^{-1}$ in the current study, skip-row planting did not have an advantage. There also was no advantage to planting two rows alternated with a skipped row planting of sorghum compared with planting all rows. Greater kernel weight and harvest index of maize and sorghum would be expected if stress were reduced during grain-fill; these were not increased by skip-row planting indicating that this practice was not effective in reducing late season stress. Therefore, skip-row planting is not likely to have an advantage for maize and sorghum production in Ethiopia unless considerably more severe late season soil water deficit stress conditions occur than encountered in this study. More severe stress does commonly occur with sorghum production in northern Ethiopia where yield was greater with planting two rows alternated with a skip row compared with planting all rows (Mesfin et al., 2010).

There may be greater advantage with skip-row planting if some form of reduced tillage or no tillage is practiced coupled with maintaining ground cover by crop residues sufficient to reduce evaporation and to reduce the potential for erosion. However, crop residues are highly valued and account for 30 to 40% of the value of the sorghum crop in Ethiopia (Wortmann et al., 2009). Therefore, fields in semi-arid areas are nearly bare of crop residue when land is prepared for planting. Any remaining crop residue is commonly consumed by termites once the rains begin. Despite much promotion, practice of conservation agriculture by smallholders is rare in semi-arid areas of Ethiopia.

The planting two rows with bean planted in the skipped row is promising for maize and sorghum production where bean or another legume of similar growth habit is adapted. Bean is an important food and market crop in Ethiopia. Other pulses may replace bean at elevations where bean is less well adapted.

## Conclusion

Tie-ridging is likely to result in improved maize yields for semi-arid areas of Ethiopia on moderately sloping land with soil of medium texture and weak aggregate stability, including much of production in the Central Rift Valley of Ethiopia. There is no indication that skip-row planting will result in increased maize or sorghum grain yield in the Central Rift Valley of Ethiopia but other information indicates that skip-row planting is appropriate where late season soil water deficits are more severe such as in northern Ethiopia and in many places throughout the Sahel. Planting bean or another pulse crop of similar growth habit in the skip-row area increases productivity relative to planting all rows of maize or sorghum alone.

## Conflict of Interest

The author(s) have not declared any conflict of interests.

## ACKNOWLEDGEMENT

This is a contribution of the Ethiopia Institute of Agricultural Research and the University of Nebraska-Lincoln Agricultural Research Division. The research was partly funded by the Hatch Act and by the U.S. Agency for International Development through INTSORMIL under the terms of Grant No. LAG-G-00-96-900009-00.

## REFERENCES

Abunyewa AA, Ferguson RB, Wortmann CS, Lyon DJ, Mason SC, Irmak S, Klein RN (2011). Grain sorghum water use with skip-row configuration in the Central Great Plains of the USA. Afr. J. Agric. Res. 6:5328-5338.

Abunyewa AA, Ferguson RB, Wortmann CS, Lyon DJ, Mason SC, Klein RN (2010). Skip-row configuration and plant population effects on sorghum grain yield and yield stability in the Central Great Plains. Agron. J. 102:296-302.

Bhatnagar GS, Chaplot PC (1991). Evaluation of intercropping of winter maize with legumes. Int. J. Trop. Agric. 9:52-55.

Brhane G, Wortmann CS (2008) Tie-ridge tillage for high altitude pulse production in northern Ethiopia. Agron. J. 100:447-453.

Brhane G, Wortmann CS, Mamo M, Gebrekidan H, Belay A (2006). Micro-basin tillage for grain sorghum production in semi-arid areas of northern Ethiopia. Agron. J. 98:124-128.

Clark LE, Knight TO (1996). Grain production and economic returns from dryland sorghum in response to tillage systems and planting patterns in the semi-arid southwestern USA. J. Prod. Agric. 9:249-256.

Gebrekidan H (2003). Grain yield response of sorghum (*Sorghum bicolor*) to tied ridges and planting methods on entisols and vertisols of Alemaya area, Eastern Ethiopia Highlands. J. Agric. Rural Dev. Trop. Sub Trop. 104:113-128.

Gusha AC (2002). Effects of tillage on soil micro relief, surface depression storage and soil water storage. Soil Till. Res. 76:105-114.

Klein R, Lyon D, Baltensperger D, Pavlista A, Shapiro C, Knezveic S, Mason S, Nelson L, Bernards M, Elmore R, Schlegel A, Vigil M, Klocke N, Melvin S. (2007). What is required to have success with rainfed skip-row corn. 64-1 In: 2007 Agronomy Abstracts. ASA Madison WI

Lal R (1977). Soil management systems and erosion control. In: Greenland D, Lal R (eds) Soil conservation and management in the humid tropics. John Wiley and Sons, Chichester NY, pp. 81-86.

Lyon DJ, Pavlista AD, Hergert GW, Klein RN, Shapiro CA, Knezevic S, Mason SC, Nelson LA, Baltensperger DD, Elmore RW, Schlegel AJ, Olson BL, Aiken RM (2009). Skip-row planting patterns stabilize corn grain yields in the central Great Plains. Crop Manage http://www.plantmanagementnetwork.org/pub/cm/research/2009/skip/ . Accessed 18 Mar 2013

Mesfin T, Tesfahunegn GB, Wortmann CS, Mamo M, Nikus O (2010). Skip-row planting and tie-ridging for sorghum production in semi-arid areas of Ethiopia. Agron. J. 102:745-750.

Mesfin T, Tesfahunegn GB, Wortmann CS, Nikus O, Mamo M (2009). Tie-ridging and fertilizer use for sorghum production in semi-arid Ethiopia. Nutr. Cycl. Agroecosyst. 85:87-94.

Milroy SP, Bange MP, Hearn AB (2004). Row configuration in rainfed cotton systems: modification of the OZCOT simulation model. Agric. Syst. 82:1-16.

Miriti JM, Esilaba AO, Bationo A, Cheruiyot H, Kihumba J, Thuranira EG (2007). Tie-riding and integrated nutrient management options for sustainable crop production in semi-arid eastern Kenya. In Bationo A, Waswa B, Kihara J, Kimetu J. (eds) Advances in integrated soil fertility management in sub-Saharan Africa: challenges and opportunities. Springer, Dordrecht, the Netherlands, pp. 435-441.

Myers RJK, Foale MA, Thomas GA, French AV, Hall B (1986). How row spacing affects water use and root growth of grain sorghum. Proc First Australian Sorghum Conf, Gatton, pp. 586-591.

Nielsen DC, Vigil MF, Caldreon FJ, Schneekloth J, Poss D, Henry WB, Benjamin JG (2007). Water extraction patterns in skip-row corn, sorghum and sunflower. 305-2. In 2007 Agronomy Abstracts. ASA, Madison, WI.

Pendke MS, Lomte MH, Gitte AU (2004). Effect of soil and water conservation practices on runoff, soil loss and yield of pigeonpea. J. Maharashtra Agric. Univ. 29:319-321.

Spackman GB, McCosker KJ, Farquharson AJ, Conway MJ (2000). Innovative management of grain sorghum in Central Queensland. Proc. Aust. Agron. Conf. Australian Soc. Accessed 3 March 2014. http://www.regional.org.au/au/asa/2001/1/a/spackman.htm

Vigil MF, Henry B, Calderon FJ, Nielsen DC, Benjamin JG, Klein B (2008). A use of skip-row planting as a strategy for drought mitigation in the west central Great Plains. Proc. Great Plains Soil Fertility Conf., Denver CO. 4-5 Mar. 2008. pp. 101-106.

Whish J, Butler G, Castor M, Cawthray S, Broad I, Carberry P, Hammer G, McLean G, Routley R, Yeates S (2005). Modeling the effects of row configuration on sorghum yield reliability in north-eastern Australia. Aust. J. Ag. Res. 56:11-23.

Workayehu T, Wortmann CS (2011). Maize-bean intercrop suppression of weeds and profitability in southern Ethiopia. Agron. J. 104:1058-1063.

Wortmann CS, Mamo M, Mburu C, Letayo E, Abebe G, Kayuki KC, Chisi M, Mativavarira M, Xerinda S, Ndacyayisenga T (2009). Atlas of sorghum (*Sorghum bicolor* (L.) Moench) production in eastern and southern Africa http://intsormil.org/smscientificpubs/Sorghum%20Atlas.pdf Accessed 6 June 2013.

# 14

# Survival, growth and biomass production of *Moringa oleifera* provenances at Gairo inland plateau and Ruvu Coastal Region in Tanzania

**Ezekiel Edward[1], Shabani A. O. Chamshama[1], Yonika M. Ngaga[1] and Mathew A. Mndolwa[2]**

[1]Faculty of Forestry and Nature Conservation, Sokoine University of Agriculture, P. O. Box 3010, Morogoro, Tanzania.
[2]Tanzania Forestry Research Institute-Lushoto Centre, P. O. Box 95, Lushoto, Tanzania.

**Survival, growth and biomass production was studied among 14 *Moringa oleifera* provenances at Gairo inland plateau-Morogoro, and Ruvu Coastal Region in Tanzania employing randomized complete block design with three replications. Growth assessment was done at six months intervals while biomass assessment was only done at 30 months. During final assessment, untransformed survival at Gairo site ranged from 65.33% for Chikwawa/Domasi (Malawi) to 98.67% for Ihumwa (Tanzania) while at Ruvu site, it ranged from 92% for PKM 2 (India) to 100% for Chikwawa/Domasi (Malawi), Mahalapye (Botswana) and Mbololo 472-029/03 (Kenya) provenances. Height ranged from 2.66 m for Makhanga (Malawi) to 5.04 m for Maun (Botswana) at Gairo site and from 4.82 m for Makhanga (Malawi) to 8.16 m for Maun (Botswana) at Ruvu site. Breast height diameter ranged from 2.80 cm for Makhanga (Malawi) to 6.07 cm for Mahalapye (Botswana) at Gairo site and from 5.18 cm for PKM 3 (India) to 8.58 cm for Maun (Botswana) at Ruvu site. Provenances Mahalapye and Maun (Botswana), Ihumwa (Tanzania), Mbololo 472-029/03 (Kenya) and PKM 1 (India) at Gairo site and Maun and Mahalapye (Botswana), Mbololo 472-029/03 (Kenya) and Ihumwa (Tanzania) at Ruvu site are recommended for planting at these and similar sites.**

**Key words:** *Moringa*, survival, growth, biomass production, Tanzania.

## INTRODUCTION

*Moringa* is the only genus in the family Moringaceae. This genus comprises 13 species, all of which are trees that grow in tropical and sub-tropical climates. *Moringa* is drought resistant and can be grown in a wide variety of poor soils, even barren ground, with soil pH between 4.5 and 9.0. The most popular species is *Moringa oleifera*, a multi-purpose tree originally from India and now found in most tropical countries (Africa, Asia and America). Some records indicate that the species is native to the sub-Himalayan tracts of India, Pakistan, Bangladesh and Afghanistan (Fahey, 2005). Fahey (2005) further indicated that the tree species is locally known by many names like horseradish tree, drumstick tree, benzolive tree, kelor, marango, mlonge, moonga, mulangay, nébéday, saijhan, sajna or ben oil tree. The species is easy to reproduce and its growth is very fast that have raised growing international interest due to its social, economic and environmental importance which can benefit humans and animals nutritionally, economically and as an energy source.

*Moringa* is an important food source in many countries. In India, *Moringa* pods are widely consumed and

**Table 1.** Sources of *M. oleifera* provenances planted at Gairo and Ruvu sites, Tanzania.

| Provenance | Collection site/country/state |
|---|---|
| Chikwawa/Domasi | Chikwawa/Domasi, Malawi |
| Makhanga | Makhanga, Malawi |
| Ngabu | Ngabu, Malawi |
| Mangochi | Mangochi, Malawi |
| Mtakataka/Dedza | Mtakataka/Dedza, Malawi |
| Mahalapye | Mahalapye, Botswana |
| Maun | Maun, Botswana |
| PKM 1 | India |
| PKM 2 | India |
| PKM 3 | India |
| Jafna/Jaffana | India |
| Honduras | Honduras |
| Mbololo 472-029/03 | Mbololo, Kenya |
| Ihumwa | Ihumwa, Tanzania |

Plantations exist to produce pods for export, fresh and tinned, to overseas consumers. In West Africa, *Moringa oleifera* leaves are commonly used to make sauces. *Moringa stenopetala* leaves are the staple food of the Konso people in Ethiopia. Studies have shown *M. oleifera* leaves to be an excellent source of vitamins, minerals and protein: perhaps more than any other tropical vegetable. Many programmes use *Moringa* leaves to fight against malnutrition and its associated diseases including blindness. Fragrant/tender flowers of *Moringa* are also used as a source of vegetables and bee forage (Mbuya et al., 1994), twigs with leaves as fodder, green leaves as mulch and solid wood as energy source (FAO, 1988; Morton, 1991; D'Souza and Kulkarmi, 1993).

*Moringa* seeds contain a cationic polyelectrolyte that has proved efficient in water treatment, as a substitute to aluminium sulphate and other flocculents (Evans, 1991; Mayer and Stelz, 1993; Olayemi and Alabi, 1994; Folkard and Sutherland, 1996). Oil extracted from the seeds is an excellent edible vegetable oil and is also useful in the cosmetics industry. The press cake left after oil extraction from the seeds can be used as a soil conditioner or as fertilizer and forage, it has the potential use as protein supplement for livestock (ruminants) (Sá nchez et al., 2006) and poultry (Evans, 1991; Olayemi and Alabi, 1994; Folkard and Sutherland, 1996).

Although *M. Oleifera* is wide spread in many African countries its original sources is unknown (Jahn, 1988a, b, c; Folkard and Sutherland, 1996; Panga, 2002). The material therefore, is most likely of narrow genetic base. Also, the knowledge about genetic improvement and proper management of this valuable multipurpose tree species is limited. Given dangers of planting materials based on a narrow genetic base, there is a need to evaluate different provenances of this species with a view of identifying the provenances that maximize productivity in various parts of Tanzania. In addition to increasing wood biomass productivity, pod, seed and leaf quality, broadening the genetic base will serve as an insurance against pests, diseases and climatic fluctuations. This study therefore evaluated *M. oleifera* provenances in terms of survival, height, diameter growth and biomass production.

## MATERIALS AND METHODS

### The study sites

The study was carried out at two contrasting sites: Gairo and Ruvu sites with established replicate trials in each. Gairo site is located in Gairo District, Morogoro Region (36° 45' E; 6° 0° S; 1 300 m a. s. l.) along the Morogoro-Dodoma highway about 130 km from Morogoro town and 140 km from Dodoma. Rainfall is poorly distributed, and varies from year to year. The average annual rainfall is around 499 mm, most of which falls between November and May (Herbert et al., 2002). The geology of the site is Usagaran system. The soil is generally classified as Haplic Lixisols (Msanya and Msaky, 1994). The soil properties of the study site are as described by Mugasha et al. (2000). The soil has low inherent fertility. The soil texture is sandy clay loam with pH in the upper 50 cm soil depth varying from 6.1-6.3, total nitrogen 0.11-0.16% and Bray I available phosphorus 0.18-3.38 ug/g. The natural vegetation found around the site consists mainly of shrubs and few scattered miombo tree species.

Ruvu site is located at Kibaha District, Coast Region (6° 33' - 6° 43'; 38° 48' S - 39v 03' E) some 60 km from Dar-es-Salaam city along Dar-es-Salaam - Morogoro highway. It is within low elevations some 80 m a. s. l. The area has mean rainfall of 900 mm per annum falling in average of 81 days (Maghembe, 1979) with irregular pattern. Heavy rains fall between March and May and light rains between November - December. Temperatures are always high tending to be the highest in January and on average the temperature ranges between 23 and 27°C having occasional minima as low as 18°C and maxima of 33°C. Soils vary substantially over short distances. They are free draining, primarily sandy, sandy loam and gravel. Soil pH varies with soil depth. The top 50 cm has pH predominantly ranging from 5 - 6.5, increasing with increase in soil depth reaching extremes of pH 9.4 at 72 cm below surface (Holmes, 1988). The common natural vegetation found at Ruvu can be described as semi deciduous open to partly closed woodland to fairly open evergreen thicket and riverine forest. Main genera include, *Pteleopsis, Strichnos, Combretum, Albizia, Brachystegia, Jubernadia, Vitex, Afzelia* and *Dalbergia*.

### Source of provenances

Fourteen (14) *M. oleifera* provenances were studied in each site (Table 1). Potted seedlings of the provenances were raised at Gairo and Ruvu nurseries using standard cultural techniques (Forest Division, 1982).

### Experimental design

Trials were planted in a randomised complete block design (RCBD) with three replications. Each plot represented a provenance planted at 5 x 5 rows in a contiguous arrangement at an espacement of 2 x 2 m. Plots measured 8 x 8 m and contained 25 trees. Distances between blocks were 4 m. Each block had one buffer row planted at the same spacing.

### Field procedures

Sites were prepared by clearing all vegetation using hand hoes followed by ploughing and pitting (pit size: 30 x 30 cm). Planting was done in January and March 2006 at Gairo and Ruvu sites respectively. Weeding was done three times during the rainy season and once during the dry season.

### Data collection

Assessments were carried out at ages of 6, 12, 18, 24 and 30 months after planting. During the first assessment, survival, root collar diameter 30 cm above ground (RCD) and height (HT) were assessed while in subsequent assessments diameter at breast height (Dbh) was also measured. Biomass production was measured at final assessment. Height was measured using calibrated height measuring pole while RCD and Dbh were measured to the nearest 0.01 cm using a veneer calliper.

### Data analysis

All statistical analyses were carried out using General Linear Model (GLM) of Statistical Analysis Systems (SAS) (SAS Inst. Inc., 1991). For all statistical analysis, a fixed effect model was fitted (equation 1) and a type III SS analysis was carried out. All data were subjected to analysis of variance (ANOVA) using plot means. Data for percentage survival was arcsine transformed prior to analysis to remove bias (Sokal and Rohlf, 1969). For significantly different provenance means, the Duncan's multiple range test (DMRT) was used for separating means (Gomez and Gomez, 1983) at 5% probability level.

The general linear model was denoted as:

$$Y = \text{replication} + \text{provenance} + \text{error} \quad (1)$$

Where, Y = the measurement

To identify the best and the worst overall performing provenance at final assessment, ordinal ranking was performed. This was done as follows; for each significant variable evaluated, provenances were assigned ranks from the best (assigned 1 point) to the worst (assigned 14 points) performing provenance. Thereafter, ranks were added, averaged, and the overall score was taken as a basis of the overall provenance ranking.

## RESULTS

### Survival

The results for tree survival for different *M. oleifera* provenances planted at Gairo and Ruvu sites are presented in Tables 2 and 3, respectively. The differences in survival were significant (P<0.05) throughout the assessment period at Gairo site while at Ruvu site provenances did not differ significantly (P>0.05).

During the last assessment occasion (30 months after planting), untransformed survival at Gairo site ranged from 61.33% for Chikwawa/Domasi (Malawi) to 98.67% for Ihumwa (Tanzania) while at Ruvu site untransformed survival ranged from 92% for PKM 2 (India) to 100% for Chikwawa/Domasi (Malawi), Mahalapye (Botswana) and Mbololo 472-029/03 (Kenya).

### Root collar diameter

The results of tree RCD for different *M. oleifera* provenances planted at Gairo and Ruvu sites are presented in Table 4. Provenances differed significantly (P<0.05) in RCD at both sites in all months of assessment. RCD ranged between 5.36 cm for Mangochi (Malawi) and 8.91 cm for Mahalapye (Botswana) at Gairo site and 7.43 cm for PKM 3 (India) and 10.67cm for Maun (Botswana) at Ruvu site during the final assessment occasion.

### Diameter at breast height

The results of tree Dbh for both sites are presented in Table 5. Provenances differed significantly (P<0.05) in Dbh at both sites in all months of assessment. Dbh ranged between 2.86 cm for Makhanga (Malawi) and 6.07 cm for Mahalapye (Botswana) at Gairo site and 5.28 cm for PKM 3 (India) and 8.58 cm for Maun (Botswana) at Ruvu site during the final assessment occasion.

### Height growth

Trends for height growth of *M. oleifera* provenances for Gairo and Ruvu sites are as shown in Table 6. Significant variation (P<0.05) in height development was observed at all assessment occasions. The mean height ranged between 2.66 m for Makhanga (Malawi) and 5.04 m for Maun (Botswana) at Gairo site and 4.82 m for Makhanga (Malawi) and 8.16 m for Maun (Botswana) at Ruvu site during the final assessment occasion (30 months after planting).

It is interesting to note that at many assessment occasions in both sites, Maun (Botswana) showed superiority in height growth even though no one single provenance maintained one rank throughout the period. Makhanga (Malawi) provenance showed poor height growth at all assessment occasions in both sites.

### Biomass production

The results for stem and foliar biomass for *M. oleifera* provenances at Gairo and Ruvu sites are presented in Table 7. Provenances differed significantly (P<0.05) in biomass production at Gairo site during the last assessment occasion while at Ruvu site provenances did not differ significantly (P>0.05) in biomass production. Stem and foliar biomass at Gairo site ranged from 5 901.08 kg ha$^{-1}$ for Mangochi (Malawi) to 24 733.60 kg ha$^{-1}$ for PKM 2 (India) and 491.02 kg ha$^{-1}$ for Ngabu (Malawi) to 4 407.18 kg ha$^{-1}$ for Maun (Botswana), respectively. At

**Table 2.** Survival and transformed survival of 6, 12, 18, 24 and 30 months old *M. oleifera* provenances planted at Gairo, Morogoro, Tanzania.

| Species/provenances | Age (months) | | | | | | | | | |
|---|---|---|---|---|---|---|---|---|---|---|
| | 6 | | 12 | | 18 | | 24 | | 30 | |
| | UTSv | TSv | UTSv | TSv | UTSv | TSv | UTSv | TSv | UTSv | TSv |
| Chikwawa/Domasi - Malawi | *61.33 (10.41) | $51.92^{d}$ | 61.33 | $51.92^{d}$ | 61.33 | $51.92^{d}$ | 61.33 | $51.92^{d}$ | 61.33 | $51.92^{e}$ |
| | | (6.35) | (10.41) | (6.35) | (10.41) | (6.35) | (10.41) | (6.35) | (10.41) | (6.35) |
| Makhanga - Malawi | 85.33 | $67.63^{c}$ | 85.33 | $67.63^{cd}$ | 84.00 | $66.53^{cd}$ | 82.67 | $65.43^{cd}$ | 78.67 | $62.59^{de}$ |
| | (2.67) | (2.10) | (2.67) | (2.10) | (2.31) | (1.82) | (1.33) | (0.99) | (2.67) | (1.92) |
| Ngabu - Malawi | 89.33 | $71.82^{bc}$ | 88.00 | $70.19^{bc}$ | 88.00 | $70.19^{bcd}$ | 88.00 | $70.19^{cd}$ | 85.33 | $67.99^{bcde}$ |
| | (4.81) | (4.42) | (4.00) | (3.38) | (4.00) | (3.38) | (4.00) | (3.38) | (4.81) | (3.82) |
| Mangochi - Malawi | 86.67 | $73.86^{abc}$ | 85.33 | $72.23^{bc}$ | 84.00 | $71.45^{bcd}$ | 80.00 | $65.55^{cd}$ | 80.00 | $65.55^{cde}$ |
| | (11.39) | (10.89) | (10.91) | (10.66) | (12.22) | (11.37) | (12.22) | (8.91) | (12.22) | (8.91) |
| Mtakataka/Dedza - Malawi | 97.33 | $82.31^{abc}$ | 97.33 | $82.31^{abc}$ | 97.33 | $82.31^{abc}$ | 97.33 | $82.31^{abc}$ | 96.00 | $78.46^{abcd}$ |
| | (1.33) | (3.85) | (1.33) | (3.85) | (1.33) | (3.85) | (1.33) | (3.85) | (0.00) | (0.00) |
| Mahalapye - Botswana | 96.00 | $78.46^{abc}$ | 96.00 | $78.46^{abc}$ | 96.00 | $78.46^{abc}$ | 96.00 | $78.46^{abc}$ | 96.00 | $78.46^{abcd}$ |
| | (0.00) | (0.00) | (0.00) | (0.00) | (0.00) | (0.00) | (0.00) | (0.00) | (0.00) | (0.00) |
| Maun - Botswana | 98.67 | $86.15^{ab}$ | 98.67 | $86.15^{ab}$ | 98.67 | $86.15^{ab}$ | 98.67 | $86.15^{ab}$ | 97.33 | $84.52^{ab}$ |
| | (1.33) | (3.85) | (1.33) | (3.85) | (1.33) | (3.85) | (1.33) | (3.85) | (2.67) | (5.48) |
| PKM 1 - India | 90.67 | $75.67^{abc}$ | 90.67 | $75.67^{abc}$ | 90.67 | $75.67^{abc}$ | 90.67 | $75.67^{abc}$ | 90.67 | $75.67^{abcd}$ |
| | (5.81) | (7.74) | (5.81) | (7.74) | (5.81) | (7.74) | (5.81) | (7.74) | (5.81) | (7.74) |
| PKM 2 - India | 98.67 | $86.15^{ab}$ | 98.67 | $86.15^{ab}$ | 98.67 | $86.15^{ab}$ | 98.67 | $86.15^{ab}$ | 97.33 | $84.52^{ab}$ |
| | (1.33) | (3.85) | (1.33) | (3.85) | (1.33) | (3.85) | (1.33) | (3.85) | (2.67) | (5.48) |
| PKM 3 - India | 93.33 | $78.29^{abc}$ | 93.33 | $78.29^{abc}$ | 93.33 | $78.29^{abc}$ | 93.33 | $78.29^{abc}$ | 93.33 | $78.29^{abcd}$ |
| | (4.81) | (6.81) | (4.81) | (6.81) | (4.81) | (6.81) | (4.81) | (6.81) | (4.81) | (6.81) |
| Jafna/Jaffana - India | 98.67 | $86.15^{ab}$ | 98.67 | $86.15^{ab}$ | 98.67 | $86.15^{ab}$ | 98.67 | $86.15^{ab}$ | 97.33 | $84.52^{ab}$ |
| | (1.33) | (3.85) | (1.33) | (3.85) | (1.33) | (3.85) | (1.33) | (3.85) | (2.67) | (5.48) |
| Honduras - Honduras | 92.00 | $73.92^{abc}$ | 90.67 | $72.29^{bc}$ | 90.67 | $72.29^{abc}$ | 89.33 | $71.01^{bcd}$ | 88.00 | $69.73^{abcde}$ |
| | (2.31) | (2.53) | (1.33) | (1.28) | (1.33) | (1.28) | (1.33) | (1.28) | (0.00) | (0.00) |
| Mbololo 472-029/03 - Kenya | 96.00 | $83.24^{abc}$ | 96.00 | $83.24^{ab}$ | 96.00 | $83.24^{ab}$ | 94.67 | $79.40^{abc}$ | 94.67 | $79.40^{abcd}$ |
| | (4.00) | (6.76) | (4.00) | (6.76) | (4.00) | (6.76) | (3.53) | (5.87) | (3.53) | (5.87) |
| Ihumwa - Tanzania | 100.00 | $90.00^{a}$ | 100.00 | $90.00^{a}$ | 98.67 | $86.15^{ab}$ | 98.67 | $86.15^{ab}$ | 98.67 | $86.15^{a}$ |
| | (0.00) | (0.00) | (0.00) | (0.00) | (1.33) | (3.85) | (1.33) | (3.85) | (1.33) | (3.85) |
| P>F Ratio | 0.0011 | 0.0038 | 0.0004 | 0.0013 | 0.0013 | 0.0032 | 0.0016 | 0.0008 | 0.0006 | 0.0032 |
| RMSE | 8.4784 | 9.2076 | 7.26 | 8.59 | 7.70 | 8.90 | 7.48 | 7.95 | 7.96 | 9.08 |
| CV | 9.24 | 11.87 | 7.90 | 11.02 | 8.33 | 11.31 | 8.19 | 10.41 | 8.84 | 12.07 |

*Means of individual species/provenance with standard error in parenthesis. Means of the same letter within the same column are not significantly different. UTSv, untransformed survival; TSv, transformed survival.

**Table 3.** Survival and transformed survival of 6, 12, 18, 24 and 30 months old *M. oleifera* provenances planted at Ruvu, Coast Region, Tanzania.

| Species/provenances | Age (months) | | | | | | | | | |
|---|---|---|---|---|---|---|---|---|---|---|
| | 6 | | 12 | | 18 | | 24 | | 30 | |
| | UTSv | TSv | UTSv | TSv | UTSv | TSv | UTSv | TSv | UTSv | TSv |
| Chikwawa/Domasi - Malawi | * 100.00 | 90.00[a] | 100.00 | 90.00[a] | 100.00 | 90.00[a] | 100.00 | 90.00[a] | 100.00 | 90.00[a] |
| | (0.00) | (0.00) | (0.00) | (0.00) | (0.00) | (0.00) | (0.00) | (0.00) | (0.00) | (0.00) |
| Makhanga - Malawi | 98.67 | 86.15[a] | 98.67 | 86.15[a] | 98.67 | 86.15[a] | 97.33 | 84.52[a] | 96.00 | 80.00[a] |
| | (1.33) | (3.85) | (1.33) | (3.85) | (1.33) | (3.85) | (2.67) | (5.48) | (2.31) | (4.87) |
| Ngabu - Malawi | 98.67 | 86.15[a] | 98.67 | 86.15[a] | 98.67 | 86.15[a] | 98.67 | 86.15[a] | 98.67 | 86.15[a] |
| | (1.33) | (3.85) | (1.33) | (3.85) | (1.33) | (3.85) | (1.33) | (3.85) | (1.33) | (3.85) |
| Mangochi - Malawi | 96.00 | 80.68[a] | 96.00 | 80.68[a] | 96.00 | 80.68[a] | 96.00 | 80.68[a] | 96.00 | 80.68[a] |
| | (2.31) | (4.87) | (2.31) | (4.87) | (2.31) | (4.87) | (2.31) | (4.87) | (2.31) | (4.87) |
| Mtakataka/Dedza - Malawi | 100.00 | 90.00[a] | 100.00 | 90.00[a] | 100.00 | 90.00[a] | 100.00 | 90.00[a] | 98.67 | 86.15[a] |
| | (0.00) | (0.00) | (0.00) | (0.00) | (0.00) | (0.00) | (0.00) | (0.00) | (1.33) | (3.85) |
| Mahalapye - Botswana | 100.00 | 90.00[a] | 100.00 | 90.00[a] | 100.00 | 90.00[a] | 100.00 | 90.00[a] | 100.00 | 90.00[a] |
| | (0.00) | (0.00) | (0.00) | (0.00) | (0.00) | (0.00) | (0.00) | (0.00) | (0.00) | (0.00) |
| Maun - Botswana | 98.67 | 86.15[a] | 98.67 | 86.15[a] | 98.67 | 86.15[a] | 96.00 | 80.68[a] | 93.33 | 77.77[a] |
| | (1.33) | (3.85) | (1.33) | (3.85) | (1.33) | (3.85) | (2.31) | (4.87) | (3.53) | (6.22) |
| PKM 1 - India | 100.00 | 90.00[a] | 100.00 | 90.00[a] | 100.00 | 90.00[a] | 97.33 | 84.52[a] | 96.00 | 80.68[a] |
| | (0.00) | (0.00) | (0.00) | (0.00) | (0.00) | (0.00) | (2.67) | (5.48) | (2.31) | (4.87) |
| PKM 2 - India | 96.00 | 80.68[a] | 96.00 | 80.68[a] | 96.00 | 80.68[a] | 94.67 | 79.40[a] | 92.00 | 79.40[a] |
| | (2.31) | (4.87) | (2.31) | (4.87) | (2.31) | (4.87) | (3.53) | (5.87) | (2.31) | (5.87) |
| PKM 3 - India | 97.33 | 82.31[a] | 97.33 | 82.31[a] | 97.33 | 82.31[a] | 97.33 | 82.31[a] | 97.33 | 82.31[a] |
| | (1.33) | (3.85) | (1.33) | (3.85) | (1.33) | (3.85) | (1.33) | (3.85) | (1.33) | (3.85) |
| Jafna/Jaffana - India | 100.00 | 90.00[a] | 98.67 | 86.15[a] | 98.67 | 86.15[a] | 96.00 | 78.46[a] | 96.00 | 78.46[a] |
| | (0.00) | (0.00) | (1.33) | (3.85) | (1.33) | (3.85) | (0.00) | (0.00) | (0.00) | (0.00) |
| Honduras - Honduras | 96.00 | 80.68[a] | 96.00 | 80.68[a] | 96.00 | 80.68[a] | 96.00 | 80.68[a] | 94.67 | 79.40[a] |
| | (2.31) | (4.87) | (2.31) | (4.87) | (2.31) | (4.87) | (2.31) | (4.87) | (3.53) | (5.87) |
| Mbololo 472-029/03 - Kenya | 100.00 | 90.00[a] | 100.00 | 90.00[a] | 100.00 | 90.00[a] | 100.00 | 90.00[a] | 100.00 | 90.00[a] |
| | (0.00) | (0.00) | (0.00) | (0.00) | (0.00) | (0.00) | (0.00) | (0.00) | (0.00) | (0.00) |
| Ihumwa - Tanzania | 98.67 | 86.15[a] | 98.67 | 86.15[a] | 98.67 | 86.15[a] | 98.67 | 86.15[a] | 98.67 | 86.15[a] |
| | (1.33) | (3.85) | (1.33) | (3.85) | (1.33) | (3.85) | (1.33) | (3.85) | (1.33) | (3.85) |
| P>F Ratio | 0.1593 | 0.188 | 0.69 | 0.75 | 0.32 | 0.38 | 0.28 | 0.20 | 0.0251 | 0.1963 |
| RMSE | 2.23 | 5.48 | 3.01 | 5.92 | 1.91 | 5.09 | 2.72 | 6.01 | 2.79 | 6.38 |
| CV | 2.26 | 6.34 | 3.05 | 6.83 | 1.93 | 5.82 | 2.78 | 7.10 | 2.87 | 7.64 |

*Means of individual species/provenance with standard error in parenthesis. Means of the same letter within the same column are not significantly different. UTSv, untransformed survival; TSv, transformed survival.

Ruvu site stem and foliar biomass ranged from 15 796.17 kg ha$^{-1}$ for Makhanga (Malawi) to 41 403.20 kg ha$^{-1}$ for Mahalapye (Botswana) and 697.10 kg ha$^{-1}$ for Makhanga (Malawi) to 1 917.21 kg ha$^{-1}$ for Maun (Botswana) respectively.

### Ordinal ranking

When the ranking of provenances in six tree variables (survival, mean height, RCD, Dbh, stem biomass and foliage biomass production) was computed, provenances were ranked as shown in Tables 8 and 9 for Gairo and Ruvu sites, respectively.

The most outstanding provenances were Mahalapye (Botswana), Maun (Botswana), Ihumwa (Tanzania), Mbololo 472-029/03 (Kenya) and PKM 2 (India) for Gairo site and Maun (Botswana), Mahalapye (Botswana), Mbololo 472-029/03 (Kenya) and Ihumwa (Tanzania) for Ruvu site.

## DISCUSSION

### Survival

General assessment at all occasions, showed good performance among provenances. All provenances in both sites except two at Gairo site had survival assessments above 80%. The significant differences in survival at Gairo site could be attributable to provenances difference in tolerance to arid/semi arid conditions of the study site (Edward et al., 2006). In this case provenances from Ihumwa (Tanzania), Mtakataka/Dedza (Malawi), Maun (Botswana), PKM 2 (India) and Jafna/Jaffana (India) appear to tolerate dry conditions than Chikwawa/-Domasi (Malawi). The excellent performance of these provenances is an indication of better adaptation to the site condition while the insignificant differences among provenances at Ruvu site implies that all provenances have adapted well in the sub-humid conditions of this

**Table 4.** Root collar diameter of 6, 12, 18, 24 and 30 months old *Moringa oleifera* provenances planted at Gairo site, Morogoro and Ruvu site, Coast Region, Tanzania.

| Species/provenances | Gairo site | | | | | Ruvu site | | | | |
|---|---|---|---|---|---|---|---|---|---|---|
| | Age (months) | | | | | | | | | |
| | 6 | 12 | 18 | 24 | 30 | 6 | 12 | 18 | 24 | 30 |
| Chikwawa/Domasi - Malawi | 3.12$^{c}$ | 4.25$^{cde}$ | 5.67$^{de}$ | 6.12$^{de}$ | 6.63$^{defg}$ | 2.69$^{def}$ | 6.45$^{cde}$ | 8.40$^{cde}$ | 8.54$^{cde}$ | 8.81$^{bcd}$ |
| | (0.07) | (0.12) | (0.20) | (0.13) | (0.53) | (0.21) | (0.33) | (0.34) | (0.12) | (0.48) |
| Makhanga - Malawi | 2.14$^{c}$ | 3.16f | 4.88$^{ef}$ | 5.28$^{e}$ | 5.73$^{fg}$ | 2.30f | 5.88$^{e}$ | 7.77$^{de}$f | 8.17$^{de}$ | 8.38$^{cd}$ |
| | (0.16) | (0.18) | (0.47) | (0.45) | (0.56) | (0.24) | (0.52) | (0.89) | (0.74) | (1.15) |
| Ngabu - Malawi | 2.41$^{de}$ | 3.51$^{ef}$ | 4.75$^{ef}$ | 5.46$^{e}$ | 6.19$^{efg}$ | 2.79$^{cdef}$ | 6.73$^{bcde}$ | 8.59$^{bcd}$ | 8.84$^{bcd}$ | 8.98$^{bcd}$ |
| | (0.21) | (0.34) | (0.20) | (0.29) | (0.22) | (0.16) | (0.08) | (0.12) | (0.04) | (0.11) |
| Mangochi - Malawi | 1.89$^{e}$ | 3.09$^{f}$ | 4.59f | 5.15$^{e}$ | 5.36g | 2.44f | 6.58$^{cde}$ | 8.65$^{bcd}$ | 8.84$^{bcd}$ | 8.93$^{bcd}$ |
| | (0.04) | (0.05) | (0.08) | (0.10) | (0.11) | (0.52) | (0.38) | (0.50) | (0.63) | (0.65) |
| Mtakataka/Dedza - Malawi | 2.82$^{cd}$ | 4.10$^{de}$ | 6.35$^{cd}$ | 6.82$^{bcd}$ | 7.48$^{bcde}$ | 2.51$^{e}$f | 7.07$^{bcd}$ | 9.36$^{abc}$ | 9.57$^{abc}$ | 10.12$^{ab}$ |
| | (0.20) | (0.03) | (0.27) | (0.11) | (0.34) | (0.22) | (0.40) | (0.46) | (0.47) | (0.44) |
| Mahalapye - Botswana | 4.01$^{a}$ | 6.11$^{a}$ | 7.94$^{a}$ | 8.31$^{a}$ | 8.91$^{a}$ | 3.40$^{abcd}$ | 7.29$^{abc}$ | 9.19$^{bc}$ | 9.34$^{abc}$ | 9.38$^{abc}$ |
| | (0.08) | (0.12) | (0.21) | (0.15) | (0.28) | (0.28) | (0.49) | (0.52) | (0.43) | (0.17) |
| Maun - Botswana | 3.71$^{ab}$ | 5.44$^{ab}$ | 7.56$^{ab}$ | 8.07$^{a}$ | 8.46$^{ab}$ | 3.82$^{ab}$ | 8.11$^{a}$ | 10.24$^{a}$ | 10.54$^{a}$ | 10.67$^{a}$ |
| | (0.28) | (0.31) | (0.43) | (0.50) | (0.33) | (0.23) | (0.32) | (0.27) | (0.41) | (0.24) |
| PKM 1 - India | 4.11$^{a}$ | 5.33$^{ab}$ | 7.16$^{abc}$ | 7.72$^{ab}$ | 7.79$^{abcd}$ | 3.93$^{a}$ | 7.04$^{bcd}$ | 8.58$^{bcd}$ | 8.70$^{bcd}$ | 8.70$^{bcd}$ |
| | (0.37) | (0.52) | (0.63) | (0.50) | (0.84) | (0.19) | (0.32) | (0.39) | (0.36) | (0.45) |
| PKM 2 - India | 3.80$^{ab}$ | 5.24$^{b}$ | 6.94$^{abc}$ | 7.46$^{abc}$ | 7.90$^{abcd}$ | 3.18$^{bcde}$ | 6.06$^{de}$ | 7.74$^{def}$ | 7.77$^{de}$ | 8.13$^{cd}$ |
| | (0.18) | (0.11) | (0.33) | (0.30) | (0.18) | (0.04) | (0.26) | (0.30) | (0.28) | (0.35) |
| PKM 3 - India | 3.98$^{a}$ | 5.26$^{b}$ | 6.86$^{abc}$ | 7.37$^{abc}$ | 7.54$^{bcde}$ | 3.24$^{abcd}$ | 6.10$^{de}$ | 7.15$^{e}$ | 7.35$^{d}$ | 7.43$^{e}$f |
| | (0.14) | (0.25) | (0.17) | (0.12) | (0.14) | (0.31) | (0.32) | (0.23) | (0.25) | (0.35) |
| Jafna/Jaffana - India | 3.86$^{ab}$ | 4.68$^{bcd}$ | 6.54$^{bcd}$ | 6.75$^{bcd}$ | 6.88$^{def}$ | 3.47$^{abc}$ | 6.32$^{cde}$ | 7.82$^{def}$ | 7.96$^{de}$ | 8.03$^{cd}$ |
| | (0.06) | (0.16) | (0.13) | (0.34) | (0.27) | (0.06) | (0.25) | (0.27) | (0.41) | (0.35) |
| Honduras - Honduras | 2.85$^{cd}$ | 4.07$^{de}$ | 6.15$^{cd}$ | 6.59$^{cd}$ | 7.03$^{cdef}$ | 2.77$^{cdef}$ | 5.68$^{e}$ | 7.35$^{f}$ | 7.43$^{de}$ | 7.77$^{cd}$ |
| | (0.21) | (0.38) | (0.55) | (0.45) | (0.63) | (0.57) | (0.37) | (0.19) | (0.20) | (0.13) |
| Mbololo 472-029/03 - Kenya | 3.07$^{c}$ | 4.69$^{bcd}$ | 7.14$^{abc}$ | 7.59$^{abc}$ | 8.32$^{abc}$ | 3.78$^{ab}$ | 8.14$^{a}$ | 9.79$^{ab}$ | 9.80$^{ab}$ | 10.08$^{ab}$ |
| | (0.16) | (0.35) | (0.37) | (0.44) | (0.29) | (0.11) | (0.04) | (0.12) | (0.09) | (0.39) |
| Ihumwa - Tanzania | 3.31$^{bc}$ | 5.01$^{bc}$ | 6.86$^{abc}$ | 7.46$^{abc}$ | 8.00$^{abcd}$ | 3.40$^{abcd}$ | 7.67$^{ab}$ | 9.66$^{abc}$ | 9.67$^{abc}$ | 10.13$^{ab}$ |
| | (0.14) | (0.11) | (0.25) | (0.22) | (0.22) | (0.07) | (0.25) | (0.24) | (0.16) | (0.26) |
| P>F Ratio | <0.0001 | <0.0001 | <0.0001 | <0.0001 | <0.0001 | <0.0001 | <0.0001 | <0.0001 | <0.0001 | 0.0008 |
| RMSE | 0.3327 | 0.46 | 0.58 | 0.55 | 0.72 | 0.38 | 0.55 | 0.62 | 0.72 | 0.84 |
| CV | 10.33 | 10.09 | 9.15 | 8.02 | 9.83 | 12.18 | 8.05 | 7.13 | 8.35 | 9.36 |

*Means of individual species/provenance with standard error in parenthesis. -Means of the same letter within the same column are not significantly different.

site.

## Root collar diameter

Significant variation (P<0.05) in RCD development was observed at all sites during the whole assessment period. At Gairo site, the study indicates the superior RCD development of Mahalapye (Botswana) and Maun (Botswana), Mbololo 472-029/03 (Kenya) and Ihumwa (Tanzania) and poor RCD development of Mangochi (Malawi) and Makhanga (Malawi). Similar results have been observed in Zimbabwe (Gadzirayi et al., 2013; Goss, 2012). At Ruvu site, the study indicates the superior RCD development of Maun (Botswana), Ihumwa (Tanzania), Mtakataka/Dedza (Malawi) and Mbololo 472-029/03 (Kenya) and poor RCD development of PKM 3 (India) and Honduras (Honduras). The differences in RCD development within a site could be attributed to variations in adaptability among provenances, but generally the provenances at Ruvu site showed good performance in RCD development, which is expected due to the sub-humid conditions of the site.

## Diameter at breast height

Significant variation (P<0.05) in Dbh was observed on all

**Table 5.** Diameter at breast height of 12, 18, 24 and 30 months old *M. oleifera* provenances planted at Gairo site, Morogoro and Ruvu site, Coast Region, Tanzania.

| Species/provenances | Gairo site | | | | Ruvu site | | | |
|---|---|---|---|---|---|---|---|---|
| | Age (months) | | | | | | | |
| | 12 | 18 | 24 | 30 | 12 | 18 | 24 | 30 |
| Chikwawa/Domasi - Malawi | $2.27^{de}$ (0.09) | $3.38^{ef}$ (0.23) | $3.52^{def}$ (0.16) | $3.87^{efg}$ (0.34) | $3.93^{ef}$ (0.33) | $5.66^{cde}$ (0.35) | $5.84^{def}$ (0.25) | $6.30^{cde}$ (0.38) |
| Makhanga - Malawi | $1.89^{f}$ (0.09) | $2.80^{f}$ (0.39) | $2.84^{f}$ (0.28) | $2.86^{h}$ (0.23) | $3.69^{f}$ (0.34) | $4.85^{e}$ (0.43) | $5.48^{ef}$ (0.71) | $6.05^{cde}$ (0.91) |
| Ngabu - Malawi | $1.95^{ef}$ (0.06) | $2.70^{f}$ (0.16) | $2.86^{f}$ (0.11) | $3.11^{gh}$ (0.11) | $4.74^{cde}$ (0.06) | $5.74^{cde}$ (0.17) | $6.23^{cdef}$ (0.25) | $6.35^{cde}$ (0.28) |
| Mangochi - Malawi | $1.83^{f}$ (0.04) | $3.02^{f}$ (0.07) | $3.11^{ef}$ (0.07) | $3.22^{gh}$ (0.16) | $5.06^{bcd}$ (0.41) | $6.31^{bc}$ (0.64) | $6.69^{bcd}$ (0.47) | $6.98^{bcd}$ (0.65) |
| Mtakataka/Dedza - Malawi | $1.96^{ef}$ (0.21) | $3.37^{ef}$ (0.07) | $3.61^{de}$ (0.09) | $3.74^{fgh}$ (0.04) | $4.87^{bcde}$ (0.43) | $6.38^{bc}$ (0.32) | $7.19^{abc}$ (0.30) | $7.81^{ab}$ (0.27) |
| Mahalapye - Botswana | $3.90^{a}$ (0.22) | $5.61^{a}$ (0.25) | $5.67^{a}$ (0.22) | $6.07^{a}$ (0.10) | $5.71^{abc}$ (0.36) | $7.23^{ab}$ (0.28) | $7.26^{abc}$ (0.40) | $7.32^{abc}$ (0.31) |
| Maun - Botswana | $3.49^{ab}$ (0.18) | $5.09^{ab}$ (0.31) | $5.32^{ab}$ (0.37) | $5.72^{ab}$ (0.28) | $6.21^{a}$ (0.51) | $7.73^{a}$ (0.27) | $8.25^{a}$ (0.27) | $8.58^{a}$ (0.29) |
| PKM 1 - India | $3.30^{b}$ (0.26) | $4.49^{bcd}$ (0.41) | $4.79^{bc}$ (0.36) | $5.01^{bcd}$ (0.64) | $5.48^{abc}$ (0.20) | $6.21^{bcd}$ (0.36) | $6.49^{bcde}$ (0.27) | $6.77^{bcd}$ (0.45) |
| PKM 2 - India | $3.29^{b}$ (0.06) | $4.52^{bcd}$ (0.12) | $4.64^{bc}$ (0.15) | $4.93^{bcd}$ (0.03) | $4.37^{def}$ (0.23) | $5.03^{de}$ (0.06) | $5.65^{def}$ (0.27) | $6.07^{cde}$ (0.18) |
| PKM 3 - India | $3.30^{b}$ (0.19) | $4.53^{bcd}$ (0.25) | $4.67^{bc}$ (0.24) | $4.84^{bcde}$ (0.26) | $4.11^{def}$ (0.33) | $5.18^{cde}$ (0.34) | $5.21^{f}$ (0.29) | $5.28^{e}$ (0.74) |
| Jafna/Jaffana - India | $2.66^{cd}$ (0.22) | $3.99^{cde}$ (0.06) | $4.01^{cd}$ (0.31) | $4.17^{defg}$ (0.16) | $4.76^{cde}$ (0.05) | $5.61^{cde}$ (0.24) | $6.07^{def}$ (0.18) | $6.27^{cde}$ (0.25) |
| Honduras - Honduras | $2.53^{cde}$ (0.15) | $3.82^{de}$ (0.28) | $4.08^{cd}$ (0.25) | $4.21^{cdef}$ (0.41) | $4.18^{def}$ (0.49) | $5.21^{cde}$ (0.31) | $5.61^{def}$ (0.24) | $5.67^{de}$ (0.25) |
| Mbololo 472-029/03 - Kenya | $3.03^{bc}$ (0.36) | $4.59^{bc}$ (0.36) | $4.74^{bc}$ (0.46) | $5.14^{abc}$ (0.44) | $6.27^{a}$ (0.13) | $7.28^{ab}$ (0.11) | $7.81^{a}$ (0.06) | $7.82^{ab}$ (0.22) |
| Ihumwa - Tanzania | $3.00^{bc}$ (0.14) | $4.59^{bcd}$ (0.33) | $4.71^{bc}$ (0.27) | $5.24^{abc}$ (0.34) | $5.81^{ab}$ (0.27) | $6.97^{ab}$ (0.40) | $7.48^{ab}$ (0.31) | $7.87^{ab}$ (0.25) |
| P>F Ratio | <0.0001 | <0.0001 | <0.0001 | <0.0001 | <0.0001 | <0.0001 | <0.0001 | 0.0001 |
| RMSE | 0.33 | 0.42 | 0.39 | 0.55 | 0.52 | 0.64 | 0.60 | 0.73 |
| CV | 11.86 | 10.27 | 9.31 | 12.51 | 10.43 | 10.43 | 9.16 | 10.69 |

*Means of individual species/provenance with standard error in parenthesis. -Means of the same letter within the same column are not significantly different.

sites during the whole assessment period. At Gairo site, the study indicates the superior Dbh development of Mahalapye (Botswana) and Maun (Botswana) and poor Dbh development of Makhanga (Malawi) and Ngabu (Malawi). These results are in agreement with those reported in Zambabwe where Mutoko provenance from Zimbabwe performed very well as compared to Malawi provenance (Gadzirayi et al., 2013). For Ruvu site, the study indicates the superior Dbh development of Maun (Botswana) and Ihumwa (Tanzania) and poor Dbh development of PKM 3 (India) and Honduras - Honduras. The differences in Dbh development could be attributed to variations in adaptability among provenances, but generally the provenances at Ruvu site showed good performance in Dbh development. Good climatic condition, that is, availability of rainfall and genetic superiority could have contributed to the good performance of these provenances at Ruvu site and those from Botswana at Gairo site.

### Height growth

This study has shown superior height growth for Maun (Botswana) and Mahalapye (Botswana) provenances at Gairo site and poor height growth of Makhanga (Malawi), Ngabu (Malawi) and Mtakataka/Dedza (Malawi) provenances. At Ruvu site, only Maun (Botswana) provenance

**Table 6.** Height growth of 6, 12, 18, 24 and 30 months old *M. oleifera* provenances planted at Gairo site, Morogoro and Ruvu site, Coast Region, Tanzania.

| Species/provenances | Gairo site | | | | | Ruvu site | | | | |
|---|---|---|---|---|---|---|---|---|---|---|
| | Age (months) | | | | | | | | | |
| | 6 | 12 | 18 | 24 | 30 | 6 | 12 | 18 | 24 | 30 |
| Chikwawa/Domasi - Malawi | $1.28^{d}$ | $1.74^{c}$ | $2.89^{e}$ | $3.05^{d}$ | $3.12^{e}$ | $2.11^{d}$ | $4.10^{de}$ | $4.74^{de}$ | $4.93^{e}$ | $5.02^{ef}$ |
| | (0.03) | (0.05) | (0.09) | (0.04) | (0.23) | (0.17) | (0.27) | (0.26) | (0.37) | (0.33) |
| Makhanga - Malawi | $1.11^{de}$ | $1.48^{c}$ | $2.57^{e}$ | $2.65^{d}$ | $2.66^{e}$ | $2.08^{d}$ | $3.92^{e}$ | $4.43^{e}$ | $4.48^{f}$ | $4.82^{f}$ |
| | (0.10) | (0.10) | (0.18) | (0.20) | (0.22) | (0.10) | (0.26) | (0.39) | (0.37) | (0.40) |
| Ngabu - Malawi | $1.22^{d}$ | $1.64^{c}$ | $2.76^{e}$ | $2.82^{d}$ | $3.02^{e}$ | $2.63^{c}$ | $4.59^{cde}$ | $5.03^{cde}$ | $5.09^{cdef}$ | $5.25^{cdef}$ |
| | (0.09) | (0.10) | (0.08) | (0.19) | (0.13) | (0.03) | (0.12) | (0.25) | (0.17) | (0.29) |
| Mangochi - Malawi | $0.89^{e}$ | $1.53^{c}$ | $2.56^{e}$ | $2.96^{d}$ | $3.15^{e}$ | $2.00^{d}$ | $4.82^{bcd}$ | $5.77^{bc}$ | $5.85^{bc}$ | $6.16^{bcd}$ |
| | (0.01) | (0.03) | (0.05) | (0.19) | (0.15) | (0.22) | (0.31) | (0.29) | (0.27) | (0.35) |
| Mtakataka/Dedza - Malawi | $1.23^{d}$ | $1.76^{c}$ | $2.88^{e}$ | $2.99^{d}$ | 3.09 | $2.07^{d}$ | $4.38^{cde}$ | $5.23^{cde}$ | $5.32^{bcde}$ | $5.81^{bcdef}$ |
| | (0.07) | (0.04) | (0.16) | (0.09) | (0.11) | (0.13) | (0.37) | (0.37) | (0.34) | (0.28) |
| Mahalapye - Botswana | $2.28^{ab}$ | $3.30^{a}$ | $4.58^{ab}$ | $4.72^{ab}$ | $4.83^{ab}$ | $3.26^{ab}$ | $5.35^{b}$ | $5.98^{b}$ | $6.20^{b}$ | $6.22^{bc}$ |
| | (0.10) | (0.06) | (0.07) | (0.08) | (0.06) | (0.12) | (0.22) | (0.10) | (0.19) | (0.62) |
| Maun - Botswana | $2.15^{b}$ | $3.22^{a}$ | $4.84^{a}$ | $4.93^{a}$ | $5.04^{a}$ | $3.38^{ab}$ | $6.38^{a}$ | $7.46^{a}$ | $7.73^{a}$ | $8.16^{a}$ |
| | (0.12) | (0.13) | (0.09) | (0.14) | (0.17) | (0.20) | (0.39) | (0.52) | (0.50) | (0.57) |
| PKM 1 - India | $2.53^{a}$ | $3.07^{a}$ | $4.14^{c}$ | $4.24^{bc}$ | $4.28^{cd}$ | $3.73^{a}$ | $5.08^{bc}$ | $5.63^{bcd}$ | $5.72^{bcd}$ | $6.06^{bcde}$ |
| | (0.06) | (0.14) | (0.27) | (0.20) | (0.26) | (0.18) | (0.12) | (0.09) | (0.02) | (0.22) |
| PKM 2 - India | $2.40^{ab}$ | $3.04^{a}$ | $4.11^{c}$ | $4.22^{bc}$ | $4.28^{cd}$ | $3.14^{b}$ | $4.43^{cde}$ | $4.93^{de}$ | $5.03^{cdef}$ | $5.26^{cdef}$ |
| | (0.04) | (0.05) | (0.12) | (0.07) | (0.10) | (0.04) | (0.19) | (0.25) | (0.24) | (0.26) |
| PKM 3 - India | $2.49^{a}$ | $2.98^{a}$ | $3.97^{c}$ | $3.99^{c}$ | $4.06^{cd}$ | $3.36^{ab}$ | $4.46^{cde}$ | $5.01^{cde}$ | $5.08^{cdef}$ | $5.12^{cde}$ |
| | (0.14) | (0.03) | (0.13) | (0.13) | (0.07) | (0.19) | (0.12) | (0.03) | (0.02) | (0.04) |
| Jafna/Jaffana - India | $2.48^{a}$ | $2.98^{a}$ | $3.78^{c}$ | $3.79^{c}$ | $3.87^{d}$ | $3.68^{a}$ | $4.48^{cde}$ | $4.88^{de}$ | $5.12^{cdef}$ | $5.16^{def}$ |
| | (0.10) | (0.06) | (0.17) | (0.23) | (0.14) | (0.01) | (0.09) | (0.13) | (0.21) | (0.29) |
| Honduras - Honduras | $1.64^{c}$ | $2.44^{b}$ | $3.44^{d}$ | $3.76^{c}$ | $3.81^{d}$ | $2.64^{c}$ | $4.37^{cde}$ | $4.99^{de}$ | $5.08^{cdef}$ | $5.33^{bcdef}$ |
| | (0.12) | (0.24) | (0.16) | (0.23) | (0.20) | (0.44) | (0.25) | (0.16) | (0.12) | (0.18) |
| Mbololo 472-029/03 - Kenya | $1.87^{c}$ | $2.60^{b}$ | $4.07^{c}$ | $4.22^{bc}$ | $4.47^{bc}$ | $3.50^{ab}$ | $5.42^{b}$ | $6.04^{b}$ | $6.11^{b}$ | $6.30^{b}$ |
| | (0.22) | (0.24) | (0.07) | (0.14) | (0.18) | (0.21) | (0.16) | (0.13) | (0.12) | (0.17) |
| Ihumwa - Tanzania | $1.79^{c}$ | $2.55^{b}$ | $4.10^{c}$ | $4.20^{bc}$ | $4.22^{cd}$ | $3.08^{bc}$ | $5.09^{bc}$ | $5.72^{bcd}$ | $5.81^{bcd}$ | $6.16^{bcd}$ |
| | (0.05) | (0.07) | (0.19) | (0.19) | (0.17) | (0.11) | (0.12) | (0.25) | (0.27) | (0.29) |
| P>F Ratio | <0.0001 | <0.0001 | <0.0001 | <0.0001 | <0.0001 | <0.0001 | <0.0001 | <0.0001 | <0.0001 | <0.0001 |
| RMSE | 0.1656 | 0.20 | 0.22 | 0.30 | 0.28 | 0.28 | 0.38 | 0.46 | 0.45 | 0.53 |
| CV | 9.14 | 8.11 | 6.13 | 8.10 | 7.21 | 9.56 | 7.85 | 8.41 | 8.24 | 9.24 |

*Means of individual species/provenance with standard error in parenthesis. -Means of the same letter within the same column are not significantly different.

showed superior height growth while Makhanga (Malawi) and Chikwawa/Domasi (Malawi) pro-venances showed poor growth in height. The differences in height growth within a site could be attributed to variations in adaptability among provenances while the between site differences in growth relate with rainfall differences between the two sites with Ruvu having higher rainfall as compared to Gairo. Generally, the provenances showed good performance in height development at Ruvu site. The growth performance of outstanding provenances in the present study compare favourably with results reported in Song Hau State farm, Cantho City, Vietnam (Manh et al., 2005) and in Zimbabwe (Gadzirayi et al., 2013; Goss, 2012). The Vietnam site has higher annual rainfall of 1800 - 2000 mm as compared to that of the study sites.

## Biomass

Most provenances grown at the Ruvu site showed high production of foliar and stem biomass than those grown at Gairo site. This could be due to sufficient availability of growth resources (rainfall) at Ruvu site unlike the Gairo site. The lowest yield in stem biomass at Gairo site were recorded from Ngabu (Malawi) and Mangochi (Malawi) while at Ruvu site were recorded for Mangochi (Malawi) and Makhanga (Malawi) provenances. The lowest yield in

**Table 7.** Biomass production of 30 months old *M. oleifera* provenances planted at Gairo site, Morogoro and Ruvu site, Coast Region, Tanzania.

| Species/Provenances | Age (months) | | | |
|---|---|---|---|---|
| | Gairo site | | Ruvu site | |
| | Stems(Kg/ha) | Foliage(Kg/ha) | Stems(Kg/ha) | Foliage (Kg/ha) |
| Chikwawa/Domasi - Malawi | 8235.60[cde] (2173.46) | 1324.10[cde] (378.86) | 21036.74[ab] (4163.94) | 738.19[ab] (107.89) |
| Makhanga - Malawi | 8481.34[cde] (803.95) | 982.04[de] (461.58) | 15796.17[b] (5022.39) | 697.10[b] (187.76) |
| Ngabu - Malawi | 6200.41[de] (1781.16) | 491.02[e] (98.81) | 20858.93[ab] (6500.60) | 1423.19[ab] (542.02) |
| Mangochi - Malawi | 5901.08[e] (1027.08) | 1035.61[de] (104.09) | 15816.98[b] (2015.65) | 972.51[ab] (31.67) |
| Mtakataka/Dedza - Malawi | 11386.00[bcde] (1053.55) | 1723.31[bcde] (598.34) | 17292.03[b] (4111.31) | 1383.46[ab] (243.10) |
| Mahalapye - Botswana | 21212.24[ab] (4618.29) | 3108.98[ab] (590.92) | 41403.26[a] (12290.79) | 1717.77[ab] (587.61) |
| Maun - Botswana | 20285.88[ab] (2910.27) | 4407.18[a] (1227.52) | 34888.16[ab] (4362.86) | 1917.21[a] (205.66) |
| PKM 1 - India | 21012.81[ab] (7635.45) | 2213.56[bcd] (129.57) | 31943.36[ab] (11629.09) | 1865.19[ab] (550.69) |
| PKM 2 - India | 24733.60[a] (1063.24) | 2562.68[bcd] (628.82) | 19061.91[b] (4741.96) | 1506.82[ab] (361.19) |
| PKM 3 - India | 16183.97[abcde] (3687.52) | 2159.61[bcd] (152.48) | 22859.98[ab] (3360.61) | 1484.22[ab] (164.91) |
| Jafna/Jaffana - India | 19981.74[ab] (2759.02) | 2740.89[bc] (552.82) | 24946.83[ab] (1697.49) | 1437.73[ab] (130.37) |
| Honduras - Honduras | 12280.77[bcde] (1835.96) | 1574.84[bcde] (685.94) | 19679.77[b] (2399.60) | 1450.30[ab] (250.07) |
| Mbololo 472-029/03 - Kenya | 17542.76[abcd] (4795.64) | 2280.33[bcd] (599.18) | 28120.39[ab] (8479.69) | 1398.37[ab] (553.10) |
| Ihumwa - Tanzania | 18859.69[abc] (3138.39) | 2987.07[ab] (971.92) | 17610.75[b] (3767.78) | 1104.62[ab] (225.14) |
| P>F RATIO | 0.0049 | 0.007 | 0.1820 | 0.3987 |
| RMSE | 6016.97 | 853.42 | 10969.71 | 612.24 |
| CV | 39.68 | 40.38 | 46.35 | 44.88 |

*Means of individual species/provenance with standard error in parenthesis. Means of the same letter within the same column are not significantly different

foliar biomass at Gairo site were recorded for Ngabu (Malawi) and Makhanga (Malawi) provenances while at Ruvu site were recorded at Makhanga (Malawi) and Chikwawa/Domasi (Malawi) provenances. This poor performance could probably be due to poor genetic adaptations influenced by climatic conditions in these areas.

## Conclusions and recommendations

The present study has shown that provenances differ significantly in survival and biomass production (Gairo site), and diameter and height growth in both sites. The best performing provenances at Gairo site were Mahalapye (Botswana), Maun (Botswana), Ihumwa (Tanzania), Mbololo 472-029/03 (Kenya) and PKM 1 (India) while at Ruvu site were Maun (Botswana), Mahalapye (Botswana), Mbololo 472-029/03 (Kenya) and Ihumwa (Tanzania). These provenances have shown promising growth throughout the study period indicating their suitability to the locality and other areas with similar soil/climatic conditions. On the other hand Chikwawa/-Domasi (Malawi), Makhanga (Malawi), Ngabu (Malawi) and Mangochi (Malawi) at Gairo and Mangochi (Malawi), Honduras (Honduras), Chikwawa/Domasi (Malawi) and Makhanga (Malawi) at Ruvu site failed to put on promising growth on these sites. The best performing provenances are recommended for planting at these and similar sites.

## ACKNOWLEDGEMENTS

We are very grateful to AFORNET for funding this research project. We are also strongly indebted to the support provided by technicians from both sides for their tireless efforts in making the work a success. Also last but not least the efforts of the late Naomi Masanika who assisted in various stages of the study cannot go

**Table 8.** Ordinal ranking of 30 months old *M. oleifera* provenances planted at Gairo site, Morogoro, Tanzania.

| Species/Provenances | Surv1 | ht | rcd | dbh | fbio | sbio | Mean | Overall rank |
|---|---|---|---|---|---|---|---|---|
| Chikwawa/Domasi - Malawi | 11 | 10 | 11 | 10 | 11 | 12 | 11 | 7 |
| Makhanga - Malawi | 10 | 13 | 13 | 14 | 13 | 11 | 12 | 8 |
| Ngabu - Malawi | 8 | 12 | 12 | 13 | 14 | 13 | 12 | 8 |
| Mangochi - Malawi | 9 | 9 | 14 | 12 | 12 | 14 | 12 | 8 |
| Mtakataka/Dedza - Malawi | 4 | 11 | 8 | 11 | 9 | 10 | 9 | 6 |
| Mahalapye - Botswana | 4 | 2 | 1 | 1 | 2 | 2 | 2 | 1 |
| Maun - Botswana | 2 | 1 | 2 | 2 | 1 | 4 | 2 | 1 |
| PKM 1 - India | 6 | 4 | 6 | 5 | 7 | 3 | 5 | 3 |
| PKM 2 - India | 2 | 4 | 5 | 6 | 5 | 1 | 4 | 2 |
| PKM 3 - India | 5 | 6 | 7 | 7 | 8 | 8 | 7 | 5 |
| Jafna/Jaffana - India | 2 | 7 | 10 | 9 | 4 | 5 | 6 | 4 |
| Honduras - Honduras | 7 | 8 | 9 | 8 | 10 | 9 | 9 | 6 |
| Mbololo 472-029/03 - Kenya | 3 | 3 | 3 | 4 | 6 | 7 | 4 | 2 |
| Ihumwa - Tanzania | 1 | 5 | 4 | 3 | 3 | 6 | 4 | 2 |

[1]Surv., Survival; ht, height; rcd, root collar diameter; dbh, diameter at breast height; fbio, foliar biomass; sbio, stem biomass.

**Table 9.** Ordinal ranking of 30 months old *M. oleifera* provenances planted at Ruvu site, Coast Region, Tanzania.

| Species/Provenances | Surv[1] | ht | rcd | dbh | fbio | sbio | Mean | Overall rank |
|---|---|---|---|---|---|---|---|---|
| Chikwawa/Domasi - Malawi | 1 | 12 | 8 | 9 | 13 | 7 | 8 | 7 |
| Makhanga - Malawi | 5 | 13 | 10 | 12 | 14 | 14 | 11 | 9 |
| Ngabu - Malawi | 2 | 9 | 6 | 8 | 8 | 8 | 7 | 6 |
| Mangochi - Malawi | 4 | 4 | 7 | 6 | 12 | 13 | 8 | 7 |
| Mtakataka/Dedza - Malawi | 2 | 6 | 3 | 4 | 10 | 12 | 6 | 5 |
| Mahalapye - Botswana | 1 | 3 | 5 | 5 | 3 | 1 | 3 | 2 |
| Maun - Botswana | 8 | 1 | 1 | 1 | 1 | 2 | 2 | 1 |
| PKM 1 - India | 4 | 5 | 9 | 7 | 2 | 3 | 5 | 4 |
| PKM 2 - India | 6 | 8 | 11 | 11 | 4 | 10 | 8 | 7 |
| PKM 3 - India | 3 | 11 | 14 | 14 | 5 | 6 | 9 | 8 |
| Jafna/Jaffana - India | 7 | 10 | 12 | 10 | 7 | 5 | 9 | 8 |
| Honduras - Honduras | 6 | 7 | 13 | 13 | 6 | 9 | 9 | 8 |
| Mbololo 472-029/03 - Kenya | 1 | 2 | 4 | 3 | 9 | 4 | 4 | 3 |
| Ihumwa - Tanzania | 2 | 4 | 2 | 2 | 11 | 11 | 5 | 4 |

[1]Surv.-Survival; ht-height; rcd-root collar diameter; dbh-diameter at breast height; fbio-foliar biomass; sbio-stem biomass.

unacknowledged.

## REFERENCES

D'souza J, Kulkarmi AR (1993). Comparative studies on nutritive values of seedlings and mature plants of *Moringa oleifera* Lam. J. Econ. Tax. 17(2):479-485.

Edward E, Chamshama SAO, Mugasha AG (2006). Growth performance of lesser-known *Leucaena* species/provenances at Gairo inlang plateau, Morogoro, Tanzania. South. Afr. For. J. 208:53-62.

Evans J (1991). Safe drinking water for developing world. Our Planet (UNEP) 3(1):12-13.

Fahey JW (2005). *Moringa oleifera:* A review of the medical evidence for its nutritional, therapeutic, and prophylactic properties. Part 1: Trees life J. 1:5. Available at: http://www.tfljournal.org/article.php/ [Accessed 20 May 2009].

FAO (1988). Non timber uses of selected arid zone trees and shrubs in Africa. Edited by F. E. Booth and G. E. Wickens, FAO Conservation Guide19, FAO, Rome, pp. 92-101.

Folkard GK, Sutherland JP (1996). *Moringa oleifera*: a tree and a litany of potential. Agrofor. Today. pp. 5-8.

Forest Division (1982). Management practices in Conifer Plantations in Tanzania: Notes on Forestry Operations. Ministry of Natural Resources and Tourism. p. 68.

Gadzirayi CT, Kubiku FNM, Mupangwa JF, Mujuru L, Chikuvire TJ

(2013). The Effect of plant spacing and cutting interval on growth of *Moringa oleifera*. JASA 2(2):131- 136.
Gomez KA, Gomez AA (1983). Statistical Procedure for Agricultural Research. John Willey and sons, New York. p. 680.
Goss M (2012). A study of the initial establishment of multi - purpose moringa (*Moringa oleifera* Lam) at various plant densities, their effect on biomass accumulation and leaf yield when grown as vegetable. Afr. J. Plant Sci. 6(3):25-129.
Herbert M, Mugasha AG, Jahn SAA, Chamshama SAO (2002). Evaluation of 19 provenances of *Calliandra calothyrsus* at Gairo and SUA Farm, Morogoro, Tanzania. South. Afr. For. J. 194:15-25.
Holmes J (1988). Ruvu Fuelwood Project. Ruvu South Forest Reserve Natural Woodlands management proposal. Unpublished. Dar-es-Salaam. pp. 232.
Jahn SAA (1988a). Flocculation experiments with waste waters in Guatemala and field notes from Costa Rica. Unpublished report for GTZ.
Jahn SAA (1988b). Using *Moringa* seeds as a coagulant in developing countries. J. Am. Water Works Assoc. 80:43-50.
Jahn SAA (1988c). Chemotaxonomy of flocculating plant materials and their application for rural water purification in developing countries. Acta Univ. Ups. Symb. Bot. Ups. 28(3):171-185. Uppsala. ISBN 91-554-2348-5.
Maghembe JA (1979). Effect of Weeding and some soil characteristics on the survival and growth of *Pinus caribaea* in plantation at Ruvu. Division of Forestry, University of Dar-es-Salaam. Record No. 8, pp. 12.
Manh LH, Dung NNX, Ngoi TP (2005). Introduction and evaluation of *Moringa oleifera* for biomass production and as feed for goats in the Mekong delta. LRRD 17(9): Available at http://www.cipav.org.co/lrrd/lrrd17/9/edit1709.htm. [Accessed 9 January 2007].
Mayer FA, Stelz E (1993). *Moringa stenopetala* provides food and low-cost water purification. Agrofor. Syst. 5:16-18.
Mbuya LP, Msanga HP, Ruffo CK, Birnie A,Tengnas B (1994). Useful Trees and Shrubs for Tanzania, Identification and Management for Agriculture and Pastoral Communities. Technical Handbook No. 6. SIDA's Regional soil Conservation Unit. Nairobi: p. 542.
Morton JF (1991). The horseradish tree, *Moringa pterygosperma-a* boon to arid lands? Econ. Bot. 45(3):318-333.
Msanya BM, Msaky JJT (1994). Land use and land evaluation systems research project. Progress Report (August 1993-March 1994). Department of Soil Science, SUA, National Soil Service, Ministry of Agriculture, Tanzania and Center for African Area Studies, Kiplo University, Japan.
Mugasha AG, Chamshama SAO, Nshubemuki L (2000). Effect of spacing on growth and yield of *Sesbania sesban* at Gairo in Morogoro region, Tanzania. TJFNC 73:53 -63.
Olayemi AB, Alabi RO (1994). Studies on traditional water purification using *Moringa oleifera* seeds. Afr. Study Monogr. 15(3):135-141.
Panga JT (2002). Effect of processing on nutritional quality of *Moringa oleifera* leaves and pods. MSc. Dissertation, SUA, Tanzania. pp. 81.
Sa´ Nchez NR, Ledin S, Ledin I (2006). Biomass production and chemical composition of *Moringa oleifera* under different management regimes in Nicaragua. Agrofor. Syst. 66(3):231-242.
SAS Institute Inc. (1991). SAS/STAT Guide for personal computers. Version 8th Edition. SAS. Inc. Cary, North Carolina, USA. pp. 672.
Sokal RR, Rohlf FJ (1969). Biometry. W.H.Freeman and Co. NY. USA.

# Camel urine, a potent tool for plant protection

**E. Nafie**

Department of Botany, College for Women (Arts, Science and Education), Ain Shams University, Cairo, Egypt.

**Cross-tolerance is the phenomenon by which a plant resistance to stress results in resistance to another form of stress. Such intriguing links has made the author to suggest that foliar camel urine (CU) application depending on its amazing constituents may mimic elicitor actions and prime tissues to pathogens in normally susceptible host plants. This study was conducted to assess anatomical and biochemical pathogenesis alterations in tomato water T1, *Fusarium oxysporum* infected T2 plants and also to evaluate the defense-related responses resulting from CU application in either nonT3 and/or T4 infected seedlings at anatomical and molecular levels. Fast and consistent increment in detoxifying enzymes peroxidase, phenylalanine ammonia-lyase, polyphenol oxidase and soluble proteins were recorded in T4 roots. In contrast, tomato roots, T2, T4 showed retard response in catalase activity. T4 accumulate highly significant level (80%) of total free and bound phenolics within its root. Root of T4 SA concomitant with coumarins had significant increment in both free and conjugated phenolic part. Anatomical microscopy shows alterations in T2, T3 as compared to T1 tomato root vascular bundle diameter. Results showed earlier reduction explained as adaptation mechanism in resisting the imposed stress regardless of its type. In contrast, T4 seedlings showed great increase in vascular bundle diameter concomitant with xylem lignin deposition and on epidermis cell walls. Tomato shoot administered CU implicate phenylpropanoid pathway in prime root to *F. oxysporum* invasion by enhancing innate immunity mechanisms, fast accumulation of $H_2O_2$ concomitant with cell wall modifications as physical barrier.**

**Key words:** Cross-resistance, signaling, defense mechanism, detoxification enzymes, free and glycoside-phenolics, phenylpropanoid pathway.

## INTRODUCTION

Plants are frequently exposed to different and simultaneous environmental stresses, which can be both biotic and/or abiotic. Exposure to abiotic stress, in some cases, enhances resistance to pathogens indicative of crosstalk between biotic and abiotic stress signaling (Bowler and Fluhr, 2000). Bowler and Fluhr (2000) and Abuqamar et al. (2009) stated that biotic and abiotic stress have similarities in plant responses to pathogens and abiotic stresses and partially share induced signaling cascades, among which are production of reactive oxygen species (ROS), accumulation of hormones such as salicylic acid, ethylene, jasmonic acid and abscisic acid. These signals ultimately induce expression of nuclear defense genes that lead to the assembly of the overall defense reaction.

Mittler et al. (2004) suggested that during plants evolution they were able to regulate ROS toxicity, through a highly balanced and tightly coordinated network of at least 152 genes which encode both ROS-producing and ROS-scavenging enzymes. For this reason, ROS molecules have been used as signaling molecules and accordingly, the interplay between the ROS-producing and ROS-scavenging pathways will determine the intensity, duration and localization of the ROS signals (Mittler et al., 2004).

Usually, cellular signaling cascade via ROS generated by plant-pathogen interactions, can also be activated by the use of elicitors, stable molecules that induce an immune defense response in plants (Spoel and Dong, 2012). Elicitor-induced plant signaling, serves as a guide to a series of intracellular events that end in the activation of transduction cascades and hormonal pathways, which trigger induced resistance and consequently activate plant immunity to environmental stresses (Spoel and Dong, 2012).

The exploitation of natural resources origin in conferring resistance to plants against abiotic and biotic stresses is gaining much attention. Urine and milk of camels (*Camelus dromedarius*) have been considered to have medicinal significance due to the active substances contained in desert plants (AL-awadi and AL-Jedabi, 2000). An inhibitory and antibiotic activity of camel urine against the growth of *Candida albicans* (yeast), *Aspergillus niger*, *Fusarium oxysporum* were proven even after it's boiling to 100°C (AL-awadi and AL-Jedabi, 2000).

The urine of camel, was very much in use as remedies for the treatment of dropsy, abdominal enlargements, anaemia, abdominal tumor, tuberculosis, haemorrhoids, leucoderma, leprosy and in mental diseases (Thakur, 2004). Al-Attas (2008), using neutron activation analysis, estimated some essential and rare elements in milk and urine of camels.

In fact, it is known that treatment of plants with elicitors, or attack with pathogens, causes a set of defense reactions such as the accumulation of defensive secondary metabolites in edible and inedible parts of plants, specific gene expression and enzymatic induction (García-Mier et al., 2013).

Plants undergo also anatomical changes in response to biotic and abiotic factors. De Micco et al. (2013) conducted a study to reveal histology alterations in *Phaseolus vulgaris* L. plants when subjected to increasing dose of X-rays (0.3, 10, 50 and 100 Gy), results showed that even at high levels of radiation, general anatomical structure was not severely perturbed. There seems to be no report in the literature on the effect of CU on physiology or anatomy of higher plants. The present study suggested that CU could encourage tomato defensive mechanisms which will prime the plant against further invasion depending on its amazing constituents and its potency to mimic elicitor's role. This work aim to first, assess tomato root responses to *F. oxysporum* challenge and CU foliar application, second to test if pretreatment aerial shoot evoke root defense mechanisms against next *Fusarium* challenge. To do this, histological concomitant with biochemical changes in treated tomato seedlings different parts were assessed.

## MATERIALS AND METHODS

Tomato seeds (*Lycopersicon esculentum* vs. *castlerock*) susceptible to *F. oxysporum* f. sp. *radicis- lycopersici* were kindly provided by Field Crops Research Institute at the National Research Centre, Giza, Egypt. Seeds were surface disinfected in 1% hydrochloric acid for 30 min and rinsed repeatedly in sterile double-distilled water prior to sowing (Fuchs et al., 1997), planted and grown in vermiculite in an environmentally controlled greenhouse at 24 ±2°C with 16/8 h of light - dark condition till desired age. Plants were fertilized twice a week, young plants, approximately with five or six fully expanded leaves were then transplanted into pots.

Camel urine was received from Maryout Research Station of Desert Research Center after collection from male camel of about seven years old in the early morning.

### Fungal inoculums' preparation

Authentic fungal culture was received from Plant Pathology Research Institute, Agriculture Research Center, Giza, Egypt. Fungus *F. oxysporum* f. sp. *Radicis lycopersici* were sub cultured in 9-cm Petri dishes on potato dextrose agar (PDA) at 25°C in the dark. Conidia were removed from 10-15-day-old cultures into distilled water and their concentration was adjusted to $10^6$ - $10^7$ conidia per ml with a cytometer before inoculation.

### Experimental design and treatments application

Upon the appearance of the fifth leaf, pots were randomly assigned to four groups of twenty replicates each, sprayed equally with either water (T1), and challenged with *F. oxysporum* (T2), sprayed with camel urine (T3), and challenged with *F. oxysporum* (T4). Challenging of T2 and T4 were carried out with pathogen two days later after foliar applications. The experiment was left till harvesting time: two, four, seven and fourteen days after inoculation.

### Preparation of tomato tissues to study enzymatic activity

Seedlings representing all treatments at the harvesting time were separated, immediately frozen in liquid $N_2$ and stored at - 80°C.

### Plant biochemical analysis

#### *Phenylalanine ammonia-lyase enzyme extraction and assay (PAL) (EC 4.3.1.5)*

The ratio of frozen tissue (g) and extraction buffer volume (ml) was 1:5. The extraction was carried out following the method proposed by Lister et al. (1996). Frozen plant material was ground at 4°C in buffer (50 mM $Na_2HPO_4/KH_2PO_4$, pH 7, polyvinylpyrrolidone (PVP) 0, 5% (Mr, 44 000), 50 mM sodium ascorbate, 18 mM mercaptoethanol, 0.1% (v/v) Triton X-100). The homogenate was filtered through four layers of cheesecloth and centrifuged at 20 000 g for 10 min. $(NH_4)_2SO_4$ was added to the supernatant (to 35% saturation), which was then centrifuged for 20 min at 20 000 g to remove the PVP. More $(NH_4)_2SO_4$ was added to this supernatant to reach a final saturation of 80%. This fraction was centrifuged at 20 000 g for 20 min, and the pellet was re suspended in extraction buffer (without PVP and Triton). This solution was used for PAL assays.

PAL activity was assayed after adapting (McCallum and Walker, 1990) method. The assay mixture consisted of 0.06 M sodium borate buffer, pH 8.8, and partially purified enzyme. The reaction was started by the addition of 11 mM L-phenylalanine. The reaction mixture was incubated at 30°C for 15 min, and the reaction was terminated by the addition of 6 N HCl and then measured at 290 nm (Lee et al., 2007). Enzyme activity was expressed as µM cinnamic acid $h^{-1}$ $g^{-1}$ FW.

### *Peroxidase enzyme extraction and assay (POD) (EC 1.11.1.7)*

Frozen plant parts were homogenized (1:5) in appropriate ice cold extraction buffer: 0.1 M potassium phosphate buffer (pH 6.8) containing 1% Triton X-100 and 0.15 g of Polyclar AT (insoluble polyvinylpolypyrrolidone, Sigma Chemical Co., St. Louis). The supernatants were used directly in the enzyme assays.

Peroxidase activity was determined following Hammerschmidt and Kuc (1982) method. The reaction mixture consisted of 0.5 ml plant extract, 1.5 ml of 0.05 M pyrogallol and 0.5 ml of 1% hydrogen peroxide. The reaction was incubated in a water bath, and then terminated after incubation for 30 min; absorbance was recorded at 420 nm. The enzyme activity was expressed as change in the absorbance of the reaction mixture, $min^{-1}$ $g^{-1}$ FW.

### *Catalase enzyme extraction and assay (CAT) (EC 1.11.1.6)*

Frozen plant parts stored at - 80°C, were extracted in 0.4 ml of buffer (potassium phosphate, 50 mM at pH 7.4, containing 10 mM DTT). Catalase activity was assayed at 25°C following the decrease in absorption at 240 nm in 10 mM sodium phosphate buffer, pH 7(Aebi, 1984). The enzyme activity was expressed as mM $H_2O_2$, $min^{-1}$ $g^{-1}$ FW.

### *Polyphenol oxidase enzyme extraction and assay (PPO) (EC 1.14.18.1)*

The extraction method proposed by Thypyapong et al. (1995) after some modifications were used. Frozen tissue was grounded in extraction buffer (100 mM Tris-HCl, pH 7.0, 100 mM KCl, 1 mM phenylmethanesulfonyl fluoride (PMSF) and 3% [w/v] PVP). The homogenates were centrifuged at 12 000g for 15 min, and the supernatant was used to measure the PPO enzyme activity. The PPO activity was determined according to Mayer (1995) with some modifications. The reaction mixture consisted of 100 mM buffer ($Na_2HPO_4$/ $KH_2PO_4$), pH 7.0 and 1.5 ml of 0.01 M catechol and reaction was initiated by the addition of 200 µl enzyme extract, the activity was expressed as changes in absorbance at 495 nm.

### Protein determination

From each treatment, 0.5 g samples were grinded in mortar with 5 ml of phosphate buffer pH 7.6 and centrifuged at 8000 g for 20 min. The supernatant was then made equal by adding phosphate buffer solution. Coomassie Brillaint Blue G-250 solution was then added and mixed thoroughly (Bradford, 1976). The total volume was 3 ml. All treatments were incubated for 5 min at room temperature and absorbance at 595 nm was recorded against the reagent blank. A standard sample (100 µg in 100 µl) was used to estimate soluble protein concentrations in the unknown treated samples as mg $g^{-1}$.

### Phenolics extraction

Tomato seedlings harvested from two, four, seven and fourteen days old plants, representing each of the four treatments were used to extract a free-phenolic fraction part, and a conjugated fraction consisting of the aglycones released after hydrolysis. The extraction method of Fawe et al. (1998) was used. Fine powder samples preserved in liquid $N_2$ were extracted in 80% acidified methanol (10 g/100 ml). The extract was then filtered with glass fiber filters (GF/C; Schlei and Schuell), and the filtrate was concentrated. The aqueous residue was adjusted to pH 2.0 and partitioned against hexane. The aqueous phase containing the phenolic constituents was further partitioned against ethyl acetate and then subjected to acid hydrolysis, 4 N HCl in an autoclave for 15 min. The hydrolysate was cooled and partitioned against ethyl acetate. The two ethyl acetate fractions obtained were dried, and the residues designated as free-phenolic fraction and conjugated-phenolic fraction, respectively, were re suspended in absolute methanol (2.5 g/ml).

### Total phenolic concentration determination

Total phenolic (TP) concentration was measured in both free and glycosidic-bound phenolic extract by using Folin Ciocalteus reagent (Sigma, St. Louis, Mo.) according to Singleton and Rossi (1965) method. Determination of total phenols concentration was done using a Jenway (UV/Vis 6405) spectrophotometer monitoring 750 nm. TP content was expressed as mg FW.

### Phytoalexins separation using high-performance liquid chromatography (HPLC)

Tomato leaves and roots from all treated plants after 48 h of infection were harvested, pooled and kept at -80°C until use (approximately 20 g). The phenolic compounds extraction was performed as described previously. The resultant free and conjugated-phenolic fractions after separation as previously described were used to determine their profiles.

The HPLC system (Thermo Separation Products, Riviera Beach, Fla.) consisted of an auto sampler (AS3000), injector (100 µl), column oven (30), pump (P3000) and diode array detector (UV6000). A reverse-phase C18 column (250 by 4.6 mm; Luna 2; Phenomenex, Torrance, Calif.), with a pre column of similar resin, was employed. Elution was performed using phosphate buffer (50 mM, pH 2.5) and methanol at a flow rate of 1 ml $min^{-1}$. A linear-gradient program was developed as follows: C18 column (time [in minutes]/methanol [percent]) = 0/0, 6/0, 31/95, 32/0. In all instances, the software was programmed to show peaks at their maximum absorbance. For each treatment, HPLC analyses were repeated twice, showing similar results.

### Tissue processing for ultra structural investigation

For optical light microscopy (OLM) investigation, roots (covering 7 and 14 day harvesting time) sections from chosen tomato plant parts were fixed in FAA (ethanol 50% + formaldehyde 5% + glacial acetic acid 10%, in water) for 48 h. After fixation, segments were

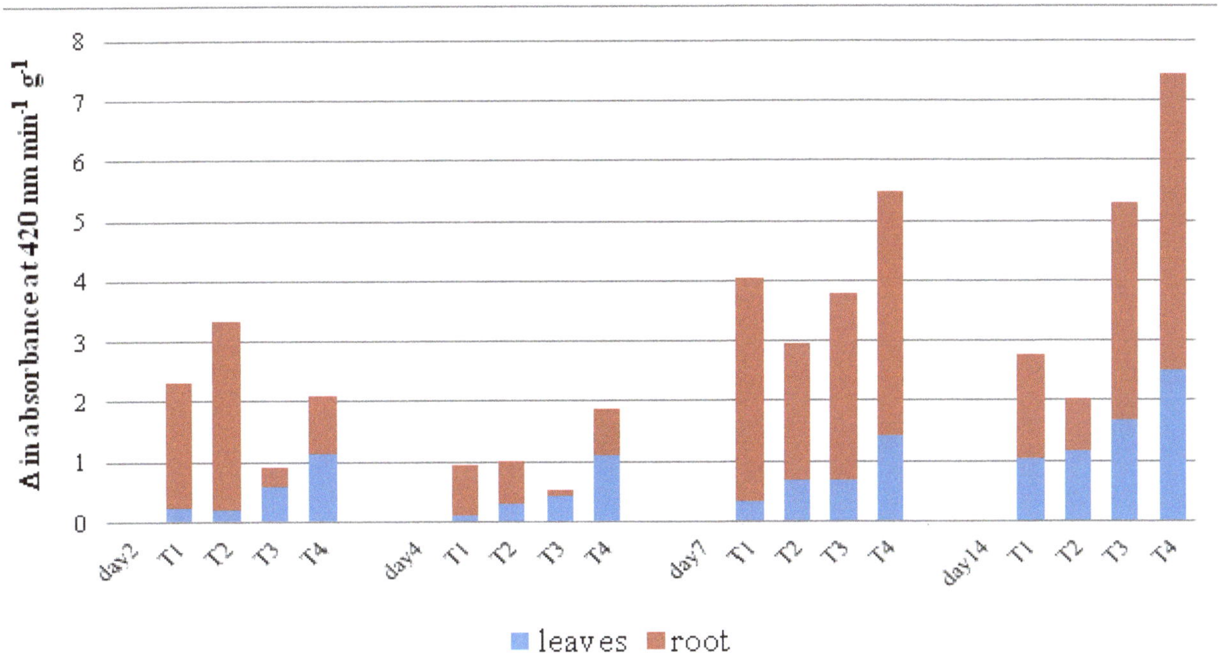


**Figure 1.** Peroxidase enzyme activity estimated in water (T1), and challenged with *F. oxysporum* (T2), sprayed with camel urine (T3), and challenged with *F. oxysporum* (T4), harvested at day two, four, seven and 14 post infection.

dehydrated through a gradient series of ethanol, cleared with xylene, and finally embedded in paraffin. Fifteen-micrometer-thick sections were cut with a microtome and transferred onto glass slides coated with egg albumin. Sections were de-paraffinized with xylene and rehydrated through a gradient series of ethanol. Sections were stained with safranin O (at ethanol 50% step) to show lignin following Balatinecz and Kennedy (1967) with slight modifications, washed well, then light green was used (at ethanol 100% step) for visualizing the plant sections anatomy. Photographs were taken using Olympus microscope BH2 with camera attachment (model U-ACAD-2). Calibrated linear ocular was used for tissue measurements.

### *Statistical analysis*

The test of least significant difference (L.S.D) at the level of 0.05% significance was used to examine differences among treatment means and interactions. Data were statistically analyzed using MSTAT-C software package according to the described method by Freed et al. (1989).

## RESULTS

### Biochemical changes

### *Camel urine application enhanced tomato resistance responses*

To assess pathogenesis and if pre CU application may prime and condition plant response to successive pathogen attack, changes in detoxification enzymes were elucidated in samples harvested at the required time.

### *Peroxidase and catalase activities*

The fungus inoculation resulted in rapid increment in T2 root enzyme activity after 48 h thereafter fluctuated, while leaves exhibited progressive levels. Peroxidase activity in T4 treated plants showed progressive local and systemic enhancement *in vivo* plant leaf and root. Root displayed higher activity than leaves as shown in Figure 1. Leaves of infected tomatoes exhibited CAT increment two days later while their roots negatively responded. T4 leaves during same time showed similarity with the control while roots enzyme sharply decreased. At the 14$^{th}$ day, enzyme activity in roots of all treated plants were increased, on the other hand leaves responded in a different manner as shown in Figure 2.

### Alterations in phenolic compounds metabolism

### *Phenylalanine ammonia-lyase and polyphenol oxidase activities*

Phenylalanine ammonia-lyase and polyphenol oxidase activities in T4 roots progressively increased till 14 day post challenge. Roots of T2 treated plants had highest level after 24 h in PAL and PPO activity thereafter

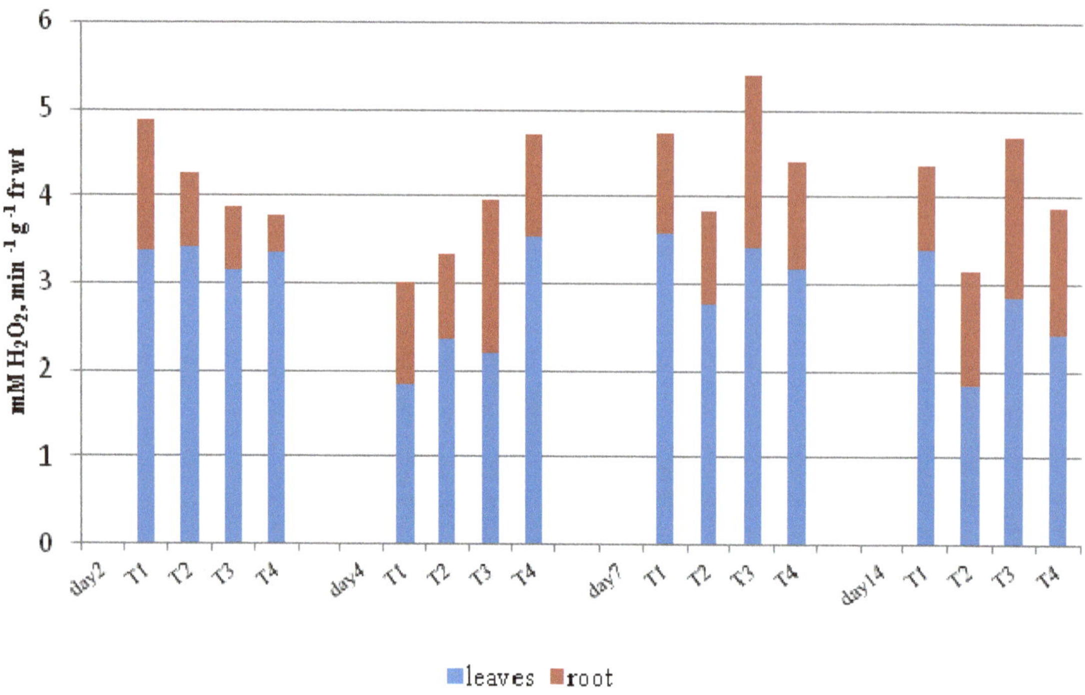


**Figure 2.** Catalase enzyme activity estimated in water (T1), and challenged with *F. oxysporum* (T2), sprayed with camel urine (T3), and challenged with *F. oxysporum* (T4)l, harvested at at day two ,four, seven and 14 post infection.

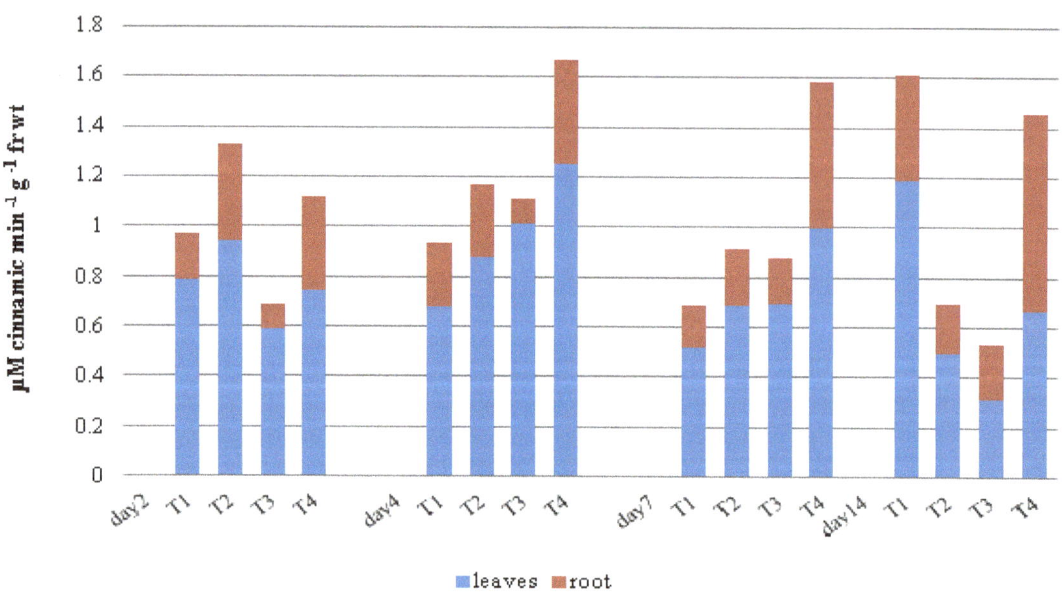


**Figure 3.** Phenylalanine ammonia-lyase enzyme activity estimated in water (T1), and challenged with *F. oxysporum* (T2), sprayed with camel urine (T3), and challenged with *F. oxysporum* (T4), harvested at day two, four, seven and 14 post infection.

generally decreased as compared to the control. Fungus inoculation caused rapid increment in leaves PAL activities then gradually ceased while PPO showed fluctuations as shown in Figures 3 and 4.

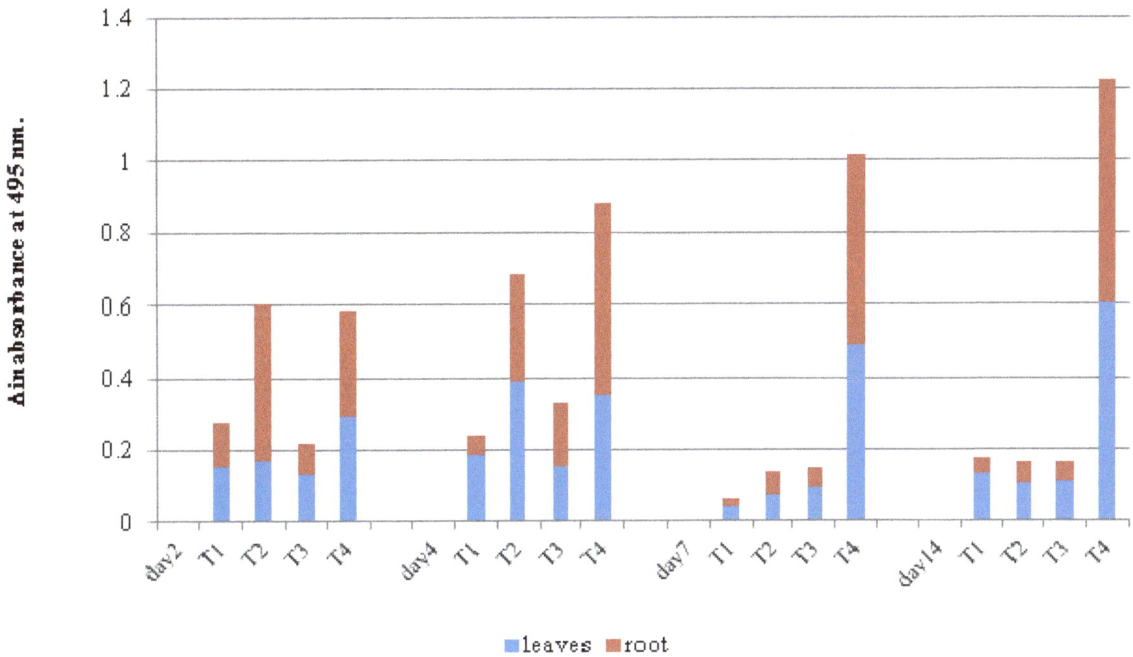


**Figure 4.** Polyphenol oxidase enzyme activity estimated in water (T1), and challenged with *F. oxysporum (*T2), sprayed with camel urine (T3), and challenged with *F. oxysporum* (T4), harvested at two day two, four, seven and 14 post infection.

### *Quantification of total phenolics in free and conjugated phenolic extracts*

Increment in total phenols throughout the study was recorded in both free and conjugated phenols in different tomato parts as shown in Table 1. Alterations in phenolic compounds recorded throughout the experiment were plant part dependent, in which T4 root recorded the highest level. T2 and T4 TP in tomato leaves showed increment reaching 43.8 and 29% respectively, and in root, 40 and 81.7% at the 14 day post infection with respect to each control.

### HPLC separation of phytoalexins

A comparative study for 48 h post harvesting of plants leaves and roots represent all treatments profile of certain fractions analyzed in conjugated and free phenolics, using reverse-phase HPLC. Four phenolic compounds were observed, and changes in their relative amount distribution between conjugate and free state were recorded as shown in Table 2.

Leaves and roots of challenged tomato plant with *F. oxysporum* T2 showed general retarded levels in free and conjugated coumarins, pyrogallol, catechol and salicylic acid amounts relative to the control T1 that was not challenged. In T4 tomato roots, coumarins, and salicylic acid were enhanced significantly in free and glycosidic-bound phenolic fraction while leaves salicylic acid enhanced bound phenolic fraction.

### *Soluble proteins*

Soluble proteins in T2 or T4 roots showed continuous increment over the control. In T2 and T4 seedlings leaves rapid and negative response was recorded and continued to increase In T4 only, results are illustrated in Figure 5.

### *Histological changes*

A part of root transverse sections from all treated tomato seedlings seven and fourteen days post-harvest are in Table 3 and Figures 6 and 7 to explore histological changes.

Tomato root section showed abundance in xylem ray with large prominent metaxylem vessels in early root as shown in Figure 6a. The structural changes associated with pathogen infection in T2 tomato seedlings root cortex and vascular bundle was reduction in both diameters, as compared to the control, T1 (Figure 6). In seedling root T3, cortex diameter was larger while vascular bundle was reduced as compared to the control T1 (Figure 6). Infection after foliar pretreatment of seedlings T4 enhanced lateral root initiation, xylem cell wall lignifications in the form of dark lesions encircling the cells as shown in Figure 6 and increment is not only in

**Table 1.** Phenolic compounds (free and conjugate) estimated in tomato leaves and roots in water (T1), challenged with *F. oxysporum* (T2), sprayed with camel urine (T3) and challenged with *F. oxysporum* (T4) harvested at two, four, 7 and 14 day post infection.

| Harvesting days | | 2 day | | | 4 day | | | 7 day | | | 14 day | | | |
|---|---|---|---|---|---|---|---|---|---|---|---|---|---|---|
| | Phenolic (mg/g FW) treatments | Free | Glycosidic-bound | Total | Free | Glycosidic-bound | Total | Free | Glycosidic-bound | Total | Free | Glycosidic-bound | Total | Total (%) |
| Leaves | T1 | 0.4c | 0.52b | 0.92d | 0.73b | 0.85b | 1.58c | 0.8c | 0.90d | 1.7d | 0.92c | 1.20c | 2.12d | |
| | T2 | 0.75b | 0.60b | 1.35c | 1.32a | 1.11b | 2.43b | 1.46a | 1.22c | 2.68c | 1.66ab | 1.39c | 3.05c | 43.80 |
| | T3 | 0.68b | 1.43a | 2.11b | 1.2a | 2.20a | 3.40a | 1.27b | 2.51b | 3.78b | 1.50b | 2.65b | 4.15b | |
| | T4 | 1.07a | 1.46a | 2.53a | 1.4a | 2.44a | 3.84a | 1.45a | 2.90a | 4.35a | 1.85a | 3.52a | 5.37a | 29.30 |
| | LSD at 5% | 0.21 | 0.45 | 0.32 | 0.27 | 0.48 | 0.63 | 0.30 | 0.26 | 0.64 | 0.28 | 0.65 | 0.73 | |
| Roots | T1 | 0.69c | 0.94b | 1.63bc | 0.95c | 1.14c | 2.09c | 0.98 c | 1.40b | 2.38 c | 1.00 c | 1.50c | 2.50c | |
| | T2 | 1.00b | 0.92b | 1.92b | 1.48b | 1.34c | 2.82b | 1.61 b | 1.53b | 3.14 b | 1.80 b | 1.70c | 3.50b | 40 |
| | T3 | 0.55c | 0.99b | 1.54c | 0.57d | 1.60b | 2.17c | 0.75 c | 1.78b | 2.53 c | 1.86 b | 2.35b | 4.21 b | |
| | T4 | 1.56a | 2.45a | 4.01a | 1.87a | 3.67a | 5.54a | 2.60 a | 4.41a | 7.01 a | 2.74 a | 4.91a | 7.65 a | 81.70 |
| | LSD at 5% | 0.26 | 0.35 | 0.23 | 0.31 | 0.22 | 0.54 | 0.48 | 1.44 | 0.50 | 0.68 | 0.57 | 0.85 | |

**Table 2.** Certain phenolic compounds estimated using HPLC (free and glycoside) in leaves and roots extracted after forty-eight hours (after infection) in either water (T1), and challenged with *F. oxysporum* (T2), sprayed with camel urine (T3), and challenged with *F. oxysporum* (T4). Phenolics estimated as µg/g FW.

| Parameter | | Free phenolics | | | | Glycosidic –bound | | | | Total | | | |
|---|---|---|---|---|---|---|---|---|---|---|---|---|---|
| | | Coumarins | Pyrogallol | catechol | salicylic acid | coumarins | pyrogallol | catechol | salicylic acid | coumarins | pyrogallol | catechol | Salicylic acid |
| Leaves | T1 | 101.63a | 50.10b | 65.25a | 13.9b | 217.36a | 101.6a | 114.21a | 7.95c | 318.99a | 151.70a | 179.46a | 21.85a |
| | T2 | 58.62b | 59.00a | 60.85a | 6.4bb | 68.50b | 27.5c | 36.60b | 2.10d | 127.12b | 86.50b | 97.45b | 8.50d |
| | T3 | 32.12c | 21.00c | 26.36b | 1.9c | 66.74b | 72.52b | 14.35c | 13.22a | 98.86c | 93.52b | 40.71c | 15.12b |
| | T4 | 33.74c | 17.13c | 22.00b | 2.1c | 32.50c | 14.31d | 12.64c | 9.62b | 66.24d | 31.44c | 34.64c | 11.72c |
| | LSD at 5% | 5.32 | 4.85 | 8.67 | 4.62 | 10.38 | 9.88 | 11.05 | 1.26 | 13.22 | 10.58 | 15.35 | 2.87 |
| Roots | T1 | 40.76b | 35.54b | 32.50a | 4.27c | 4.50c | 10.20c | 29.18b | 1.75c | 45.26c | 45.74b | 61.68b | 6.02c |
| | T2 | - | 30.90b | 14.63c | 0.30d | 6.31c | 9.70c | 24.44c | 4.23c | 6.31d | 40.60bc | 39.07d | 4.53c |
| | T3 | 38.62b | 50.00a | 36.00a | 8.44b | 59.68b | 39.63a | 40.00a | 11.00b | 98.3b | 89.63a | 76.00a | 19.44b |
| | T4 | 78.00a | 10.60c | 27.52b | 10.66a | 65.33a | 23.45b | 23.62c | 17.10a | 143.33a | 34.05c | 51.14c | 27.76a |
| | LSD at 5% | 10.24 | 8.65 | 4.32 | 2.08 | 4.65 | 5.34 | 3.98 | 4.21 | 9.25 | 8.62 | 7.36 | 5.14 |

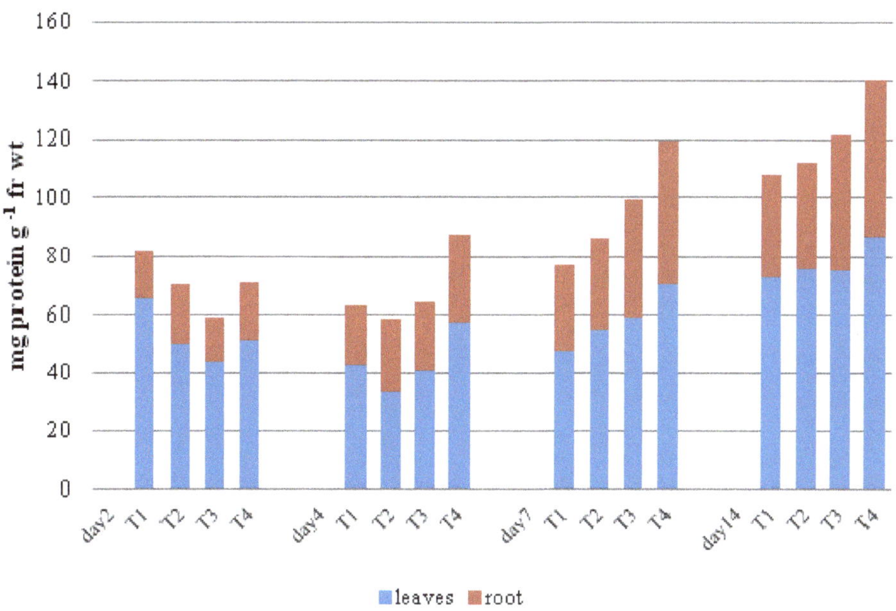


**Figure 5.** Soluble proteins estimated in water (T1), and challenged with *F. oxysporum* (T2), sprayed with camel urine (T3), and challenged with *F. oxysporum* (T4), harvested at day two, four, seven and 14 post infection.

**Table 3.** Tomato seedlings root, cortex and stele cross section diameter (µm) in either water (T1), and challenged with *F. oxysporum* (T2), sprayed with camel urine (T3), and challenged with *F. oxysporum* (T4), harvested at 7 and 14 days post infection.

| Treatments | Root | | Cortex | | Stele | |
|---|---|---|---|---|---|---|
| | 7 day | 14 day | 7 day | 14day | 7 day | 14 day |
| T1 | 21.6 | 26.0ab | 4.0b | 5.0b | 12.0b | 13.0b |
| T2 | 20.0 | 24.0b | 3.4b | 4.8b | 11.4b | 12.6b |
| T3 | 21.0 | 24.60.b | 5.0a | 6.0a | 11.0b | 12.4b |
| T4 | 23.6 | 27.8a | 3.4b | 5.0b | 14.4a | 16.4a |
| LSD at 5% | NS | 2.44 | 0.85 | 0.79 | 0.2.10 | 0.57 |

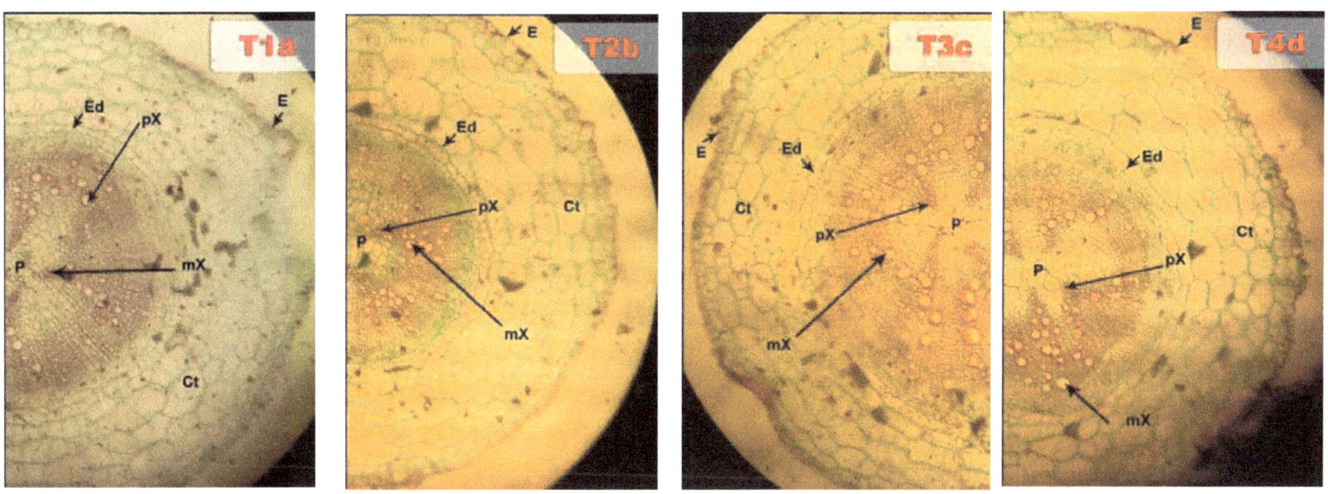


**Figure 6.** Light Microscopic images showing cross sections of root from 7-day post infection tomato seedlings. Different treatments were: T1a, control water non infected; T2b, *F. oxysporum* infected; T3c, urine camel foliar spraying; T4d, urine camel sprayed and infected. Ct, tomato seedlings cortex; E, epidermis; Ed, endodermis; mX, metaxylem; P, pith; Ph, phloem; pX, protoxylem.

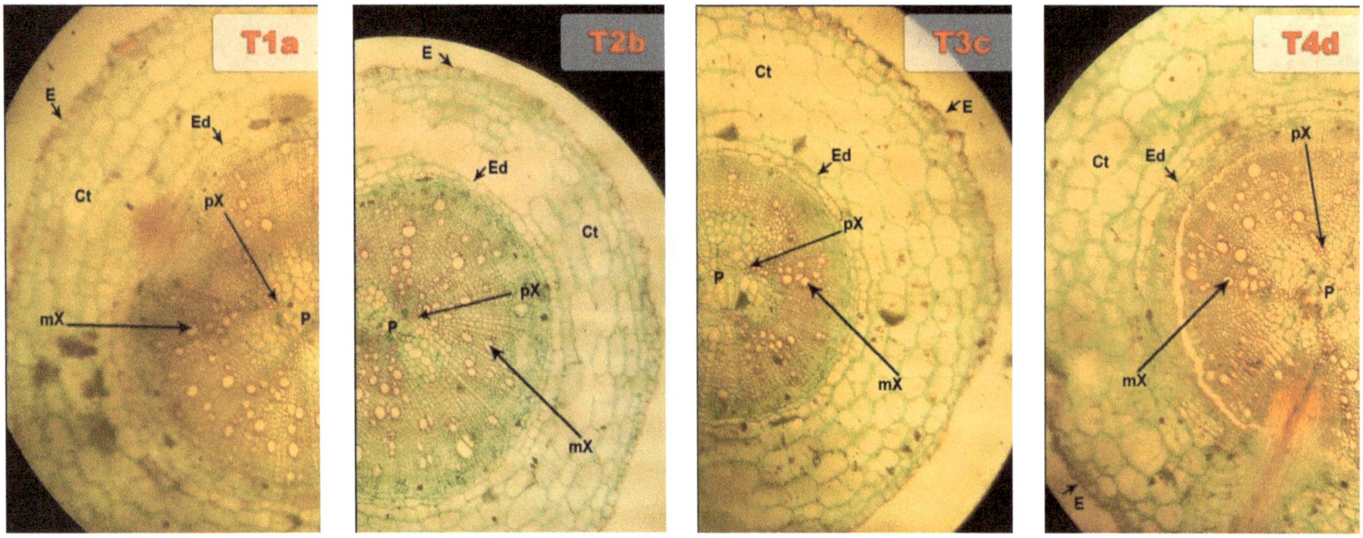


**Figure 7.** Light Microscopic images showing cross sections of root from 14-days post infection tomato seedlings. Different treatments were: T1a, control water non infected; T2b, *F. oxysporum* infected; T3c, urine camel foliar spraying; T4d, urine camel sprayed and infected. Ct, tomato seedlings cortex; E, epidermis; Ed, endodermis; mX, metaxylem; P, pith; Ph, phloem; pX, protoxylem.

xylem cell width but also in number. Root age follow the same trends (Figure 7).

## DISCUSSION

In nature, plants often deal with simultaneous or subsequent abiotic and biotic stresses which can influence the primary induced defense responses (Abuqamar, 2009). Activation of plant defense mechanisms is associated with ecological fitness costs.

Hence, plants need regulatory mechanisms to effectively and efficiently adapt to changes in their complex environment. Share between different signaling pathways provides the plant with such a powerful regulatory potential (Abuqamar, 2009).

Presented results show that tomato plants responded to exogenous, foliar applications of the CU solution. The amazing solution was found to sensitize susceptible tomato plants to react more rapidly and more efficiently to *F. oxysporum* f. sp. *Radicis lycopersici* attack.

Peroxidase has often been used as an enzymatic marker in studies of defense-related processes (Karthikeyan et al., 2006). The enhanced peroxidase activity shown by the *pschit*, transgenic tobacco lines, was linked to their increased pathogen resistance (Dana et al., 2006). Tomato tissues responded to CU application and this point to CU recognition. Tomato leaves tissues point to fast and consistent POD increments thereafter CU systemic transporting, preferentially enhanced the roots peroxidase. In seedlings roots, peroxidase activity increments inT4 implicate wall-bound phenolics accumulation in lignifications with pathogen restriction in systemic parts as a result of pre CU foliar application (Karthikeyan et al., 2006). Root CAT depletion (in the present work) may be explained by salicylic acid binding affinity to iron-containing enzymes (Klessig et al., 2000) resulting in $H_2O_2$ accumulation to directly award off pathogen spreading.

PAL is a key enzyme in the biosynthesis of phenyl propane unit. Tomato plant pretreated by camel urine as compared to water spray before challenging both implicated total phenolics and PAL activity increment in roots within the plant strategy to lignin deposition which had been confirmed by the histological part resulted in physical protection enhancement. Increased PAL activity level in response to pathogen or elicitor like raw cow's milk spray has been reported (Arun et al., 2010).

Karthikeyan et al. (2006) stated that infected chickpea and palm seedlings treated with biocontrol agents resulted in increased accumulation of phenolics and PAL. Tomato T2 roots suggested that immediate transient activation in PAL after infection could re-establish the notion that plant recognize its enemy and respond to invasion by PAL expression.

PPOs are a group of copper containing enzymes that catalyze oxidation of hydroxy phenols to their quinone derivatives, which have antimicrobial activity (Thipyapong and Stiffens, 1997). Recently, study on elicitation of resistance and defense related enzymes by amino acids and raw cow's milk (RCM) in Pearl millet against downy mildew disease (Arun et al., 2010) recorded increased PPO in both RCM and amino acids treated plants.

The prominent increase in PPO roots of T4 treatments exhibited the highest enzyme levels, involved camel urine spray in conditioning root defense responses against pathogen toxic metabolites which may activate phenol-oxidizing enzymes where *F. oxysporum* was found to produce such a toxic metabolite.

Emerging evidence suggests that hormones signaling pathways, as well as ROS, play key roles in the crosstalk between biotic and abiotic stress signaling (Abuqamar, 2009). They act synergistically or antagonistically with each others to regulate plant responses to pathogens and abiotic stress factors (Abuqamar, 2009). SA quantity in roots of T4 treated seedlings, showed that SA could be a signaling molecule forming a feedback amplification cycle in concert with ROS (Abuqamar, 2009). In this way, SA induction is not required but the endogenous SA present amplifies the effects of ROS initial levels. This study contributes to defining the role of SA during the camel urine application. Asselbergh et al. (2007) found that in tomato, increase in SA-regulated defense gene expression, faster accumulation of $H_2O_2$ and the associated cell wall modifications explain the increase in resistance to *Botrytis cinerea*. In the present work, root of the pretreated seedlings with camel urine and pathogen challenged record enhancement conjugated SA concomitant with catalase inhibition and peroxidase increment so, shoot perception to camel urine and the cascade signaling pathways cross-talk *via* specific systematic signals like SA to prime root to other stress by enhanced defense gene expression, fast accumulation of $H_2O_2$ concomitant with cell wall modifications as physical barrier after lignin deposition. Cvikrová et al. (2006) conducted a study to identify the defense-related alterations in phenylpropanoid pathway, they explain the rise of phenolic contents in the reaction zones by scoring predominately derivatives of cinnamic acid, in particular *p*-coumaric acid, content of which increased by about 2.3- and 3.0-fold.

Many of the pathogen-induced phenylpropanoids like coumarins are considered phytoalexins because they exhibit antimicrobial properties (Cvikrová et al., 2006). Immediate excellent increment after two days post infection in total coumarins in T4 seedlings roots share increments in total antioxidant defenses. Further physiological and molecular research has to be done in order to clearly define the signaling pathways involved in camel urine induced resistance. Current accumulation of proteins in CU treated tomato may provide a storage form of nitrogen that is re-utilized when stress is over, while increasing soluble proteins in infected and pretreated may be due to synthesis structural proteins in particular synthesis of those proteins which are involved in modification of cell wall especially in roots to collaborate defense against pathogen infection.

In the present work, root of seedlings received both foliar application and pathogen challenged showed no tissue injury or disorganization in vascular tissues. Enhanced lignifications on epidermis cell wall and enhanced thickness inside xylem cell wall which were correlated with peroxidase activity and related to camel urine unique constituents especially Ca and Se ions as compared to water treated seedling. Selenium ions as one of CU constituents (Al-Atta, 2008) has been shown to protect plants from both herbivore and pathogen infection (Hanson et al., 2004).

In the light of the above recorded results, the present study may suggest that tomato aerial parts can recognize and respond to CU, evoke cascade signaling pathways, systematically crosstalk with roots using SA and sensitize tomato underground parts to respond faster and to a greater extent to *F. oxysporum* attack. A better understanding of these mechanisms will make it possible to genetically engineer crops disease resistance, leading to greater food production.

## Conflict of Interests

The author(s) have not declared any conflict of interests.

## ACKNOWLEDGEMENTS

The author wish first to acknowledge the Botany Department, Ain Shams University for permitting this work to be done in the Botany Garden, second to thank Faculty of Agriculture, Ain Shams University, for providing *F. oxysporum* cultures, finally, I wish to express my thanks to Dr. Azza Elshafey, Professor of Plant Physiology, Botany Department, Faculty of Science, Ain Shams University, for her scientific support, encouragement and revising the manuscript.

### REFERENCES

Abuqamar S, Luo, H, Laluk K, Mickelbart M, Mengiste T (2009). Crosstalk between biotic and abiotic stress responses in tomato is mediated by the *AIM1* transcription factor. Plant J. 58(2):347-60. Epub 2008 December 25. Erratum in: Plant J. 2009 December 60(5):929.

Aebi H (1984). Catalase *in vitro*. Meth. Enzy.105:121-126.

Al Atta (2008). Determination of essential elements in milk and urine of camel and in *Nigella sativa* seeds. Arabian J. Chem. 1(2):123-129.

AL-awadi A, AL-Jedabi A (2000). Antimicrobial agents in camel's urine. Microbiol. Viruses (9B) 8-11 November, pp. 265-281.

Arun K, Mali P, Manga V (2010). Changes of some phenolic compounds and enzyme activities on infected pearl millet caused by *Sclerospora graminicola*. Int. J. Plant Physiol. Biochem. 2(1):6-10.

Asselbergh B, Curvers K, Franca S, Audenaert K, Vuylsteke M, Van Breusegem F, Hofte M (2007). Resistance to *Botrytis cinerea* in *sitiens*, an abscisic acid-deficient tomato mutant, involve timely production of hydrogen peroxide and cell wall modifications in the epidermis. Plant Physiol. 144:1863-1877.

Balatinecz J, Kennedy R (1967). Maturation of ray parenchyma cells in pine. For. Prod. J. 17:57-64.

Bowler C, Fluhr R (2000). The role of calcium and activated oxygens as signals for controlling cross-tolerance. Trends Plant Sci. 5: 241–246.

Bradford M (1976). A rapid and sensitive method for quantization of microgram quantities of protein utilizing the principle of protein-dye binding .Anal. Biochem. 72:248-254.

Cvikrová M, Malá J, Hrubcová M, Eder J ( 2006). Soluble and cell wall-bound phenolics and lignin in *Ascocalyx abietina* infected *Norway pruces.* Plant Sci. 170:563-570.

Dana M, Pintor-Tor J, Beatriz Cubero B (2006). Transgenic Tobacco plants over expressing chitinases of fungal origin show enhanced resistance to biotic and abiotic stress agents. Plant Physiol. 142:722-730.

De Micco V, Arena C, Aronne G (2013). Anatomical alterations of *Phaseolus vulgaris* L. mature leaves irradiated with X-rays.

Fawe A, Abou-Zaid J, Menzies G, Bélanger R (1998). Silicon-mediated accumulation of flavonoid phytoalexins in cucumber. Phytopathol. 88:396-401.

Freed R, Einmensmith SP, Guetz S, Reicosky D, Smail VW, Wolberg P (1989). User's guide to MSTAT-C, an analysis of agronomic research experiments. Michigan State University, USA.

Fuchs J, Moenne-Loccoz Y, Defago G (1997). Nonpathogenic *Fusarium oxysporum* strain *Fo47* induces resistance to Fusarium Wilt in tomato. Plant Dis. 81:492-496.

García-Mier L, Guevara-González RG, Mondragón-Olguín VM, del Verduzco-Cuellar BR, Torres-Pacheco I (2013). Agriculture and bioactives: achieving both crop yield and phytochemicals. Int. J. Mol. Sci.14:4203-4222.

Hammerschmidt R, Kuc J (1982). Lignification as a mechanism for induced systemic resistance in cucumber. Physiol.Pl. Pathol. 20:61-71.

Hanson B, Lindblom S, Loeffler M, Pilon-Smits E (2004). Selenium protects plants from phloem-feeding aphids due to both deterrence and toxicity. New Phytol. 162:655-662.

Karthikeyan M, Radhika K, Mathiyazhagan S, Bhaskaran R, Samiyappan R, Velazhahan R (2006). Induction of phenolics and defense-related enzymes in coconut (*Cocos nucifera* L.) roots treated with biocontrol agents. Braz. J. Plant Physiol. 18 (3).

Klessig FD, Durner J, Nod R, Navarre AD, Wendehenne D, Kumar D, Ma Zhou J, Shah J, Zhang S, Kachroo P, Trifa Y, Pontier D, Lam E, Silva H (2000). Nitric oxide and salicylic acid signaling in plant defense. Proceedings of the National Academy of Sciences of the United States of America. 97(16):8849-8855.

Lee R, Kim, Y, Jung J, Avice C, Ourry A, Kim H (2007). Peroxidases and lignification in relation to the intensity of water-deficit stress in white clover (*Trifolium repens* L.). J. Exp. Bot. 58(6):1271-1279.

Lister C, Lancaster J, Walker J (1996). Phenylalanine Ammonia-Lyase activity and its relationship to anthocyanin and flavonoid levels in New Zealand-grown apple cultivars. Am. Soc. Hortic. Sci.121:281-285.

Mayer AM, Harel E, Shaul RB (1965). Assay of catechol oxidase a critical comparison of methods. Phytochem. 5:783-789.

McCallum J, Walker J (1990). Phenolic biosynthesis during grain development in wheat: changes in phenylalanine ammonia-lyase activity and soluble phenolic content. J. Cereal Sci. 11:35-49.

Mittler R, Vanderauwera S, Gollery M, van Breusegem F (2004). Reactive oxygen gene network of plants Trends. Plant Sci.9:490-498.

Singleton V, Rossi J (1965). Colorimetry of total phenolics with phosphomolybdic-phosphotungstic acid reagents. Am. J. Enol. Vitic. 16:144-158.

Spoel SH, Dong X (2012). How do plants achieve immunity? Defense without specialized immune cells. Nat. Rev. Immunol.12:89–100.

Thakur A (2004).Therapeutic use of urine in early Indian medicine, Indian J. History Sci. *39*(4):415-427.

Thipyapong P, Stiffens J (1997). Tomato Polyphenol oxidase differential response of the PPO F Promoter to injuries and wound signals. Plant Physiol. 115:409-418.

Thypyapong P, Hunt M, Steffens J (1995). Systemic wound induction of potato (*Solanum tuberosum*) polyphenol oxidase. Phytochem 40:673-676.

# Comparative effects of NPK fertilizer, cowpea pod husk and some tree crops wastes on soil, leaf chemical properties and growth performance of cocoa (*Theobroma cacao* L.)

**Adejobi, K. B., Akanbi, O. S., Ugioro, O., Adeosun, S. A., Mohammed, I., Nduka, B. A. and Adeniyi, D. O.**

Cocoa Research Institute of Nigeria, P. M. B. 5244, Ibadan, Nigeria.

**A nursery experiment was carried out in Akure (rain forest zone), south-western Nigeria to study relative effect(s) of some organic wastes as fertilizers on growth performance, soil and leaf chemical composition of cocoa seedlings (*Theobroma cacao* L.) in the nursery. The experiment comprised of five treatments: Cowpea Pod Husk (CPH) (2.5 t/ha), cocoa pod husk ash (CPHA) (2.5 t/ha), kola pod husk (KPH) (2.5 t/ha), NPK15-15-15 (2.5 t/ha) and control (no fertilizer application). Each treatment was applied to 2.5 kg of soil filled polythene bags containing cocoa seedlings. The experiment was arranged in completely randomized design (CRD) with three replications. The organic wastes increased significantly (P>0.05) the plant height, stem diameter, leaf area, number of leaves, fresh root and shoot weights and dry root and shoot weights of cocoa seedlings. The treatments also increased significantly (P>0.05) soil and leaf N, P, K, Ca, Mg, Na, soil pH and organic matter (OM) content relative to the control. Kola pod husk (KPH) was the most effective in improving cocoa growth, leaf and soil chemical composition.**

**Key words:** Cocoa, growth, organic matter, NPK, cowpea pod husk, cocoa pod husk ash, kola pod husk.

## INTRODUCTION

Cocoa is one of the most important tropical crops (FAOSTAT, 2006). West Africa contributes about 70% of the world's cocoa production. The crop significantly contributed to the economies of countries in this sub-region, as well as economics of many other countries in Central America and South East Asia. Nigeria is the fourth largest producer of cocoa in the world with an estimated production of 485,000 metric tons in 2006 (FAOSTAT, 2006). Cocoa is therefore a major commodity crop cultivated in Nigeria and is a major raw material used in the production of cocoa powder (for beverage drink), various chocolate based products, biscuits and confectioneries. Processed cocoa bean is also used to make sweets, sweetening products, cocoa butter (used in making chocolate), perfume, and in pharmaceuticals. Locally, cocoa bean is used in cooking soup that has resemblance of okra and in treating various abdominal problems or ailments (Opeke, 2005). The production of cocoa in Nigeria has witnessed a downward trend since the early 1970s due to numerous factors like ageing trees, ageing farmers, wrong application of recommended agronomic techniques by farmers, effects of pests and diseases and deficiencies in macro and micro nutrients in the soils (Adejobi et al., 2011a).

Previous studies have attributed this yield decline essentially to soil nutrients imbalance (Ojeniyi et al., 1981). One way of combating this problem is the use of fertilizer. However, African farmers use very little fertilizer (8 kg/ha) compared to their counterparts in other parts of the agrarian world; hence, Africa's soils are increasingly depleted of nutrients (IFDC, 2008/2009). This is particularly true with cocoa farmers in Nigeria.

Ogunlade et al. (2009) reported that more than 85% of cocoa farmers in Nigeria do not use fertilizers on cocoa. Reasons for this low usage of fertilizers vary from lack of farmers' knowledge of the nutrients status of their soils to scarcity and high cost of fertilizers where available. The need to pay attention to soil fertilization is now almost as important as the control of capsids and black-pod disease in cocoa. Ayanlaja (2002), Adejobi et al. (2011a, b, c, d) and Moyin-Jesu (2008) reported the use of organic residues such as animal manures, urban refuse, agro-industrial processing wastes, animal dungs, refuse dump compost, pit latrine compost, foot of the hill compost, mulching, passive refuse dump in home gardens and alley cropping with appropriate nitrogen fixing shrubs, have been found capable of increasing and balancing soil nutrients with consequential increase in yield and crop performance.

The main objective of this study therefore was to examine the influence of different organic wastes on soil, leaf chemical composition and growth performance of cocoa seedlings.

## MATERIALS AND METHODS

The experiment was carried out at Federal college of Agriculture, Akure between 2010 and 2011. Akure is located in the sub-humid region with distinct dry and wet seasons. The annual rainfall ranges from 1100 to 1300 mm per annum, temperature ranges between 24 and 30°C and relative humidity is about 85%.

### Soil sampling and analysis before planting

Soil samples were randomly collected from 0 to 15 cm depth on the site. The soils were bulked, air dried and made to pass through a 2 mm sieve for chemical analysis. The soil pH (1:1 soil/water) was measured using pH meter. Organic matter was determined by the Murphy blue coloration and determined on a spectronic 20 at 882 um (Murphy and Riley, 1962). Soil potassium (K), calcium (Ca) and Magnesium (Mg) were extracted with IMNH$_4$ OAC, PH$_7$ and were determined with flame photometer; Mg was determined with an atomic absorption spectrophotometer. The total nitrogen (N) was determined by the Microkjedahl method (AOAC, 1990).

### Processing of the organic residues used for the experiment

Cocoa pod husk ash (CPHA) and kola pod husk (KPH) were both obtained from cocoa and kola processing departments of Cocoa Research Institute of Nigeria (CRIN), Ibadan. Cowpea pod husk (CPH) was obtained from near-by farm in Ibadan, Oyo State while NPK 15-15-15 fertilizer was obtained from Mikky Farm Limited, Akure, Ondo State. Cocoa pod husk, cowpea pod husk and kola pod husk were sun dried for 32 h. Only cocoa pod husk was bunt to ash and allowed to cool for another 32 h, bagged and kept in a dry place. Kola and cowpea pod husk were ground with heavy mortal, bagged and kept in a dry place.

### Chemical analysis of the organic material used

Two (2) grams each of the processed forms of the organic material used were analysed for nutrient composition using the standard procedure as described by Udo and Ogunwale (1986).

### Nursery experiment

Mature, disease-free and ripe cocoa pods were harvested from cacao plantation of the Cocoa Research Institute of Nigeria (CRIN). The pods were broken and the beans were hand- scooped for planting. The bulked soil taken from the site (0 to 15 cm depth) of the experiment was sieved to remove stones and plant debris and 2.5 kg of the sieved soil was placed into a polythene bag (25 × 13 cm). There were 5 treatments: 2.5 t/ha cowpea pod husk (CPH), 2.5 t/ha cocoa pod husk ash (CPHA), 2.5 t/ha kola pod husk (KPH), 2.5 t/ha NPK 15-15-15 and the control (no fertilizer application). Two cocoa beans were sown per polythene bag arranged in completely randomized design (CRD) and later thinned to one seedling per polythene bag. The amount of the treatments were applied using spot method a month after sowing, the parameters such as plant height, number of leaves, leaf area, stem girth and number of branches were recorded from 8 to 32 weeks after planting. At 32 weeks after planting in the nursery, the seedlings were carefully removed from the polythene bags for the measurement of shoot and root lengths, fresh shoot and root weights; then oven dried and both dry shoot and root weights were taken before they were finally analysed for N, P, K, Ca and Mg contents.

At the time of taking the shoot weight, soil samples were taken from each of the polythene bag, air dried and sieved for analysis of major elements (soil N, P, K, Ca, Mg, pH and OM) as described earlier.

## RESULTS AND DISCUSION

The result of the initial physico-chemical properties of the soil used for the experiment is presented in Table 1. The soil which was classified as an Affisol belonging to Akure series (Soil Survey Staff, 1999) had pH ($H_2O$) of 5.40, Organic matter (0.52%), total N (0.11%), available P (6.05 mg/kg), exchangeable K, Ca, and Mg being 1.20, 1.42 and 0.95 mole/kg, respectively. The values for organic matter, N, P, and Mg were generally low and fell below the critical level required for optimal performance of most tree crops in Nigeria (Egbe et al., 1989). With low N, P, K, Ca, Mg and organic matter, it is quite obvious that the soil is inherently low in fertility and therefore expected to show positive response to soil amendment. The insufficient levels of the major nutrients in the soil showed that the soil is depleted and would not be able to meet the nutritional needs of the cocoa plants unless external nutrients supply is made for the soil to be able to support optimum growth of cocoa plants. The soil particle size distribution indicated that the overall mean sand, silt

**Table 1.** Soil physiochemical composition before planting cocoa.

| Soil properties | Value |
|---|---|
| **Physical properties** | |
| Sand | 76.02% |
| Silt | 16.25% |
| Clay | 7.73% |
| Textural class | Sandy loam |
| **Chemical properties** | |
| Soil pH ($H_2O$) | 5.40 |
| Organic matter | 0.52% |
| Organic carbon | 0.25% |
| Nitrogen | 0.11% |
| Available P | 6.05 mg/kg |
| **Exchangeable bases** | |
| $K^+$ | 1.20 cmol/kg |
| $Ca^{2+}$ | 1.42 cmol/kg |
| $Mg^{2+}$ | 0.95 cmol/kg |
| $Mn^{2+}$ | 0.89 cmol/kg |
| **Exchangeable acidity** | |
| $Al^{3+}$ | 1.39 cmol/kg |
| $H^+$ | 0.12 cmol/kg |
| ECEC | 6.97 |

and clay contents of the soil were 76.02, 16.25 and 7.73%, respectively. The clay + silt values were generally below 32% estimated to be adequate for soils considered to be ideal for tree crop production especially cocoa plant (Egbe et al., 1989).

Table 2 presents data on the nutrient composition of the organic materials used for raising the cocoa seedlings; cowpea pod husk (CPH) contained 4.02% OM, 22.93 mg/kg P, 8.25 mg/kg Mg, 4.91 mg/kg Ca and 4.19 mg/kg Na. Kola pod husk (KPH) on the other hand had 2.6% N, 3.21% OM, 6.51 mg/kg P, 1.09 mg/kg Mg, 2.66 mg/kg Ca and 2.61 mg/kg Na. On the contrary, cocoa pod husk ash (CPHA) contained 2.0% OM, 1.02% N, 4.02 mg/kg P, 5.31 mg/kg K, 1.08 mg/kg Mg. 3.60 mg/kg Ca and 3.06 mg/kg Na. Cocoa pod husk ash had high K with low N and P. The low value of N in CPHA might be as a result of volatilization during the burning process since the carbon present in the material has been partially destroyed by burning. This is consistent with the findings of Ajayi et al. (2007) and Odedina et al. (2003) that cocoa pod ash contained N, P, K, Ca and Mg. The high pH of the organic materials especially the CPHA, is an indication that the soil used for the conduct of the experiment which is confirmed to be acidic will benefit positively from their addition, and hence, moderate the acidity of the soil. This finding is in agreement with the earlier results of Ayeni et al. (2008a, b) and Ajayi et al. (2007a, b) that CPHA increased soil pH due to its liming effects on the soil. The growth parameters of cocoa seedlings as influenced by different organic fertilizers application are presented in Table 3.

The organic fertilizer materials positively and significantly affected the growth parameters of cocoa seedlings such as plant height, stem diameter, number of leaves per plant and leaf area relative to control. Kola pod husk produced the highest plant height, number of leaves per plant and leaf area respectively relative to control and other fertilizer materials; this was closely followed in descending order by NPK 15-15-15, cocoa pod husk ash, cowpea pod husk and control (KPH > NPK 15-15-15 > CPHA >CPA> control). Generally, the values of KPH with respect to these parameters were either higher or comparable to the in-organic fertilizer (NPK 15-15-15) and other organic material.

Fresh and dry root weight of cocoa seedlings (Table 4) showed that NPK fertilizer and organic materials of plant origin were comparable in their values. However, values due to organic fertilizers of plant origin were higher compared to that of inorganic origin (NPK 15-15-15 fertilizer). This might be due to presence of other vital nutrient elements presence in the organic fertilizer mate-rials (Ca, Mg, organic carbon and other micronutrients) that are required for good seedling growth which are absent in the NPK 15-15-15 fertilizer. Similar results were obtained for both fresh and dry shoot weight of cocoa seedlings with kola pod husk having the highest shoot weight (13.13 g) relative to control (6.33 g). The mean weight differences recorded for KPH and NPK 15-15-15 were not significantly ($p = 0.05$) different from each other although the highest response was recorded with KPH. CPH and CPHA recorded similar mean values of 10.00 and 9.00 g, respectively for dry shoot weight while KPH was significantly ($p \geq 0.05$) higher relative to other materials and control.

The effects of the treatments on the chemical properties of the soil as presented in Table 5 shows that all the organic materials and most importantly the cocoa pod husk ash increased the soil pH significantly ($p \leq 0.05$) compared to NPK and control, respectively. This findings is in agreement with the result of Nottidge et al. (2007) that affirmed the role of ash as a liming material and effective source of nutrients for crops such as vegetables, maize and cocoa (Odedina et al., 2003; Ayeni et al., 2008).

The soil N contents ranged between 0.11 to 0.99 g/kg soil. NPK 15-15-15 significantly increased soil nitrogen content relative to control. The effect of the treatments on soil N status shows the soil was significantly ($p \leq 0.05$) and positively affected by all the treatments. The effect of kola pod husk on soil N was more pronounced followed by CPHA and CPH, respectively. The difference between NPK 15-15-15 and KPH in respect to soil was not significant (Table 5). The effects of the applied organic

**Table 2.** Chemical analysis of the organic manures used for the experiment.

| Treatment | pH ($H_2O$) | C/N ratio | OM (%) | N (%) | P (mg/kg) | K (mg/kg) | Mg (mg/kg) | Ca (mg/kg) | Na (mg/kg) |
|---|---|---|---|---|---|---|---|---|---|
| CPH | 7.02 | 6.00 | 4.02 | 2.63 | 22.93 | 3.89 | 8.25 | 4.98 | 4.19 |
| CPHA | 7.20 | 9.50 | 2.00 | 1.02 | 40.21 | 5.31 | 1.08 | 3.60 | 3.06 |
| KPH | 6.99 | 5.60 | 3.21 | 2.68 | 6.51 | 3.29 | 1.09 | 2.66 | 2.61 |

CPH, Cowpea pod husk; CPHA, cocoa pod husk ash; KPH, kola pod husk.

**Table 3.** The growth parameters of cocoa seedlings between 4 to 24 weeks after planting under different organic fertilizer application.

| Treatment | Plant height (cm) | Number of leaves | Stem girth (cm) | Leaf area ($cm^2$) |
|---|---|---|---|---|
| CPH | $27.81^b$ | $9.55^b$ | $2.24^a$ | $47.09^b$ |
| CPHA | $27.82^b$ | $10.46^a$ | $2.19^a$ | $45.91^b$ |
| KPH | $30.56^a$ | $11.38^a$ | $2.19^a$ | $53.64^a$ |
| NPK 15-15-15 | $27.87^b$ | $11.27^a$ | $2.16^a$ | $52.26^a$ |
| Control | $19.96^c$ | $6.86^c$ | $1.13^b$ | $28.96^c$ |

CPH, Cowpea pod husk; CPHA, cocoa pod husk ash; KPH, kola pod husk.
Treatment means within each column followed by the same letter are not significantly different from each other using Duncan multiple range test at 5% level.

**Table 4.** The yield parameters of cocoa seedlings under different organic fertilizer application.

| Treatment | Fresh root weight (g) | Dry root weight (g) | Fresh shoot weight (g) | Dry shoot weight (g) |
|---|---|---|---|---|
| CPH | $6.30^a$ | $5.60^a$ | $12.86^c$ | $10.00^b$ |
| CPHA | $7.30^a$ | $3.80^b$ | $14.3^{ab}$ | $9.00^b$ |
| KPH | $7.53^a$ | $3.95^b$ | $18.66^a$ | $13.13^a$ |
| NPK 15-15-15 | $6.86^a$ | $2.85^b$ | $15.90^b$ | $12.83^a$ |
| Control | $3.37^b$ | $2.88^b$ | $10.00^c$ | $6.33^c$ |

CPH, Cowpea pod husk; CPHA, cocoa pod husk ash; KPH, kola pod husk. Treatment means within each column followed by the same letters are not significantly different from each other using Duncan multiple range test at 5% level.

**Table 5.** Soil chemical analysis after the experiment under different organic fertilizer application.

| Treatment | Soil pH ($H_2O$) 1:1 | Organic carbon (g/kg) | Organic matter (%) | N (%) | P (mg/kg) | K (mg/kg) | Mg (mg/kg) | Ca (mg/kg) | Na (mg/kg) |
|---|---|---|---|---|---|---|---|---|---|
| CPH | $7.05^a$ | $2.05^a$ | $3.59^a$ | $0.20^b$ | $12.66^c$ | $1.11^a$ | $2.21^a$ | $4.10^a$ | $1.13^c$ |
| CPHA | $7.38^a$ | $1.53^b$ | $2.15^b$ | $0.27^b$ | $40.00^a$ | $1.88^a$ | $1.98^b$ | $3.19^b$ | $1.09^a$ |
| KPH | $7.28^a$ | $2.78^a$ | $2.46^b$ | $0.62^a$ | $16.00^c$ | $1.31^a$ | $2.01^a$ | $3.11^a$ | $0.66^b$ |
| NPK 15-15-15 | $5.03^b$ | $1.79^b$ | $0.85^d$ | $0.99^a$ | $20.00^b$ | $1.40^a$ | $1.10^b$ | $2.93^b$ | $0.60^b$ |
| Control | $5.95^b$ | $1.00^b$ | $0.56^d$ | $0.11^c$ | $12.00^d$ | $1.20^a$ | $0.58^c$ | $1.42^c$ | $0.50^b$ |

CPH, Cowpea pod husk; CPHA, cocoa pod husk ash; KPH, kola pod husk. Treatment means within each column followed by the same letters are not significantly different from each other using Duncan multiple range test at 5% level.

material on soil P revealed that CPHA gave significantly higher mean values relative to control and NPK 15-15-15. This might not be unconnected to the higher P present in the material as revealed by the analysis of the materials (Table 2). Similarly, all the organic materials improved soil K, Mg, Ca, and Na, respectively relative to the control. CPHA recorded the highest value of 1.88 mg/kg K followed by NPK 15-15-15 (1.40 mg/kg). CPH recorded the least soil K value. There is no significant difference among all the treatments applied. The amount of Mg, Ca and Na were also positively influenced with organic fertilizers addition irrespective of sources. This result is

**Table 6.** The leaf chemical composition under different organic manure application.

| Treatment | N (%) | P (mg/kg) | K (mg/kg) | Ca (mg/kg) | Mg (mg/kg) | Na (mg/kg) |
|---|---|---|---|---|---|---|
| CPH | 1.98[a] | 1.44[a] | 5.94[b] | 2.08[a] | 2.11[a] | 2.40[a] |
| CPHA | 1.48[b] | 1.24[b] | 7.08[a] | 1.08[b] | 1.91[b] | 1.99[b] |
| KPH | 1.49[b] | 1.31[b] | 6.01[a] | 0.73[c] | 2.06[a] | 1.99[b] |
| NPK 15-15-15 | 1.93[a] | 1.20[b] | 5.48[b] | 2.06[a] | 1.88[b] | 1.08[b] |
| Control | 1.00[c] | 0.27[c] | 0.81[c] | 0.83[c] | 0.92[c] | 1.00[b] |

CPH, Cowpea pod husk; CPHA, cocoa pod husk ash; KPH, kola pod husk. Treatment means within each column followed by the same letter are not significantly different from each other using Duncan multiple range test at 5% level.

consistent with the findings of Odedina et al. (2003) who reported that cocoa pod husk ash significantly increased soil OM, N, P, K, Ca and Mg, respectively. Adejobi et al. (2011a, b), in their work on the effects of organo-mineral fertilizer and cocoa pod husk ash in the soil, leaf chemical composition and growth of coffee concluded that combined application of organo-mineral fertilizer and cocoa pod husk ash increased soil N, P, K, Ca, Mg and pH. The nutrients element composition of the cocoa leaf as affected by different organic manure application is shown in Table 6. The leaf N and P composition of cocoa seedlings was either comparable or higher than the NPK 15-15-15 in leaf nutrient composition. The mean values in the fertilizer treated seedlings were significantly higher compared with that of control. Similar trend as obtained in leaf N and P compositions was recorded with Ca. CPH gave a significantly higher value relative to control.

The difference in values recorded for CPH and NPK 15-15-15 were comparable, though CPH produced a higher value, the difference was not significant. The leaf Mg contents was higher in CPH relative to control and other applied fertilizer material. The low leaf chemical composition value noticeable with the control is a clear indication that the soil is inherently low in soil fertility and basic nutrients for cocoa seedlings. Hence, application of organic fertilizer amendments is quite necessary for enhanced production.

## Conclusion

The use of both chemical and organic fertilisers significantly enhanced cocoa growth parameters, fresh and dry matter yield and leaf and soil chemical com-position. However, addition of organic materials such as CPH, CPHA and KPH as nutrient sources produced a promising effects on cocoa seedlings comparable to inorganic fertilizer; hence, they are advised for the purpose of cocoa seedlings establishment.

## REFERENCES


Adejobi KB, Famaye AO, Adeniyi DO, Orisajo SB, Adeyemi EA (2011a). Effect of cocoa pod husk and goat dung on nutrient content and growth performance of cashew (*Anacardium occidentale*). Adv. Environ. Biol. 5(7):1536-1542.

Adejobi KB, Famaye AO, Adeniyi DO, Akanbi OS, Orisajo SB (2011b). Comparative effect of organo-mineral fertilizer and cocoa pod husk ash on the soil, leaf chemical composition and growth performance of cacao (*Theobroma cacao* L.) in South-western Nigeria. Obeche J. 29(1):212-217.

Ajayi CA, Awodun MA, Ojeniyi SO (2007). Comparative effect of cocoa pod husk ash and NPK fertilizer on soil and nutrient content and growth of kola seedlings. Int. J. Soil Sci. 2(2):148-153.

AOAC (1990) Official methods of Analysis 12th ed. Association of Official Analytical Chemistry, Washington, D.C., USA.

Ayanlaja SA (2002). Soil nutrient management research on the humid forest zone of Nigeria for cocoa production between 1940 and 2000 missing gaps. 17th WCSS. Thailand Symposium No.16 paper, No.1503.

Ayeni LS, Adetunji MT, Ojeniyi SO (2008a). Comparative Nutrient Release from cocoa pod ash, poultry manure and NPK 20:10:10 fertilizer and their combinations. Incubation Study. Nig. J. Soil Sci. 18:114-123.

Ayeni LS, Adetunji MT, Ojeniyi SO (2008b). Integrated application of NPK fertilizer, cocoa pod ash and poultry manure effect on maize performance, plant and soil nutrient content. Int. J. Pure Appl. Sci. 2(2):34-41.

FAOSTAT (2006). Food and Agricultural Organization. World Food Balance sheet. www.faostat. fao.org.

Moyin-Jesu EI (2008). Comparative evaluation of different organic fertilizers on the soil fertility, leaf mineral composition and growth of dike nut (*Irvingia gabonensis*). Emirate J. Food Agric. 20(2):1-9.

Murphy J, Riley JP (1962). A modified single solution method for the determination of phosphate in waters. Anal. Chem. Acta 9:00-82.

Ogunlade MO, Iloyanomon CI (2009). Leaf litter fall and soil nutrient status under cocoa plantation of different ages in Ibadan, Nigeria. Nig. J. Soil Sci. 19(1):25- 28.

Ojeniyi SO (1981). Effect of long term NPK application on secondary and micronutrient contents of *Coffea canephora* Pierre. Plant Soil 60:477-480.

Udo EJ, Ogunwale JA (1986). Laboratory Manual for the analysis of Soil, Plant and Water samples. 2nd edition.

# 17

# The effect of salinity (NaCl) on germination and early seedling growth of *Lathyrus sativus* and *Pisum sativum* var. *abyssinicum*

**Berhanu Abraha Tsegay**[1] **and Berhane Gebreslassie**[2]

[1]Department of Biology, Bahir Dar University, Ethiopia.
[2]Department of Biology, Woldia University, Ethiopia.

**High salt level of a germination medium may induce a reduction, delay and even complete inhibition of germination due to osmotic effect and/or ion toxicity. The objective of this study was to investigate the effect of salinity due to NaCl on germination and early seedling growth of two crops, *Pisum sativum* var. *abyssinicum* and *Lathyrus sativus*. Seeds of these two crops were treated with NaCl induced saline germinating media prepared in Petri dishes. Fifty (50) surface sterilized seeds per Petri dish were sown in five salt treatments (0, 5, 7, 9 and 15 dSm$^{-}$). Each treatment was replicated four times. Germination percentage, shoot length and root length of both crops decreased with an increase in salinity level. Although both crops are low salt tolerant legumes, *P. sativum* var. *abyssinium* was found to be less tolerant than *L. sativus*. This study could be strengthened by further work under field conditions and also at mature vegetative and reproductive stages of the crops.**

**Key words:** Germination time, water uptake percentage, salt tolerance, seedling biomass.

## INTRODUCTION

Germination is the initial stage of a plant's life cycle and determines where and when a crop can be established. It is a complex metabolic process that oxidizes the lipids and carbohydrates within the seed and breaks down storage proteins in order to obtain energy and amino acids necessary for plant development (Almansouri et al., 2001). Seed germination and early seedling growth under saline conditions are considered as major factors limiting the establishment of crops (Kitajima and Fenner, 2000). Interaction between seedbed environment and seed quality is also important (Khajeh-Hosseini et al., 2003). Plant available water is restricted in soils containing excess sodium chloride, resulting in partial dehydration of cell cytoplasm. Such plasmolysis affects the metabolism of cells and functions of macromolecules and, ultimately, results in cessation of growth (Le Rudulier, 2005).

The effect of salinity on germination can be either by creating osmotic potential which prevents the uptake of water or by the toxic effects of ions on embryo viability (Houle et al., 2001). Salts absorb and retain water with such strength that it is not freely available in the soil, causing an increase of soil solution osmotic pressure. Salt stress may cause significant reductions in the rate and final germination percentage (FG %), which in turn may lead to an uneven stand establishment and a reduction in yields.

Rapid, uniform and a high germination percentage for legumes is a prerequisite for successful stand establishment and yield (Demir and Ermis, 2003). The specific ions likely to be most abundant and cause greatest problems are sodium ($Na^+$) and chloride ($Cl^-$).

The legume family is the third largest of flowering plants (Morris et al., 2003; Lewis et al., 2005). Economically, legumes represent the second most important family of crop plants after Poaceae (grass family), accounting for approximately 27% of the world's crop production (Graham and Vance, 2003). Legumes, in developing countries, are largely produced as subsistence crops by small hold farmers. Ethiopia is a good producer of food legumes after Egypt in Africa (Lewis et al, 2005). However, about 10,608 ha of Ethiopia's total land is affected by salinity in semi-arid and arid regions (Geressu, 2011), which may reduce yield. Moreover, irrigated lands in the semi-arid parts of the country are increasingly becoming saline and turning to a new scenario of hampering food production for the fast growing population.

Reclaiming salt affected land is always costly and time consuming (Turan et al., 2007). Therefore, in countries like Ethiopia where legumes are principally grown for human consumption; the focus should remain on the development of salt tolerant cultivars to attain food security of the rapidly growing population. For this purpose, screening of suitable crops and cultivars is necessary. Identification of salt tolerant crops and cultivars is important so that such unproductive land can be used. Although, to our knowledge, there is no previous data available on *L. sativus* and *P. sativum* var. *abyssinicum* response to salinity in Northern Ethiopia. The objective of this study was therefore to investigate the effect of salinity on germination and early seedling growth of *P. sativum* var. *abyssinicum* and *L. sativus*.

## MATERIALS AND METHODS

This study was conducted from October 2012 - March 2013 in the Laboratory of Botanical Sciences at the Department of Biology, Bahir Dar University, Ethiopia. Sterilized, hand selected seeds of both *L. sativus* (from Adet, Amhara Region) and *P. sativum* var *abyssinicum* (from Maichew, Tigray Region) were used for the study. Before beginning the experiment, solutions were made by dissolving sodium chloride in distilled water at five different concentrations (0, 5, 7, 9 and 15 dSm⁻ EC) (Li, 2008) and left for 48 h in order to dissolve. The salt solutions were prepared every six days so that it is relatively fresh for the germinating seeds.

Following the solution preparation, Petri dishes were washed and disinfected with alcohol and air dried. Germination chambers were prepared from the sterilized glass Petri dishes and filter papers (Whatman No. 2). The filter papers were cut into two pieces of equal size with labeled seed compartments and concentrations of salt solutions; and made ready for use. Treatments were arranged in a completely randomized design (CRD) and each treatment combination was replicated four times.

Dry weight of the hand selected seeds were recorded before the seeds were surface sterilized with 5% sodium hypochlorite solution for 10 min and rinsed with sterile distilled water three times. Fifty seeds were sown in each Whatman filter paper bedded dishes (Kaya et al., 2005). The seeds were watered with the appropriate salt solutions and left for 24 h where after, the fresh weight was recorded to calculate the uptake of water by the seeds. This was done as follows (Gairola et al., 2011):

Water uptake percentage =

$$\frac{\text{Seed fresh weight} - \text{Seed dry weight}}{\text{Seed fresh weight}} \times 100$$

After the measurement of water uptake, the seeds were re-sown into the Petri dishes. They were then watered daily with the appropriate salt solutions. All the germination and early seedling growth parameters were evaluated using the method used by Li (2008) with some modifications. Counting germinated seeds started 24 h after sowing every day for 14 days. A seed was considered to be germinated when plumule and radical emerge from the seeds. In all treatments, a continuous assessment in seedling growth were carried out during the subsequent days for both study crops, until day 14.

Germination rate (GR), which is important for state of readiness for early seedling growth under laboratory condition was evaluated as follows:

Germination rate = $(X_n - (X_n\text{-}1))/Y_n$

where, $X_n$ is the number of germinated seeds at the $n^{th}$ day and $y_n$ is the number of days from sowing until the $n^{th}$ harvesting time.

Germination Percentage (GP %): (GP) =

$$\frac{\text{Number of germinated seeds}}{\text{Total number of seeds sown}} \times 100$$ (Kandil et al., 2012).

Mean germination time (MGT): MGT = $\frac{\sum dn}{\sum n}$ (Kandil et al., 2012)

Where n is the number of seeds which germinated on day 'd', and 'd' is the number of days counted from the beginning of germination.

Germination index (GI): was calculated as the product of number of days after sowing and number of germinated seeds divided by the total number of sown seeds.

GI = $\frac{\sum diNi}{S}$ (Li, 2008)

where 'di' is number of days after sowing of seeds under a particular treatment, 'Ni' is number of germinated seeds, 'S' is total number of sown seeds of the study legume crops.

Seedling height reduction (SHR): the delay in root length and shoot length expressed in percentage was calculated as:

$$\frac{\text{Plant height at control} - \text{plant height at salt treat condition}}{\text{Plant height at control}} \times 100$$

(Morris et al., 2003)

Relative NaCl injury rate was calculated as the difference between germination percentage in control and germination percentage in salt treated seeds dividing by the germination percentage in the untreated seeds. The emergence of radicle (root) and shoot (plumule) of *P. sativum* and *L. sativus* from each Petri dish were assessed every day after sowing. And then the salt tolerance rate was calculated by using the standard formula used by Kaymakanova (2009):

$$\text{Salt tolerance (ST)} = \frac{\text{Seedling dry weight of treated}}{\text{Seedling dry weight in control}} \times 100$$

**Data analysis**

All collected data for germination and early seedling growth were organized, analyzed and interpreted using SPSS 20 for Windows and Microsoft excel softwares. Variance were statistically computed by analysis of variance (ANOVA) for the complete random design using SPSS version 20, and comparisons of means were made using the Tukey HSD significant difference test (HSD) at $P<0.05$. Graphics and tables were used to show the distribution of the response of the seeds to the different treatments.

## RESULTS AND DISCUSSION

### Water uptake percentage

Entry of water into seeds is greatly influenced by the nature of the seed coat (pericarp). Seeds of *L. sativus* and *P. sativum* var. *abyssinicum* respond differently though they are treated with the same salt concentrations. The results in Table 1 indicate that water uptake percentage increased in seeds of *L. sativus* and *P. sativum* treated with higher salt concentrations as compared to those treated with distilled water. In both seeds, the ability to absorb water is influenced by the concentration of salts and index of reduction increases with increasing salinity level from 9 to 15 dSm⁻.

Seeds of *P. sativum* showed a high percentage of water uptake increment in all NaCl solution treatments except at the 7 dSm⁻ as compared to that of *L. sativus* seeds (Table 1). The overall water uptake percentage was not, however, statistically significant ($p < 0.05$) between the crop types and among the different salt concentrations except at 5 dSm⁻ for *L. sativus* (Table 1). This is as a result of large sample size.

### Germination percentage

The interaction between the crops and salt concentrations was significant, meaning that the crops respond differently at specific salt concentration. This is shown by the significant differences ($P < 0.001$) in germination percentage of the two crop types given in all treatments (Table 1). Increased salt concentration caused decrease in germination percentage, but the extent of reduction under high concentration stress was much greater than that under low salt stress (Table 1).

Under the highest salt concentration (15 dSm⁻), no seed germinated for either crop. Germination percentage of *L. sativus* was significantly ($p < 0.05$) higher than *P. sativum* var. *abyssinicum* at all salt concentrations (Table 1). This suggests that seeds of *L. sativus* could germinate well at a relatively higher concentration of NaCl than *P. sativum* var. *abyssinicum* seeds.

### Germination rate

Differences among NaCl treatment were statistically significant ($P<0.0001$) for germination rate of the study seeds. The germination rate decreased as salt concentration increased to a 9 dSm⁻ and delayed for the highest salt dosage (Table 1). Since the higher salt concentration limited the water absorption, it slows down the germination rate. Higher germination rate was recorded for *L. sativus* seeds at 0, 5, and 7 dSm⁻ salt concentrations as compared to *P. sativum* var. *abyssinicum*. No seeds germinated at the highest salt concentration (15 dSm⁻) for both species. Since higher salinity limited water absorption, it prevents nutrient assimilation, as a result, germination rate declined with increasing salinity. The results of this study were similar to the findings of Akhtar and Hussain (2008) and Kaydan and Yagmur (2008) who worked on different plant species. For example, Akthar and Hussain (2008) reported that germination of three grasses (*Bothriochlo apertusa, Dichanthium annulatum* and *Panicum antidotale*) significantly declined even at 5 dSm⁻ level of treatment.

### Mean germination time

Salinity had a considerable increasing effect on germination time of *P. sativum* var. *abyssinicum* and *L. sativus* up to a certain level (Table 1). The increasing effect was more for *L. sativus* as compared to *P. sativum* var. *abyssinicum.* The highest average mean germination time was obtained from seeds treated with the 5 dSm⁻ salt concentration. In this study, mean germination time for *P. sativum* var. *abyssinicum* seeds at 5 dSm⁻ was higher than the control. This difference is significant at $p< 0.05$. There was also a significant difference ($P < 0.001$) in mean germination time of the two crops at 0, 5 and 7 dSm- salt concentrations. It means that the higher the salt concentration the longer the germination time until seeds develops tolerance and starts to germinate by overcoming the germination time delay (Patto, 2009).

*L. sativus* seeds took much time for germination than *P. sativum* var. *abyssinicum* seeds though both of them increased in mean germination time with increased salt solution treatment up to 5 dSm⁻. That means seeds of *L. sativus* can tolerate salinity effect up to 9 dSm⁻ than *P. sativum* var. *abyssinicum* seeds. This shows that *L. sativus*

**Table 1.** Means comparison for effects of salinity (NaCl) on water uptake, percent germination, germination index, germination rate and mean germination time of *L. sativus* and *P. sativum* var. *abyssinicum*.

| NaCl ($dSm^{-}$) | Water uptake (%) | | Germination (%) | | Germination index | | Germination rate | | Mean germination time (day) | |
|---|---|---|---|---|---|---|---|---|---|---|
| | *L. sativus* | *P. sativum* | *L. sativus* | *P. sativum* | *L. sativus* | *P. sativum* | *L. sativus* | *P. sativum* | *L. sativus* | *P. sativum* |
| 0 | $84.83^{ab}$ | $55.34^{ab}$ | $99.9^{a}$ | $84.40^{bc}$ | $4.31^{a}$ | $3.92^{ab}$ | $13.22^{a}$ | $12.86^{a}$ | $2.91^{ab}$ | $1.98^{c}$ |
| 5 | $91.95^{ab}$ | $73.89^{ab}$ | $81.90^{b}$ | $38.55^{d}$ | $5.45^{a}$ | $2.73^{bc}$ | $10.34^{b}$ | $5.44^{c}$ | $3.63^{a}$ | $2.26^{bc}$ |
| 7 | $87.72^{ab}$ | $71.90^{ab}$ | $65.95^{c}$ | $18.85^{ef}$ | $5.24^{a}$ | $1.11^{d}$ | $8.66^{b}$ | $2.82^{de}$ | $3.50^{a}$ | $0.74^{d}$ |
| 9 | $86.45^{ab}$ | $81.30^{ab}$ | $22.10^{e}$ | $5.90^{gf}$ | $1.37^{cd}$ | $0.58^{d}$ | $3.53^{cd}$ | $0.92^{ef}$ | $0.91^{d}$ | $0.38^{d}$ |
| 15 | $89.11^{ab}$ | $84.70^{ab}$ | $0.00^{g}$ | $0.00^{g}$ | $0.00^{d}$ | $0.00^{d}$ | $0.00^{f}$ | $0.00^{f}$ | $0.00^{d}$ | $0.00^{d}$ |

Superscripts within the means of each column (a-f) with different letters indicate significant difference among means ($p<0.05$, using Tukey HSD test).

seeds are capable of germinating if the adverse effect of salinity is not extreme during their dormancy period. Delayed germination causes increased irrigation cost and irregular and weak seedling growth in the establishment of legume crops. Relevant results were reported by Gunjaca and Sarcevic (2000) and Almansouri et al. (2001). They reported that increasing osmotic potential decrease water uptake and slow down germination time.

## Germination index

The germination index of both crops decreased significantly with increasing NaCl concentration (Table 1). The reduction gets stronger particularly at the higher level of salt concentration when compared with the control. Thus, germination index and NaCl concentration were negatively correlated. *L. sativus* was less affected by salt treatments as it has long, hard and penetrating root system than *P. sativum* var. *abyssinicum*. On the other hand, *P. sativum* var. *abyssinicum* seeds were more sensitive to the same salinity levels. In this crop, increase in salt concentration caused higher decreases in germination index values as compared to *L. sativus* (Table 1).

Generally, increase in salt concentration decreased the germination index of the study crops and this is in line with the findings of Khayatnez and Gholamin (2011) in *Zea mays* that showed decreased germination index as salt concentration increased. Previous reports from other workers also corroborate these results, for instance, Khan et al. (2009) reported on hot pepper. In this study, germination index of *P. sativum* var. *abyssinicum* decreased significantly with increasing NaCl concentration (Table 1). The reduction gets stronger particularly at 9 $dSm^{-}$ NaCl concentration as compared to the control. Total decline in germination was observed at the highest salinity level (15 $dSm^{-}$ NaCl). An increased germination index is indicative of decreased phytotoxicity and thus of a more mature germinated seeds (Khayatnezhad and Gholamin, 2011). However, results obtained using the germination index should be interpreted with care because they are affected by the type of seed used and the source of salinity. Application of damaged seeds and non-stabilized salts ($CaCl_2$) to germinating seeds may lead to immobilization of plant nutrients and cause phytotoxicity (Khan et al., 2009).

## Effects of salinity on early seedling growth

Generally, germination and early seedling growth declined with increasing salinity levels, although reduction in root length was higher than reduction in shoot length. With higher salt concentration seedling growth rate declined in both crops. Root elongation was sensitive than shoot under the stresses on both seeds. It was affected at the lowest salt stresses (5 and 7 $dSm^{-}$) and was completely inhibited at the highest salt (15 $dSm^{-}$) concentrations (Table 2). As the osmotic pressure decrease at the germination environment, water cannot carry most of the water soluble nutrients to the root in order to increase osmotic pressure of roots to get enough water. Similar results were also found in older seedlings of chick pea by Li (2008). Thus, salinity had a deleterious effect and even total decline in seedling growth rate as the level of salinity increased beyond the tolerance range which could have resulted from a lower absorption of salt component by seed, and

**Table 2.** Mean comparisons for effects of salinity (NaCl) on root length, root height reduction, root fresh weight, root dry weight, shoot length and shoot height reduction of *L. sativus* and *P. sativum* var. *abyssinicum*.

| NaCl (dS/m⁻) | Root length (cm) | | Root height reduction (%) | | Root fresh weight (g) | |
|---|---|---|---|---|---|---|
| | *L. sativus* | *P. sativum* | *L. sativus* | *P. sativum* | *L. sativus* | *P. sativum* |
| 0 | $10.26^{a}$ | $8.03^{b}$ | $0.00^{d}$ | $0.00^{c}$ | $7.4^{a}$ | $6.7^{a}$ |
| 5 | $4.13^{c}$ | $2.01^{c}$ | $57.37^{c}$ | $74.34^{bc}$ | $6.5^{a}$ | $4.55^{b}$ |
| 7 | $1.47^{cd}$ | $0.55^{d}$ | $57.37^{c}$ | $92.84^{ab}$ | $3.7^{bc}$ | $4.55^{b}$ |
| 9 | $1.47^{cd}$ | $0.55^{d}$ | $93.32^{ab}$ | $92.84^{ab}$ | $3.7^{bc}$ | $1.83^{cd}$ |
| 15 | $1.47^{cd}$ | $0.00^{d}$ | $93.32^{ab}$ | $100^{a}$ | $3.69^{bc}$ | $1.83^{cd}$ |

| NaCl (dS/m⁻) | Root dry weight (g.) | | Shoot length (cm) | | Shoot height reduction (%) | |
|---|---|---|---|---|---|---|
| | *L. sativus* | *P .sativum* | *L. sativus* | *P. sativum* | *L. sativus* | *P. sativum* |
| 0 | $2.85^{ab}$ | $3.48^{a}$ | $10.12^{a}$ | $10.29^{b}$ | $0.00^{e}$ | $0.00^{a}$ |
| 5 | $2.51^{abc}$ | $2.28^{bc}$ | $8.93^{b}$ | $7.79^{b}$ | $12.70^{de}$ | $24.06^{cde}$ |
| 7 | $2.51^{abc}$ | $1.82^{c}$ | $5.31^{c}$ | $6.15^{cd}$ | $45.58^{bcd}$ | $40.72^{bcd}$ |
| 9 | $1.68^{cd}$ | $0.98^{de}$ | $3.80^{cd}$ | $4.51^{de}$ | $71.06^{ab}$ | $54.58^{bc}$ |
| 15 | $0.00^{e}$ | $0.00^{e}$ | $0.00^{e}$ | $.0.00^{d}$ | $79.82^{ab}$ | $100^{e}$ |

Superscripts (a-e) with different letters indicate significant difference among means within the columns (p<0.05, using Tukey HSD).

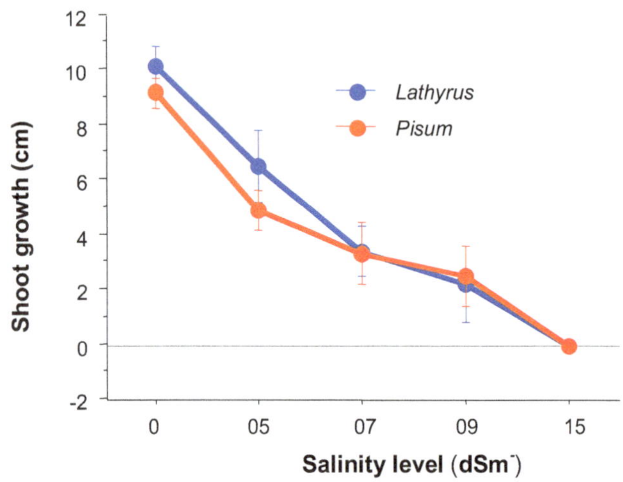


**Figure 1.** Seedling length of *L. sativus* and *P. sativum* var. *abyssinicum* grown under saline conditions at different stress levels. Error bars ± standard deviation.

germination process is also less responsive to high tissue sodium concentrations than seedlings growth (Figure 1).

The results of the present study showed that *P. sativum* var. *abyssinicum* and *L. sativus* roots during early seedling growth were more markedly inhibited than shoots under same salt stress conditions (Table 2). When compared with shoot, root elongation is more sensitive to the stress and is injured more severely (Table 2) because they are the first organs to face the stress. The injurious effect caused by salt stress was much greater on *P. sativum* var. *abyssinicum* than *L. sativus* (Table 2).

Sowing *P. sativum* var. *abyssinicum* under 7 dSm⁻ salt stress had recorded lower root elongation than *L. sativus* seeds with significant difference between the two legumes ($p<0.01$). This species was highly affected as salinity stress level increases. However, both study crops at higher salinity levels, that is 9 dSm⁻ produced the shortest roots with significant differences between them. In addition, results showed that maximum root length was obtained from sown *L. sativus* in the control treatment (Table 2).

Means comparison indicated that a significant reduction of leaf number was observed between seed types when salinity stress was increased from 0 to 9 dSm⁻ while there

**Table 3.** Means comparison for effects of salinity (NaCl) on shoot dry weight, leaf number, relative NaCl injury rate, shoot fresh weight, seedling biomass and salt tolerance of *L. sativus* and *P. sativum* var. *abyssinicum*.

| NaCl (dSm$^{-}$) | Shoot dry weight | | Leaf number | | Relative injury rate | |
|---|---|---|---|---|---|---|
| | ***L. sativus*** | ***P. sativum*** | ***L. sativus*** | ***P. sativum*** | ***L. sativus*** | ***P. sativum*** |
| **0** | $3.42^{ab}$ | $2.95^{bc}$ | $1.93^{a}$ | $1.68^{ab}$ | $00^{f}$ | $8^{ef}$ |
| **5** | $4.12^{a}$ | $3.52^{ab}$ | $1.48^{b}$ | $0.79^{cd}$ | $18^{de}$ | $56^{c}$ |
| **7** | $2.24^{cd}$ | $1.73^{de}$ | $0.99^{c}$ | $0.79^{cd}$ | $34^{d}$ | $78^{b}$ |
| **9** | $2.44^{cd}$ | $1.40^{e}$ | $0.37^{de}$ | $0.33^{e}$ | $78^{b}$ | $93^{ab}$ |
| **15** | $0.00^{f}$ | $0.00^{f}$ | $0.00^{e}$ | $0.00^{e}$ | $100^{a}$ | $100^{a}$ |

| NaCl (dS/m$^{-}$) | Shoot fresh weight | | Seedling biomass | | Salt tolerance | |
|---|---|---|---|---|---|---|
| | ***L. sativus*** | ***P. sativum*** | ***L. sativus*** | ***P. sativum*** | ***L. sativus*** | ***P. sativum*** |
| **0** | $7.40^{a}$ | $5.03^{a}$ | $1.00^{a}$ | $0.79^{ab}$ | $100^{a}$ | $100^{a}$ |
| **5** | $8.8^{a}$ | $7.40^{b}$ | $0.46^{bc}$ | $0.28^{cd}$ | $84.19^{a}$ | $83.86^{a}$ |
| **7** | $4.83^{b}$ | $3.36^{bc}$ | $0.27^{cd}$ | $0.10^{d}$ | $55.10^{b}$ | $49.01^{b}$ |
| **9** | $5.06^{b}$ | $2.88^{c}$ | $0.27^{cd}$ | $0.21^{cd}$ | $59.95^{b}$ | $39.18^{b}$ |
| **15** | $0.00^{f}$ | $0.00^{f}$ | $0.00^{d}$ | $-0.00^{d}$ | $-0.00^{c}$ | $-0.00^{c}$ |

Superscripts (a-f) with different letters indicated significant difference among means within columns ($p<0.05$, using Tukey HSD).

was no significant difference between salinity levels of 0, 5 and 7 dSm$^{-}$. Increasing salinity to 9 dSm$^{-}$ NaCl decreased shoot fresh weight, shoot dry weight, leaf number, seedling biomass, salt tolerance and increased relative sodium chloride injury of the two crops (Table 3). Shoot fresh weight was significantly influenced ($P<0.05$) by salinity levels. The highest shoot fresh weight was obtained from 5 dSm$^{-}$ salinity level while the lowest weight was at 15 dSm$^{-}$. Plant growth was affected positively up to 5 dSm$^{-}$ salt levels. At this level shoot fresh weight was higher than the control. However, shoot fresh weight significantly decreased as salinity level increased above 5 dSm$^{-}$ (Table 3). This result was supported by the findings of Kaya et al. (2005) who showed that shoot fresh weight increased in low NaCl levels for some plants.

Salinity stress significantly ($P<0.05$) affected shoot dry weight as the salt concentration dosage increased. Shoot dry weight significantly decreased at salt levels over 7 dSm$^{-}$. The highest shoot dry weights obtained (4.12 and 3.52 g for *L. sativus* and *P. sativum* var. *abyssinicum*, respectively) were from 5 dSm$^{-}$ salinity level, which were statistically different from the control. The present findings are in agreement with the results of other researchers on Alfalfa and Ryegrass (Mohammadi et al., 2008; Pessarakli and Kopec, 2009).

## Leaf number and seedling biomass

Means comparison indicated that a significant reduction of leaf number and seedling biomass was observed between seed types when salinity stress was increased from 0 to 9 dSm$^{-}$. At the higher salt concentration (9 dSm) the leaf number was extremely affected and made seedlings less thrived for the stress. Leaves were totally delayed at the highest concentration (15 dSm$^{-}$) for both crops (Table 3). Seedling biomass of the crops also reduced as salt concentration increased (Table 3). Rahman et al. (2008) reported such a reduction and the reasons were believed to be due to slow or less mobilization of reserve foods, suspending the cell division, enlarging and injuring hypocotyls with increase in salt concentration in Gina cultivar. When the two species were compared, seedling biomass was higher for *L. sativus* than *P. sativum* var. *abyssinicum*.

## Relative sodium chloride (NaCl) injury rate

The injury rate at 15 dSm$^{-}$ was 100% for both species. Effects were significant at all salt concentrations except at 5 and 7 dSm$^{-}$ for *L. sativus* and at 7 and 9 dSm$^{-}$ for *P. sativum* var. *abyssinicum* (Table 3). Relative NaCl injury rate was found to be higher for *P. sativum* var. *abyssinicum* as compared to *L. sativus* seeds at the same salt treatments. There was a direct relationship between the salinity levels and injury rate on both crops. Injury rate increased with increasing concentration of salinity level. This was similar to earlier studies made by Hadush and Gebreslassie (2012) on *L. sativus* landraces.

## Salt tolerance

The present study indicates that salt tolerance in early

seedling stage was not correlated with seed germination. There were germinated seeds at the highest concentration (15 dSm$^{-}$) but died upto three days after germination (Table 3). This shows germination stage is more saline tolerant, which could result from a lower absorption of salt components by the seed. Germination process is also less responsive to high tissue sodium concentrations than seedlings growth. The result agrees with that of Mahdavi and SANAVY (2007) for *Schinopsis quebracho* where seeds at germination stage were relatively tolerant to salinity that at a later seedling stage. There was no tolerance at highest salinity (15 dSm$^{-}$) in both species. The difference in response to salinity shown by the two species at early stage of development may be related to their morphological and physiological variation. Results by Munns and James (2000) showed that crop plants could be salt tolerant at germination but turned salt sensitive during vegetative development.

## Conclusion

Salt (NaCl) stress through enhancement of osmotic pressure leads to the reduction of germination percentage, germination rate, germination index and increase in mean germination time of *P. sativum* var. *abyssinicum* and *L. sativus* seeds. Salinity disturbed the metabolic and physiological processes starting from the imbibitions stage. *L. sativus* seeds were different from *P. sativum* var. *abyssinicum* for all the studied germination and seedling growth traits. Germination percentage decreased significantly with increasing salt concentration (≥ 7 dSm$^{-}$). Dry and fresh weight of seedlings decreased as seedling length declined with increasing salinity levels because root number, shoot number, root length and shoot length decreased significantly. The injurious effect caused by salinity stress was much greater on *P. sativum* var. *abyssinicum* than *L. sativus*. Roots of the early seedlings of *P. sativum* var. *abyssinicum* and *L. sativus* were more markedly inhibited than shoots under salt stress conditions. Salt tolerance in early seedling stage was not correlated with later developments. The results indicated the availability of salinity tolerant crops (to medium salinity) that could be useful in future planning for saline environments.

### REFERENCES


Akhtar P, Hussain F (2008). Salinity tolerance of three range grasses at germination and early growth stages. Pak. J. Bot. 40:2437-2441.

Almansouri MJ, Kinet M, Lutts S (2001). Effect of salt and osmotic stress on germination in durum Wheat (*Triticum durum* Desf.). Plant Soil 231:243-254.

Demir I, Ermis S (2003). Effect of controlled hydration treatment on germination and seedling growth under salt stress during development in tomato seeds. Euro. J. Hortic. Sci. 68:53-58.

Gairola KC, Nautiyal AR, Dwivedi AK (2011). Effect of temperatures and germination media on seed germination of *Jatropha Curcas* L. Adv. Biores. 2:66-71.

Geressu K (2011). The Response of Some Haricot Bean (*Phaseolus vulgaris*) Varieties for Salt Stress during Germination and Seedling Stage. Curr. Res. J. Biol. Sci. 3:282-288.

Graham PH, Vance CP (2003). Legumes: Importance and constraints to greater use. Plant Physiol. 131:872-877.

Gunjaca J, Sarcevic H (2000). Survival analysis of the wheat germination data. 22nd International Conference. Information Technology Interfaces ITI 2000. June, 2000, Pula, Croatia.

Hadush T, Gebresilassie B (2012). The effect of salinity (NaCl) on germination of selected Grass pea (*Lathyrus sativus* L.). Landraces of Tigray. Asian J. Agric. Sci. 4:96-101.

Houle GL, Morel L, Reynolds CE, Seigel J (2001). The effect of salinity on different developmental stages of an endemic annual plant, *Asterlaurenti anus*. Am. J. Bot. 88:62-67.

Kandil AA, Sharief AE, Ahmed SRH (2012). Germination and Seedling Growth of Some Chickpea Cultivars (*Cicer arietinum* L.) under salinity stress. J. Basic Appl. Sci. 8:561-571.

Kaya MD, Kaya G, Kolsarici O (2005). Effects of NaCl concentration on germination and emergence of some *Brassica* species. J. Agric. Sci. 11:448-452.

Kaydan D, Yagmur M (2008). Germination, seedling growth and relative water content of shoot in different seed sizes of triticale under osmotic stress of water and NaCl. Afr. J. Biotechnol. 7:2862-2868.

Kaymakanova M (2009). Effect of salinity on germination and seed physiology in bean (*Phaseolus vulugaris* L.). Biotechnol. 23:326-329.

Khajeh-Hosseni MA, Powell A, Bingham IJ (2003). The interaction between salinity Stress and seed vigour during germination of soybean seeds. Seed Sci. Technol. 31:751-725.

Khan HA, Ayub CM, Pervez MA, Bilal RM, Shahid MA, Ziaf K (2009). Effect of seed priming with NaCl on salinity tolerance of hot pepper (*Capsicum annuum* L.) at seedling stage. Soil Environ. 28:81-87.

Khayatnezhad M, Gholamin R (2011). Effects of salt stress levels on five maize (Zea mays L.) cultivars at germination stage. Afr. J. Biotechnol. 10:12909-12915.

Kitajima K, Fenner M (2000). Ecology of seedling regeneration. J. Plant Cell Environ. 25:239-250.

Le Rudulier D (2005). Osmoregulation in Rhizobia: The key role of compatible solutes. Grain Legume 42:18-19.

Lewis G, Schrirer B. Mackinder B (2005). *Legumes of the World*; Royal Botanical Gardens. Kew, UK.

Li Y (2008). Effect of salt stress on seed germination and seedling growth of three Salinity plants. Pak. J. Biol. Sci. 11:1268-1272.

Mahdavi B, Sanavy SAMM (2007). Germination and seedling growth in grass pea (*Lathyrus sativus*) cultivars under salinity conditions. Pak. J. Biol. Sci.10:273-279.

Mohammadi H, Poustini K, Ahmadi A (2008). Root nitrogen remobilization and ion status of two alfalfa (*Medicago sativa* L.) cultivars in response to salinity stress. J. Agron. Crop Sci. 194:126-134.

Morris B, Charles S, Son S (2003). Legumes. In: Solomon HK (ed.). *Encyclopedia of Food and Culture*. New York.

Munns R, James RA, Läuchli A (2006). Approaches to increasing the salt tolerance of wheat and other cereals. J. Exp. Bot. 57:1025-1043.

Patto MC (2009). Genetics, genomics and breeding in crop plants. In: *Cool season Legumes*. Science Publication. Plymouth, UK.

Pessarakli M, Kopec DM (2009). Screening various Ryegrass cultivars for salt stress tolerance. J. Food Agric. Environ. 7:739-743.

Rahman M, Soomro U, Zahoor-ul-Hag M, Gul Sh (2008). Effects of NaCl salinity on Wheat (*Triticum aestivum* L.) cultivars. World J. Agric. Sci. 4:398-403.

Turan MA, Turkmen N, Taban N (2007). Effect of NaCl on stomatal resistance and proline, chlorophyll, Na, Cl and K concentration of lentil plants. J. Agron. 6:378-381.

# Yield and water use efficiency (WUE) responses of forage sorghum ratoon crop under varying salinity and irrigation frequency

**A. R. Saberi**

Scientific Board of Agricultural and Natural Science Research Center of Golestan Province, Gorgan, Iran

**Water stress is associated with low availability of water or with osmotic effects arising from salinity. Besides affecting crop yields, salinity may also influence biochemical composition and nutrient concentration of forage crops. To determine the effect of salinity on biochemical composition and nutrient concentration of forage sorghum [*Sorghum bicolor* (L.) *Moench*], two varieties of sorghum, Speedfeed and KFS4 were grown under rain shelter at salinity levels of 0, 5, 10, 15 dS $m^{-1}$ and irrigated when the leaf water potential reached -1 (control), -1.5 and -2 MPa. The factorial treatment combinations were arranged in a randomized complete block design with three replications. Stem dry matter, leaf dry matter and eventual dry matter yields of ratoon forage sorghum decreased with increasing salinity and irrigation interval. The reduction in plant biomass under stress conditions was found to be associated with increase in water use efficiency (WUE). Since the saline applications continued until the second harvest (first ratoon crop) as soil salinity increased, dry matter yields were reduced dramatically. No viable plants were obtained in the second ratoon crop for KFS4 variety at 15 dS $m^{-1}$ salinity and -2 MPa irrigation frequency treatments. Based on all parameters evaluated in the first ratoon crop, there was no concrete evidence to suggest that Speedfeed was superior in performance, but in the second ratoon, forage yields suggest that this variety can be a good alternative in planning for forage production.**

**Key words:** Salinity, irrigation frequency, ratoon crop, forage sorghum.

## INTRODUCTION

The productivity of plants is strongly influenced by environmental stresses. Scientists consider soil salinity as one of the major factors that reduce plant growth in many regions of the world (McCarty and Dudeck, 1993; Murdoch, 1987). Sodium chloride (NaCl) is the predominant component contributing to salinity in soils (Jungklang et al., 2003). To overcome this problem, the search for salt tolerant forage sorghums has increased (Harivandi et al., 1992). Salt tolerant plants have the capability to minimize these detrimental effects by a series of morphological, physiological and biochemical processes (Jacoby, 1999). Under the variable saline environments, plants have developed different adaptive mechanisms (Rhodes et al., 2002 and Borsani, et al., 2003).

Sorghum has tillering characteristics that enable it to completely regenerate the above-ground portions of the plant. Individual sorghum plants have been kept alive for as long as 7 years where the climate is mild enough to avoid winterkill and disease and insect protection have been provided. These important features of sorghum have allowed producers to seek a second crop within the same growing season. This decision, however, must be made at or soon after harvest of the main crop, and re-growth should be managed in the same manner as the main crop (Teutsch, 2006). A new generation of salt-tolerant forage varieties would allow for landscape development in saline environments and where salt water usage is a problem, or where fresh water is limited or not available for irrigation. Forage sorghum development in these areas often requires the use of brackish water from affected wells or other secondary sources. The effect of salinity and water deficit on ratoon crops of sorghum is not known. Hence, studies are needed to improve understanding on yield re-growth of forage sorghum under influence of salt and water stresses. In addition, many factors need to be considered when addressing the suitability of irrigation water with respect to salinity. To our knowledge, there are no published studies on drought and salt water tolerance among these forage sorghum varieties. Proper utilization of highly salt tolerant forage sorghum varieties will benefit the growing forage sorghum industry. The specific objectives of this study were: to determine yield, biomass partitioning and water use efficiency (WUE) responses in the first ratoon crop of the two forage sorghum varieties to different levels of salinity and irrigation frequency.

## MATERIALS AND METHODS

The factorial experiment was conducted under rain shelter at University Putra Malaysia (02°N 59.476′ 101°°E 2.867′, 51 m altitude), from January 2009 to December 2009. The climatic conditions recorded under rain shelter were 31°C mean temperature, 88% humidity, 4.5 mm evaporation and 71% light at 12 am. Two selected (Fouman et al., 2003) salt tolerant varieties namely Speedfeed and KFS4, of forage sorghum [*Sorghum bicolor* (L.) Moench] were subjected to the salinity levels of 0, 5, 10 and 15 dS $m^{-1}$ of NaCl concentrations, and irrigated when the leaf water potential reached -1 (control),-1.5 and -2 MPa.

The treatments were arranged in a randomized complete block design with three replications. Polybags (40 × 45 cm) were filled with a mixture of top soil, peat moss and sand at the ratio of 3:2:1 (v/v), respectively. The plants were irrigated with non-saline water for seedling establishment and with saline water starting from the $2^{nd}$ week after germination according to the treatments. The soil mixture which had a pH of 6.52 was filled into polyethylene bags of size 40 x 45 cm. Each bag of soil was also thoroughly mixed with 62 g of $CaCO_3$, 9 g complete fertilizer (15% N, $P_2O_5$, $K_2O$), 1 g of triple super phosphate (45% $P_2O_5$) and 2.4 g of urea (46% N). Soil field capacity (FC) and permanent wilting point (PWP) were measured before and after completion at the experiment. Soil moisture was determined by gravimetric method (Aslam et al., 2008).

The amount of water required for the irrigation of each treatment was calculated using the following equation: $V = SMD \times A$ where; V= volume of water to be applied (liter); A= polybag area = $\pi r^2$; SMD = ($\theta_{FC}$ - $\theta_i$) D Bd /100: SMD = soil moisture deficit; $\theta_{FC}$ = gravimetric soil moisture content at field capacity (%); $\theta_i$ = Soil moisture content before irrigation (%); D = rooting depth (cm); Bd = bulk density (in this soil 1.5 g $cm^{-3}$).

### Parameters measured

#### *Determination of shoot dry weight in main crop*

After harvesting, main crop plants were allowed to re-grow and 49 days after harvest, plants had grown to approximately 170 cm in height. Shoots were harvested at the cutting height of 15 cm from soil surface the same as for the main crop. The plant samples were carefully washed to remove all soil particles. Samples were then dried in an oven at 70°C for 3 days until constant weight was achieved. The dry weight of shoots (g $plant^{-1}$) was recorded for each treatment.

#### *Determination of shoot dry weight in first ratoon crop*

After harvesting main crop plants were re-growth and 49 days later than harvest, plants had reached to 170 cm height approximately, shoots were harvested at a cutting height of 15 cm from soil surface, then dried and weighed the same way as main crop.

#### *Determination of shoot dry weight in second ratoon crop*

At the end of the experiment (10% flowering stage) when plants had re-growth that reached 150 cm height approximately, fifty-day-old shoots were harvested, then dried and weighed the same as main crop.

The data were analyzed using Procedure ANOVA in the SAS. Treatment means were compared using least significant differences (LSD) at the 95% ($P \leq 0.05$) probability level.

## RESULTS

### Yield and yield components at harvest in first ratoon crop

Irrigation frequency and salinity, significantly ($P<0.01$) affected; growth, yield and yield components of forage sorghum (Table 1). Under influence of salinity at 15 dS $m^{-1}$ forage mass was reduced, 33.7% as compared to none saline treatment, while the decreases under salinity at 5 and 10 dS $m^{-1}$ were 13.9 and 22.3%, respectively. Comparison between varieties showed no significant differences between KFS4 and Speedfeed.

The highest forage dry mass was obtained with the more frequent irrigation, but decreased 4.3% with moderate irrigation. The impact of irrigation schedule however was strongly evident in plants when irrigation was delayed from -1.5 to -2 Mpa and as a result the forage yield decreased by 18.5 %.

**Table 1.** Effect of different levels of irrigation frequency and salinity on leaf, stem and forage dry weight of forage sorghum varieties for the first ratoon crop.

| Treatment | Leaf dry weight (g plant$^{-1}$) | Stem dry weight (g plant$^{-1}$) | Forage dry weight (g plant$^{-1}$) |
|---|---|---|---|
| **Variety** | | | |
| KFS4 | 8.74$^{a}$ | 18.67$^{a}$ | 27.42$^{a}$ |
| Speedfeed | 8.51$^{a}$ | 18.38$^{a}$ | 26.85$^{a}$ |
| LSD $_{0.05}$ | 0.58 | 1.68 | 2.26 |
| **Irrigation frequency at** | | | |
| LWP -1.0 (MPa) | 9.38$^{a}$ | 20.07$^{a}$ | 29.38$^{a}$ |
| LWP -1.5 (MPa) | 8.77$^{a}$ | 19.32$^{a}$ | 28.09$^{a}$ |
| LWP -2.0 (MPa) | 7.73$^{b}$ | 16.19$^{b}$ | 23.93$^{b}$ |
| LSD $_{0.05}$ | 0.75 | 2.17 | 2.76 |
| **Salinity (dS m$^{-1}$)** | | | |
| 0 | 10.25$^{a}$ | 22.75$^{a}$ | 32.89$^{a}$ |
| 5 | 8.93$^{b}$ | 19.36$^{b}$ | 28.30$^{b}$ |
| 10 | 8.13$^{b}$ | 17.40$^{b}$ | 25.53$^{b}$ |
| 15 | 7.20$^{c}$ | 14.60$^{c}$ | 21.80$^{c}$ |
| LSD $_{0.05}$ | 0.90 | 2.58 | 3.19 |
| **F value** | | | |
| VxI | 2.11** | 1.53$^{*}$ | 0.61** |
| VxS | 0.13$^{ns}$ | 0.15$^{ns}$ | 0.10$^{ns}$ |
| IxS | 0.57$^{ns}$ | 0.67$^{ns}$ | 0.45$^{ns}$ |
| VxIxS | 0.80$^{ns}$ | 0.60$^{ns}$ | 0.72** |
| **Error and CV** | | | |
| Error (MS) | 216.26 | 12.62 | 22.71 |
| CV (%) | 14.35 | 19.17 | 17.56 |

**, * and ns are significant at 0.01, 0.05 level and non significant, respectively.

In addition to reduction in yield, stress resulted in changes in composition of leaves and stems of the forage sorghums. The decreases of leaves under leaf water potential of -1.5 to -2 Mpa were 6.5 and 17.5%, respectively, while under salinity treatments the decreases were 12, 20.6 and 29.7%, respectively. Accumulations in dry stem weights under water stress were 3.7 and 19.3%, while under salinity the values were 14.9, 23.5 and 35.8%, respectively. Comparison between yield and yield components of the two varieties showed no significant differences between KFS4 and Speedfeed variety (Table 1).

### Water use efficiency at harvest in first ratoon crop

The water use by plants showed very complicated responses; and it was significantly dependent on salinity, variety and interaction of irrigation and variety as well as interaction of salinity and irrigation (Table 2). Unlimited water use lead to higher yields, however this caused some wastage. There was a significant interaction effect of irrigation and salinity levels on water use efficiency (WUE).

WUE in the control treatment was 34.5%, while the WUE with the irrigation frequency at leaf water potential of -2 MPa was 19.3% more than in the control treatment. Irrigation when applied at longer intervals had higher WUE. Irrespective of variety, the more frequently watered plants accumulated greater dry matter to eventually produce higher dry forage yield than other irrigation frequencies (Table 2).

The results derived from the irrigation study showed that despite the possibility of greater surface evaporation with light frequent irrigations, differentials of the sorghum varieties and other indicators of plant water stress were found to be improved with low frequent irrigation. Irrespective of variety, intermediate irrigation regime had

**Table 2.** Effect of irrigation frequency on yield of first ratoon crop and water use efficiency (WUE)

| Irrigation schedule | Days after treatment | | | | | | | Dry forage yield (g plant$^{-1}$) | Total water used (liters) | WUE (g plant$^{-1}$ liter$^{-1}$) |
|---|---|---|---|---|---|---|---|---|---|---|
| | 1 | 11 | 19 | 28 | 33 | 38 | 45 | | | |
| I1 | 1 | 0.98 | 1.17 | 1.27 | 1.09 | 1.08 | 1.26 | 29.38 | 7.85 | 3.73 |
| I2 | 1 | 0 | 1.59 | 0 | 1.70 | 0 | 2.01 | 28.09 | 6.30 | 4.43 |
| I3 | 1 | 0 | 0 | 2.36 | 0 | 0 | 2.48 | 23.93 | 5.84 | 4.07 |
| LSD $_{0.05}$ | | | | | | | | 2.76 | 0.61 | 0.43 |

I1, I2 and I3 are irrigation frequencies applied when the leaf water potential reached -1, -1.5 and -2 MPa, respectively. * Salinity levels averaged for these data.

**Table 3.** Mean forage yields in ratoon crops of sorghum varieties at different levels of irrigation frequency and salinity.

| Treatment | Main crop (dry forage) (g plant$^{-1}$) | First ratoon crop (dry forage) (g plant$^{-1}$) | Second ratoon crop(dry forage) (g plant$^{-1}$) | Total (dry forage) (g plant$^{-1}$) |
|---|---|---|---|---|
| **Variety** | | | | |
| KFS4 | 42.25 | 27.42 | 2.14 | 71.81 |
| Speedfeed | 39.41 | 26.85 | 11.55 | 78.50 |
| LSD $_{0.05}$ | 2.66 | 2.26 | 0.61 | 4.52 |
| **Irrigation frequency at** | | | | |
| LWP -1.0 (MPa) | 45.12 | 29.38 | 7.91 | 81.91 |
| LWP -1.5 (MPa) | 38.88 | 28.09 | 6.88 | 74.11 |
| LWP -2.0 (MPa) | 38.48 | 23.93 | 5.74 | 69.45 |
| LSD $_{0.05}$ | 3.25 | 2.76 | 0.75 | 5.54 |
| **Salinity (dS m$^{-1}$)** | | | | |
| 0 | 45.73 | 32.89 | 8.38 | 87.23 |
| 5 | 43.61 | 28.30 | 7.15 | 79.46 |
| 10 | 39.79 | 25.53 | 6.47 | 72.36 |
| 15 | 34.17 | 21.80 | 5.37 | 61.57 |
| LSD $_{0.05}$ | 3.76 | 3.19 | 0.87 | 6.40 |
| **F value** | | | | |
| V×I | $4.37^{*}$ | $0.61^{**}$ | $16.93^{**}$ | $0.97^{ns}$ |
| V×S | $0.03^{ns}$ | $0.10^{ns}$ | $8.41^{**}$ | $0.20^{ns}$ |
| I×S | $0.29^{ns}$ | $0.45^{ns}$ | $1.48^{ns}$ | $0.61^{ns}$ |
| V×I×S | $0.70^{ns}$ | $0.72^{**}$ | $1.73^{ns}$ | $0.35^{ns}$ |
| **Error and CV** | | | | |
| Error (MS) | 31.43 | 22.71 | 2.91 | 91.05 |
| CV (%) | 13.73 | 17.56 | 18.94 | 12.69 |

**, * and ns are significant at 0.01, 0.05 level and non significant, respectively.

higher WUE than the most frequently irrigated regime. Though, when comparing the two varieties, variety KFS4 significantly produced higher dry forage in the main crop; however, in the second ratoon crop, the total dry weight (submission of three cutting) of Speedfeed was higher than KFS4 (Table 3 and Figure 1).

The results of this study showed that the yield of forage sorghum was significantly increased with more frequent irrigation. More frequent irrigation is usually associated with increase in salt concentrations in the soil as water is extracted by the crop. Typically, salt concentrations are lowest following irrigation and higher just before the next

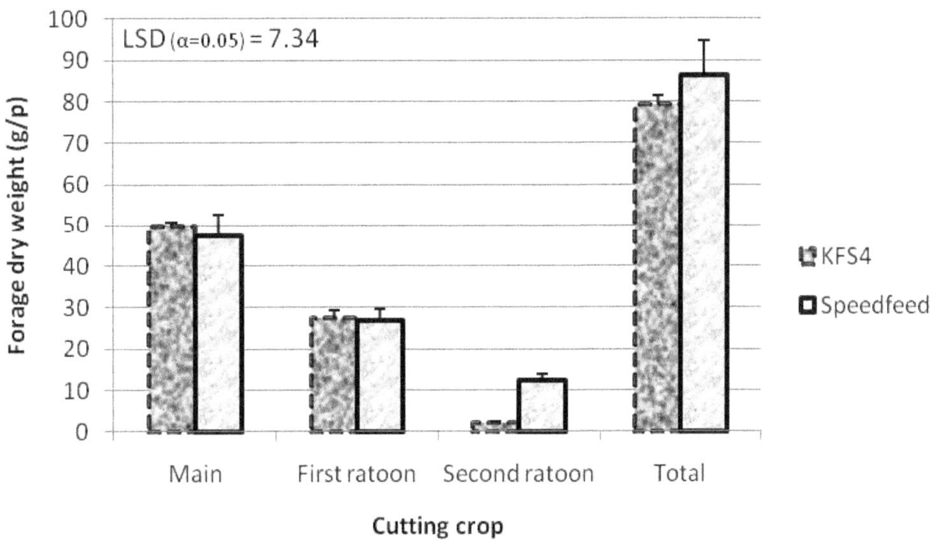


**Figure 1.** Forage yield in ratoon crops of the two sorghum varieties.

irrigation. Increasing irrigation frequency maintains higher constant moisture content in the soil, and thus more of the salts are kept in solution which aids the leaching process.

## DISCUSSION

The effect of salinity and irrigation frequency as reflected by a lower yield may be the result of a combination of osmotic and specific ion effects of Cl and Na. As the salt concentration and water stress increased, dry matter yield diminished the first ratoon with reductions of 13.9, 22.3 and 33.7% as compared to none saline treatment, respectively (Table 1). The highest reduction (33.7%) was from the most salt stressed (15 dS $m^{-1}$) against 18.6% for the most water stressed plants, and significant decreases were also recorded for other salt levels. This is consistent with reports in other monocots including rice, wheat and maize (Krishna et al., 1993; Shabala, et al., 1998). KFS4 variety which in the main crop study exhibited a higher yield in terms of dry matter production did not show any significant differences with Speedfeed variety in the first ratoon crop. As the saline applications continued into the first ratoon crop, there were significant losses in yield under the stress treatments, however with three cuttings characteristic of Speedfeed led to higher total production of Speedfeed (Figure 1 and Table 3). No viable plants were obtained in the second ratoon crop for KFS4 variety at 15 dS $m^{-1}$ salinity and -2 MPa irrigation frequency treatments, because KFS4 was genetically known to be a one ratoon crop.

Salinity stressed plants certainly faced osmotic challenges. This is in agreement with several previous reports (Munns and Tester, 2008; Lee et al., 2004), which concur that osmotic adjustment is the main response for survival and growth of plants under salinity stress. Also under saline conditions a positive association between photosynthetic rate and yield has been reported in sorghum (Faville et al., 1999).

When plants are under water stress, they firstly reduce and then stop leaf expansion (Hsiao and Jing, 1997). In this study, the WUE obtained for the infrequently watered plants were higher than earlier reports on five forage sorghum cultivars grown on dry land in Texas (Gulzar et al., 2003) or values previously reported for the same crop in the northern part of Sudan (Mustafa and Abdel Magid, 1982). Although larger volumes of water were used in this investigation as compared to the two previous reports, the higher WUE obtained here could be attributed to the reduction in irrigation with the infrequent watering.

## Conclusion

There were significant losses in yield in the second ratoon crop under the stress treatments. Irrigation can overcome effect of salinity as increasing irrigation frequency maintains constant moisture content in the soil. Thus, more of the salts are then kept in solution which aids in the leaching process. Irrigation may be intensified in saline soils to mitigate the effect of salinity on plant growth. However, there is a critical level of salinity after which irrigation cannot mitigate the effect of salinity. The critical level of salinity for KFS4 was 15 dS $m^{-1}$, while for Speedfeed it was 10 dS $m^{-1}$. The results obtained in this study would serve as a useful guide for managing forage sorghum in saline and water stressed field conditions. On the other hand, WUE in forage sorghums can be increased with reduced irrigation frequency, and this will enable larger field areas to be irrigated with the savings in water used. The present findings suggest that in semiarid environments (where saving water is very important), sorghums should be irrigated infrequently but heavily, if the aim is to get high WUE forage sorghum.

The final concluding remarks describe characteristics, advantages and limitations of the varieties studied: KFS4 variety under stress condition gave better vegetative growth performance especially in main crop dry matter production as compared to Speedfeed. While considering second ratoon crop characteristics, Speedfeed was superior to KFS4 variety.

## Conflict of Interests

The author(s) have declared that there is no conflict of interests.

### REFERENCES

Aslam M, Haji Kh, Ahmad H, Muhammad A, Ejaz A, Muhammad A (2008). Effect of available soil moisture depletion levels and topping treatments on growth rate total dry biomass in chickpea. J. Agric. Res. 46(3):229-243.

Borsani O, Valpuesta V, Botella MA (2003). Developing salt tolerant plants in a new century: a molecular biology approach. Plant Cell Tissue Org. Cult. 73:101-115.

Faville MJ, Silvester WB, Green TGA, Jermyn WA (1999). Photosynthetic characteristics of three asparagus cultivars differing in yield. Crop Sci. 39:1070-1077.

Fouman A, Heravan EM, Nakano Y (2003). Evaluation forage sorghum varieties for salt tolerance. Proceedings of 7$^{th}$ International Conference on Development of Drylands. 14-17 September 2003, Tehran, IRAN.

Gulzar S, Khan MA, Ungar IA (2003). Salt tolerance of a coastal salt marsh grass. Commun. Soil Sci. Plant Analyt. 34:2595-2605.

Harivandi MA, Bulter JD, Wu L (1992). Salinity and turfgrass culture. In Turfgrass Agronomy, Monograph, No. 32.'. (Eds DV Waddington *et al.*) pp. 207–229. (ASA, CSSA, and SSSA, Madison, WI).

Hsiao TC, Jing J (1997). Leaf and root expansive growth in response to water deficits – In: Cosgrove, D,J., Knievcl, Growth. The American Society of Plant Physiologists, Rockville. pp. 180-192.

Jacoby B (1999). Mechanism involved in salt tolerance of plants. In Handbook of Plant and Crop Stress, ed. M. Pessarakli, Marcel Dekker, Inc., New York, pp. 97-124.

Jungklang J, Usui K, Matsumoto H (2003). Differences in physiological responses to NaCl between salt-tolerant (*Sesbania rostrata* Brem. and Oberm.) and non-tolerant (*Phaseolus vulgaris* L.). Weed Biol. Manag. 3:21-27.

Krishna S, Raj BT, Mawson EC, Yeung T, Thorpe A (1993). Utilization of induction and quenching kinetics of chlorophyll a fluorescence for in vivo salinity screening studies in wheat (*Triticum aestivum* vars. Kharchia-65 and Fielder), Can. J. Bot. 71(1):87-92.

Lee G, Duncan RR, Carrow RN (2004). Salinity tolerance of seashore paspalum ecotypes, Shoot growth responses and criteria, Hortic. Sci. 39(2):1138-1142.

McCarty LB, Dudeck AE (1993). Salinity effects on bentgrass germination. Hortic. Sci. 28:15-17.

Munns R, Tester M (2008). Mechanisms of salinity tolerance. Ann. Rev. Plant Biol. 59:651-681.

Murdoch CL (1987). Water the limiting factor for golf course development in Hawaii. *USGA:* Green section record 25:11-13.

Mustafa MA, Abdel MEA (1982). Effects of irrigation interval, urea and gypsum on N, P and K uptake by forage sorghum on highly saline-sodic clay. Exp. Agric. 18(2):177-182.

Rhodes D, Nadolska-Orczyk A, Rich PJ (2002). Salinity, osmolytes and compatible solutes. In: Lauchli, A., Luttge, U. (Eds.), *Salinity: Environment-Plants-Molecules.* Boston, Kluwer Academic Publ C, pp.181-204, doi: 10.1007/0-306-48155-3_9

Shabala SN, Shabala SI, Martynenko AI, Babourina O, Newman IA (1998). Salinity effect on bioelectric activity, growth, $Na^+$ accumulation and chlorophyll fluorescence of maize leaves: a comparative survey and prospects for screening, Australian J. Plant Physiol. 25:609-616.

Teutsch C (2006). Warm-season annual grasses for summer forage. Publication 418-004. Communication and Marketing, College of Agriculture and Life Sciences, Virginia Polytechnic Institute and State University. Accessed at: http://pubs.ext.vt.edu/418/418-004/418-004.pdf

# Germination tests of seeds of argan tree (*Argania spinosa* (l.) skeels) of two sources (Tindouf and Mostaganem) in the semi-arid western Algerian.

**Benaouf Zohra[1], Miloudi Ali[2] and Belkhodja Moulay[3]**

[1]Faculty of Biology, Laboratory of Ecology, Houari Boumediene University of Sciences and Technology (U.S.T.H.B), El Alia, BP 32, Bab Ezzouar, 16111 Algiers, Algeria.
[2]Laboratory of Plant Biology - Faculty of Science, University of Mascara 29000 Algeria.
[3]Laboratory of Plant physiology - Faculty of Science, University of Oran 31000 Algeria.

**The Argan tree (*Argania spinosa*) is a drought-tolerant, and forest species observed in arid and semi-arid zone in Algeria, has specific ecological characteristics and many interests (forest, field, and fruit). The natural reproduction of the tree has become difficult; we have assessed the propagation method by seedlings. In our experiments, we used two seed sources collected from Tindouf and Mostaganem. Based on the results of the regeneration of the Argan tree seedlings from the laboratory, we report that soaking of seed for at least four days will certainly contribute to the success of germination. Sterilization prevents microbial contamination and improves germination. According to the results, the germination tests revealed a very high rate of germination (95%) for seeds pre-soaked in water for 96 and 120 h at 25 and 30°C. The analysis of morphological characteristics of plants under greenhouse showed that there was growth in the root system of the argan seedlings and the aerial part improved quantitatively and qualitatively. Seedlings that acclimatized are two years and above, and had a well developed and lignified air device with a large leaf, which promotes their growth. The success rate for seedlings of 12 months was very low.**

**Key words:** Argan tree, domestication, germination, pre-soaked, acclimatization, seedlings.

## INTRODUCTION

In south Algeria (Tindouf), forest based on the endemic argan tree (*Argania spinosa* L. Skeels), Propagation by seeds is the most common method used to reproduce the argan forest species. All the reforestation projects of argan in Algeria adopt this method. The most successful example of this is the Mostaganem project in coastline, which began to bear fruit after six years, and all nurseries in Tindouf which aim to rehabilitate the argan tree. The sustainability of this agroforestry system is now threatened by over-exploitation and overgrazing in Tindouf (Touaref bouaama, Targanat and Markala) by the local population and the nomads of the Polisario. This

**Figure 1.** Components of the argan seed.

results in serious degradation of the soils and a decrease in production of oil (traditional extraction) and food betailles (camels and sheep). It is therefore a national concern for the authorities involved to protect our rare and vulnerable inherited Argan forests against anthropogenic pressure. They should allow the establishment of an executive order by the Wali of Wilaya, Tindouf bearing draft reserve natural creation (Decree No. 04/96 of 1 2/06/2004) on the protection of argan plant species in the Wilaya of Tindouf.

Perfectly suited to its environment, this endemic tree can grow on poor, shallow soils, and owing to its deep rooting system, it is considered as having a strong effect against erosion and desertification, which are the main environmental problems. The argan tree is monoecious and allogamous and exhibits high genetic diversity (Msanda, 1993). This diversity can both be preserved for ecological purposes and used through domestication. Recent initiatives have promoted the domestication of multi-purpose agroforestry species for their ability to alleviate poverty and mitigate environmental degradation (Leakey and Simons, 1998). This approach is relevant to *Argania spinosa* which could be domesticated for oil and fodder production. What is required is a package of technical procedures from the selection of the best genotypes to the production of cultivars in nurseries for successful integration into agroforestry systems. Currently, due to the failure of natural regeneration and reforestation, the only possibility of rejuvenating mature argan forest is through coppicing. Seedling production and use for plant production in nurseries could allow for the conservation of the genetic diversity. However, according to the literature, seed germination is difficult. Through selection and mass production of trees with desirable characteristics, biotechnology could improve argan tree productivity, as well as overall production (Sasson, 1993). Vegetative propagation offers the opportunity to multiply selected genotypes and to provide a significant step towards 'domestication'. But the germination is a very complex biological phenomenon which requires a good understanding and control of the factors behind. For multiplication of argan, it was found that the technique of soaking seeds in water is quite sufficient for good germination (Nouaïm and Chaussod, 1993). To overcome this constraint related to the biological nature of the seed, we performed a pre soaking in water of seeds at varying durations before planting. The purposes of the present study were: to examine the factors affecting variation in the germination of argan seeds, acclimatization of plantlets produced by germination, and to improve the success rates of nursery production and transplantation

## MATERIALS AND METHODS

### Propagation by seeds

The seeds used in this study came from ripe fruits collected from randomly chosen trees located in Tindouf forest (south of Algeria), and nursery of Mostaganem (coastline). The fruits were dried and their pulp was removed by hand to obtain nuts. Just before germination, they were disinfected with hydrogen peroxide for 20 min (Figure 1).

### Tests for germination

The technique used was by soaking argan seeds in water at different times (72, 96 and 120 h), then the seeds were placed in the oven to heat thresholds considered (30, 28, and 25°C). Under these conditions, we want to know the ability of seeds' germination of two sources: Tindouf (natural area of argan) and Mostaganem (introduced species) (Figure 2). We conducted a test with 20 seeds for each treatment.

#### *First germination test with a temperature of 30°C*

The argan seeds were previously disinfected with bleach for 20 min; then they were pre-soaked in water and placed on cotton in Petri dishes: Treatment 1: The seeds are soaked in water for 72 h; Treatment 2: the seeds were soaked in water for 96 h; Treatment 3, the seeds were soaked in water for 120 h.

#### *The germination tests with temperatures of 28 and 25°C*

The argan seeds were disinfected prior to pure bleaching for 20

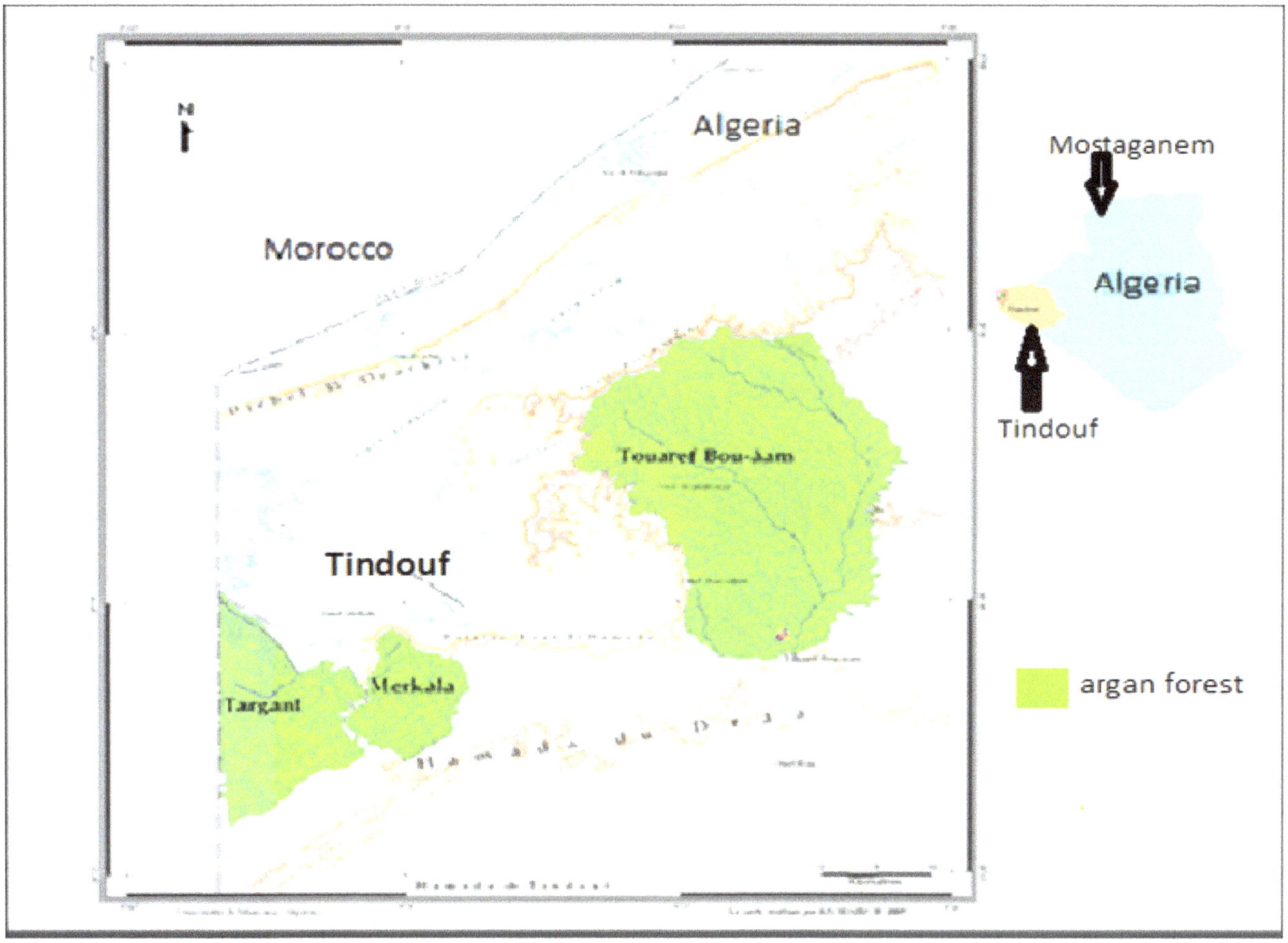


**Figure 2.** The whereabouts of argan in Algeria.

**Figure 3.** Reforestation project of argan in Mostaganem.

min, and then they pre-soaked in water treatment for 96 and 120 h only. This is because soaking for 72 h gives us very low germination than when placed on cotton in boxes kneaded (Figure 3).

**Figure 4.** Seedlings of argan in plastic bag.

**Figure 5.** Seedlings of argan in pots.

***Treatment 1***

The seeds were soaked in water for 96 h.

***Treatment 2***

The seeds were soaked in water for 120 h.

**Acclimatization of seedlings**

In order to know the behavior of the argan tree under the influence of abiotic conditions, we found it useful to perform planting in two different settings: Under greenhouse and open field. The choice of these two environments was done so that we can follow the growth of seedlings after planting argan and to compare the environmental conditions.

***On emissions***

***Transplanting seedlings argan in plastic bag***

In the laboratory we installed in plastic bags peat sand filled to two thirds and one-third. The experimental conditions were characterized in the laboratory by a thermal range of 18 to 24°C, exposure to sunlight and a relative humidity of ambient air of about 65 to 75%. Seedlings of the argan tree were watered every 48 h (Figure 4).

***Transplanting seedlings argan in pots***

Young seedlings argan obtained under the above conditions (under glass) were removed from the bags with their lumps around the roots, and then they were transplanted into pots. The substrate used consists of peat mixed with sand in equal proportion (1/2 and peat 1/2 sand) (Figure 5).

**In the field**

The argan tree seedlings were monitored on the ground after planting. They have the following characteristics.

***Origin***

These two types of argan seedlings that aged between one and two years are from the nursery of Mostaganem (littoral). We also noted that these seedlings had varying dimensions (height, diameter, branch), even those of the same age.

***Planting period***

Generally, planting periods are chosen depending on the rainfall. In Algeria, plantations are generally done from October to March because the soil is in good condition during that time and the humidity is high (about 90%). In our case, the argan tree seedlings were sown in 28/02/2008.

***Planting method***

There were 24 argan seedlings of one year and 20 seedlings of 2 years. The length of the seedlings ranged from 06-73 cm, with a number of sheets of about 150 units. A plot at the experimental farm was prepared to serve the planting site. Planting distances were 1 m between seedlings and 2 m between the lines. The first watering was done immediately after planting, then it was repeated once a month for six months after planting (the amount of irrigation water is 3 L per plant).

**Statistical analysis**

Both parametric (t test) and non-parametric tests (Mann-Whitney-Wilcoxon test, based on sum of ranks) were performed on experimental data for comparison of treatments (Conover, 1980).

## RESULTS

### Germination rate

Germination is a complex biological phenomenon that requires control and identification of factors causing it. In our case, we are interested in the argan fruit covered with a very hard shell. So we practice several treatments before sowing seeds at different temperatures to find the

**Figure 6.** Germination of argan seeds

optimal conditions for germination (Figure 6).

## Germination tests at a temperature of 30°C

### *Pre soaking in water for 72 h*

The seed germination started on the fourth day at a rate of 10% for those in Mostaganem against those germinated in Tindouf. Raw sproutswe observed as early as 10 days at a rate of 5%. Then, the germination rate increased slowly and reached maximum rates of 95 and 80%, respectively at duration of 20 days (Figure 7).

### *Pre soaking in water for 96 h*

The first sprouts were obtained from the second day at a rate of 10% for both seeds. Then, the germination rate increased to a maximum of 80% on the 12th day (Figure 8).

### *Pre- soaking in water for 120 h*

Seeds pre-soaked in water for 120 h express a relatively high germination rate (15%) from the third day of sowing: 15% for Mostaganem seeds and 10% for those of Tindouf. Germination rate increased rapidly up to 75% in seeds of Mostaganem and 80% for seeds of Tindouf on the 13$^{th}$ day of sowing (Figure 9).

## Testing the germination temperature of 28°C

Based on the fact that the best results on germination of argan were obtained from pre soaking in water for 96 h and 120 h, we found it useful to continue the remainder of the tests with the two treatments. Seed treatment consisting of pre-soaking in water for 75 h gave very poor results.

### *Pre soaking in water for 120 h*

Seeds pre-soaked in water for 120 h express variable responses to germination (Figure 10). Indeed for the two batches of seeds, germination begins at the third day of planting at a rate of 15% for seeds of Mostaganem and 10% for those of Tindouf. Thereafter, germination evolves to a maximum of 90% for Mostaganem seeds and 95% for those of Tindouf.

### *Pre soaking in water for 96 h*

There was a significant difference in the behavior of the two types of seeds. Indeed, the seeds of Tindouf germinated on the sixth day and those of Mostaganem, on the fourth day, at 10 and 5% rates, respectively. As a result, germination increased to 95% for seeds of Mostaganem and 90% for those of Tindouf at day 18 (Figure 11).

## Test germination temperature of 25°C

### *Pre soaking in water for 120 h*

The first sprouts were obtained from the second day at a rate of 10 to 45% for Mostaganem seeds. Seeds of Tindouf gave a rate of 5% at the 2nd day of planting. Germination evolves rapidly to a maximum of 95% for Mostaganem seeds and 90% for Tindouf seeds on the 18th day (Figure 12).

### *Pre soaking in water for 96 h*

There was an early germination of the second day of planting at a rate of 5% for both types of seed. As a result, the levels increase slowly and reached a maximum of 95% on the 20$^{th}$ day of sowing (Figure 13).

## Acclimatization step

### *Morphological characteristics of the argan tree seedlings*

Plants produced from seed had a tap root system with a fast and powerful development under these conditions: 15 days after sowing is enough for the taproot to appear through the hole at the bottom of the bags.

### *Main characters observations and measurements:*

#### i) Number of leaves and thorns

Young argan seedlings had a strong capacity for growth and development. Therefore, the leaf system was significant in the first month, with an average of 30 leaves per seedling, and about 70 leaves per seedling after 9 months (Figure 14).

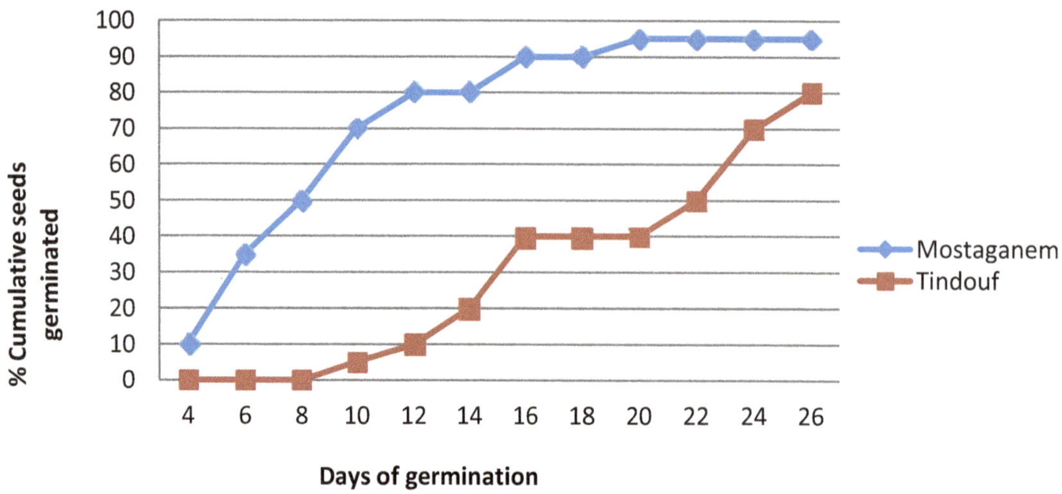


**Figure 7.** Daily argan seed germination rate pre-soaked for 72 h and subjected to a temperature of 30°C.

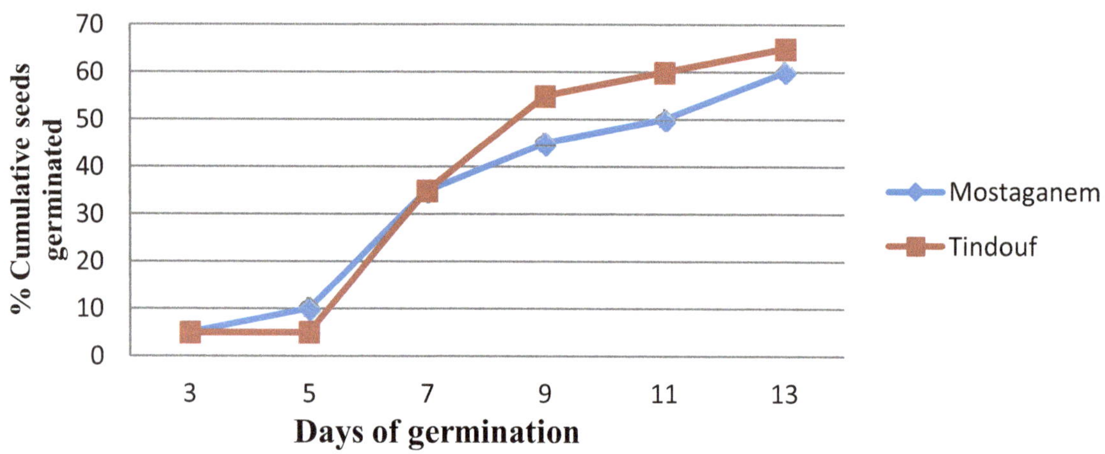


**Figure 8.** Daily argan seed germination rate, pre-soaked for 96 h and subjected to a temperature of 30°C.

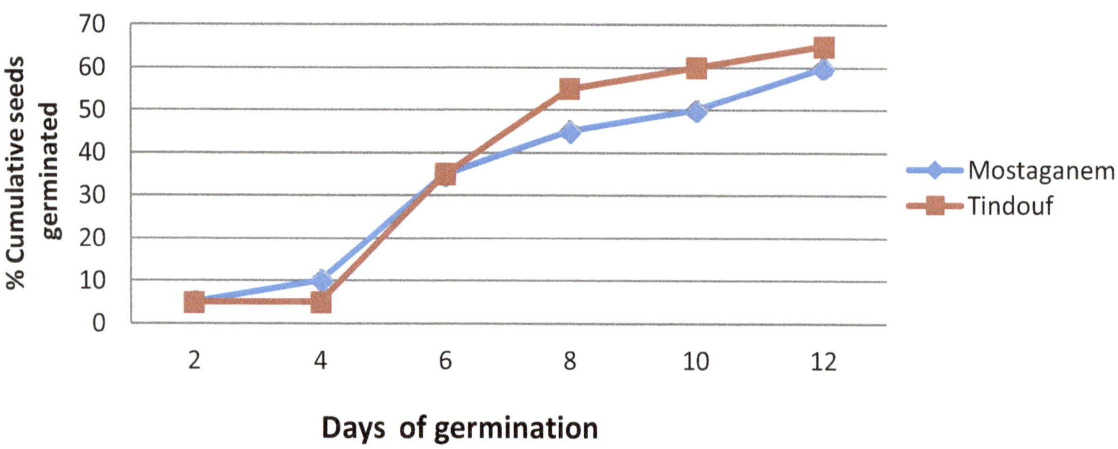


**Figure 9.** Daily argan seed germination rate, pre-soaked for 120 hours and subjected to a temperature of 30°C.

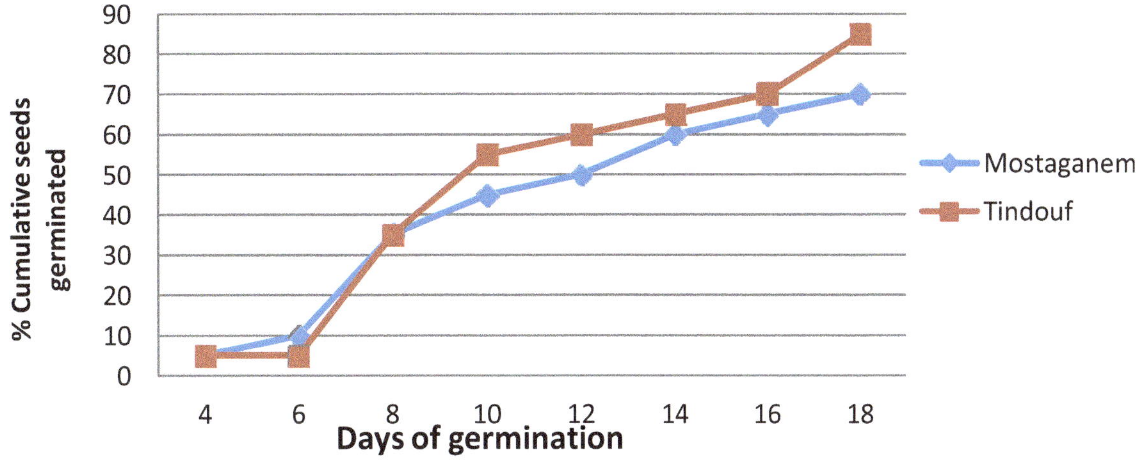


**Figure 10.** Daily argan seed germination rate, pre-soaked for 120 h and subjected to a temperature of 28°C.

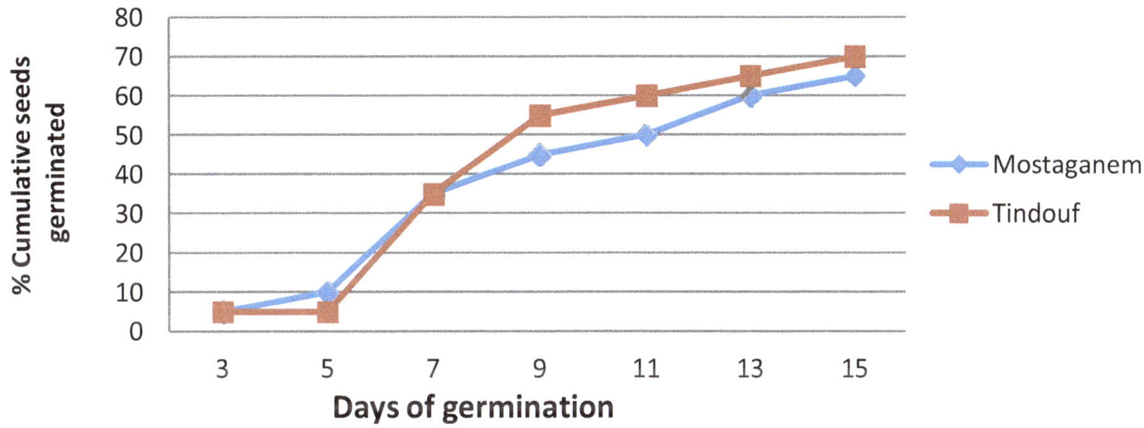


**Figure 11.** Daily argan seed germination rate, pre-soaked for 96 h and subjected to a temperature of 28°C.

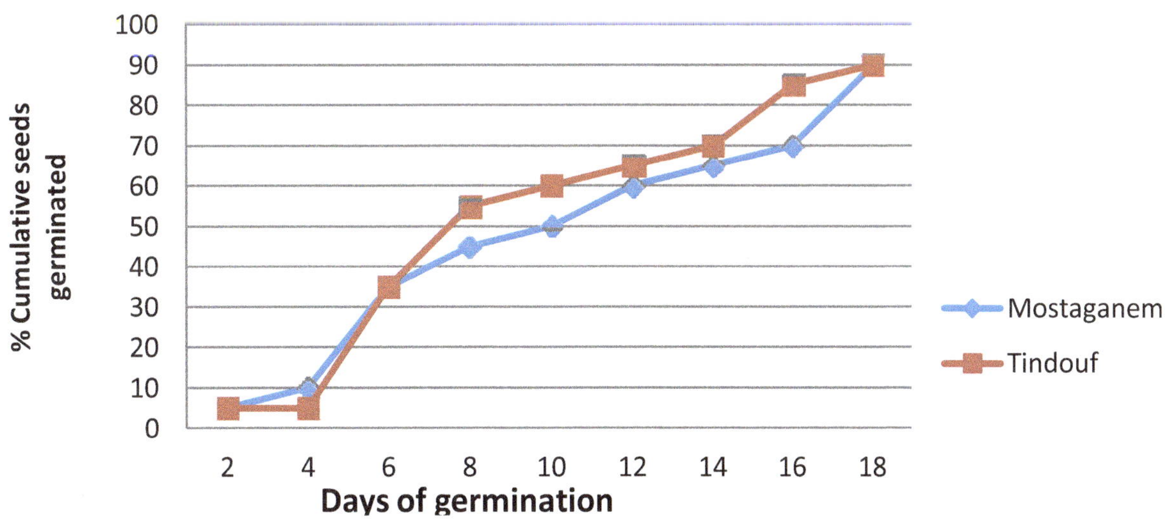


**Figure 12.** Daily argan seed germination rate, pre-soaked for 120 h and subjected to a temperature of 25°C.

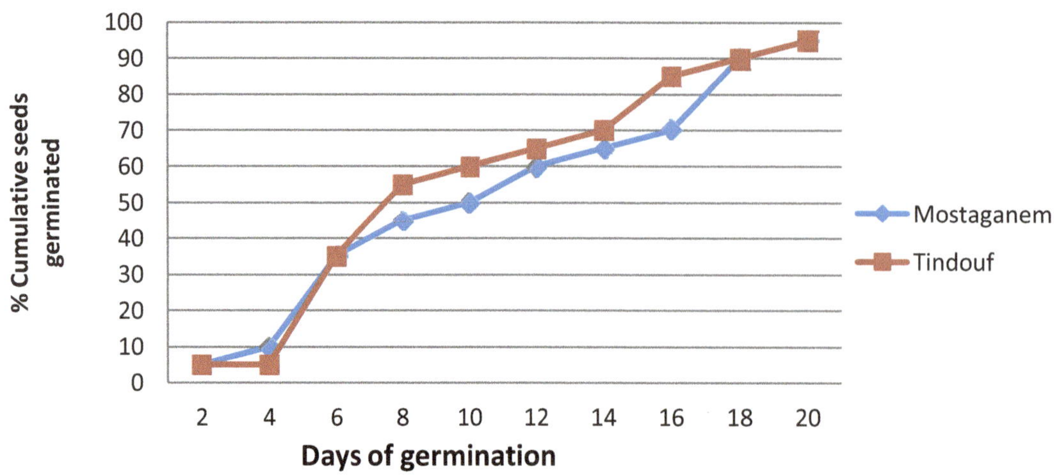


**Figure 13.** Daily argan seed germination rate, pre-soaked for 96 hours and subjected to a temperature of 25°C.

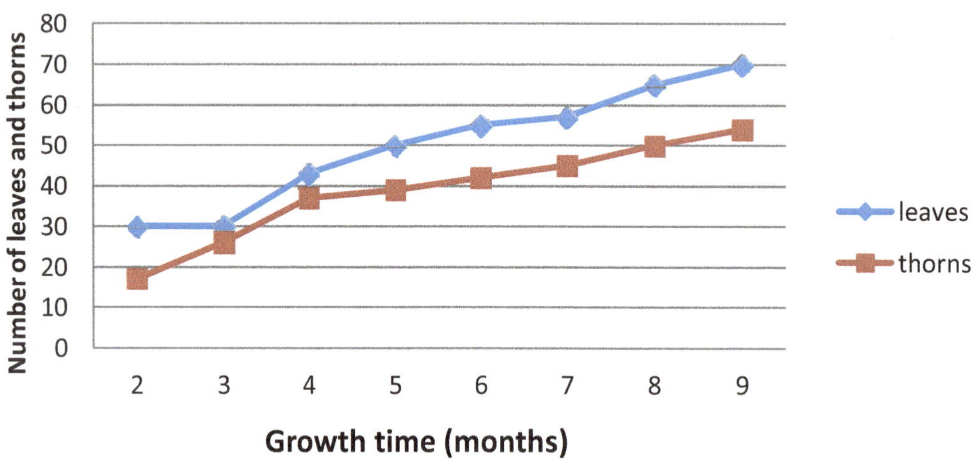


**Figure 14.** Monitoring the evolution of leaves and thorns of argan seedlings for 9 months.

Thorns appeared from the 20th day of the plantation, and increased with time. Thus, we noted that older seedlings of 2, 7, and 9 months haved a varying number of thorns: 17, 42, and 54, respectively.

### ii) Growth in height

Height is the average height of seedlings, measured from collar to the end bud. The results show that height growth was faster in 14 days old seedlings with a height of 0.8 cm and got to 6.5 cm after a month. There is slow height growth in a month or more with two months old seedlings that have height of 18 cm and 9 months old seedlings with a height of 40 cm (Figure 15).
Results on the root system show that the length of the main root was significant; the secondary and tertiary roots were thin and brittle. After 14 days of planting, the length of the main root was 3 cm, then it underwent rapid growth to reach a length of 8 cm in a month. There was continuous growth in the root system with normal development but at a variable rate. For 4 and 5 months, the length of the primary root was respectively 25 and 29 cm. However, we observed a slowdown of root growth after 6 and 9 months; their lengths we slightly higher than the previous (32 and 35 cm). Similarly, we note that the growth in height of stem is small compared to the roots.

### In the field

#### *Height of stems*

Growth in height of stem varied in the two types of

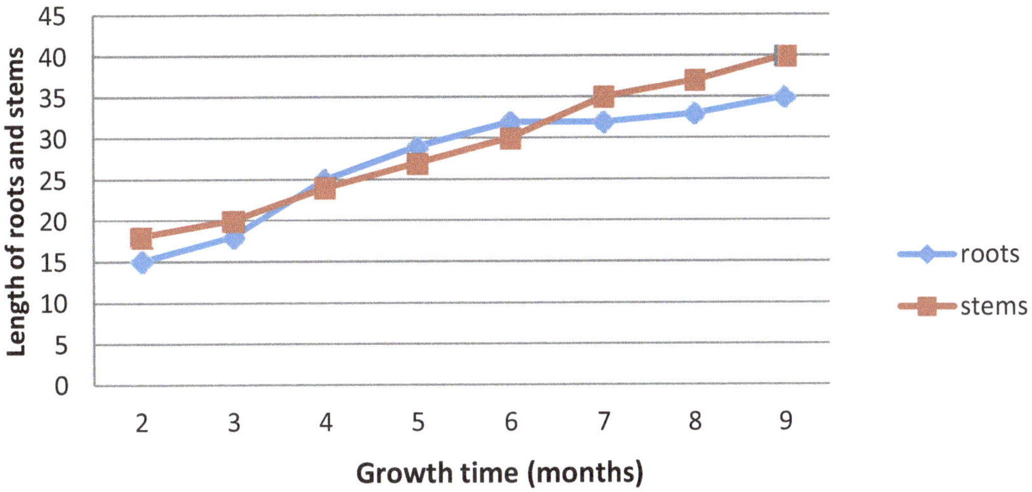


**Figure 15.** Monitoring the evolution of heights of stems, and roots of argan seedlings of 9 months.

seedlings. After the first week of planting, we observed that stem height has quadrupled in older seedlings of 2 years compared to one year. Then the seedlings were growing two batches of irregular height up to 74 cm (seedlings older than 2 years) and about 12 cm (seedlings older than one year) in the first six months of planting.

At the age of 12 months after planting, shoot growth slowsed down: height reachesd 18 cm respectively in one year seedlings and 32 cm for that of 2 years.

### *Leaf drying*

Leaf desiccation is when the foliage or leaves of few seedlings dry up and become detached from the main stem. After a week when transplantation was done, the rate of drying of the leaves is very low for the two types of seedlings.

### *Number of thorns*

From the observations, we find that the number of thorns increased with the age of seedlings; older seedlings of one year hadve 10-28 thorns each against those of two years, with 40-260 thorns each. So after a month of planting, these have the same values like previous thorns.

### *Branch of the stems*

Young argan seedlings have an axis of stem that is rarely branched. As argan tree is vigorous, it acquires ramifications of the first order. At 2 years, the seedlings have remarkable branching of approximately 2 to 15 branches per seedling up to the 3rd month of the planting.

### *Rate of recovery after transplantation*

The number of successful transplanted seedlings after one year, compared to results of those originally planted is seen in Table 1.

The aim of this study was to highlight the biological conditions related to promoting good regeneration of the argan tree. Our experimental tests on regeneration of seedlings argan of two years have encountered a number of climate and technical field problems, which impact negatively the development of seedlings. Then, the highest success rate was found in 24 months old seedlings (80%) compared to seedlings of 12 months (50%). This high success rate (80%) shows that the age of argan tree seedlings is significant for a good recovery.

## DISCUSSION

Many authors have reported the difficulty involved the germination of the seeds of the argan tree and they recommended either scarification or acid treatment. Our experiment showed that in seeds pre-soaked in water for 120 h at 30 and 28°C, raw sprouts were obtained from the third day at a rate of 15%, as against 25 and 20°C, in which germination starts on day 3, but at a rate of 10%. Regarding the rate of germination, it was significant in seeds pre-soaked in water for 96 and 120 h in the thermal range of 30 and 25°C. Under the same conditions, in seeds pre-treated before sowing, there was optimal rate of 85% obtained at 30°C.

Regarding the origin of the seeds, the results show

**Table 1.** Success rate of transplanted argan seedlings.

| Age (month) | Seedlings transplanted | Live seedlings | Success rate (%) |
|---|---|---|---|
| 12 | 24 | 12 | 50 |
| 24 | 20 | 16 | 80 |

some difference in the rate and the beginning of the germination. The seeds begin to germinate in Tindouf 2 to four days late compared to those of Mostaganem. The delay in germination may be related to the different morphology of the seeds, compared to those of Mostaganem. We hold that the seeds of Tindouf have a relatively hard and thin shell.

Regarding the relationship between the germination time, temperature, and nature of the pretreated seeds source, the results indicate the existence of a correlation between the duration used for pre-soaking seeds in water and germination temperature. Moreover, the duration used in pre-soaking argan seed is very short when the optimum temperature for germination was high (30°C). The variation in germination parameters of Argan seeds we measured is reported by many authors working in the field. For example, Come (1975) reported that seed will sprout if the embryo has the possibility of imbibing; Mazliak (1982) concluded that temperature remains a limiting factor that affects directly germination by acting on the speed chemical reactions. In addition, Nouaim and Chaussod (1993) argue that the argan tree can regenerate from seeds, but a lot of failures were observed.

These authors report that a simple dipping of argan seeds in water for three or four days encourage a high percentage of germination. In our case, soaking seeds in water for 96 and 120 hbefore sowing has positive effect on early and high rates of germinated seeds (80, 85 and 95%) from 25 to 30°C. Compared to some work done in Algeria on the possibilities of multiplication of argan seeds in a thermal range of 25 to 30°C, we observe that the germination varies with sources of seeds. For example, for seeds from Tindouf, germination rate reached 50% (Slimani, 1996). This rate increaseds to 70% (Kechairi and Lakhdari, 2002) and 80% (Miloudi, 2006) for seed harvested from Oggaz Station, Mascara. For seeds from Mostaganem, germination rate reached 90% (Baoui, 2001) and 55, 70 and 80% in the thermal range of 25 to 30°C (Miloudi, 2006).

Furthermore, Renard (1975) notes that among the seeds used for the test (same experimental conditions), some do not germinate due to the hardness of the integument and others due to the endogenous inhibitors or dormancy. Under these conditions, it is essential to achieve optimal conditions for germination, confirming the general rule on the germination of halophytes (Grouzis et al., 1976), and Glycophytes (Francoit et al., 1986; Belkhodja and Soltani, 1992).

## Conclusion

Failure of argan tree regeneration is often attributed to multiplication difficulties. However, our experimental results have shown that this tree can be propagated by seeds. Due to its allogamous reproduction (Msanda, 1993), different multiplication methods could be used, according to the objective. Reforestation by seedlings is the best method because it maintains the great genetic variability of the species and confers ecological resistance. It reduces production costs in forest nurseries. The aim of our work was not to define ideal methods for propagation, but to check the feasibility of this technique. Our results represent regene-ration of argan tree by seeds. This technique will be used for future production of argan trees in agroforestry systems. This points out the imbalance of traditional agroforestry system, which is now clearly threatened. The only way to avoid irreversible damage to the environment and to promote sustainable management of the argan forest is to improve agricultural income. It seems now possible to optimize argan tree based agroforestry systems, through the production of high quality exotic oil, for which there is already a large demand, exceeding present supplying capacities.

'Planting trees that yield a good profit' is the best way to fight against desertification. Leakey and Simons (1998) showed that increasing the quality, number and diversity of domesticated trees could provide a wide array of non-timber forest products (NTFPs). NTFPs could enhance the capacity of agroforestry to fulfill its ultimate potential as a way to alleviate poverty and to mitigate deforestation and land depletion. Domestication of trees for agroforestry to produce NTFPs can therefore benefit both the farmer and the environment. Vegetative propagation, enabling argan tree domestication, is a real chance for development (Sasson, 1993). We showed that argan tree domestication is technically feasible. However, dissemination of this knowledge is necessary for a true sustainable development.

## Conflict of Interest(s)

The author(s) have not declared any conflict of interests.

## ACKNOWLEDGEMENTS

We are thankful to the R. S.B.G. laboratory, the professors in USTHB University and Mascara University, Faculty of Biology

## REFERENCES


Baoui B (2001). Test multiplication and generative transplantation argan plants (*Argania spinosa* L. Skeels) in the plateau of Mostaganem. State Engineer in Agronomy memory. Univ of Mostaganem. p. 80.

Belkhodja M, Soltani N (1992). Replies bean Vicia faba L to salinity: a study of the germination of some lines to growth. determined.Bull. Soc. Bot. FR. pp. 357-368.

Come D (1975). Some problems of terminology on seeds and their germination (Chaussâte nine Germination desemences, bordas, Paris, Bruxelles, Montréal. 1975, pp. 11-26.

Conover WJ (1980) Practical Non-parametric Statistics (2$^{nd}$ edition). John Wiley & sons, New York, 493p.

Francoit L, Maas E, Ovonan T (1986). Effect of salinity on grain yield, vegetative growth and germination of semi dwarf and durum wheat in an arid environment. Agro. J. pp. 1053-1058.

Grouzis M, Berger A, Heim G (1976). Polymorphism and germination in three annual species of the genus Salicornia ; OEcol. Plant. pp. 41-52.

Kechairi R, Lakhdari A (2002). Contribution to the study of the argan (Argania spinosa L. Skeels) and multiplication by seed testing laboratory in memory of engineering status in Bio. Univ of Mascara. p.120.

Leakey RRB, Simons AJ (1998). The domestication and commercialization of indigenous trees in agroforestry for the alleviation of poverty. Agroforestry Systems 38:165-176.

Mazliak P (1982). Physiologie II Plant Growth and Development. Collection methods Hermaun Paris. p. 465.

Miloudi A (2006). The physiological and biochemical responses of the argan (*Argania spinosa* L. Skeels) natural abiotic factors. PhD thesis, University of Oran ES SENIA. pp. 1-2, 30-90.

Msanda F (1993). Ecologie et cartographie des groupements végétaux d'Anzi (Anti-Atlas Occidental, Maroc) et contribution à l'étude de la diversité génétique de l'arganier (Arganai spinosa (L.) Skeels). Thèse Univ. Joseph Fourrier, Grenoble (France) 116 p.

Nouaim R, Chaussod G (1993). Argan (*Argania spinosa* (L.) Skeels). The flamboyant newsletter network members tropical trees n° 27, Septembre1993. pp. 50-64.

Renard H (1975). Germination techniques, practical criteria and significance. Chemical, biochemical criteria, physiological and molecular. Laboratory of the University of Avigon, France. pp. 24-35.

Sasson A (1993). Biotechnologies in Developing Countries: ²Present and Future. UNESCO Publishing, Paris. 764p.

Slimani H (1996). Contribution to the study of the argan (*Argania spinosa* L. Skeels) the germination of the Inland halophyte *Hordeum jubatum*. Can. J. Bot. 1420-1425.

# Correlation and path coefficient analysis of seed yield and yield components in lentil (*Lens culinaris* Medik.) genotype in Ethiopia

**Fikru Mekonnen[1], Firew Mekbib[1], Shiv Kumar[2], Seid Ahmed[3] and T. R. Sharma[4]**

[1]School of Plant Sciences, College of Agriculture and Environmental Sciences, Haramaya University, P.O. Box, 138, Dire Dawa, Ethiopia.
[2]International Center for Agricultural Research in the Dry Areas (ICARDA), Rabat, Morocco.
[3]International Center for Agricultural Research in the Dry Areas (ICARDA), P O Box 5689, Addis Ababa, Ethiopia.
[4]Department of Agricultural Biotechnology, Chaudhary Sarwan Kumar Himachal Pradesh Agricultural University, Palampur 176 062, India.

**Knowledge of correlation among different traits and further partitioning of the correlation coefficients into direct and indirect effects on yield is one of the approaches to understand the nature and extent of the relationship among characters. The objective of this study was to determine the degree of associations between seed yield and yield components of lentil. In this study, 228 genotypes from diversified origin were considered. The experiment was conducted at three locations using randomized complete block design and augmented design. Significant strong positive phenotypic and genotype correlations were observed between seed yield and biomass, seed weight per plant, number of seed per plant and number of pods per plant. These four yield-contributing characters had also strong positive correlations with each other. The path coefficient analysis is in harmony, with the phenotypic and genotypic correlation, that is, seed weight per plant, followed by number of pods per plant, biomass yield and 100 seed weight have a considerable positive direct effect on lentil seed yield. However, days to 50% flowering, days to 90% maturity and rust disease severity score had negative phenotypic correlation and negative direct effect on seed yield.**

**Key words:** Genotype correlation coefficient, lentil, path analysis, phenotype correlation coefficient.

## INTRODUCTION

Lentil is a short and slender annual cool-season food legume, which was, domesticated early in the Fertile Crescent of the Near East (Sarker et al., 2010). Lentil provides sufficient amounts of the most essential amino acids to meet the nutrient requirements of humans. It is also a cash crop fetching a lot of money in domestic markets as compared to all other food legumes and major cereals (Geletu et al., 1996). Because of its significant economic roles and social contribution, lentil production has recently been expanding both in stressed and non-stressed environments (Asnake and Geletu, 2006). The global lentil productivity is about 0.9 t/ha, while in Africa it is about 0.6 t/ha (FAO, 2013) and in Ethiopia about 1.2 t/ha (CSA, 2012). In many traditional lentil-producing countries, most of the lentil production area is covered by the local landraces that are vulnerable to a range of biotic and abiotic factors and produced lower yield and seed quality (Sarker et al., 2003).

Characterizing crop genotypes for agronomic morphological traits have immense importance in crops breeding. Some of the traits are easy to select due to its visual nature but its scientific understanding is important for its manipulation. To facilitate selection in breeding for high yield, it is important to examine the association of various yield components and give more attention to those having greatest influence on yield. Gomez and Gomez (1984) showed that in order to determine important components of crop performance among their variables, it is essential to examine association between primary and secondary traits. Falconer and Mackey (1996) also indicated correlation coefficients measures the degree of association between characters and relative influence of various characters on yield and amongst themselves. Since breeding is commonly aimed at multiple traits, it is essential to know the correlated responses in primary traits like seed yield through selection of secondary traits (Borojevic, 1990).

Knowledge of correlation among different traits and further partitioning of the correlation coefficients into direct and indirect effects is one of the approaches to understand the nature and extent of such relationship among the characters in both local and introduced genotypes. As more traits are considered for correlation study, the indirect associations between traits become more complex, less obvious and somewhat perplexing. The path coefficient analysis, a method developed by Wright (1921) and later elaborated by Dewey and Lu (1959), provides an effective means of partitioning direct and indirect effect of association. Path coefficient is simply a standardized partial regression coefficient, measures the direct influence of one variable upon another, and permits the separation of correlation coefficient into components of direct and indirect effects This technique allows identifying major yield contributing characters, and specific traits producing a given correlation (Rao et al., 1997).

Researchers in Ethiopia and elsewhere have reported association yield and yield components traits of lentil germplasm, and the information was not comprehensive and exhaustive (Erskine et al., 1989; Ramgiry et al., 1989; Geletu et al., 1996; Abebe et al., 2001; Tigest, 2003; Edossa et al., 2010; Roy, et al., 2013). Besides, the value of correlation coefficients and the contribution of different components vary in different environments and, in different population and the magnitude of correlation coefficient can often be influenced by the choice of individuals upon which the observations are made (Jatasra and Paroda 1978; Dabholkar, 1992). It is therefore, advisable to consider large number of genotypes and use the correlation to establish an index in deciding the direction of selection. Hence, understanding of such an important association is necessary to conduct effective selection activities. An understanding of the genetic relationships of different yield components and their direct and indirect effect is imperative for Ethiopian lentil, in relation to lentil from exotic origin. The objective of this study was to determine the degree and nature of associations among seed yield and yield components in local and exotic lentil genotypes under the Ethiopia conditions.

## MATERIALS AND METHODS

### Description of the study site

The field experiment was conducted on selected hot spot locations for rust. The locations included Sirinka Agricultural Research Center (SRARC) in the northeastern part of Ethiopia for two seasons (2011 and 2012), Chefe Donsa in the central part of Ethiopia and Sinana Agricultural Research Center (SARC) in southeastern parts of Ethiopia during the 2011/12 cropping season (Table 1 and Figure 1).

### Plant materials

The experiments consist of 228 genotypes collected from Ethiopian Biodiversity Institute (EBI) and DebreZeit Agricultural Research Station (DZARC), Ethiopia, and the International Center for Agricultural Research in Dry Areas (ICARDA). Name of genotype number of genotype, source of origin, and breeding status of the genotypes were presented in Table 1. In 2010/2011 cropping season, 158 genotypes were planted for morphological evaluation at SRARC. In the 2011/12 cropping season, 228 genotypes including recombinant inbreed lines (RIL) were included (Table 1) in the study across three locations: Sirinka, Chefe Donsa and Sinana.

### Experimental layout and design

A randomized complete block design (RCBD) with three replications, at SRARC in the 2010/11 cropping season, and an augmented design with five blocks were used in 2011/12 cropping season over the three locations. The genotypes were planted in July in a two rows plot size of 0.8 $m^2$. The row to row distance was 20 cm. The distance between two plots was 50 cm and the distance between two blocks was 100 cm. Eight checks were replicated within each block. Planting was done in the first week of July at SRARC, in August at Chefe Donsa and in mid-September at SARC. The recommended agronomic packages at each location were applied for raising a successful crop.

### Data collection

Data were recorded on 10 important phonological, yield and yield

**Table 1.** List of genotypes used in this study and their origin.

| Source of origin | No. of genotypes | Name of genotype | Breeding status of the genotypes |
|---|---|---|---|
| Tigray | 8 | Acc. no. 219957,235383, 237503,237504, 241785, 242604, 243447 | Landrace |
| Amhara | 54 | Acc. no. 36003, 36025, 36028, 36039, 36041, 36061, 36071, 36085, 36088, 36089, 36097, 36103, 36104, 36105, 36137, 36139, 36150, 36162, 36165, 36168, 207258, 207274, 207287, 207309, 212745, 215248, 215249, 223221, 228242, 229179, 229182, 229183, 231247, 235013, 235015, 235016, 235017, 236484, 236486, 236487, 237502, 238978, 238979, 241784, 241786, 243433, 243436, 243440, 243443, 244606, 244610, 244615, 244619, 244623 | Landrace |
| Oromya | 29 | Acc. no. 36001, 36007, 36009, 36013, 36015, 36019, 36023, 36029, 36033, 36042, 36048, 36058, 36110, 36120, 36131, 203141, 215806, 216877, 228809, 230521, 230833, 230834, 230837, 231248, 235698, 236438, 236892, 237027, 238971 | Landrace |
| SNNP | 2 | Acc. no. 36147 and Acc. No.,228243 | Landrace |
| Somali | 1 | Acc. no. 230832 | Landrace |
| DZARC | 6 | /ILL4225 x ILL4605/ /ILL 6821/ Alemaya, /ILL 1 x ILL 1169//ILL 6027/ ADAA, /ILL 7978/ Teshale, / Alemaya x FLIP88-41L/ DERASH, /ILL 7981/ *Aleme Tena* and P160/ILL 2704/ / Chekol/ | Commercial varieties |
| Unknown | 10 | Acc. no. 36134, 207260, 211062,211078, 211110, 220120, 211131, 233349, 233973, 241782 | Landrace |
| ICARDA | 22 | X2003S 222/ILL 213/, X2003S 238 /ILL 4605/, X2006S 128/ILL 5480/, L-9-12, X2002S 219 /ILL 6821/, X2006S 129 /F2/, X2005S 215 /ILL 6002/, X2006S 133/FLIP87-21L/ /ILL 4349 x ILL4605//ILL 6211/, X2006S 130/FLIP 93-46L/ /ILL 7547/, FLIP-2004-7L, X2003S 223, X2003S 195/ILL 7115/, X2006S 130/FLIP 96-46 L//ILL 7978/, X2002S 221/FLIP 96-47 L//7979/, X2002S 221 /7980/, X2003S 233 /ILL 8009/, X2006S 134/ILL8174/, 2006S 122 /FLIP 2003-43L/ /ILL 7010 x ILL 1939/ /ILL 9932/, /FLIP 2003-56L/ /ILL 9945/X2006S 127, /ILL 2573 x ILL 7537/ /FLIP 2003-62 L/ /ILL 9951/X2006S 122, X2002S 219 /shehor-74/ /ILL 7554/, 2003S 236 | Parent |
| DZARC | 11 | EL-142 /ILL 5071/, EXOTIC #DZ/2008 AK, R-186XFLIP-86-38L-24, ILL-358 X ILL-2573-2-2000, 87S-93549XEL-1O3-4, 87s-93549XEL-03-5, Chekol X R-186-1, R-186X FLIP-86-38L, Chekol x R-186-2, EL-142 X R-186-2, EL-142 X R-186-3 | Breeding line |
| ICARDA | 15 | ILL-590/NEL 590/, FLIP-2006-60L, FLIP-97-68L, FLIP-04-26L, ILL-28501, FLIP-2006-20L, FLIP-87-68L, /ILL 6037 x ILX 87062/ FLIP2005-24L/ ILL-10045, FLIP-97-16L/ILL 8078, ILL-10680, FLIP-2004-37L, FLIP-84-95L /ILL 5722, FLIP-97-61L, L-830, Precoze/ILL 4605/ | Breeding line |
| ICARDA | 70 | RIL1- RIL70 | RIL |

Unknown: Originated from Ethiopia but site of collection not mentioned RIL: Recombinant inbred line.

components and reaction to rust disease. The data were recorded in randomly selected plants on plant and plot basis (Table 2).

#### Data analysis

The data were subjected to statistical analysis using GenStat Release 15.1 statistical software (VSN International Ltd., 2012). The correlation and path coefficient was analyzed based on the row data value for each agro-morphological trait. The phenotypic and genotypic correlation coefficients were calculated for data from RCBD by the following formula as suggested by Miller et al. (1959). Only the phenotypic correlation coefficients were calculated for augmented design using the formula adopted by Singh and Chudhary (1977) on Genstat 15 Release 4.2 software. Path coefficient analysis was carried out to partition the phenotypic and genotypic correlation coefficients into direct and indirect effects of yield attributing traits (independent characters) on grain yield (dependent character) using the general formula suggested by Wright (1921) and worked out by Dewey and Lu (1959). For augmented design, path coefficient analysis was calculated using the formula and syntax developed by Singh and Chudhary (1977) on Genstat 15 Release 4.2 software.

## RESULTS AND DISCUSSION

### Phenotypic and genotypic correlation

Significant positive phenotypic correlation was observed between number of pods per plant and seed weight per plant ($r_p$ = 0.9), seed yield and above ground biomass ($r_p$ = 0.7), seed weight per plant ($r_p$ = 0.6), number of seed per plant ($r_p$ = 0.5) and number of pods per plant ($r_p$ = 0.5). In addition, numbers of seeds per pod, plant height and 100 seed weight have significant ($P \leq 0.05$) positive association with seed yield at Sirinka (Table 3).

Highly significant strong positive correlations were observed between number of pods per plant and seed weight per plant ($r_p$ = 0.9), seed weight per plant and

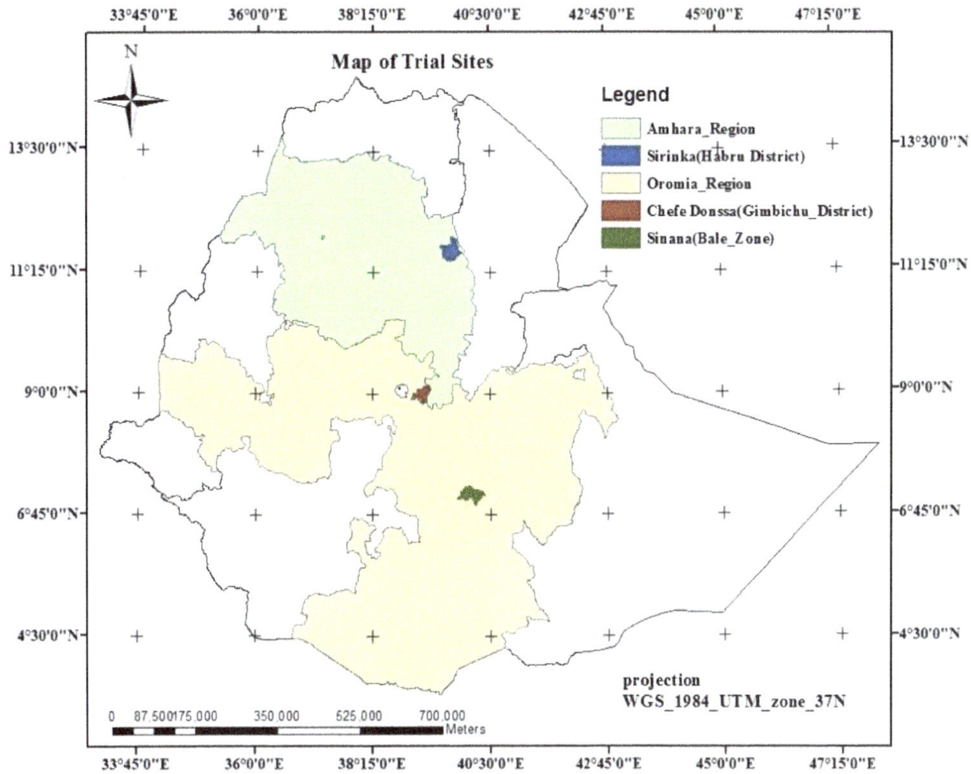


**Figure 1.** Geographical position of the experimental sites.

**Table 2.** List of agro-morphological, phonological traits and rust severity score collected for the study.

| Traits | Description |
|---|---|
| Number of pods/plant (NPPP) | Mean of pods from ten randomly selected plants |
| Number of seeds/ Pod (NS) | Average number of seeds per pods counted from 10 randomly sampled pods of 10 different plants. |
| Seed weight per plant (SWPP) | The average matured seed weight from 10 randomly selected plants. |
| Number of seeds per plant (NSPP) | Mean of number of seeds counted from 10 randomly selected plants |
| 100-seed weight (SW) | Weight of 100 seeds taken from the bulk harvest in gram (g) |
| Plant height (PH) | The height from the ground surface to the top of the main stem at physiological maturity in centimeter (cm) |
| Days to flowering (DFF) | Days from planting to flowering of 50% plant |
| Days to maturity (DM) | Days from planting to physiological maturity of 75% plant |
| Biomass per plant (BI) | Above ground biomass of the mean yield of two harvestable rows in gram |
| Seed yield per plant (SY) | Mean seed yield in gram of two harvestable rows in gram |
| Rust disease severity score after flowering (RDSAF) | The RILs and the accession were scored for rust severity quantitative scale developed by Chen (2007) had 1–9 scale, whereby 1 = 0–10% leaf area infected, 3 = 11–30% leaf area infected, 5 = 31–50% leaf area infected, 7 = 51–70% leaf area infected, 9 = more than 70% leaf area infected. |

number of seeds per plant ($r_p$ = 0.8), number of pods per plant and number of seeds per plant ($r_p$ = 0.8). Whereas intermediate significant positive correlations were observed between days to 50% flowering and days to 90% maturity ($r_p$ = 0.6), seed weight per plant and above ground biomass ($r_p$ = 0.6), plant height and above ground biomass ($r_p$ = 0.6) and number of seeds per plant and above ground biomass ($r_p$ = 0.5) at Sirinka in 2010/11 (Table 3).

Significant ($P \leq 0.05$) negative phenotypic correlation

**Table 3.** Genotypic correlation in above diagonal and phenotypic correlation in below diagonal of yield and yield component characters in lentil genotypes at Sirinka in 20110/11.

| Trait | DFF | DM | NPPP | NS | SWPP | NSPP | PH | SW | BI | AY |
|---|---|---|---|---|---|---|---|---|---|---|
| DFF | | 0.8** | 0.1 | -0.3** | -0.1 | 0.3* | 0.3* | -0.1 | 0.5** | 0.2 |
| DM | 0.6** | | -0.1 | -0.8** | -0.3** | 0.1 | -0.1 | 0.1 | 0.3** | 0.04 |
| NPPP | 0.05 | -0.2* | | 0.1 | 0.8* | 0.8** | 0.3** | -0.3** | 0.4** | 0.4** |
| NS | -0.2** | -0.5** | 0.1* | | 0.4** | 0.5** | 0.1 | -0.07 | 0.2** | 0.3** |
| SWPP | -0.1 | -0.2** | 0.9** | 0.2** | | 0.8** | 0.2 | 0.20* | 0.6** | 0.6** |
| NSPP | 0.06 | -0.04 | 0.8** | 0.4** | 0.8** | | 0.3** | -0.3** | 0.4** | 0.** |
| PH | 0.2* | -0.2* | 0.4** | 0.001 | 0.3** | 0.3** | | -0.14 | 0.6** | 0.3** |
| SW | -0.06 | 0.2* | -0.07 | -0.03 | 0.2** | -0.2** | -0.1* | | 0.2* | 0.2* |
| BI | 0.3** | 0.2* | 0.5** | 0.13* | 0.6** | 0.5** | 0.6** | 0.2** | | 0.7** |
| AY | 0.02 | -0.07 | 0.5** | 0.2** | 0.5** | 0.5** | 0.3** | 0.2** | 0.66** | |

DFF = days to 50% flowering, DM = days to 90% maturity, NPPP = number of pods per plant, NS= number of seeds per pod, SWPP = Seed weight per Plant, NSPP = Number of Seeds per Plant, PH = Plant height in cm, SW = 100-seed weight in gram, BI = above ground biomass, SY = seed yield,
** and * =Significant at 0.01and 0.05 probability levels.

**Table 4.** Pearson's phenotype correlations of yield and yield component characters in lentil genotypes at Sirinka 2011/12.

| Trait | DF | DM | NPPP | NS | NSPP | SW | SWPP | PH | BI | SY |
|---|---|---|---|---|---|---|---|---|---|---|
| DF | - | | | | | | | | | |
| DM | 0.70** | - | | | | | | | | |
| NPPP | 0.002 | -0.16** | - | | | | | | | |
| NS | 0.20** | 0.17 | 0.03 | - | | | | | | |
| NSPP | 0.03 | -0.15* | 0.62** | 0.32** | - | | | | | |
| SW | -0.02 | 0.22 | -0.19** | -0.02 | -0.22** | - | | | | |
| SWPP | 0.07 | -0.01 | 0.51 | 0.19** | 0.83** | 0.11 | - | | | |
| PHH | 0.20** | -0.06 | 0.31 | 0.10 | 0.47** | -0.11 | 0.41** | - | | |
| BI | 0.26** | 0.14* | 0.39** | 0.19** | 0.57** | 0.03 | 0.65** | 0.57** | - | |
| SY | 0.05 | -0.10 | 0.43** | 0.13* | 0.64** | 0.04 | 0.72** | 0.41** | 0.77** | - |

DFF = days to 50% flowering, DM = days to 90% maturity, NPPP = number of pods per plant, NS= number of seeds per pod, SWPP = Seed weight per Plant, NSPP = Number of Seeds per Plant, PH = Plant height in cm, SW = 100-seed weight in gram, BI = above ground biomass, SY = seed yield,
** and * =Significant at 0.01and 0.05 probability levels.

coefficients were observed between days to 90% maturity and number of seeds per pods ($r_p$ = -0.5), plant height and number of pods per plant and 100 seed weight and number of seeds per plant (Table 3).

Similarly, significant phenotypic correlation was observed between seed yield and number of pods per plant, number of seeds per pods, number of seeds per plant, seed weight per plant and plant height at Sirinka, Chefe Donsa and Sinana in 2011/12 cropping season (Tables 4, 5 and 6). However, negative significant correlations were observed between seed yield and days to 50% flowering, and rust disease score with seed yield at Chefedonsa and Sinana locations (Tables 5 and 6). Above ground biomass showed significant positive phenotypic correlation with all characters except rust disease severity score (Tables 4, 5 and 6). Significant positive correlations were observed between days to 50% flowering and days to 90% maturity, and between days to 90% maturity and 100 seed weight in all the three environments; significant phenotypic correlations were observed between number of pods per plant and seed weight per plant at Sirinka ($r_p$ = 0.51) and Chefe Donsa ($r_p$ = 0.85), and number of pods per plant and number of seeds per plant at Sirinka ($r_p$ = 0.62), Chefe Donsa ($r_p$ = 0.94) and Sinana ($r_p$ = 0.92). Seed weight per plant and number of seeds per plant ($r_p$ = 0.83) at Sirinka, seed weight per plant with plant height ($r_p$ = 0.41) at Sirinka and number of seeds per plant with plant height ($r_p$ = 0.47) at Sirinka. Highly significant phenotypic correlation was observed between seed weight per plant and above

**Table 5.** Pearson's phenotype correlations of yield and yield component characters in lentil genotypes at Chefe Donsa 2011/12.

| Traits | DF | DM | NPPP | NS | NSPP | SW | SWPP | PH | BI | SY |
|---|---|---|---|---|---|---|---|---|---|---|
| DF | - | | | | | | | | | |
| DM | 0.54** | - | | | | | | | | |
| NPPP | -0.15* | -0.19** | - | | | | | | | |
| NS | -0.26** | -0.38** | 0.10 | - | | | | | | |
| NSPP | -0.21** | -0.26** | 0.94** | 0.40** | - | | | | | |
| SW | 0.15* | 0.45** | -0.22** | -0.25** | -0.26** | - | | | | |
| SWPP | -0.12 | -0.06 | 0.85** | 0.18** | 0.83** | 0.08 | - | | | |
| PH | 0.27** | 0.21** | 0.02 | -0.05 | 0.01 | -0.02 | 0.01 | - | | |
| BI | 0.30** | 0.39** | 0.07 | 0.05 | 0.06 | 0.14* | 0.17** | 0.44** | - | |
| SY | -0.15* | -0.18* | 0.45** | 0.22** | 0.47** | -0.09 | 0.54** | 0.25** | 0.51** | - |

DFF = days to 50% flowering, DM = days to 90% maturity, NPPP = number of pods per plant, NS= number of seeds per pod, SWPP = Seed weight per Plant, NSPP = Number of Seeds per Plant, PH = Plant height in cm, SW = 100-seed weight in gram, BI = above ground biomass, SY = seed yield,
** and * =Significant at 0.01and 0.05 probability levels.

**Table 6.** Pearson's phenotype correlations of yield and yield component characters in lentil genotypes at Sinana in 2011/12.

| Trait | DF | DM | NPPP | NS | NSPP | SW | PHH | BI | SY | RDSBF | RDSAF |
|---|---|---|---|---|---|---|---|---|---|---|---|
| DF | - | | | | | | | | | | |
| DM | 0.2** | - | | | | | | | | | |
| NPPP | -0.14* | 0.2** | - | | | | | | | | |
| NS | -0.03 | 0.07 | 0.14* | - | | | | | | | |
| NSPP | -0.10 | 0.2** | 0.9** | 0.5** | - | | | | | | |
| SW | 0.15* | 0.3** | 0.2** | 0.01 | 0.2** | - | | | | | |
| PH | 0.09 | -0.01 | -0.02 | 0.11 | 0.05 | 0.09 | - | | | | |
| BI | 0.02 | 0.11 | 0.03 | 0.06 | 0.07 | 0.13* | 0.2** | - | | | |
| SY | -0.09 | 0.2** | 0.5** | 0.03 | 0.4** | 0.5** | 0.2** | 0.3** | - | | |
| RDSBF | -0.2** | -0.3** | -0.2** | 0.01 | -0.2** | -0.3** | 0.04 | 0.06 | -0.3** | - | |
| RDSAF | -0.06 | -0.4** | -0.3** | 0.02 | -0.2** | -0.5** | 0.00 | -0.12* | -0.5** | 0.40 | - |

RDSBF= rust disease severity score before flower, RDSAF= rust disease severity score after flower, DFF = days to 50% flowering, DM = days to 90% maturity, NPPP = number of pods per plant, NS= number of seeds per pod, SWPP = Seed weight per Plant, NSPP = Number of Seeds per Plant, PH = Plant height in cm, SW = 100-seed weight in gram, BI = above ground biomass, SY = seed yield, ** and * =Significant at 0.01and 0.05 probability levels.

ground biomass ($r_p$ = 0.65) at Sirinka (Table 4). Negative phenotypic correlations were observed between days to 90% maturity and rust disease score before and after flowering ($r_p$ = 0.3 and 0.4) at Sinana (Table 6). Negative phenotypic correlations were observed between days to 90% maturity and number of pods per plant, number of seed per plant and seed weight per plant at Sirinka and ChefeDonsa (Tables 4 and 5). Hundred seed weight also showed significant negative phenotypic correlation with number of seeds per plant at Sirinka and ChefeDonsa but not at Sinana (Tables 4, 5 and 6).

Seed yield per plant positively and significantly correlated with biomass ($r_g$ = 0.7), seed weight per plant ($r_g$ = 0.6), number of seeds per plant ($r_g$ = 0.5), number of pods per plant ($r_g$ = 0.4) (Table 3). Significant positive genotypic correlation coefficient were also observed among yield component traits including number of pods per plant with seed weight per plant ($r_g$ = 0.8) and number of seeds per plant ($r_g$ = 0.8), seed weight per plant with number of seeds per plant ($r_g$ = 0.8), days to 50% flowering with days to 90% maturity ($r_g$ = 0.8). However, negative and significant genotypic correlation coefficients were observed between days to 90% maturity and number of seeds per pod ($r_g$ = -0.8), number of pods per plant with 100 seed weight ($r_g$ = - 0.3), number of seeds per plant and 100 seed weight ($r_g$ = -0.3), days to 50% flowering and number of seeds per pod ($r_g$ = -0.3).

## Estimates of direct and indirect effects of yield component traits on seed yield

In the present study, seed yield (SY) was considered as

**Table 7.** Phenotypic direct effects on main diagonal (bold) and indirect effects of different agronomic traits on seed yield of lentil genotypes at Sirinka in 2010/11.

| Traits | DF | DM | NPPP | NS | SWPP | NSPP | PHH | SW | PSS |
|---|---|---|---|---|---|---|---|---|---|
| DF | **-0.06** | -0.12 | 0.03 | 0.03 | 0.11 | -0.04 | 0.05 | 0.03 | -0.02 |
| DM | 0.04 | **-0.19** | -0.03 | 0.03 | 0.12 | -0.01 | -0.01 | 0.11 | -0.02 |
| NPPP | 0.01 | 0.02 | **0.31** | 0.00 | 0.36 | -0.22 | 0.05 | -0.04 | 0.00 |
| NS | 0.01 | -0.02 | 0.00 | **0.21** | 0.15 | -0.13 | 0.01 | -0.01 | -0.01 |
| SWPP | 0.01 | -0.04 | 0.23 | 0.07 | **0.48** | -0.22 | 0.04 | 0.05 | -0.003 |
| NSPP | 0.01 | -0.003 | 0.25 | 0.10 | 0.39 | **-0.27** | 0.05 | -0.03 | -0.002 |
| PHH | 0.02 | 0.01 | 0.10 | 0.02 | 0.10 | -0.08 | **0.17** | -0.03 | 0.002 |
| SW | 0.01 | -0.09 | -0.05 | -0.01 | 0.10 | 0.04 | -0.02 | **0.22** | -0.001 |
| PSS | -0.02 | 0.06 | 0.00 | -0.03 | -0.03 | 0.01 | 0.01 | -0.01 | **0.05** |

Contribution of residuals in the variability = 0.57. DFF = days to 50% flowering, DM = days to 90% maturity, NPPP = number of pods per plant, NS = number of seeds per pod, SWPP = seed weight per plant, NSPP = number of seeds per plant, PH = plant height in cm, SW = 100-seed weight in gram, BI = above ground biomass, SY = seed yield, **, * = Significant at 0.01 and 0.05 probability levels.

effect dependent on nine independent variables, which were considered as causes. The independent characters were: days to 50% flowering (DFF), days to 90% maturity (DM), number of pods per plant (NPPP), number of seeds per pods (NS), number of seeds per plant (NSPP), seed weight per plant (SWPP), plant height (PH), 100 seed weight (SW) and rust disease severity score.

The path analysis revealed that seed weight per plant (0.48) followed by number of pods per plant (0.31) and 100 seed weight (0.22) had exerted positive direct effects on seed yield at Sirinka in 2010/11 (Table 7). In addition, numbers of seeds per pod and plant height have positive direct effects on seed yield. However, number of seeds per plant (-0.27) and days to 90% maturity (-0.19) had negative direct effects on seed yield. With the exception of seed weight per plant and number of seeds per plant, the indirect effects of all traits on seed yield via other traits were small and negligible. The contribution of all traits in explaining the variability of seed yield was lower than the residual effect (0.57). More than 50% of the variability of seed yield at Sirinka 2011/12 was accounted for by seed weight per plant and number of pods per plant followed by the interaction of seed weight per plant and number of pods per plant, the rest of the traits had very little effect on the variability of seed yield.

Result of the path analysis are consistent with the phenotypic and genotype correlation coefficients which showed that seed weight per plant and number of pods per plant had highly significant genotypic and phenotypic correlation coefficients with seed yield (Table 3). Number of seeds per plant had significant positive correlation with seed yield and had a substantially high negative direct effect (-0.27). The positive indirect effects of number of seeds per plant via seed weight per plant, number of pods per plant and number of seeds per pod, might counter-balance the final positive correlation with seed yield (Table 7). Path coefficient analysis at Sirinka in 2011/12 cropping season, gave essentially similar pattern of relationship between yield and yield components with that of Sirinka in 2010/11. Seed weight per plant, plant height and numbers of pods per plant had strong positive direct effect on seed yield. These characters had positive indirect influence via seed weight per plant, days to 90% maturity, number of pods per plant, number of seeds per plant and 100 seed weight on seed yield (Table 8).

At Chefe Donsa, seed weight per plant (0.62), plant height (0.32), and number of pods per plant (0.28) had positive direct effect on seed yield. Number of seeds per pod, number of seeds per plant, and 100 seed weight had significant positive indirect effect on seed yield via seed weight per plant and plant height. The strong direct effects of these three characters on seed yield is consistent with the phenotypic correlation coefficient of these traits with seed yield ($r_p$ = 0.54, 0.25 and 0.45, respectively) (Table 5). Number of seeds per plant had negative direct effect on seed yield but it had a significant phenotypic correlation coefficient ($r_p$= 0.47) with seed yield. This may be due to the indirect positive effect of this character on seed yield via seed weight per plant, number of pods per plant, number of seeds per pod, 100 seed weight and days to 50% flowering and 90% maturity that counter balanced the final correlation in the positive direction (Table 9). Rust disease score, days to 50% flowering and 90% maturity and 100 seed weight in gram had considerable negative direct and indirect effects on seed yield via days to 50% flowering, days to 90% maturity, number of pods per plant, number of seeds per pod, plant height and 100 seed weight (Table 9). The direct and indirect effect of the casual variables considered in this study were higher than the residual effects

**Table 8.** Phenotypic direct effects on main diagonal (bold) and indirect effects of different agronomic traits on seed yield of lentil genotypes at Sirinka in 2011/12.

| Trait | DF | DM | NPPP | NS | SWPP | NSPP | PHH | SW |
|---|---|---|---|---|---|---|---|---|
| DF | **-0.04** | -0.023 | -0.001 | -0.001 | 0.027 | -0.001 | 0.018 | -0.0001 |
| DM | -0.021 | **-0.045** | -0.006 | -0.001 | -0.01 | -0.01 | -0.004 | -0.002 |
| NPPP | 0.001 | 0.005 | **0.056** | 0.001 | 0.303 | 0.027 | 0.033 | 0.001 |
| NS | -0.006 | -0.008 | 0.002 | **-0.01** | 0.116 | 0.014 | 0.012 | 0.0002 |
| SWPP | -0.002 | 0.0004 | 0.026 | -0.001 | **0.641** | 0.038 | 0.044 | -0.001 |
| NSPP | 0.001 | 0.005 | 0.033 | -0.002 | 0.533 | **-0.046** | 0.051 | 0.002 |
| PHH | -0.007 | 0.002 | 0.017 | -0.001 | 0.256 | 0.021 | **0.11** | 0.001 |
| SW | -0.001 | -0.01 | -0.011 | 0.001 | 0.072 | -0.01 | -0.01 | **-0.007** |

Contribution of residuals in the variability = 0.4195. DFF = days to 50% flowering, DM = days to 90% maturity, NPPP = number of pods per plant, NS = number of seeds per pod, SWPP = seed weight per plant, NSPP = number of seeds per plant, PH = plant height in cm, SW = 100-seed weight in gram, BI = above ground biomass, SY = seed yield, **, *Significance at 0.01 and 0.05 probability levels.

**Table 9.** Phenotypic direct effects on main diagonal (bold) and indirect effects of different agronomic traits on seed yield of lentil genotypes at Chefe Donsa in 2011/12.

| Trait | DF | DM | NPPP | NS | SWPP | NSPP | PHH | SW | RDSAF |
|---|---|---|---|---|---|---|---|---|---|
| DF | **-0.09** | -0.1 | -0.01 | -0.02 | -0.08 | 0.04 | 0.08 | -0.02 | 0.03 |
| DM | -0.05 | **-0.19** | -0.02 | -0.03 | -0.05 | 0.05 | 0.06 | -0.05 | 0.06 |
| NPPP | 0.02 | 0.04 | **0.28** | 0.01 | 0.52 | -0.16 | 0 | 0.03 | -0.01 |
| NS | 0.02 | 0.08 | 0.01 | **0.07** | 0.11 | -0.06 | -0.01 | 0.03 | -0.03 |
| SWPP | 0.01 | 0.02 | 0.07 | 0.01 | **0.62** | -0.14 | 0.006 | -0.01 | 0.01 |
| NSPP | 0.02 | 0.06 | 0.07 | 0.03 | 0.51 | **-0.17** | 0.002 | 0.03 | -0.02 |
| PHH | -0.02 | -0.04 | 0.001 | 0.003 | 0.01 | 0 | **0.32** | 0.01 | 0.01 |
| SW | -0.02 | -0.09 | -0.02 | -0.02 | 0.03 | 0.05 | -0.01 | **-0.11** | 0.06 |
| RDS | 0.02 | 0.08 | 0.01 | 0.02 | -0.04 | -0.03 | -0.03 | 0.05 | **-0.13** |

Contribution of residuals in the variability = 0.4905. DFF = days to 50% flowering, DM = days to 90% maturity, NPPP = number of pods per plant, NS= number of seeds per pod, SWPP = Seed weight per Plant, NSPP = Number of Seeds per Plant, PH = Plant height in cm, SW = 100-seed weight in gram, BI = above ground biomass, SY = seed yield, ** and * =Significant at 0.01 and 0.05 probability levels.

(0.49) which explain its effect on seed yield. More than 50% of seed yield response was explained by seed weight per plant followed by plant height at Chefe Donsa in 2011/12.

At Sinana in the 2011/12 cropping season, the highest positive direct effect were shown by number of pods per plant (0.49), followed by 100 seed weight in gram (0.28) (Table 10). Plant height (0.18) showed the third largest positive direct effect on seed yield. In addition to its direct effect, the indirect influence of number of pods per plant via number of seeds per plant was substantially important for final contribution to seed yield ($P_r$=0.4). However, rust disease severity score (-0.25), number of seeds per plant (-0.171) and days to 50% flowering (-0.12) showed negative direct effects on seed yield.

In general, from path analysis result of the four environments, we can deduce that seed weight per plant, number of pods per plant and plant height had consistently strong direct effects on seed yield. These traits had also average strong phenotypic correlation coefficient values ($r_p$ = 0.64, 0.47 and 0.29, respectively) with seed yield. However, days to 50% flowering, days to 90% maturity, rust disease severity score and number of seeds per plant had negative direct effect on seed yield at the four environments. Phenotypic correlation coefficients also revealed that these traits had negative association with seed yield of lentil except for number of seeds per plant with an average value of $r_p$ = -0.023, -0.26 and -0.055, respectively. Despite, negative direct effect of number of seeds per plant on seed yield, it had an average strong phenotypic correlation coefficient ($r_p$= 0.54) with seed yield. This may be due to the significant positive indirect effect of this character via number of pods per plant, seed weight per plant, number of seeds per pod, 100 seed weight in gram, days to 50% flowering and 90% maturity, and plant height that substantially step

**Table 10.** Phenotypic direct effects on main diagonal (bold) and indirect effects of different agronomic traits on seed yield of lentil genotypes at Sinana in 2011/12.

| Trait | DF | DM | NPPP | NSPP | PHH | SW | RDSAF |
|---|---|---|---|---|---|---|---|
| DF | **-0.12** | -0.012 | -0.094 | 0.023 | 0.014 | 0.05 | 0.025 |
| DM | -0.03 | **0.042** | 0.093 | -0.04 | 0.009 | 0.09 | 0.121 |
| NPPP | 0.02 | -0.008 | **0.493** | -0.16 | -0.002 | 0.06 | 0.07 |
| NSPP | 0.02 | -0.009 | 0.451 | **-0.17** | 0.014 | 0.06 | 0.061 |
| PHH | -0.01 | -0.002 | -0.007 | -0.01 | **0.18** | 0.03 | 0.011 |
| SW | -0.02 | -0.014 | 0.104 | -0.03 | 0.017 | **0.28** | 0.125 |
| RDSAF | 0.01 | 0.02 | -0.14 | 0.042 | -0.008 | -0.14 | **-0.25** |

Contribution of residuals in the variability = 0.5398. DFF = days to 50% flowering, DM = days to 90% maturity, NPPP = number of pods per plant, NS= number of seeds per pod, SWPP = Seed weight per Plant, NSPP = Number of Seeds per Plant, PH = Plant height in cm, SW = 100-seed weight in gram, BI = above ground biomass, SY = seed yield, ** and * = Significant at 0.01and 0.05 probability levels.

down the final correlation in the positive direction (Tables 8, 9 and 10).

## DISCUSSION

Seed yield is a complex quantitative character governed by polygene, and is highly influenced by the environment. To improve yield, study of association of yield and yield components is of paramount importance. To increase yield, the breeder has to give sufficient attention to yield components that are favorably correlated with seed yield. Studies on correlation provide an opportunity for critically assessing the relationship of the component characters with seed yield. At Sirinka in 2010/11 cropping season, the magnitudes of genotypic correlation coefficients between all traits were higher than their phenotype correlation coefficients except for number of pods per plant, signifying that the association among various characters were of genetic causes rather than environmental effects. A similar trends were reported by Singh et al. (1999) and Tyagi and Khan (2011), genotypic correlation were of higher magnitude than their phenotypic counter parts in lentil.

Significant positive correlations were observed between seed yield and biomass, number of pods per plant, seed weight per plant, number of seed per plant and plant height at the four environments (Tables 3, 4 and 5). Other investigators also reported that seed yield was positively correlated with number of pods per plant, number of seeds per pods, seed weight per plant and biological yield in lentil (Dixit and Dubey, 1984; Balayan and Singh, 1986; Esmail et al., 1994; Bhattacharya, 1999; Naji et al., 2003; Singh et al., 2003). In another study, Abo-Shetaia et al. (1997) showed similar report both with highly significant phenotypic and genotypic association between lentil seed yield and number of pods per plant, and number of seeds per plant, the genotype correlation coefficient of seed yield with number of seeds per pod being negative in their study.

Number of pods per plant, number of seed per plant, seed weight per plant and above ground biomass had strong positive correlations among each other. Kumar and Sapra (1984) and Singh et al. (2003) reported that seed weight per plant was directly linked with number of pods per plant. Sinha and Singh (2002), Tigest (2003) and Ketema (2007) reported that number of seeds per plant was strongly and positively associated with number of pods per plant in lentil. This implies that genotype with larger number pods per plant produce higher number of seeds per plant. The relationship between seed yield and number of pods per plant, number of seeds per plant, seed weight per plant and above ground biomass suggested that selection based on these component characters would result in maximum yield in lentil. From this, we can infer that selection for either character increases the other traits and has to be considered as selection index for seed yield of lentil.

In support of our recommendation, Hamdi et al. (1991) and Khattab (1999) observed that pods per plant, seeds per plant and seed weight and biomass yield were the most important characters that contributed to grain yield. In addition to these, number of seeds per pods, plant height and 100 seed weight had significant (*p≤0.05*) positive phenotypic association with seed yield in all location except with 100 seed weight at Chefe Donsa. Vir et al. (2001) and Tyagi and Khan (2011) had reported such positive correlation between seed yield with plant height and 100-seed weight in lentil. Sinha and Singh (2002) and Sing et al. (2003) also reported in agreement with our findings that seed yield was significantly correlated with plant height, number of seeds per pods and seed weight. Hundred seed weight has negative phenotypic association with number of seeds per plant, number of pods per plant and number of seeds per pods at all the environments except at Sinana. Such observation was also made by others (Tyagi and Sharma, 1985; Sharma et al., 1993; Tyagi and Khan, 2011). However, as opposed to our finding, Sinha and Singh (2002) reported that 100 seed weight was strongly and positively associated with number of seeds per plant and number of pods per plant in lentil.

Days to 50% flowering and 90% maturity and rust disease severity score showed significant negative phenotypic correlation with seed yield at all locations, except at Sinana where positive association was observed between days to 90% maturity and seed yield. At Sirinka and Chefe Donsa where terminal drought is a common phenomena, earliness is favorable characters for better performance in such location however, at Sinana, incidence of terminal drought is not common, late maturing genotype may better perform for seed yield. Others also reported that seed yield had negative correlation with time to flowering, days to 90% maturity and seed weight in moisture deficit environments (Kumar and Sapra, 1984; Mia et al., 1986; Esmail et al., 1994; Tyagi and Khan, 2011). In another high potential environment, Manara and Manara (1988) found positive correlation of seed yield with days to 90% maturity, suggesting that seed yield could be increased by selecting late maturing genotypes with greater number of pods per plant whenever there is no terminal drought. These findings are also strengthened by Tyagi and Sharma (1985) and Dutta et al. (1993) who reported that above ground biomass and seed yield were negatively correlated with earliness.

Two characters may show correlation just because they are correlated with a common third one. In such cases, it becomes necessary to study a method that takes into account the causal relationship between the variables in addition to the degree of such relationship. The phenotypic and genotypic correlations were further subjected to path coefficient analysis, which involves partitioning of the correlation coefficients into direct and indirect effects via alternative characters (Falconer and Mackay, 1996). Since correlations provide only limited information ignoring complex interrelationships among traits, further partitioning of genetic correlations into direct and indirect effects using the path-coefficient analysis provides better picture of the relationship of predictor variables with the response variable (Rao et al., 1997). Seed yield being a complex outcome contributed by different component traits, it was considered to be the resultant variable, while days to 50% flowering, days to maturity, number of pods per plant, number of seeds per pods, pod weight per plant, number of seeds per plant, seed weight per plant, plant height and 100 seed weight were causal variables.

Path coefficient analysis for seed yield over the two seasons at Sirinka revealed that seed weight per plant, number of pods per plant, and plant height had consistently strong direct effects on seed yield. Therefore, for selecting high yielding genotypes, the breeder should give more emphasis to plants with higher seed weight per plant, and more number of pods per plant. Phenotypic and genotype correlation coefficients of seed weight per plant and number of pods per plant were also high with seed yield at Sirinka 2010/11. Dixit and Dubey (1984), Balayan and Singh (1986), Tikka et al. (1997), Hamdi et al. (2003), Verma et al. (2004) and Tyagi and Khan (2011) also reported that number of pods per plant, plant height and seed weight per plant were the highest direct contributors towards better seed yield in their respective studies. In addition, Jain et al. (1991) and Begum and Begum (1996) reported that a combination of two or three variables, viz. plant height, number of branches per plant and pods per plant were found to be better than other combination of characters for the improvement of seed yield in lentil.

However, days to 50% flowering and days to 90% maturity, which had negative phenotypic correlation coefficients with seed yield, had also substantial negative direct effects on seed yield. In spite of the negative direct effect of number of seeds per plant it had high genotypic and phenotype correlation coefficient with seed yield, and this association was due to positive indirect effect of this character via seed weight per plant, number of pods per plant, days to 50% flowering and 90% maturity, plant height and 100 seed weight that step down the final association in the positive direction (Table 7). Verma et al. (2004) also reported that number of seeds per plant has significant correlation with seed yield but it had a negative direct effect on seed yield under rain fed condition. The path analysis residual values, at Sirinka in 2010/11 was high indicating that some other factors, which have not been considered need to be included in the analysis to fully account for the variation in seed yield (Annex 1). The contribution of all traits in the variability of seed yield was higher than the residual effects (0.42) in 2011/1 (Table 8). More than 50% of seed yield variability was contributed by the effect of seed weight per plant (Annex 1, 2 and 3).

At Chefe Donsa and Sinana 2012, almost similar results were obtained in terms of both magnitude and direction. The highest positive direct effects were shown by number of pods per plant (0.49), followed by 100 seed weight (0.281) (Table 9). Plant height (0.18) showed the third largest positive direct effect on seed yield. These results were consistent with results from other studies (Khattab, 1999; Bhattacharya, 1999; Singh et al., 2003; Çokkizgin, 2007).

Days to 90% maturity, rust disease score and days to 50% flowering had negative direct effects on seed yield at the four environments, except positive direct effects of days to 90% maturity at Sirinka in 2010/11. The same traits had also negative phenotypic correlation coefficients with seed yield of lentil. This was obvious for a dry land site like Sirinka where terminal drought is a common phenomena, genotypes which had long flowering period and late maturing genotypes with their physiological seed sinking process would be disturbed and fail to bring better seed yields. Whereas genotypes possessing early flowering and maturity attributes do well in these kinds of environments by escaping the terminal drought.

However, at Sinana where we do not have the incidence of terminal drought, late maturing genotype perform better in seed yield. Similar observation were reported by Kumar and Sapra (1984) and Tyagi and Khan (2011). However, Dutta et al. (1993) and Kumar et al. (2004) observed that days to 50% flowering and 90% maturity had the highest direct effects on seed yield, and could justify that increased seed yield were due to a long crop growing season that may affect an increase sink sites, not to impaired grain development during high temperature in the latter part of the season.

## Conclusion

This study clearly showed that for selecting high yielding genotypes, the breeder should give more emphasis to plants with higher seed weight per plant, more number of pods per plant and higher number of seeds per plant, and plants with higher above ground biomass. The path coefficients were consistent with phenotypic and genotype correlation coefficients in that seed weight per plant and number of pods per plant showed not only significant positive genotypic and phenotypic correlation coefficients with seed yield but also higher direct effects. On the other hand, days to 50% flowering and 90% maturity and rust disease severity score showed both negative phenotypic correlation coefficient and negative direct effects on seed yield except for the positive direct effect of days to 90% maturity with seed yield at Sinana. From phenotypic, genotypic and path coefficients it could be conclude that traits like seed weight per plant, number of pods per plant, above ground biomass, number of seeds per plant and plant height had a considerable role as selection criteria to improve lentil seed yield.

## Conflict of Interests

The author(s) have declared that there is no conflict of interests.

### REFERENCES


Abebe T, Kusmenoglu I, McPhee KE, Muehlbauer FJ (2001). Characterization of core collection of lentil germplasm for phenology, morphology, seed and straw yields. Genet Res Crop Evol. 48:143-152.

Abo-Shetaia AM, El-Gawad AAA, Khattab AM, Mokhtar SA (1997). Physiological and yield attributes as selection criteria in lentil crop. Ann. Agric. Sci. 42:323-336.

Asnake F, Geletu B (2006). Breeding lentil for wider adaptation. In: Kemal A, Gemechu K, Seid A, Malhotra R, Beniwal S, Makkouk K, and Halila MH, (eds). Forage and Food Legumes of Ethiopia: Progress and Prospects. Proceedings of the workshop on Food and Forage Legumes, 22-26 September 2003, Addis Ababa, Ethiopia. Sponsors: EIAR and ICARDA. International Center for Agricultural Research in the Dry Areas (ICARDA), Aleppo Syria, pp. 80-86.

Balayan HS, Singh S (1986). Character association in lentil. Lens Newsletter 13:1-3.

Begum S, Begum S (1996). Morphological study and character associations in germplasm of lentil (Lens culinaris Medik). Bangladeshi. J. Bot. 25:79-82.

Bhattacharya A (1999). Lentil yield as affected by yield components under irrigated and nonirrigated conditions. Legume Res. 22:222-226.

Borojevic S (1990). Principle and methods of plant breeding. Elsevier Science Publishers, Elsevier. pp. 310.

Chen W (2007). Stemphylium blight disease scoring in field condition. USDA-ARS, Washington State University, Pullman. pp. 73.

Çokkizgin A (2007). A Research On Determination Of Botanical And Agronomic Properties Of Local Genotypes Of Some Lentils(Lens Culinaris Medik.) Selected From South And Southeastern Anatolian Regions in Turkey. PhD Thesis, Department of field crops, institute of natural and applied sciences university of çukurova. pp.140.

CSA (2012). Central Statistical Agency Agricultural Sample Survey Report on: Area and Production of Crops, Statistical Bulletin, Addis Ababa. pp. 445.

Dewey DR, Lu KH (1959). A correlation and path coefficient analysis of crested wheat grass seed production. Agro. J. 51:515-518.

Dixit P, Dubey DK (1984). Path analysis in lentil (Lens culinaris Medik.). Lens Newsletter. 11:15-17.

Dutta RK, Shaikh MAQ, Chowdhury SI, Muslimuddin M (1993). Physiology of flowering and pod development in lentil in relation to photoperiod and temperature. Lens Newsletter. 20:52-56.

Edossa F, Kassahun T, Endashaw B (2010). A comparative study of morphological and molecular diversity in Ethiopian lentil (*Lens culinaris* Medikus.) landraces. Afr. J. Plant Sci. 4:242-254.

Erksine W, Adham Y, Holly L (1989). Geographic distribution of variation in quantitative traits in a world lentil collection. Euphytica. 43:97-103.

Esmail AM, Mohamed AA, Hamdi A, Rabie EM (1994). Analysis of yield variation in lentil (*Lens culinaris* Medik.). Ann. Agric. Svi. 32:1073-1087.

Falconer DS, Mackay FC (1996). Introduction to Quantitative Genetics. 4$^{th}$ ed., Longman Group Limited, Malaysia. pp. 70-80, 464.

FAO (2013). FAO statistical yearbook. Food and Agriculture Organization of the United Nations, Rome, 2013: World Food and Agriculture. pp. 158.

Geletu B, Million E, Yadeta A (1996). Improved cultivars and production technology of lentil in Ethiopia. Research Bulletin No. 3. Debre Zeit Agricultural Research Center, Alemaya University, Debre Zeit, Ethiopia.

Gomez KA, Gomez AA (1984). Statistical Procedure for agricultural research. 2nd. ed. John Wiley and Sons, New York. pp. 375-378.

Hamdi A, Erskine W, Gates Phenotype (1991). Relationship among economic characters in lentil. Euphytica. 57:109-116.

Hamdi AA, El-Ghareib A, Shafey SA, Ibrahim MAM (2003). Direct and indirect relationships among lentil characters. J. Agric. Res. 81:224-229.

Jain SK, Sharma HL, Mehra RB, Khare JP (1991). Multiple correlation and regression analysis in lentil. Lens Newslett. 18:11-13.

Jatasra DS, Paroda RS (1978). Character associations in three generation of wheat under different environmental conditions. Cereal Res. Commun. 6:85-91.

Ketema D (2007). Genetic Variation for Agronomic and Root Characters in Ethiopian Lentil (*Lens culinaris* Medikus) Landraces Grown Under Moisture Stress. M.Sc Thesis, Haramaya University, Ethiopia. pp. 84.

Khattab SAM (1999). Association and path analysis in lentil under different irrigation regimes. Egypt. J. Agron. 20:13-25.

Kumar B, Sapra RL (1984). Factor analysis in Lens culinaris Medik. Lens Newsletter. 11:17-19.

Kumar R, Sharma SK, Malik BP, Sharma S (2004). Path coefficient analysis of seed yield components in lentil (*Lens culinaris* Medik.). Legume Res. 27:305-307.

Manara NTF, Manara W (1988). Morphological and developmental trait association in lentil (*Lens culinaris*). Lens Newsletter. 15: 34-36.

Mia MW, Mian MAK, Rahman MM (1986). Performance of exotic germplasm in Bangladesh. Lens Newsletter. 13:12-13.

Miller PA, William JC, Robinson HF (1959). Variety by Environment Interaction in cotton variety tests. Agro. J. 51:132-134.

Naji Z, Kafawin O, Halila H, Saoub Heritability (2003). Genotype by Environmental interaction, growth rate and correlation for some lentil (Lens Culinaris) genotypes grown under arid conditions in Jordan. Dirsat Agric. Sci. 30:374-383.

Ramgiry SR, Paliwal KK, Tomar SK (1989). Variability and correlation of grain yield and other qualitative characters in lentil. Lens Newslett. 16:19-21

Rao SA, Khan MA, McNeeilly T, Khan AA (1997). Cause and effect relations of yield and yield components in rice (*Oryza sativa* L.). J. Genet. Breed. 51:1-5.

Roy S, Islam MA, Sarker A, Malek MA, Rafii MY. Ismail MR (2013). Determination of genetic diversity in lentil germplasm based on quantitative traits. Australia J. Crop Sci. 7:14-21.

Sarker A, Erskine W, Singh M (2003). Variation in shoot and root characteristics and their association with drought tolerance in lentil landraces. Genet. Res. Crop Evol. 52:87-95.

Sarker A, Singh M, Rajaram S, Erskine W (2010). Adaptation of Small-Seeded Red Lentil (*Lens culinaris* Medik. subsp. *culinaris*) to Diverse Environments. Crop Sci. 50:1250-1259.

Sharma B, Tyagi MC, Asthana AN (1993). Progress in breeding bold seeded lentil in India. In: W.Erskine and M.C. Saxena (eds.). Lentil in Asia. ICARDA/ICAR, 11-15$^{th}$ March, 1991, New Delhi, India. pp. 22-38.

Singh A, Kumar P, Ramkrishna R, Tiwari SK, Yadav VK (2003). Dissecting association in lentil grown in tropical zone. Environ. Ecol. 21:186-190.

Singh KB, Chaudhary BD (1977). Biometrical Methods in Quantitative Genetic Analysis. Kalyani Publishers, New Delhi, India, pp. 300-304.

Sinha RP, Singh RN (2002). Correlation of seed yield with other traits in lentil. J. App. Bio. 12:1-3.

Tigest D (2003). Genetic variability and associations among yield and yield related characters in exotic lentil lines (*Lens culinaris* Medik.). M.Sc thesis. Alemaya University, Ethiopia. p. 88.

Tikka SBS, Asama BN, Gupta VK (1997). Interrelationships of Quantitative Characters with Seed Yield in Lentil (*Lens culinaris* Medik). Indian J. Heredity. 9:16-20.

Tyagi MC, Sharma B (1985). Association among economic traits in lentil. Lens Newsletter. 12:10-12.

Tyagi SD, Khan MH (2011). Correlation, path-coefficient and genetic diversity in lentil (Lens culinaris Medik) under rainfed conditions. Int. Res. J. Plant Sci. 2:191-200. http://www.interesjournals.org/IRJPS.

Verma AK, Mahto RN, Bhattacharya A (2004). Path analysis in lentil (*Lens culinaris* Medik). J. Res. Birsa Agric. University. 16:135-138.

Vir O, Gupta VP, Vir O (2001). Association among yield and yield contributing characters in *microsperma* X *macrosperma* derivatives of lentil. Crop Improv. 28:75-80.

VSN International (2012). GenStat for Windows 15th Edition. VSN International, Hemel Hempstead, UK. Web page: GenStat.co.uk

Wright S (1921). Correlations and causations. J. Agric. Res. 20:557-587.

**Annex 1.** Contribution of direct and indirect effects of agro-morphological traits on seed yield variability of lentil at Sirinka 2011 cropping season.

| Trait | DFF | DM | NP | NSS | SWPP | TNSPP | PH | SW |
|---|---|---|---|---|---|---|---|---|
| DFF | 0.003 | | | | | | | |
| DM | -0.014 | 0.035 | | | | | | |
| NP | 0.003 | 0.011 | 0.095 | | | | | |
| NSS | 0.004 | -0.009 | 0 | 0.046 | | | | |
| SWPP | 0.012 | -0.043 | 0.221 | 0.062 | 0.233 | | | |
| TNSPP | -0.005 | 0.002 | -0.135 | -0.054 | -0.214 | 0.074 | | |
| PH | 0.005 | 0.004 | 0.032 | 0.006 | 0.035 | -0.028 | 0.028 | |
| SW | 0.003 | -0.039 | -0.024 | -0.005 | 0.045 | 0.017 | -0.01 | 0.048 |

Contribution of residuals in the variability = 0.5713.

**Annex 2.** Contribution of direct and indirect effects of agro-morphological traits on seed yield variability of lentil at Sirinka 2012 cropping season.

| Trait | DFF | DM | NP | SWPP | TNSPP | PH | SW |
|---|---|---|---|---|---|---|---|
| DFF | 0.00163 | | | | | | |
| DM | 0.00188 | 0.00206 | | | | | |
| NP | 0.00011 | 0.00055 | 0.00309 | | | | |
| SWPP | -0.00221 | 0.00054 | 0.03364 | 0.41112 | | | |
| TNSPP | 0.00008 | 0.00049 | 0.00304 | 0.04878 | 0.00209 | | |
| PH | -0.00144 | 0.00036 | 0.00365 | 0.0565 | 0.00463 | 0.01215 | |
| SW | 0.00001 | 0.00014 | 0.00015 | -0.00099 | 0.00014 | 0.00017 | 0.00005 |

Contribution of residuals in the variability = 0.4195.

**Annex 3.** Contribution of direct and indirect effects of agro-morphological traits on seed yield variability of lentil at Chefe Donsa 2012 cropping season.

| Trait | DFF | DM | NP | NS | SWPP | TNSPP | PH | SW | RDSAF |
|---|---|---|---|---|---|---|---|---|---|
| DFF | 0.009 | | | | | | | | |
| DM | 0.019 | 0.037 | | | | | | | |
| NP | 0.002 | 0.006 | 0.006 | | | | | | |
| NS | 0.003 | 0.011 | 0.001 | 0.005 | | | | | |
| SWPP | 0.014 | 0.019 | 0.081 | 0.015 | 0.385 | | | | |
| TNSPP | -0.007 | -0.018 | -0.024 | -0.009 | -0.169 | 0.027 | | | |
| PH | -0.015 | -0.024 | 0.001 | -0.001 | 0.004 | -0.001 | 0.105 | | |
| SW | 0.003 | 0.020 | 0.004 | 0.004 | -0.006 | -0.010 | 0.002 | 0.013 | |
| RDSAF | -0.006 | -0.023 | -0.002 | -0.004 | 0.012 | 0.007 | 0.008 | -0.013 | 0.018 |

Contribution of residuals in the variability = 0.4905.

**Annex 4.** Contribution of direct and indirect effects of agro-morphological traits on seed yield variability of lentil at Sinana 2012 cropping season.

| Trait | DFF | DM | NP | NSPP | PH | SW | RSBF | RSAF |
|---|---|---|---|---|---|---|---|---|
| DFF | 0.015 | | | | | | | |
| DM | 0.003 | 0.002 | | | | | | |
| NP | 0.023 | -0.008 | 0.243 | | | | | |
| NSPP | -0.006 | 0.003 | -0.154 | 0.029 | | | | |
| PH | -0.003 | -0.001 | -0.002 | -0.005 | 0.032 | | | |

**Annex 4.** Contd.

| | | | | | | | | |
|---|---|---|---|---|---|---|---|---|
| SW | -0.011 | -0.008 | 0.059 | -0.019 | 0.009 | 0.079 | | |
| RSBF | -0.001 | -0.001 | 0.005 | -0.002 | 0.000 | 0.004 | 0.001 | |
| RSAF | -0.006 | -0.010 | 0.069 | -0.021 | 0.004 | 0.070 | 0.005 | 0.061 |

Contribution of residuals in the variability = 0.4905.

# Seed germination and abnormality of cotyledon of *Peganum harmala* populations in Mongolia

**Amartuvshin NARANTSETSEG**

Institute of Botany, Mongolian Academy of Sciences, Ulaanbaatar 210351, Mongolia.

**The present study assessed whether seed germination of *Peganum harmala* L. is related to population size and whether factors that control three and four cotyledons and difference in cotyledon shape are different and whether abnormality of cotyledons is related to genetic drift. The percentage of final seed germination was highest in Dzungarian Gobian population and followed by Trans-Altai Gobian population and lowest was in East Gobian population (Tukey HSD test, $p<0.01$). Most seedlings of *P. harmala* have two cotyledons in our study areas but there were abnormal cotyledons which are three, four and different shape dicotyledons. Three and four cotyledons showed low frequency in biggest (limited by river basins) and bigger (limited by oasis) populations and high frequency in small population (limited by well) whereas pattern of different shape dicotyledon frequency does not follow difference of population size. The different shape dicotyledons showed higher frequency in region with lower elevation, as compared to regions with higher elevation. The present results suggest that percentage of seed germination of *P. harmala* and frequency of more than two cotyledons might depend on population size, while frequency of different shape dicotyledons might be related to habitat difference which is high soil salinity. Also, our results suppose more than two cotyledons might be related to genetic drift.**

**Key words:** Abnormality of cotyledons, genetic drift, population size, seed germination.

## INTRODUCTION

Seed germination is the last stage of embryogenesis and cotyledons are formed during embryogenesis of angiosperms (Goldberg et al., 1994; Gomez et al., 2006; Chandler, 2008). Previous studies described effects of temperature, water supply, light and salinity on seed germination of *Peganum harmala* L. These studies were done in the desert region, east of Cairo and the semi-desert Mediterranean coastal (Hammouda and Bakr, 1969) and Pakistan (Hussain and Nasrin, 1985). Menges (1991) reported that percentage of seed germination is related to population size.

The cotyledon functions primarily during seed germination and senescence shortly after the seedling emerges from the soil (Goldberg et al., 1994). Angiosperms have mono- or dicotyledons but more and less than two cotyledons have been reported in over 15

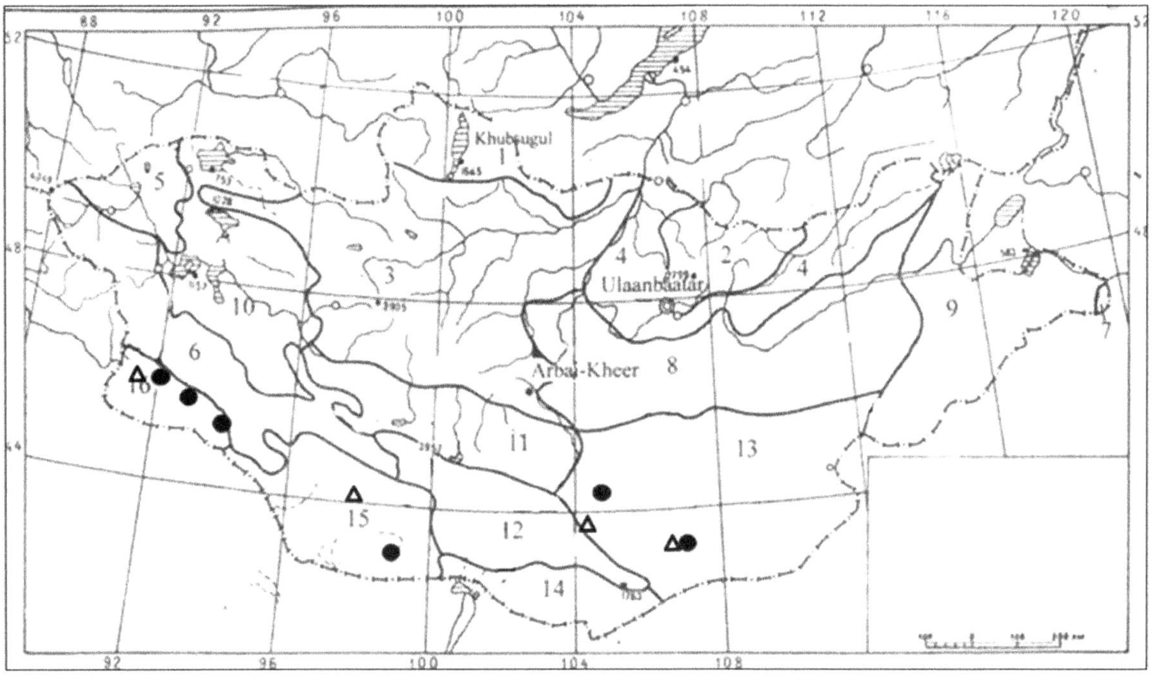


**Figure 1.** Seed samplng points of *P. harmala* in Mongolia (the botany-geographic region according to Grubov, 2001). Δ- weather station; seed sampling point. 1- Khubsugul, 2- Khentii, 3- Khangai, 4- Mongol-Dahuria, 5- Khovd, 6- Mongolian Altai, 7- Great Khingan, 8- Middle Khalkh, 9- Eastern Mongolia, 10- Great Lake Depression, 11- Valley of Lakes, 12- Gobi-Altai, 13-East Gobi, 14- Alasha Gobi, 15- Trans-Altai Gobi, 16- Dzungarian Gobi.

families (Conner and Agrawal, 2005).

The earliest defects were observed at the transition from the globular to the heart stage of embryogenesis with the formation of more than two cotyledons (Al-Hammadi et al., 2003). Previous studies supposed that cotyledon number range is related to hormonal asymmetric distribution and maternal effects (Liu et al., 1993; Mayer et al., 1993; Hadfi et al., 1998; Al-Hammadi et al., 2003; Orlova et al., 2006; Swarup et al., 2004), some kind of selection (Taylor and Mundell, 1999), lack of genetic variation (Conner and Agrawal, 2005), etc. Experimental studies reported that arabinogalactan proteins (AGPs) play functional roles in embryo development and cotyledon formation on *Nicotiana tabacum* (Qin and Zhao, 2007). Chandler (2008) reported that cotyledon bifurcation and lobing can result in supernumerary cotyledons. External nutrient (organic or mineral) supply during embryogenesis is important for embryo development (Qin and Zhao, 2007). However researchers still argue on the factor that could cause abnormality of cotyledon, most results supposed that genetic reasons might be related to abnormality of cotyledons.

Genetic variation was significantly reduced in endangered species when compared with non endangered species (Frankham, 1996). The effect of genetic drift is larger in small populations and smaller in large populations (Small et al., 2007). Dzungarian Gobi desert isolates distribution of *P. harmala* in Mongolia from a main distribution range in Middle Asia. In Mongolia, its distribution also isolated by deserts and then belongs to endangered species in Mongolia (Shiirevdamba et al., 1997).

The purpose of the present study was to assess whether seed germination of *P. harmala* is related to population size and whether factors that control more than two cotyledons and difference in cotyledon shape are different and whether abnormality of cotyledons is related to genetic drift.

## MATERIALS AND METHODS

Seed samples were taken from three sites in the Dzungarian Gobi (46° 06' 10.33N, 91° 35'14.49E, alt: 1191 m in Khovd river basin; 45° 53' 03.28N, 91° 47'01.18E, alt: 1163 m in Unch river basin and 45° 34' 34.89N, 93° 04'33.38E, alt: 1498 m near Shiir us spring) and one in Thras-Altai Gobi (43° 14' 679N, 099° 00' 411E, alt: 971 m in Ekhiin gol oasis) and two in East Gobi (44° 46' 27.12N, 105° 15'32.55E, alt: 1200 m near well of herder camp between the sums Tsogt-Ovoo and Khongor and 43° 10' 48.11N, 107° 10'29.36E, alt: 1060 m near well in Khanbogd sum, Southgobi aimag).

Climatic data is given based on database of Institute of Metereology and Hydrology of Mongolia. Nearest weather stations with seed sampling points were used for the present study: The nearest one with sampling points was "Baitag" station in Dzungarian Gobi and "Tooroi" station in Trans-Altai Gobi and "Dalanzadgad" and "Hanbogd" stations in East Gobi (Figure 1). Average temperature in July (°C) was 20.9, 23.4 and 21.2-23.8 and annual precipitation (mm) was 71.9, 49.5 and 84.0-127.1 in Dzungarian Gobi, Trans-Altai Gobi and East Gobi. Dzungarian Gobi

**Figure 2.** Diversity of *P. harmala* cotyledons in Mongolia. a- three cotyledons, b- four cotyledons, c- different shape dicotyledons.

region belongs to the province Dzungaria, Mts. Tien Shan and East Gobi and Trans-Altai Gobi belong to the province Mongolia, according to botany-geographic divisions (Grubov, 1963; Takhtajan, 1978).

The seed collection at the Institute of Botany was used for seed germination and observation of seedling morphology. In total, 6 seed samples were examined (Figure 1) which was collected between 1983 and 2001, from above-mentioned regions. When plants shed seeds, capsules were harvested then dried them paper bags at room temperature. The seeds were sampled randomly in each population.

Seed maturity of *P. harmala* was evaluated by weight of dry seeds as reported Harrington (1972). Weight of a thousand seeds was measured 10 times, using the analytic scale Shimadzu AY220 (d-0.1 mg). Seed germination was determined at 25±1°C for 10 days in the seed germinator, without dormancy breaking treatments, using Petri dishes and moist blotter by distilled water. Frequency of cotyledon number and shape was counted on 1000 embryos in each seed sample.

Maximum percentage of final seed germination was compared among regions, climatic variability and population sizes, because of only one site in Trans-Altai Gobi.

Population sizes (N) corresponded to the red-listed categories of "critically endangered" and "endangered", "vulnerable" respectively (Brook et al., 2002). Main distribution range of *P. harmala* is in the Dzungaria, Mts. Tien Shan (Central Asia).

In Mongolia, population of this species in the Dzungaria is isolated from other populations by Gobi-Altai mountain system and Trans-Altai Gobian population is isolated from East Gobian population by Alasha Gobi desert. Distribution of *P. harmala* is limited to river basins, area of oasis and well in the Dzungarian Gobi, Trans-Altai Gobi and East Gobi regions. Hence, population size of *P. harmala* in river basins, oasis and near well was evaluated biggest (in the Dzungarian Gobi), bigger (in Trans-Altai Gobi) and small (in East Gobi), respectively.

Geographic differences of weight of a thousand seeds, final germination, frequency of different shape dicotyledons, three and four cotyledons were estimated by Tukey HSD test.

## RESULTS

Mean of a thousand seed weight of *P. harmala* was 2.67, 2.43 and 2.6 g in the regions Dzungarian Gobi, Trans-Altai Gobi and East Gobi. Percentage of final seed germination was 95, 72 and 61% in the Dzungarian Gobi, Trans-Altai Gobi and East Gobi, respectively. To compare percentage of final seed germination with climatic variability, high and low percentage of seed germination was found in region of low and high air temperature in July but regional difference of its percentage does not follow regional difference of annual precipitation. The percentage of final seed germination was highest in Dzungarian population and followed by Trans-Altai Gobian population and lowest in East Gobian population (Tukey HSD test, $p<0.01$).

Most seedlings of *P. harmala* have two cotyledons in our study areas but there were abnormal cotyledons which are three (Figure 2a), four cotyledons (Figure 2b) and different shape dicotyledons (Figure 2c).

In comparison, frequency of abnormal seedlings among populations, 1.6, 2.7 and 6.8% of all seedlings in Dzungarian, Trans-Altai Gobian and East Gobian populations have three cotyledons but 0.5 and 1.5% in Dzungarian and East Gobian populations have four cotyledons. Different shape dicotyledons were found (9.6, 14 and 7.4%) in Dzungarian, Trans-Altai Gobian and East Gobian populations. Frequency of three cotyledons in East Gobian population was significantly higher than in other populations. However, frequency of three cotyledons was not significantly different between Dzungarian and Trans-Altai Gobian populations, Dzungarian population showed low frequency, when compared with Trans-Altai Gobian population. Also, frequency of four cotyledons was not significantly different between Dzungarian and East Gobian populations but its frequency was low in Dzungarian population, when compared with East Gobian one. Frequency of different shape dicotyledons was significantly higher in Trans-Altai Gobian population than in Dzungarian and East Gobian populations. However, frequency of different shape dicotyledons was not significantly different between Dzungarian and East Gobian populations, Dzungarian population showed higher frequency, when compared with East Gobian populations (Figure 3).

Different shape dicotyledons showed low and high frequency in region with high and low altitude but frequency of three and four cotyledons does not follow. To compare frequency of abnormality of cotyledons with climatic variability, high frequency of three and four cotyledons was found in regions with 21.2-23.8°C of average temperature in July while low frequency was found in regions with 20.9-23.4°C. Different shape dicotyledons showed high frequency in region with 49.5 mm of annual precipitation and low frequency in regions with 71.9-127.1 mm whereas frequency of three and four

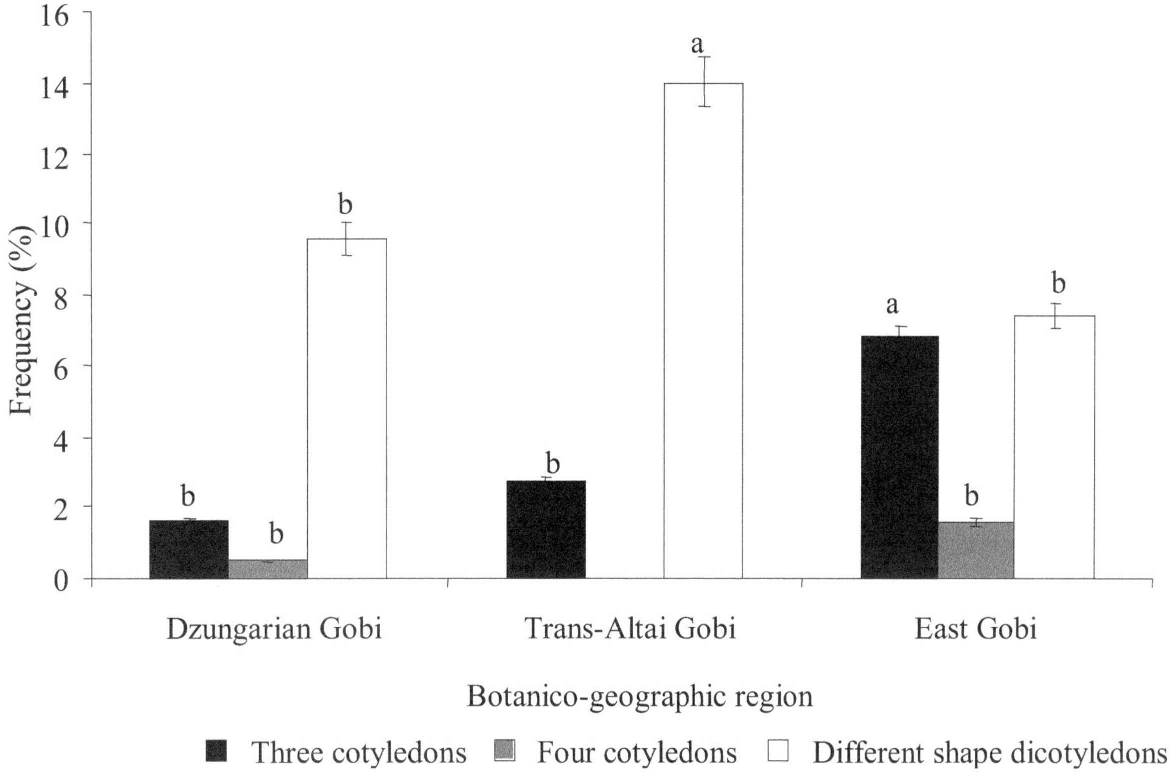


**Figure 3.** Cotyledon number and shape differences of *P. harmala* among the botany-geographic regions in Mongolia (Tukey HSD test, $p<0.05$).

cotyledons does not follow regional difference of annual precipitation.

Highest frequency of three and four cotyledons was found in East Gobian population with lowest final seed germination but that was lowest in Dzungarian population with highest final seed germination. The highest frequency of different shape dicotyledons was found in Trans-Altai Gobian population with averaged final seed germination whereas it was significantly low in Dzungarian and East Gobian populations with highest and lowest percentage of final seed germination. These patterns show that frequency of three and four cotyledons increased with decreasing percentage of final seed germination but frequency of different shape dicotyledons was not correlated with percentage of final seed germination.

Three and four cotyledons showed low frequency in biggest (limited by river basins) and bigger (limited by oasis) populations and high frequency in small population (limited by well) whereas pattern of different shape dicotyledon frequency does not follow difference of population size.

## DISCUSSION

Seed germination is the most important factor for *P. harmala* growth and distribution. *P. harmala* seeds germinated well at about 10 mm of rainfall (Hammouda and Bakr, 1969). Daily amount of precipitation in our study sites is insufficient for seed germination of this species, resulting in low amount of annual precipitation. But, river, oasis and well water supply could be sufficient for seed germination.

*P. harmala* exhibits 40 and 80-90% of final seed germination at 25 and 30°C in room temperature (Hussain and Nasrin, 1985). The present experiment indicated that seeds of *P. harmala* germinated 60-95% at 25°C. To compare results of previous and present experiments, seed germination of *P. harmala* is higher in Mongolia than in Cairo, Mediterranean coastal and Pakistan at same temperature. It seems seeds of *P. harmala* in Mongolia could adapt to coldness, better than above mentioned regions. However, negative correspondence between percentage of final seed germination and experimental temperature was found, seeds of *P. harmala* showed highest percentage of germination in Dzungarian Gobi with lowest temperature in July, when compared with other regions. Hence, regional difference of annual precipitation and air temperature in July cannot be main limiting factors for seed germination of this species.

However, ground water supply is sufficient for seed germination of *P. harmala*, water supplying area must be different among river, oasis and well. The results indicate that seed germination of *P. harmala* is low and high in small and bigger populations, as reported by Menges (1991). Hence seeds of *P. harmala* cannot potentially germinate in Mongolian desert, because of area limitation.

Previous studies reported more and less than two cotyledons in the family Aceraceae, Juglandaceae, Rubiaceae, Pedaliaceae, Protaceae, Ranunculaceae, Papaveraceae, Brassicaceae, Fabaceae, Geraniaceae, Chenopodiaceae, Onagraceae, Solanaceae, Scrophulariaceae, Salicaceae and Euphorbiaceae (Gates, 1910; Went, 1944; Harrison, 1964; Dessureaux, 1967; Magsar and Tsagaanmaam, 1984; Rajora and Zsuffa, 1986; Taylor and Mundell, 1999; Graz, 2001; Conner and Agrawal, 2005; Chandler, 2008). The present results show three, four cotyledons and different shape dicotyledons in *Peganum* genus (Peganaceae), in addition to above mentioned families.

Experimental study on tomato showed that abnormality of cotyledons is formed during embryogenesis (Al-Hammadi et al., 2003). Qin and Zhao (2007) showed that βGlcY affects cotyledon formation of *Nicotiana tabacum* L. Recent studies explain abnormality of cotyledons as a lack of genetic variation (Conner and Agrawal, 2005) but researchers still argue on factors and genetic reasons of abnormality of cotyledons (Chandler, 2008).

Frequency of more than two cotyledons showed same pattern with regional difference of average temperature in July but frequency of different shape dicotyledons showed negative pattern with regional difference of annual precipitation and elevation. The correspondences suppose that frequencies of more than two cotyledons and different shape dicotyledons might be signs of different factors, showing frequency of more than two cotyledons corresponding to heating that cause water evaporation while frequency of different shape dicotyledons was with length of vegetation period that depends on elevation.

Frequency of more than two cotyledons negatively corresponded with percentage of final seed germination. Seed maturation was not different among regions, because of seed weight indifference (Tukey HSD test, $p>0.05$). Correspondence between frequency of more than two cotyledons and population size indicated that frequency of more than two cotyledons decreases towards the main distribution range of *P. harmala*. This results suppose that more than two cotyledon might be related to genetic drift. Several genes such as DRN/ESR1 and DRN-LIKE/ESR2 have a role in cotyledon development, with mutant showing syncotyly and pleiocotyly (Chandler, 2008).

Cytokinin and arabinogalactin proteins (AGPs) play an important role during the stage of embryogenesis at which cotyledon number is determined (Chaudhury et al., 1993; Qin and Zhao, 2007). The present study discuss canalization of cotyledon number and bifurcation or lobbing as cause of more than two cotyledons formation (Conner and Agrawal, 2005; Chandler, 2008).

More than two cotyledons means shape of cotyledon leaves on a seedling of *P. harmala* is similar while different shape dicotyledons means shape of two cotyledon leaves on a seedling is different. The different shape dicotyledons showed higher frequency in region with lower elevation (in oasis), as compared to in regions with higher elevation. High value of average temperature in July and low value of annual precipitation indicated drought condition in Trans-Altai Gobi. Hence, *P. harmala* in oasis used more ground water for growth during drought season. However, soil salinity in oasis is higher than other dry regions, resulting in ground water evaporation (Pankova et al., 2004; Pankova, 2008). Safina (1977) reported that calyx lobe of *P. harmala* was found in high saline habitats (in oasis) but it was complete in low saline habitats and showed it as polymorphism.

Water stress induced by drought affects β-amylase activity of cucumber cotyledons in non-saline soil condition (Todaka et al., 2000) but our results suggest that different shape dicotyledons in *P. harmala* might depends on soil salinity which explains osmotic and toxicity effects, in saline soil condition.

The present results suggest that percentage of seed germination of *P. harmala* and frequency of more than two cotyledons might depend on population size, while frequency of different shape dicotyledons might be related to habitat difference which is high soil salinity. Also, our results suppose more than two cotyledons might be related to genetic drift.

## Conflict of Interests

The author(s) have not declared any conflict of interests.

## ACKNOWLEDGEMENTS

The author thanks Professor G. Tserenbaljid and academician Ch. Dugarjav (Institue of Botany) for advice and comment on this study and the writing of the manuscript.

### REFERENCES

Al-Hammadi ASA, Sreelakshmi Y, Negi S, Siddiqi I, Sharma R (2003). The polycotyledon mutant of tomato show enhanced polar auxin transport. Plant Physiol. 133:113-125.

Brook BW, Tonkyn DW, O'Grady JJ, Frankham R (2002). Contribution of inbreeding to extinction risk in threatened species. Conserv. Ecol. 6(1):16.

Chandler JW (2008). Cotyledon organogenesis. J. Exp. Bot. 59(11):2917-2931.

Chaudhury AM, Letham S, Craig S, Dennis ES (1993). Amp1-a mutant with high cytokinin levels and altered embryonic pattern, faster vegetative growth, constitutive photomorphgenesis and precocious flowering. Plant J. 4(6):907-916.

Conner JK, Agrawal AA (2005). Mechanisms of constraints: the contributions of selections and genetic variance to the maintenance

of cotyledon number in wild Radish. J. Evol. Biol. 18:238-242.
Dessureaux L (1967). Selection for pleiocotyly in alfalfa. Can. J. Genet. Cytol. 9:658.
Frankham R (1996). Relationship of genetic variation to population size in wildlife. Conserv. Biol. 10:1500-1508.
Gates RR (1910). Abnormalities in Oenothera. Missouri Botanical Garden Annual Report, 21: 175-187.
Goldberg RB, de Paiva G, Yadegari R (1994). Plant embryogenesis: Zygote to Seed. Sci. 266:605-614.
Gomez LD, Baud S, Gildey A, Li Y, Graham IA (2006). Delayed embryo development in the ARABIDOPSIS TREHALOSE-6-PHOSPHATE SYNTHASE1 mutation is associated with altered cell wall structure, decreased cell division amd starch accumulation. Plant J. 46 69-84.
Graz FP (2001). Threecotyledons on *Schinziophyton rautanenii* seedlings. South Afr. J. Bot. 67:69-70.
Grubov VI (2001). Key to the vascular plants of Mongolia. Volume I and II., "Science" Publisher, Plymouth.
Grubov VI (1963). The plants of Central Asia. Fasc.1, Academy Press, Moscow-Leningrad.
Hadfi K, Steph V, Neuhaus G (1998). Auxin-induced developmwntal pattern in Brassica juncea embryos. Dev. 125(5):879-887.
Hammouda MA, Bakr ZY (1969). Some aspects of germination of Desert seeds. Phyton, 13(3-4):183-201.
Harrington JF (1972). Seed storage and longevity. In: Kozlovski T, (eds) Seed biology. Academic Press, New York, 3:145-245.
Harrison BJ (1964). Factors affecting the frequency of tricotyly in *Antirrhinum majus*. Nature, 201:424.
Hussain F, Nasrin R (1985). Germination study of the seed of *Peganum harmala*. Pak. J. Agric. Res. 6(2):113-118.
Liu C, Xu Z, Chua NH (1993). Auxin polar transport is essential for the establishment of bilateral symmetry during early plant embryogenesis. Plant Cell 5(6):621-630.
Magsar D, Tsagaanmaam D (1984). Three cotyledon of *Adonis mongolica* Sim. Proceed. Inst. Bot. 10:57-62 (in Mongolian).
Mayer U, Buttner G, Jurgens G (1993). Apical-basal pattern formation in the Arabidopsis embryo: studies on the role of the gnom gene. Development, 117(1):149-162.
Menges ES (1991). Seed germination percentage increases with population size in a fragmented Prairie species. Conserv. Biol. 5(2):158-164.
Orlova I, Marshall-Colyn A, Schnepp J, Wood B, Varbanova M, Fridman E, Blackeslee JJ, Peer WA, Murphy AS, Rhodes D, Pichersky E, Dudareva N (2006). Reduction of bensoid synthesis in Petunia flowers reveals multiple pathways to bensoid acid and enhancement in auxin transport. Plant Cell 18(2):3458-3475.
Pankova EI, Mandakhbayar Z, Golovanov DL (2004). Changes in soil salinization in the Ekhiin gol oasis (Mongolia) according to monitoring data of 1977 and 2001. Euroasian Soil Sci. 37(9):911-926.
Pankova EI (2008). Environmental conditions and soils of natural oases in the Alashan Gobi Desert, Mongolia. Eurasian Soil Sci. 41(8):827-836.
Qin Y, Zhao J (2007). Localization of arabinogalactin-proteins in different stages of embryos and their role in cotyledon formation of *Nicotiana tabacum* L. Sex Plant Reprod. 20:213-224.
Rajora OP, Zsuffa L (1986). A typical seedlings of *Populus* L: their genetic significance and value in breeding. Silvae Genetica, 35:122-124.
Safina LK (1977). *Peganum harmala* L. "NAUKA" Press, Alma-Ata (in Russian).
Shiirevdamba Ts, Shagdarsuren O, Erdenejav G, Amgalan Ts, Tsetsegmaa Ts (1997). Mongolian Red Book. "ADMON" Printing, Ulaanbaatar (text in Mongolian, summary in English).
Small KS, Brudno M, Hill MM, Sidow A (2007). Extreme genomic variation in a natural population. Proceedings of the National Academy of Sciences of the United States of America, 104(13):5698-5703.
Swarup R, Kargul J, Marchant A, Zadik D, Rahman A, Mills R, Yemm A, May S, Williams L, Millner P, Tsurumi S, Moore I, Kerr ID, Bennett MJ (2004). Structure function analysis of the presumptive *Arabidopsis* auxin permease AUX1. Plant Cell 16(11):3069-3083.
Takhtajan AL (1978). The floristic regions of the world. "NAUKA" Press, Leningrad (in Russian).
Taylor NL, Mundel NE (1999). Registration of multiple-cotyledon red clover genetic marker stock: L38-1485. Crop Sci. 39:1259.
Todaka D, Matsushima H, Morohashi Y (2000). Water stress enhances β-amylase activityin cucumber cotyledons. J. Exp. Bot. 51(345):739-745.
Went RR (1944). Morphological observations on the tomato plant. Bull. Torrey Bot. Club 71:77-92.

# 22

# Influence of boron on seed germination and seedling growth of wheat (*Triticum aestivum* L.)

**Habtamu Ashagre, Ibrahim A. Hamza, Urgecha Fita and Worku Nedesa**

College of Agriculture and Veterinary Sciences, Department of Plant Sciences, Ambo University, P. O. Box 19, Ambo, Ethiopia.

**A laboratory experiment was conducted at Department of Plant Science, Ambo University, Ethiopia, to study the effect of boron availability on seed germination and seedling growth of wheat (*Triticum aestivum* L. var *Danda'a*). Seeds were sown in petri dishes with varying concentrations of boron (0, 0.25, 0.5, 1, 2, 4, 8, and 16 mg/L) at room temperature (24 ± 2°C) in complete randomized design (CRD) with four replications. Germination percentage and rate, shoot and root lengths, shoot and root fresh and dry weights, root number, root - shoot ratio and seedling vigor index were found to decrease beyond 0.25 mg/L, and phytotoxicity increased significantly (except on shoot at 0.25 mg/L) with increase in the concentration of boron in the germinating medium.**

**Key words:** Boron, germination, phytotoxicity, seedling growth, tolerance, vigor, wheat.

## INTRODUCTION

Boron (B) is a micronutrient required by plants in a very small quantity (Abd El-Wahab, 2008), and its availability in soil and irrigation water is an important determinant of agricultural production (Tanaka and Fujiwara, 2007). In soil solution, boron exists primarily as boric acid ($H_3BO_3$), which can be easily leached under high rainfall conditions (Yan et al., 2006) leading to plant deficiencies. On the contrary, under low rainfall conditions, boron cannot be sufficiently leached and therefore may accumulate to toxic levels for plant growth (Reid, 2007). This occurs very often in arid and semi arid regions where parent material and groundwater may contain high concentrations of boron. The accumulation of boron in top soil due to evaporation of groundwater reaches toxic levels that can reduce crop yields (Tanaka and Fujiwara, 2007). Boron is often found in high concentrations in association with saline soils and saline well water (Dhankhar and Dahiya, 1980). Of all the potential sources, irrigation water is the most important contributor to high levels of soil boron (Chauhan and Power, 1978). In assessing the potential toxicity of B rich irrigation water, the physical and chemical characteristics of the soil must be considered (Rauf et al., 2007). Boron can be regenerated through the mineralisation of soil organic matter, or through weathering processes of soil minerals (Peryea et al., 1985).

Plants exposed to excess of boron have reduced vigor, retarded development, leaf burn (chlorotic and necrotic patches in older leaves), and decreased number, size, and weight of fruits (Nable et al., 1997). Boron toxicity is an important agricultural problem that limits crop productivity in different regions of the world, and can occur in B-rich soils or in soils exposed to B-rich irrigation waters, fertilizers, sewage sludge, or fly ash (Luis et al., 2012).

Wheat (*Triticum aestivum* L.) is a staple food for more

than 35% of the world population and is also the first grain crops in most of the developing countries (Metwali et al., 2011). Bread wheat is the main food of people in many countries and about 70% calories and 80% protein of human diet is supplied from its consumption (Taregh et al., 2011). In Ethiopia, the total area put to wheat production was estimated to be 1.68 million ha with an average yield of 1827 kg/ha (CSA, 2011). Abiotic stress, especially drought stress and boron toxicity, are worldwide problems which seriously constrain global crop production (Schnurbusch et al., 2010; Pan et al., 2002). It is one of the major causes of crop loss worldwide, which commonly reduces average yield of many crops by more than 50% (Wang et al., 2003; Bayoumi, 2008). In countries like Australia, B toxicity impacts heavily on wheat production (up to 11% yield reduction) in affected areas (Moody and Rathjen, 2007).

Shoot biomass reduction of wheat was observed owing to high B addition to soil (Wimmer et al., 2003). Wheat with high B concentrations showed leaf edge burning and necrosis compared with the control treatments (Sonmez et al., 2009). Genotypic variation in susceptibility to B toxicity has been reported (Torun et al., 2006). Paull et al., (1988) reported wide range of intra-specific variation in response to B occurs in a number of crops, including bread wheat (*Triticum aestivum* L). In addition, seed germination and seedling growth are the most important phases in the life cycle of plant, and are highly responsive to the existing environment. Hence, the current experiment was conducted to investigate the effect of boron on germination and seedling growth parameters of wheat (*Triticum aestivum* L. var *Danda'a*).

## MATERIALS AND METHODS

A laboratory experiment was conducted in October, 2013 at the Department of Plant Sciences, Ambo University, to investigate the effect of boron on germination and seedling growth of wheat. The experiment was arranged in completely randomized design with four replications. Cultivar *Danda'a* was treated with eight levels of boron (0, 0.25, 0.5, 1, 2, 4, 8, and 16 mg/L) for the experiment, deionized water was used for the control treatment. Boric acid ($H_3BO_3$) was used as a source of boron. Seeds were surface sterilized with 30% hydrogen peroxide solution for 5 min, and rinsed with deionized water. Twenty seeds were uniformly placed per Petri dish (9.5 cm diameter) using a forceps after the Petri dish were sterilized with 98% ethanol, and rinsed with deionized water. Filter papers were well soaked by adding 7 ml with the respective solutions (7 treatment solutions and the control) at an interval of 48 h as described by Naveed et al. (2001). All the Petri dishes were covered with lids and kept at room temperature (24±2°C). Germination continued for 10 days, and germinated seeds were counted daily. Germination was considered to have occurred when radicles attained a length of 2 mm. After 10 days, parameters such as percent germination and rate of germination were calculated according to ISTA (1999); and root and shoot lengths of seedling were measured using a scale. Root and shoot dry weights were recorded after oven drying for 72 h at 60°C. The seedling vigor index (SVI) was determined as Hosseein and Kasra (2011):

Seedling Vigor Index = % Germination × Seedling dry weight (g)

Tolerance index (T.I.) was determined by Iqbal and Rahmati (1992) method:

T.I. = (Mean root length in treatment solution / Mean root length in distilled water) × 100

The percentage of phytotoxicity on shoot and root of seedlings was calculated following the formula given by Chou and Lin (1976):

$$\text{Phytotoxicity of Shoot (\%)} = \frac{\text{Shoot length of control} - \text{Shoot length of treatment}}{\text{Shoot length of control}} \times 100$$

$$\text{Phytotoxicity of Root (\%)} = \frac{\text{Root length of control} - \text{Root length of treatment}}{\text{Root length of control}} \times 100$$

Relative water content (RWC) of seedling was calculated as per the formula used by (Shalaby et al., 1993):

$$\text{RWC} = \frac{\text{Fresh weight} - \text{Dry weight}}{\text{Fresh weight}} \times 100$$

Statistical analysis of the data was performed using one-way ANOVA using SAS statistical software (Version 9). Based on the ANOVA results, mean separations were performed by Duncan's multiple range test at 5% level.

## RESULTS AND DISCUSSION

### Seed germination and seedling growth

The germination percentage at 0.25mg/l showed no difference compared to the control (Table 1). A significant decrease in germination was observed at boron concentrations higher than 0.5 mg/L. At 8 and 16 mg /L, wheat seed failed to germinate, indicating that germination is totally inhibited at such high concentra-tions of boron (Table 1). The consistent decrease in percentage and rate of seed germination beyond 0.25 mg/L in the present study is in line with the findings of Yau and Saxena (1997) and Muhammad et al., (2013) who stated that high boron concentration reduced germination percentage of durum wheat and maize, respectively.

The shoot and root lengths, shoot and root fresh and dry weights, and seedling dry weight decreased significantly with the increase in boron concentration (Tables 1 and 2). However, the highest shoot length (5.12 cm) and root length (5.31 cm) were obtained with 0.25 mg/L boron

**Table 1.** Effect of boron on germination, and shoot and root lengths of wheat.

| Boron Conc. (mg/L) | Germination (%) | Germination rate | Shoot length (cm) | Root length (cm) |
|---|---|---|---|---|
| 0 | 96.25[a] | 5.39 (3.25)[a] | 14.3 (4.78)[ab] | 18.8(5.31)[a] |
| 0.25 | 96.25[a] | 4.97(3.15)[a] | 17.1 (5.12)[a] | 12.7(4.56)[b] |
| 0.5 | 85[ab] | 2.82 (2.68)[ab] | 12.6(4.55)[b] | 6.1 (3.46)[c] |
| 1 | 77.5[bc] | 2.25( 2.5)[b] | 7.4 (3.72)[c] | 2.2(2.46)[d] |
| 2 | 68.75[c] | 1.64 (2.27)[bc] | 3.4(2.73)[d] | 1.4 (2.14)[de] |
| 4 | 38.75[d] | 0.8 (1.77)[c] | 0.95(1.78 )[e] | 0.88(1.76)[e] |
| 8 | 0[e] | 0 (1.00)[d] | 0(1.00 )[f] | 0(1.00)[f] |
| 16 | 0[e] | 0 (1.00)[d] | 0(1.00)[f] | 0(1.00)[f] |
| SEm (±) | 10.4 | 0.39 | 0.36 | 0.35 |
| CV (%) | 18 | 17.5 | 11.7 | 12.9 |

Means with similar letters in each column are not significant at 5% level of probability. Data in parenthesis are square root transformed.

**Table 2.** Effect of boron on fresh and dry weights of wheat seedling.

| Boron Conc. (mg/L) | Shoot fresh weight (g) | Shoot dry weight (g) | Root fresh weight (g) | Root dry weight (g) | Seedling fresh weight (g) | Seedling dry weight (g) |
|---|---|---|---|---|---|---|
| 0 | 0.09 (1.306)[ab] | 0.008(1.089)[a] | 0.22(1.373)[a] | 0.007(1.081)[a] | 0.31(1.56)[a] | 0.015(1.39)[ab] |
| 0.25 | 0.11(1.332)[a] | 0.011(1.106)[a] | 0.06(1.240)[ab] | 0.009(1.092)[a] | 0.17(1.41)[ab] | 0.020(1.14)[a] |
| 0.5 | 0.07(1.260)[b] | 0.009(1.092)[a] | 0.03(1.177)[abc] | 0.007(1.082)[a] | 0.10(1.32)[bc] | 0.015(1.12)[ab] |
| 1 | 0.03(1.1 47)[c] | 0.004(1.059)[b] | 0.025(1.137)[bc] | 0.006(1.073)[ab] | 0.055(1.23)[cd] | 0.009(1.10)[b] |
| 2 | 0.009(1.097)[cd] | 0.002(1.037)[c] | 0.019(1.136)[bc] | 0.003(1.049)[bc] | 0.03(1.17) [cd] | 0.004(1.06)[c] |
| 4 | 0.006(1.065)[d] | 0.002(1.029)[c] | 0.014(1.095)[bc] | 0.003(1.040)[c] | 0.02(1.14)[de] | 0.004(1.06)[c] |
| 8 | 0(1.0)[e] | 0(1.0)[d] | 0(1.0)[c] | 0(1.0)[d] | 0(1.0)[e] | 0(1.0)[d] |
| 16 | 0(1.0)[e] | 0(1.0)[d] | 0(1.0)[c] | 0(1.0)[d] | 0(1.0)[e] | 0(1.0)[d] |
| SEm(±) | 0.03 | 0.014 | 0.13 | 0.017 | 0.107 | 0.018 |
| CV (%) | 3 | 1.3 | 11.2 | 1.6 | 8.8 | 1.7 |

Means with similar letters in each column are not significant at 5% level of probability. Data in parenthesis are square root transformed.

concentration and control, respectively; and the lowest shoot and root lengths were with 8 mg/L and 16 g/L concentrations that caused a complete failure of germination. Boron inhibits root growth primarily through limiting cell elongation rather than cell division (Brown et al., 2002). Nable et al. (1997) also reported that shoot and root growth reduced when exposed to high B levels.

Shoot and root fresh and dry weights, and seedling fresh and dry weights decreased significantly with increase in boron concentration as compared to control and 0.25 mg/L (Table 2). Fresh weight and dry matter yield of the plants decreased significantly with increasing levels of applied boron (Ayvaz et al., 2012; Alpaslan and Gunes, 2001). Muhammad et al. (2013) and Turan et al. (2006) reported that shoot and root fresh and dry weights of maize and wheat decreased with the increase in the concentration of boron, respectively. Boron at 0.25 mg/L concentration resulted in the highest shoot fresh weight (1.33 g), shoot (1.1 g) and root (1.09 g) dry weights, and seedling dry weight (1.14 g). The significant increase at low concentration of boron could be due to its involvement in cell elongation or cell division and meristematic growth (Khan et al., 2006). Bonilla et al. (2004) and Farr (2010) reported that low concentrations of exogenous boric acid stimulated seed germination and seedling growth, while high concentrations showed an inhibitive effect on these parameters (Ölçer and Kocaçalışkan, 2007). Metwally et al. (2012) reported a gradual reduction in fresh and dry matter yield of shoots and roots, with increasing boron concentration in sand culture media.

### Root number and root- to-shoot ratio

Root number and root-to-shoot ratio showed no significant difference up to 0.5 and 0.25 mg/L of boron concentrations, respectively. Further increase in boron concentrations significantly reduced both root and root-to shoot ratio (Figure 1). Cokkizgin (2013) reported similar

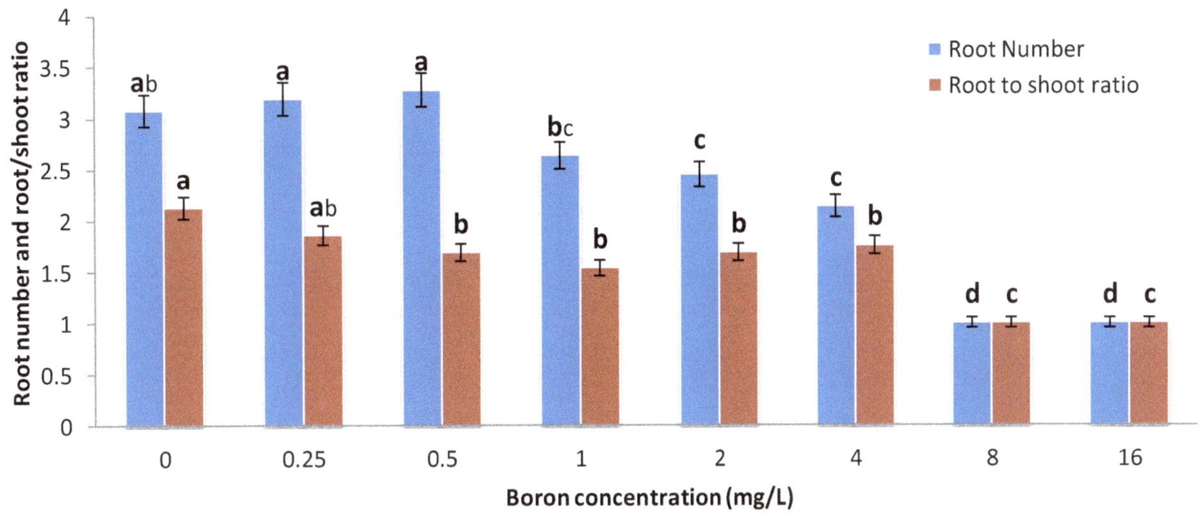


**Figure 1.** Effect of boron on root number and root /shoot ratio in wheat.

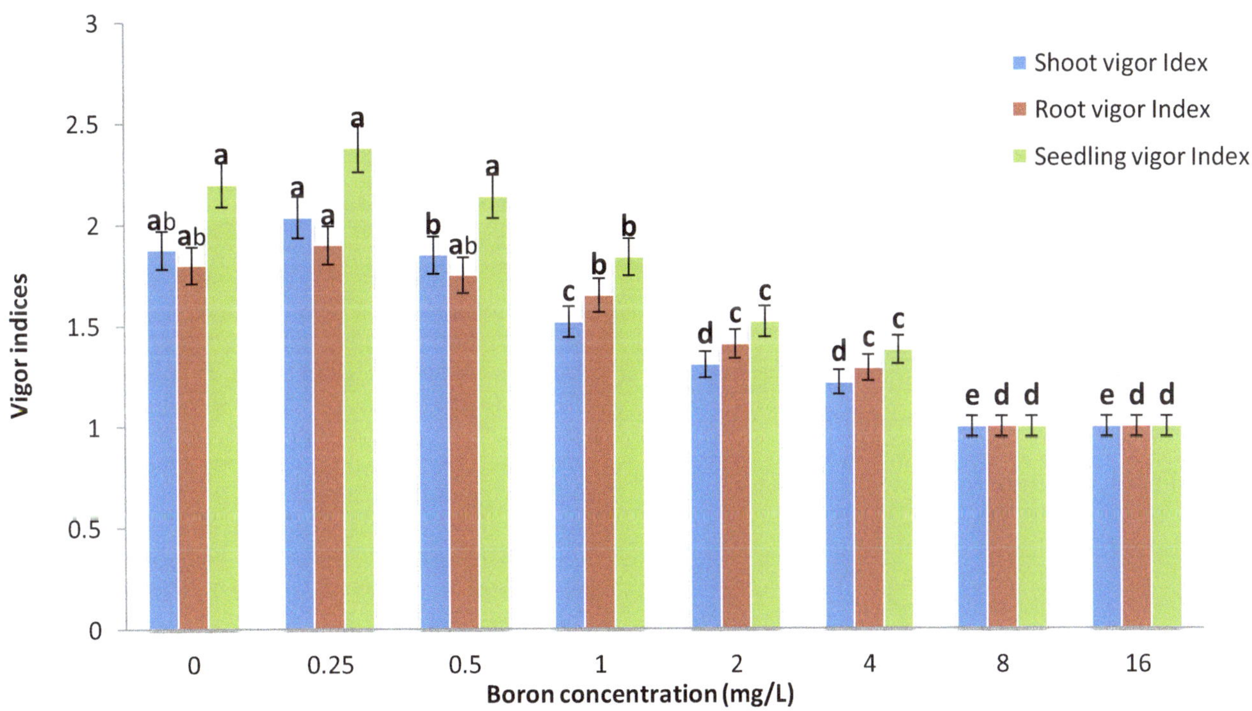


**Figure 2.** Effect of boron on vigor indices of Wheat.

findings on bean, and the increase in B concentration inhibited the secondary root emergence (Rehman et al., 2012).

### Seedling vigour, tolerance and phytotoxicity

Boron concentrations showed a significant effect on seedling vigour index, shoot and root vigour indices, and tolerance index (Figure 2). The highest value for each trait was noticed in control treatment and 0.25 mg/L. Boron concentrations above 0.5 mg/L decreased vigour indices and tolerance index. Similar findings were reported by Mirshekari (2012) and Cokkizgin (2013), who observed a restricted, seedling vigor index of dill and

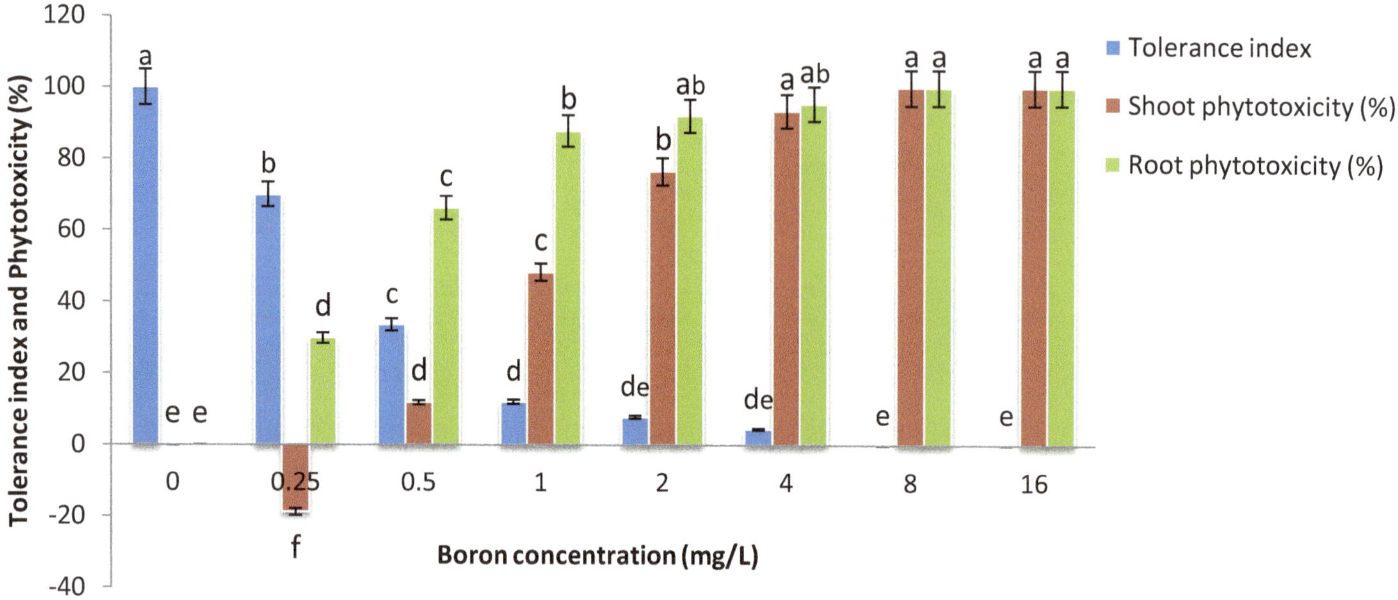


**Figure 3.** Effect of boron on tolerance index and phytotoxicity of wheat.

*Phaseolus vulgaris* at high level of boron concentrations, respectively. Ivanova et al., (2010) also reported a decrease in rapeseed seedling vigor indices with increasing micronutrient concentrations.

Boron showed a significant effect ($p<0.05$) on shoot and root phytotoxicity (Figure 3). Phytotoxicity of shoot and root increased with the increase in boron concentration. Lesser shoot and root toxicity was recorded with control treatment and lower concentration (0.25 mg/L), while it increased at higher concentration. At 0.25 mg/L B concentration shoot growth was highly promoted compared to control, hence showed a negative toxicity. Maximum phytotoxicity of boron on shoot and root was observed with ≥8 mg/L. The finding of our study is in agreement with the recent reports of Shaikh et al. (2013) and Habtamu et al. (2013) who reported that micronutrient toxicity of shoot and root decreased at lower concentration. On the other hand, it has the characteristics of promoting seedling growth at low concentrations based on crop types and varieties; however, its toxicity increased with increased in concentrations.

The tolerance index of wheat seed declined significantly with the increase in boron concentration. The maximum value of the tolerance index was obtained in the control treatment (100%) followed by 0.25 mg/L (70.1%), while a zero tolerance was attained for boron concentrations ≥8 mg/L. This result is in agreement with the recent findings of Shaikh et al. (2013) and Habtamu et al. (2013) who reported that increasing micronutrient concentrations decreased wheat and tomato tolerance index, respectively.

### Relative water content

Increased boron concentrations ≥8 mg/L caused significant decrease in shoot and root relative water content of seedlings (Figure 4). In boron concentrations between 0 and 4 mg/L, the relative shoot and root water content decreased from 91.4 to 60.6%, and 85.8 to 61.4%, respectively. The highest values of relative water content of shoot and root were observed in the control, while the lowest in boron concentrations higher than ≥4 mg/L. High water content on shoot and root implies the higher dilution effect of boron that reduces its toxicity. This result was in agreement with Kinfemichael (2011) who found a drastic decrease seedling shoot and root relative water content at higher salinity level.

### Conclusion

High boron concentrations caused a decrease in germination and germination rate, shoot and root lengths, shoot and root fresh and dry weights, vigor, tolerance and toxicity indices and relative water content of shoot and root in wheat. Low boron concentration (0.25 mg/L) showed the highest shoot fresh weight, shoot and root dry weights, seedling dry weight, and the lower shoot toxicity index. At higher boron concentrations, a deleterious effect on germination and seedling growth traits of wheat (*Triticum aestivum* L. var. *Danda'a*) was observed.

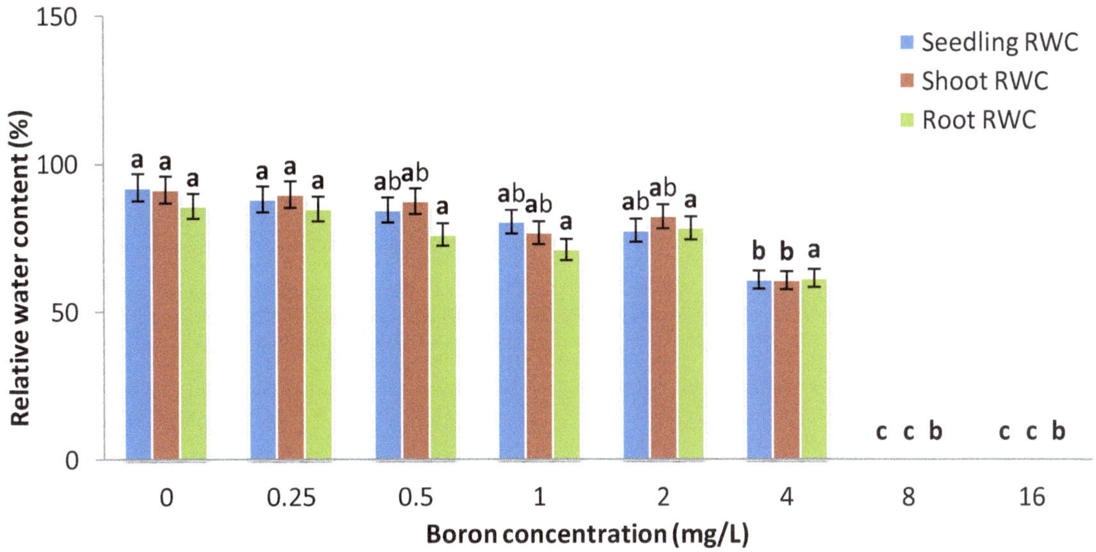


**Figure 4.** Effect of boron on shoot and root relative water content of wheat.

## REFERENCES


Abd El-Wahab AM, Mohamed A (2008). Effect of some trace elements on growth, yield and chemical constituents of *Trachyspermum ammi* L. plants under Sinai conditions. Res. J. Agric. Biol. Sci 4:717-724.

Alpaslan M , Gunes A (2001). Interactive effects of boron and salinity stress on the growth, membrane permeability and mineral composition of tomato and cucumber plants. Plant and Soil 236:123-128,

Ayvaz, M, Koyuncu M, Guven A, Fagerstedt KV (2012). Does boron affect hormone levels of barley cultivars? EurAsian J. Biosci. 6:113-120.

Bayoumi TY, Manal HE, Metwali EM (2008). Application of physiological and biochemical indices as a screening technique for drought tolerance in wheat genotypes. African J. Biot. 7(14):2341-2352.

Bonilla I, El-Hamdaoui A, Bolanos L (2004). Boron and calcium increase *Pisum sativum* seed germination and seedling development under salt stress. Plant and Soil 267:97-107.

Brown PH, Bellaloui N, Wimmer M, Bassil ES, Riuz J, Hu H, Pfeffer H, Dannel F, Romheld V (2002). Boron in plant biology. Plant Biol. 4:205-227.

Chou CH, Lin HJ (1976). Autointoxication mechanism of *Oriza sativa* L. phytotoxic effects of decomposing rice residues in soil. J. Chem. Ecol. 2:353-367.

Chauhan RPS, Powar SL (1978). Tolerance of wheat and pea to boron in irrigation water. Plant and Soil 50:14-149.

Cokkizgin A (2013). Boron ($H_3BO_3$) Toxicity in bean (*Phaseolus vulgaris* L.) germination. Ann. Res. Rev. Biol. 4(1):325-336.

CSA (2011). Statistical Abstract of 2009/10. CSA, Addis Ababa, Ethiopia.

Dhankhar DP, Dahiya SS (1980). The effect of different levels of boron and soil salinity on the yield of dry matter and its mineral composition in Ber (*Zizyphus rotundifola*). Int. Symp. Salt Affected Soils. Carnal, India, pp. 396-403.

Farr HJ (2010). Early Growth Tolerance to Boron and Salt in Wheat and Barley. M. Sc. thesis, Curtin Univ., Agri. Tech., Australia, p.95.

Habtamu A, Derara A, Tesfaye F (2013). Effect of copper and zinc on seed germination, phytotoxicity, tolerance and seedling vigor of tomato (*Lycopersicon esculentum* L. cultivar Roma VF). Int. J. Agric. Sci. Res. 2(11):312-317.

Hosseein AF, Kasra M (2011). Effect of hydropriming on seedling vigour in basil (*Ocimum basilicum* L.) under salinity conditions. Adv. Envi. Biol., 5(5), 828-833.

Iqbal MZ, Rahmati K (1992).Tolerance of *Albizia lebbeck* to Cu and Fe application. Ekologia (CSFR) 11:427-430.

ISTA (1999). International for seed testing rules. International seed testing association, Zurich, Switzerland.

Ivanova EM, Kholodova VP, Kuznetsov VIV (2010). Biological effects of high copper and zinc concentrations and their interaction in rapeseed plants. Rus. J. Plant Physiol. 57(6):806-814.

Khan R, Gurmani AH, Gurmani AR, Zia MS (2006). Effect of boron application on rice yield under wheat rice system. Int. J. Agri. Biol. 8:805-808.

Kinfemichael GA (2011) The response of some haricot bean (*Phaseolus vulgaris*) varieties for salt stress during germination and seedling stage. Cur. Res. J. Biol. Sci. 3(4):282-288.

Luis M Cervilla, Bego˜na Blasco, Juan J Rios, Miguel A Rosales, Eva S´anchez-Rodr´ıguez, MariaM. Rubio-Wilhelmi, Luis Romero, Juan M Ruiz (2012). Parameters Symptomatic for Boron Toxicity in Leaves of Tomato Plants. J. Bot. pp. 1-17.

Metwally A, El-Shazoly R, Hamada AM (2012). Effect of boron on growth criteria of some wheat cultivars. J. Biol. Earth Sci. 2(1):B1-B9

Metwali MR Ehab, Manal HE, Tarek YB (2011). Agronomical traits and biochemical genetic markers associated with salt tolerance in wheat cultivars (*Triticum aestivum* L.). Australian J. Basic Appl. Sci. 5(5):174-183.

Mirshekari B (2012). Seed priming with iron and boron enhances germination and yield of dill (*Anethum graveolens*). Turkish J. Agri. Fore. 36:27-33.

Moody D, Rathjen AJ (2007). Importance of boron tolerance in wheat. http://www. grdc.com.au/director/events/researchupdates.cfm?item_id=AC8FE77498FBCC5F1347F9451CC147D3&page Number=18.

Muhammad HRS, Tasveer ZB, Uzma Y (2013). Boron irrigation effect on germination and morphological attributes of *Zea mays* cultivars (*Cv.*Afghoee & Cv.Composite). Int. J. Sci. Engi. Res. 4(8):1563-1569.

Nable RO, Ba˜nuelos GS, Paull JG (1997). Boron toxicity. Plant and Soil 193(12):181-198.

Naveed KM, Lqbal HF, Tahir A, Ahmad AN (2001). Germination potential of chickpea (*Cicer arietinum* L.) under saline condition. Pak. J. Bio. Sci. 4:395- 360.

Ölçer H, Kocaçalışkan İ (2007). Excess boron reduces polyphenol oxidase activities in embryo and endosperm of maize seed during germination. J. Biosci. 62:111-115.

Pan XY, Wang YF, Wang GX, Cao QD, Wang J (2002). Relationship between growth redundancy and size inequality in spring wheat population mulched with clear plastic film. Acta Phytoecol. Sinica, 26:177-184.

Paull JG, Cartwright B, Rathjen AJ (1988). Responses of wheat and barley genotypes to toxic concentrations of soil boron. Euphytica 39:137-144.

Peryea FJ, Bingham FT, Rhoades JD (1985). Mechanisms for boron regeneration. Soil Sci. Soc. Am. J. 49:840-843.

Rauf M, Munir M, Ul-Hassan M, Ahmed M, Afzai M (2007). Performance of wheat genotypes under osmotic stress at germination and early seedling growth stage. African J. of Biot., 8:971-975.

Rehman AU, Farooq M, Nawaz A, Iqbal S, Rehman A (2012). Optimizing the boron seed coating treatments for improving the germination and early seedling growth of fine grain rice. Int. J. Agric. Biol. 14:453-456.

Reid R (2007). Identification of boron transporter genes likely to be responsible for tolerance to boron toxicity in wheat and barley. Plant Cell Physiol. 48:1673-1678.

Schnurbusch T, Hayes J, Sutton T (2010). Boron toxicity tolerance in wheat and barley: Australian perspectives. Breed. Sci. 60:297-304.

Shaikh IRS, Rafique AS, Shaikh AA (2013). Phytotoxic effects of heavy metals Parveen Rajjak (Cr, Cd, Mn and Zn) on Wheat (*Triticum aestivum* L.) seed germination and seedlings growth in black cotton soil of Nanded. India. Res. J. Chem. Sci. 3(6):14-23.

Shalaby EE, Epstein E, Qualset OC (1993) Variation in salt tolerance among some wheat and triticale genotypes. J .Agron. Crop. Sci. 171:298-304.

Sonmez O, Aydemir S, Kaya C (2009). Mitigation effects of mycorrhiza on boron toxicity in wheat (*Triticum durum*) plants. New Zealand J. Crop Horti. Sci. 37(2):99-104.

Tanaka M, Fujiwara T (2007). Physiological roles and transport mechanisms of boron: perspectives from plants. Eur. J. Physiol. 456(4):671-677.

Taregh G, Mostafa valizadeh, Hossein S (2011). Effect of drought on germination indices and seedling growth of 12 bread wheat genotypes. Adv. Environ. Biol. 5(6):1034-1039.

Turan MA, Taban N, Taban S (2009). Effect of calcium on the alleviation of boron toxicity and localization of boron and calcium in cell wall of wheat. Not. Bot. Hort. Agrobot. Cluj 37(2):99-103.

Torun AA, Yazici A, Erdem H, Çakmak I (2006). Genotypic variation in tolerance to B toxicity in 70 durum wheat genotypes. Turk. J. Agri. Fore. 30:49-58.

Wang XY, Vinocur P, Altman A (2003). Plant responses to drought, salinity and extreme temperatures: towards genetics engineering for stress tolerance. Planta. 218:1-14.

Wimmer MA, Muhling KH, Lauchli A, Brown PH, Goldbach HE (2003). The interaction between salinity and B toxicity affects the subcellular distribution of ions and proteins in wheat leaves. Plant, Cell Environ. 26:1267-1274.

Yan X, Wu P, Ling H, Xu G, Xu F, Zhang Q (2006). Plant nutriomics in China: An overview. Ann. Bot. 98:473-482.

Yau SK, Saxena MC (1997). Variation in growth development and yield of durum wheat in response to high soil boron I: Average effects. Aust. J. Agri. Res. 48:945-949.

# Sustainability of maize-wheat cropping system by different legume intercropping and nitrogen level treatments on light distribution, soil temperature and crop productivity

R. K. Naresh[1], S. S. Tomar[1], U. P. Shahi[2], S. P. Singh[2], Purushottam[3], B. Singh[4] and Parvinder Kumar[5]

[1]Department of Agronomy, Sardar Vallabhbhai Patel University of Agriculture and Technology, Meerut (U.P), India.
[2]Department of Soil Science, Sardar Vallabhbhai Patel University of Agriculture and Technology, Meerut (U.P), India.
[3]Department of Pathology & Microbiology, Sardar Vallabhbhai Patel University of Agriculture and Technology, Meerut (U.P), India.
[4]Department of Horticulture, Sardar Vallabhbhai Patel University of Agriculture and Technology, Meerut (U.P), India.
[5]Department of Agricultural Meterology C. C. S. H. A. U., Hissar, Haryana, India

**Agriculture designs for cropping system with legume intercropping for increased resource use efficiency, profitability, productivity and reduced adverse environmental impact are urgently required. A three-year field experiment consisting of six cropping treatments applied during both seasons was conducted during 2008-2009 to 2010-2011 to study the effect of nitrogen and legume intercropping with maize for sustainability of maize-wheat cropping systems. Results indicate that the photosynthetic active radiation was at noon reaching nearly 1940 µmol $m^{-2}$ $s^{-1}$ for maize and 1620 µmol $m^{-2}$ $s^{-1}$ for intercropped legumes. Intercropping can increase light interception, shading and reduce evaporation as compared to sole maize. Maize intercropped had higher values of stomatal conductance and leaf temperature than pure crop. The grain yield and yield attributing characters of wheat crop increased significantly under intercropping treatments when compared with wheat only. Wheat yield significantly increased up to 160 kg N $ha^{-1}$. However, there was no significant increase in yield of maize beyond 120 kg N $ha^{-1}$. Sole maize-wheat rotation showed a decline in soil organic carbon by 3.7%, while black gram and cowpea intercropping with maize in paired rows (2:2 row ratio) followed by wheat had increased contents of per cent organic carbon in soil as 0.63 and 0.67, respectively, as compared to initial values of 0.54%. Plots treated with intercrops/farm yard manure (FYM) during the rainy season sustained the wheat yield while the control plot showed a decline in wheat yield by 4-9%.**

**Key words:** Sustainability, intercropping, soil temperature, crop productivity.

## INTRODUCTION

Traditionally, soil fertility was maintained through the application of farmyard manure. Manure was largely produced from weeds, crop residue and fodder trees that were fed to tethered animals. Currently, farmyard manure is the primary source of nutrients for maize fields, although the use of fertilizers is growing in importance. Since farmers apply all the manure available on their farms, it is the increased use of fertilizer that is likely to enable increases in maize production in the future. The agricultural scientists has identified chemical fertilizer as major contributing factor to accelerating agricultural growth and has considered it as one of the priority inputs. Fertilizer application per unit area in maize cultivation is lower than in other crops for two reasons. First, most of the maize is produced in the rainy season where transportation of fertilizer is costly. Second, most of the farmers are subsistent, who often do not have cash to buy expensive fertilizer. Multiple cropping (intercropping or mixed cropping) plays an important role in agriculture because of the effective utilization of resources, significantly enhancing crop productivity when compared with monoculture crops (Li et al., 1999). Introduction of a grain legume in cereal-based cropping system aims at increased productivity and profitability to achieve food and nutritional security and sustainability (Swaminathan, 1998). Intercropping is widely accepted as a sustainable practice due to its yield advantage, high utilization efficiency of light and water, pest and disease suppression (Willey, 1979; Zhu et al., 2000). According to Saleem et al. (2011) and Solanki et al. (2011), the intercropping system provides higher cash return to smallholder farmers than growing the monocrops. Most studies on intercropping have focused on resource utilization, including water light (Shackle and Hall, 1984; Donald, 1985) and nutrients (Dhiman et al., 2007), resulting in substantial yield advantage when compared with sole cropping. In the intercrop, the degree of resource complementarily, the total yield and the relative yield between the individual species is determined by both inter- and intra-specific competition, which again is influenced by the availability of environmental resources, the relative frequency of the species and the density of components (Hauggaard et al., 2006). However, the intercropped species might utilize the growth resources more efficiently than sole crops, and resources may thus support a greater number of plants. A number of mechanisms exist by which intercrops utilize resources such as light, water and nutrients more efficiently than the equivalent sole crops (Anil et al., 1998; Ghanbari et al., 2010; Undie et al., 2012).

Human efforts to produce ever-greater amounts of food leave their mark on our environment. Persistent use of conventional farming practices based on extensive tillage, and especially when combined with *in situ* burning of crop residues, have magnified soil erosion losses and the soil resource base has been steadily degraded (Montgomery, 2007). Now, people understand that agriculture should not only be about high yield, but also sustainable (Naresh et al., 2011). Farmers concerned about the environmental sustainability of their crop production systems combined with ever-increasing production costs have begun to adopt and adapt improved management practices which lead to the ultimate vision of sustainable conservation agriculture. Conservation agriculture addresses a concept of the complete agricultural system, combining three basic principles: (1) reduction in tillage, (2) retention of adequate levels of crop residues and cover of soil surface and (3) use of economically viable crop rotations. These conservation agriculture principles are applicable to a wide range of crop production systems. Obviously, specific and compatible management components will need to be identified through adaptive research with active farmers' involvement in contrasting agro-climatic production systems. The objective of this study was to identify the effect of nitrogen addition and different legume intercropping with maize for sustainability of maize-wheat cropping systems.

## MATERIALS AND METHODS

An experiment was conducted for maize-wheat system with legume intercropping in farmers participatory mode in the jurisdiction of Sardar Vallabhbhai Patel University of Agriculture and Technology, Meerut (Uttar Pradesh), India, during 2008-09 to 2010-11.Treatments included: Sole maize (60 cm) and wheat control grown with FYM @ 10 t/ha, maize intercropped with blackgram/cowpea in paired rows (30/90 cm) in 2:2 row ratio, normal planted maize (60 cm) intercropped in 1:1 row ratio with pigeon pea/blackgram with pure stand of maize as checked under irrigated condition followed by wheat during *rabi* season with three nitrogen levels (80,120 and 160 kg N/ha), applied to maize as well as to wheat. The climate of the area is semi-arid, with an average annual rainfall of 805 mm (75-80% of which is received during July to September), minimum temperature of 4°C in January, maximum temperature of 41 to 45°C in June, and relative humidity of 67 to 83% throughout the year. In general, the soils of the experimental sites was sandy loam soil texture with medium fertility (86.5 sand, 9.2 silt and 4.3% clay) with pH 7.4 and had 0.54% organic carbon at the time of initiation of the study. Maize and intercrop were sown simultaneously during the 1st fourth night of June in each of the three years. Intercropping treatments were randomly allocated to main plots and N levels to subplots to evaluate treatments in split plot design. Nitrogen as per treatment and a uniform dose of $P_2O_5$ @ 60 kg $ha^{-1}$ to all the treatments were applied through urea and single super phosphate, respectively. 1/3 nitrogen along with full dose of phosphorus was applied as basal while remaining nitrogen was equally top-dressed at knee high and tesseling stages. During *rabi* season wheat crop was sown during the 2nd fourth night of November in each of the three years.

At maturity, seed and stover/straw samples were analyzed for total N to determine the N uptake. After completion of three sequences, soil samples were analyzed for total N to calculate N-fixation and N balance. Contribution of atmospheric nitrogen through legumes was computed by monitoring the changes in total nitrogen as well as that removed by the crop sequences by using the equation:

NF = NR - NA + N/number of leguminous crops

Where, NF = Nitrogen fixed by legume crop (kg/ha), N = change in total N + addition depletion of N in soil (kg/ha), NR = total N removed by sequence (kg/ha), NA = total N applied through fertilizer or manure (kg/ha).

Soil moisture dynamics were measured at 0-20 cm of soil depth during the study in all treatments; using TDR probes (three replications per treatment). The measurement system for the TDR was based on a cable tester (Tektronix 1502C) coupled to a handheld computer (Husky FS/2) (Thomsen, 1994). Soil moisture content during the growing period was calculated for all treatments. The soil layer above thermometers was used to record soil temperatures. The thermometers were buried in the soil horizontally (at 20 cm depth), between two plants in each of the maize and intercrop rows in a middle row. The measurement of temperature in the afternoon was made on relatively clear days.

The fraction of PAR intercepted was calculated by taking ten readings in rapid succession above the canopy and ten readings below the canopy at the soil surface using a Ceptometer. The soil surface measurements were taken by placing the Ceptometer at right angles to the plant rows. Stomatal conductance and leaf transpiration were measured with a portable photosynthesis system (LI-6400 Licor, USA). Measurements were made on the flag leaf for maize and on the central mature leaflet for legume crops.

## RESULTS AND DISCUSSION

### Photosynthetic active radiation (PAR)

A characteristic pattern of light interception was found for each cropping systems. There was significant difference in light interception between the maize pure stands as compared to the legume intercrop. In light interception by the maize-wheat monoculture increased linearly, reaching about 80% interception for light above ground and for soil moisture is less than in many other intercropping systems. The study showed that maize-wheat intercropping systems had significantly higher energy gain (more radiation intercepted) than pure crop systems. Lower PAR values were observed in Figure 1 in intercropped legumes indicating significant interception by maize leaves during the morning because of planting techniques. The greatest variation between systems occurred at 10 and 16 h when about 960 and 610 µmol $m^{-2}$ $s^{-1}$ were intercepted. The highest values were recorded at noon reaching nearly 1940 µmol $m^{-2}$ $s^{-1}$ for maize and 1620 µmol $m^{-2}$ $s^{-1}$ for intercropped legumes. However, decreases of 1°C in canopy temperature in inter-cropped canopy between 8 and 10 h was observed. The highest canopy temperature and vapour pressure deficit (VPD) were recorded in maize in the afternoon showing high transpiration conditions during the experimentation. Maize stomatal conductance and leaf transpiration in Figure 2 were also positively affected by intercropping. Between 12 and 16 h values recorded were significantly higher for maize intercropping than for sole crop and highest differences were computed at 14 h when stomatal conductance and leaf transpiration of intercropped were 0.218 mol $m^{-2}$ $s^{-1}$ and 7.09 m mol $m^{-2}$ $s^{-1}$, respectively. In the sole system, these variables reached 0.110 mol $m^{-2}$ $s^{-1}$ and 5.17 m mol $m^{-2}$ $s^{-1}$, respectively.

### Soil moisture content (SM)

Soil moisture content was reduced dramatically in the sole crop of maize due to high evapotranspiration potential; on the contrary, soil moisture content was increased dramatically in the legume intercropping due to low evapotranspiration potential for the growth period (Figure 3). However, comparing maize SM and legume intercropping, the patterns of soil water distribution in the soil profiles differed supporting the null hypothesis. Legume intercropping provided better soil cover as compared to sole maize, so water evaporation at soil surface was low and soil moisture was high as compared to sole maize. Distribution of root systems among species and cropping system influenced the soil moisture down the soil profile. Comparing the soil moisture content of the soil layer, the maize + cowpea intercropping tended to display the lowest differences followed by maize + pigeonpea and the highest difference followed by maize + blackgram, showing intermediate and greater differences (Figure 4). The measured soil moisture contents in the sole maize system were lower than those in the intercropping systems. In intercropping system, water uptake from soil surface layers increased due to increased root density in the upper layers, thus decreasing water dissipated by evaporation. A similar result was reported by Ghanbari et al. (2010).

### Growth and yield parameters

Various inter-cropping treatments did vary significantly the growth parameters of maze, namely plant height and dry mater accumulation at 90 days after sowing (DAS) (gm/plant). These parameters also vary significantly with successive increase in nitrogen levels from 80-160 kg N/ha. Mean data for plant height (Table 1) reveals that plant height increased by 8.2 and 14.1% at 80 and 160 kg N/ha respectively, measured under lowest level of nitrogen (80 kg N/ha). During the three years, dry matter produced at 90 DAS was maximum at 160 kg N/ha but remained at par with that accumulated at the preceding level of 120 kg N/ha. Mean data for dry mater accumulation at 90 DAS indicated that dry matter increased by 14.7 and 17.6% at 120 and 160 kg N /ha, respectively over the least dry matter of 102.4 gm/plant obtained under 80 kg N/ha (Table 1).

Intercropping treatments during rainy season had significant variations in spikelets per spike in succeeding wheat crop. Spikelets per spike obtained in the plot

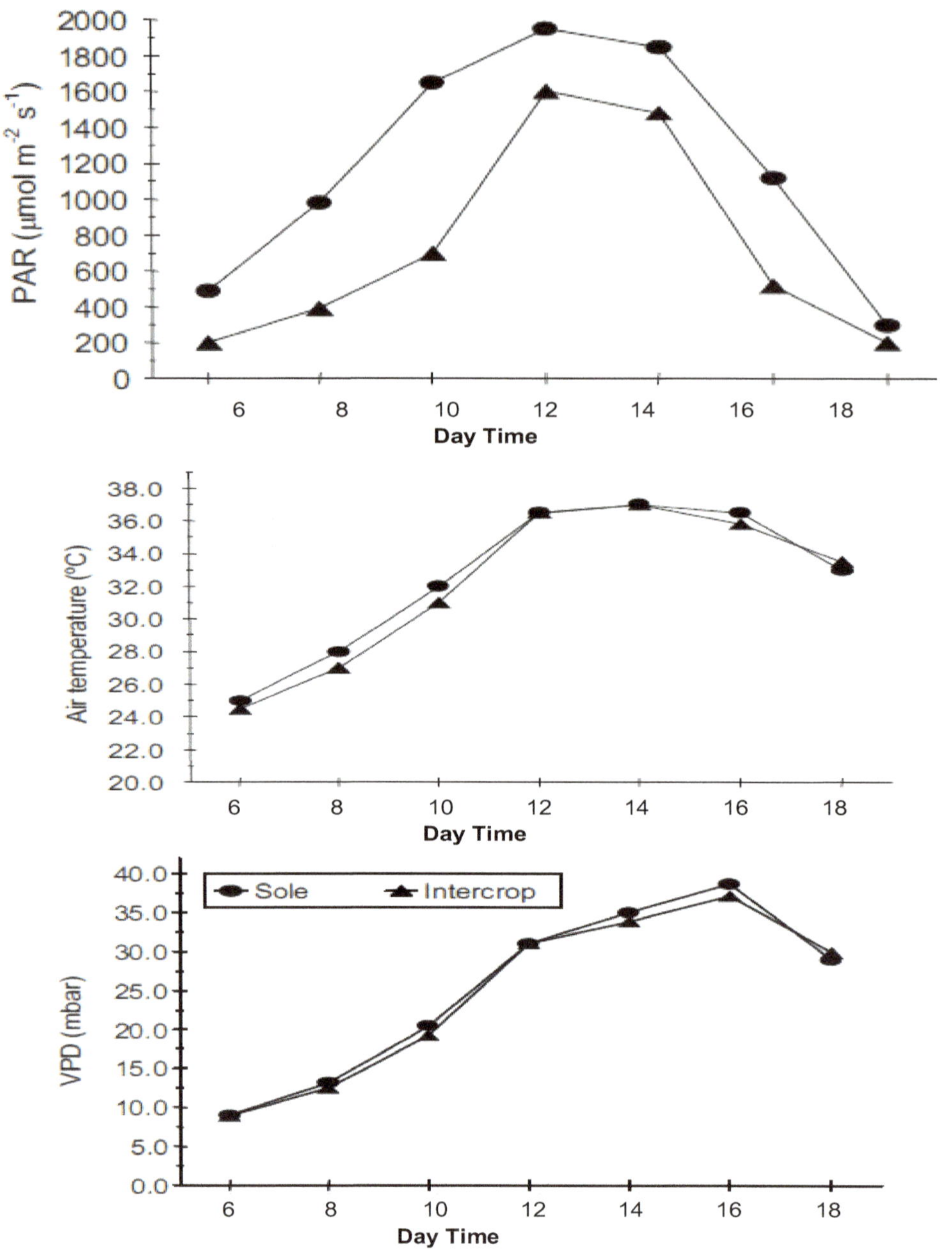


**Figure 1.** Diurnal fluctuation of photosynthetic active radiations (PAR), air temperature (T) and vapour pressure deficit (VPD) in maize and legumes intercropped.

treated either with FYM or intercrops were significantly superior to that in control plots during respective years (Table 2). Mean data for spikelets per spike calculated for three years indicated that maximum spikelets per spike of 16.4 were counted in the treatment when the crop was raised in the plot where maize + cowpea intercropping was completed, resulting in increase tillers by 3.48% as compared to the control plot (11.8 spikelets/spike). During the three years, increasing levels of nitrogen significantly increased the spikelets per spike till up to 120 kg N/ha. Mean data for spikelets per spike (Table 2) revealed that they increased by 13.8 and 19.4% at 120 and 160 N/ha, respectively, over the least spikelets (12.5 per spike) counted at 80 kg nitrogen level. Shelling percent of maize remained unaffected due to different intercropping treatments in all the three years (Table 2). However, the shelling percent of maize significantly improved with increasing levels of nitrogen during the three years of the study. Mean data for shelling percent (Table 1) revealed that it was improved by 2.20 and 2.60 at 120 and 160 kg N/ha, respectively, over that calculated under 80 kg nitrogen level (71.20). Different intercropping treatments applied in rainy season brought out significant variation in the number of grains per spike and 1000

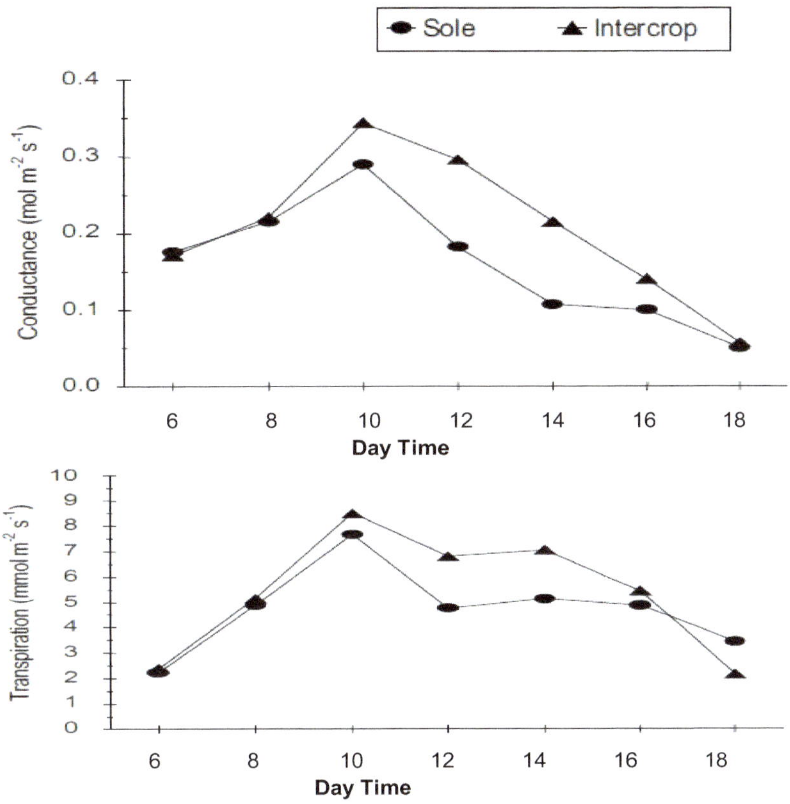


**Figure 2.** Diurnal fluctuation of stomatal conductance and transpiration of maize sole crop and intercropped with legume.

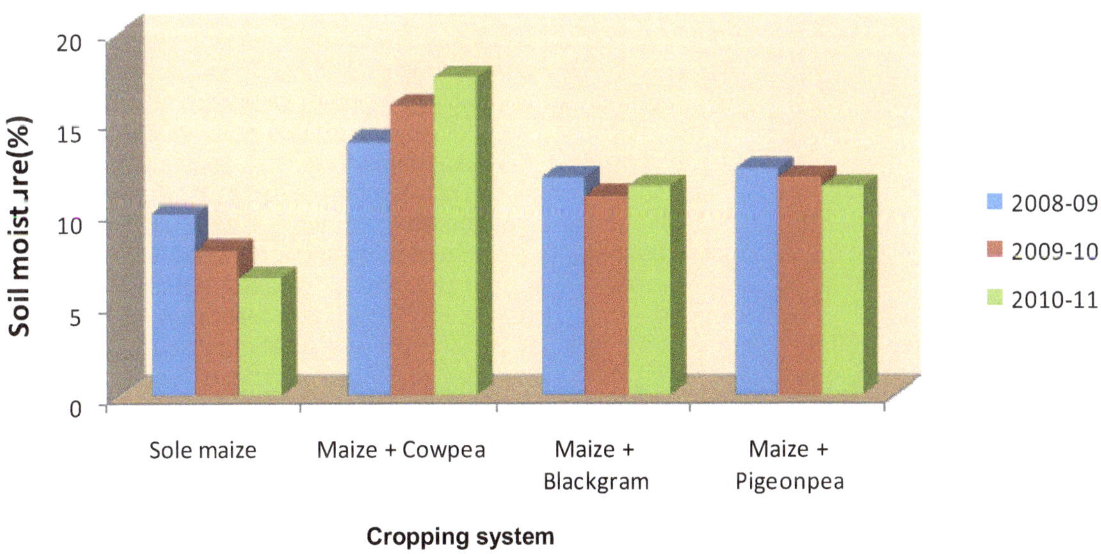


**Figure 3.** Variation of soil moisture (SM) at different cropping systems.

grain weight in wheat during the years of study.

The number of grains/spike and 1000-grain weight gradually reduced from 56.6 to 42.6 and 42.6 to 35.7 g, respectively, during initial and final years of the study in the control plot where sole maize during rainy season was raised. While in the case of intercropping treatments, numbers of grains/spike and 1000 grain weight under different intercropping treatments were significantly

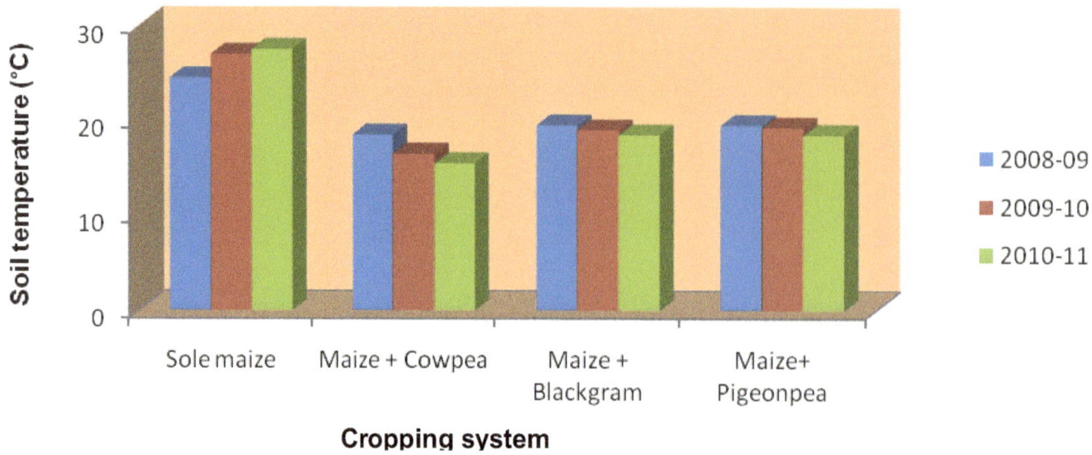


**Figure 4.** Variation of soil temperature (ST) at different cropping systems.

superior over the control plots but were at par during the third year of the study. Increasing level of nitrogen significantly increased the grains/spike and 1000 grain weight up to 120 kg N/ha. On mean basis, grains/spike increased by 30.8 and 34.0% at 120 and 160 kg N/ha, respectively, over that observed (35.7grains) under initial level of nitrogen. Similarly, 1000 grain weight were increased by 14.2 and 16.4% at 120 and 160 kg N /ha, when compared with those counted (36.2) at initial nitrogen level. These results were in close conformity with that of Solanki et al. (2011) and Undie et al. (2012).

### Yield and economic analysis

Maize yield was significantly affected by different intercropping treatments, where pigeonpea intercropping in contrast to pure stand of maize reduced the maize yield significantly. However, the grain yield of subsequent wheat crop was significantly increased due to different intercropping treatments applied to maize. During the first year of the study, wheat yields obtained under the plots treated with FYM (40.35 q/ha) or intercropped with blackgram in 2:2 row ratios (39.64 q/ha) produced significantly superior yield as compared to the control plot (37.82 q/ha). During the following years, the control plot showed a gradual decline in wheat yield by 4.49 and 8.89%, respectively, during 2009-10 and 2010-11. During these years, wheat yields obtained under the plots either treated with FYM or intercrops were significantly higher than that of yield obtained under the respective control plots. This increase in wheat yield might be attributed to nitrogen benefit to soil due to blackgram and cowpea intercropping. Wheat yield significantly increased up to 160 kg N/ha. However, there was no significant increase in yield of maize beyond 120 kg N/ha (Table 3). B : C ratio analysis indicated that the B : C ratio of 2.03 was calculated when maize in paired rows was intercropped with blackgram in 2:2 row ratios (30/90 cm wide beds), followed by wheat. Cowpea intercropping in a similar fashion was the next best treatment in sustaining productivity of wheat as compared to the control. These results are in conformity with the findings of Shah et al. (1991).

### Nutrient studies

Organic carbon contents varied with different inter-cropping treatments (Table 3). After completion of the study, the control plot, having pure maize-wheat rotation, showed a decline in soil organic carbon of 3.70%, while blackgram and cowpea inter-cropping with maize in paired rows in 2:2 row ratio followed by wheat increased content of organic carbon in soil as 0.63 and 0.67%, respectively, as compared to initial values of 0.54%. Total nitrogen content of soil also increased with increasing level of nitrogen (Table 4) in all the intercropping treatments. The highest total nitrogen (1399 kg/ha) was found under maize + cowpea (paired rows in 2:2 row ratios (30/90cm)) wheat treatment with net positive N balance of 139 kg/ha less than 160 kg N/ha level. This increase in N availability might be due to the decaying of nodules in legumes. These results are in line with the findings of Saleem et al. (2011) and Naresh et al. (2011).

From the above, it can be concluded that there are opportunities for improving the productivity of maize-wheat system with legumes intercropping in the western Uttar Pradesh, India. Intercropping is the best cropping system because in this system, light interception, soil moisture, soil temperature and yields were higher as compared to sole crops. Microclimatic variations in intercropping system had caused favourable environmental conditions for growth and high yield as compared to sole crops. The legume-maize inter-cropping exploited more soil moisture than a sole maize crop; the beneficial effects of the inter-cropping appeared to greatly compensate for the interception and uptake

**Table 1.** Growth and yield attributing characters of maize as affected by legume intercropping and N levels.

| Intercropping treatments | Growth and yield attributes | | | | | | | | | | | |
|---|---|---|---|---|---|---|---|---|---|---|---|---|
| | Plant height (cm) at harvest | | | | Dry matter at 90 DAS (g $plant^{-1}$) | | | | Shelling (%) | | | |
| | 08-09 | 09-10 | 10-11 | Mean | 08-09 | 09-10 | 10-11 | Mean | 08-09 | 09-10 | 10-11 | Mean |
| $T_1$ Sole maize (60 cm narrow beds)- wheat (control) | 260.7 | 258.3 | 269.8 | 262.9 | 120.7 | 116.4 | 123.6 | 120.2 | 74.1 | 73.9 | 72.3 | 73.4 |
| $T_2$ Sole maize (60 cm narrow beds with FYM @ 10t/ha-wheat | 277.9 | 271.2 | 276.3 | 275.1 | 132.7 | 128.9 | 136.2 | 132.6 | 71.8 | 72.9 | 73.6 | 72.8 |
| $T_3$ Maize + blackgram paired row in 2:2 row ratio (30/90 cm wide beds)-wheat | 253.7 | 249.6 | 267.4 | 256.9 | 106.2 | 112.7 | 119.4 | 112.8 | 71.7 | 73.2 | 72.8 | 72.6 |
| $T_4$ Maize + cowpea paired row in 2:2 row ratio (30/90 cm wide beds) –wheat | 251.6 | 257.9 | 258.3 | 255.9 | 98.7 | 114.3 | 121.6 | 111.5 | 73.1 | 72.6 | 72.9 | 72.9 |
| $T_5$ Maize + pigeonpea alternate rows in 1:1 row ratio (30/30 cm flat beds ) – wheat | 248.7 | 253.5 | 256.7 | 253.0 | 118.5 | 123.4 | 126.8 | 122.9 | 72.3 | 71.8 | 72.6 | 72.2 |
| $T_6$ Maize + blackgram alternate rows in 1:1 row ratio (30/30 cm flat beds ) – wheat | 242.2 | 251.8 | 257.6 | 250.5 | 92.3 | 109.6 | 117.9 | 106.6 | 71.3 | 72.6 | 72.3 | 72.1 |
| C D at 5% | 6.8 | 9.3 | 8.6 | - | 3.2 | 5.8 | 4.7 | - | NS | NS | NS | - |
| **Nitrogen levels (kg/ha)** | | | | | | | | | | | | |
| 80 | 229.7 | 240.3 | 249.2 | 239.7 | 95.2 | 103.5 | 108.4 | 102.4 | 71.2 | 71.5 | 70.8 | 71.2 |
| 120 | 254.1 | 265.4 | 267.9 | 262.5 | 109.3 | 123.4 | 127.7 | 120.1 | 72.3 | 73.4 | 72.8 | 72.8 |
| 160 | 267.3 | 274.1 | 271.6 | 271.0 | 112.6 | 128.6 | 131.5 | 124.2 | 72.4 | 73.7 | 73.2 | 73.1 |
| C D at 5% | 3.4 | 4.09 | 2.41 | - | 3.6 | 6.11 | 3.84 | - | 0.73 | 0.69 | 0.58 | - |

**Table 2.** Yield attributing characters of wheat as affected by legume intercropping and N level.

| Intercropping treatments | Yield attributes | | | | | | | | | | | |
|---|---|---|---|---|---|---|---|---|---|---|---|---|
| | Spikelets $spike^{-1}$ | | | | Number of grains $spike^{-1}$ | | | | 1000- Grain weight in (g ) | | | |
| | 08-09 | 09-10 | 10-11 | Mean | 08-09 | 09-10 | 10-11 | Mean | 08-09 | 09-10 | 10-11 | Mean |
| $T_1$ Sole maize (60cm narrow beds)- wheat (control) | 11.7 | 11.4 | 12.3 | 11.8 | 42.6 | 43.9 | 44.6 | 43.7 | 35.7 | 36.6 | 37.8 | 36.7 |
| $T_2$ Sole maize (60cm narrow beds with FYM @ 10t/ha-wheat | 12.8 | 12.1 | 13.6 | 12.8 | 44.7 | 45.3 | 45.8 | 45.3 | 36.1 | 38.3 | 36.7 | 37.0 |
| $T_3$ Maize+blackgram paired row in 2:2 row ratio (30/90 cm wide beds )-wheat | 14.3 | 14.7 | 15.6 | 14.9 | 52.6 | 53.1 | 55.3 | 53.7 | 41.3 | 42.7 | 42.4 | 42.1 |
| $T_4$ Maize+ cowpea paired row in 2:2 row ratio (30/90 cm wide beds )-wheat | 14.8 | 15.3 | 16.4 | 15.5 | 53.8 | 54.1 | 56.6 | 54.8 | 42.7 | 41.9 | 42.6 | 42.4 |
| $T_5$ Maize + pigeonpea alternate rows in 1:1 row ratio (30/30 cm flat beds ) – wheat | 13.2 | 13.8 | 14.6 | 13.9 | 48.7 | 51.0 | 53.4 | 51.0 | 41.3 | 41.6 | 42.4 | 41.8 |

**Table 2.** Contd.

| | | | | | | | | | | | | |
|---|---|---|---|---|---|---|---|---|---|---|---|---|
| $T_6$ Maize +blackgram alternate rows in 1:1 row ratio (30/30 cm flat beds ) – wheat | 11.9 | 12.3 | 12.8 | 12.3 | 48.5 | 49.0 | 52.5 | 50.0 | 39.5 | 40.8 | 41.3 | 40.5 |
| C D at 5 % | 1.08 | 1.27 | 0.93 | - | 4.06 | 6.59 | 4.35 | - | 0.94 | 1.29 | 1.65 | - |
| **Nitrogen levels (kg/ha)** | | | | | | | | | | | | |
| 80 | 11.1 | 13.3 | 13.2 | 12.5 | 31.3 | 36.2 | 39.7 | 35.7 | 35.3 | 36.8 | 36.4 | 36.2 |
| 120 | 13.4 | 14.7 | 15.5 | 14.5 | 50.6 | 51.5 | 52.8 | 51.6 | 41.3 | 42.5 | 42.8 | 42.2 |
| 160 | 14.8 | 15.2 | 16.4 | 15.5 | 52.7 | 53.4 | 56.3 | 54.1 | 42.9 | 43.2 | 43.7 | 43.3 |
| C D at 5 % | 1.54 | 0.65 | 0.97 | - | 2.61 | 2.11 | 3.84 | - | 2.06 | 1.52 | 1.44 | - |

**Table 3.** Grain yield and net monetary returns, and soil organic carbon of maize-wheat cropping system as affected by legume intercropping and N levels.

| Intercropping treatments | Grain yield (q/ha ) | | | | | | | | B:C ratio | O.C. (%) |
|---|---|---|---|---|---|---|---|---|---|---|
| | Maize and intercrops | | | | Wheat | | | | | |
| | 08-09 | 09-10 | 10-11 | Mean | 08-09 | 09-10 | 10-11 | Mean | | |
| $T_1$ Sole maize (60cm narrow beds)-wheat (control) | 39.13 | 40.05 | 40.95 | 40.04 | 37.82 | 36.12 | 34.46 | 36.13 | 1.92 | 0.52 |
| $T_2$ Sole maize (60cm narrow beds with FYM @ 10t/ha-wheat | 42.60 | 43.45 | 45.60 | 43.88 | 40.35 | 40.19 | 41.40 | 40.65 | 1.87 | 0.58 |
| $T_3$ Maize+blackgram paired row in 2:2 row ratio (30/90 cm wide beds )-wheat | 41.85 (2.85)* | 42.75 (3.85)* | 43.15 (4.45)* | 42.58 (3.72) | 39.64 | 38.00 | 37.90 | 38.51 | 2.03 | 0.63 |
| $T_4$ Maize+ cowpea paired row in 2:2 row ratio (30/90 cm wide beds )-wheat | 40.42 (3.65)* | 41.40 (4.35)* | 42.65 (5.25)* | 41.49 (4.42) | 37.95 | 39.10 | 38.35 | 38.47 | 2.01 | 0.67 |
| $T_5$ Maize + pigeonpea alternate rows in 1:1 row ratio (30/30 cm flat beds ) – wheat | 34.50 (3.24)* | 36.65 (4.45)* | 37.60 (5.80)* | 36.25 (4.49) | 38.45 | 38.30 | 37.75 | 38.17 | 1.98 | 0.51 |
| $T_6$ Maize + blackgram alternate rows in 1:1 row ratio (30/30 cm flat beds ) – wheat | 38.46 (2.82)* | 39.35 (3.75)* | 40.85 (5.65)* | 39.55 (4.07) | 39.40 | 38.30 | 37.20 | 38.30 | 1.96 | 0.61 |
| C D at 5 % | 2.54 | 2.08 | 3.15 | - | 1.92 | 2.16 | 2.35 | - | - | 0.049 |
| Nitrogen levels (kg/ha) | | | | | | | | | | |
| 80 | 24.60 (2.60) | 27.35 (2.95) | 28.20 (3.15) | 26.72 (2.90) | 28.25 | 35.20 | 32.15 | 31.87 | 1.83 | 0.56 |
| 120 | 40.35 (3.85) | 41.85 (4.25) | 42.45 (4.85) | 41.55 (4.32) | 42.80 | 38.95 | 39.85 | 40.53 | 1.99 | 0.59 |
| 160 | 41.35 (3.60) | 42.80 (4.45) | 43.75 (4.95) | 42.63 (4.33) | 45.80 | 41.15 | 41.65 | 42.87 | 2.08 | 0.61 |
| C D at 5 % | 1.65 | 2.82 | 3.73 | - | 2.14 | 2.32 | 2.23 | - | - | N.S. 0.54** |

*Figure in parentheses indicate yield of intercrops, **Initial value.

**Table 4.** Balance sheet of total nitrogen in maize + intercrops-wheat cropping system (kg/ha).

| Treatments | Nitrogen levels (Kg/ha) | Mean N removed by crops | Soil N at initiation + N added –N removed | Estimated N after cropping (3 years ) | Net soil N balance (Kg/ha) |
|---|---|---|---|---|---|
| $T_1$ Maize (narrow beds) - wheat (control) | 80 | 205 | 1037 | 1218 | -56 |
| | 120 | 259 | 1176 | 1291 | +24 |
| | 160 | 298 | 1255 | 1337 | +56 |
| $T_2$ Maize ( narrow beds )with FYM @ 10t/ha-wheat | 80 | 229 | 1010 | 1206 | -47 |
| | 120 | 297 | 1137 | 1295 | +19 |
| | 160 | 367 | 1249 | 1367 | +89 |
| $T_3$ Maize+blackgram paired rows in 2:2 row ratio (30/90 cm wide beds )-wheat | 80 | 248 | 1135 | 1287 | +16 |
| | 120 | 305 | 1205 | 1327 | +63 |
| | 160 | 348 | 1321 | 1388 | +121 |
| $T_4$ Maize+ cowpea paired row in 2:2 row ratio (30/90 cm wide beds )-wheat | 80 | 253 | 1174 | 1304 | +43 |
| | 120 | 318 | 1216 | 1343 | +71 |
| | 160 | 353 | 1345 | 1399 | +139 |
| $T_5$ Maize + pigeonpea alternate rows in 1:1 row ratio (30/30 cm flat beds ) – wheat | 80 | 247 | 1024 | 1232 | -41 |
| | 120 | 323 | 1145 | 1301 | +47 |
| | 160 | 364 | 1299 | 1384 | +119 |
| $T_6$ Maize +blackgram alternate rows in 1:1 row ratio (30/30 cm flat beds ) – wheat | 80 | 251 | 1046 | 1240 | -23 |
| | 120 | 341 | 1129 | 1298 | +52 |
| | 160 | 378 | 1221 | 1346 | +93 |

losses near the legume crops canopy. Intercropping improves water relations as compared to the sole crops and intercropped maize is more competitive than cowpea in terms of use of resource uptake, mainly soil moisture. In intercropping system, water uptake from soil surface layers increased due to increased root density in the upper layers, thus decreasing water dissipated by evaporation. However, it is evident from the results that legumes and legume-maize intercropping are more effective in improving soil moisture and decrease evaporation from the soil surface due to shading and increase the soil moisture potentially available for transpiration and growth as compared to sole maize.

A yield advantage in intercropping is achieved only when components crops do not compete for the same resources over the same time and space. In these experiments, the degree of below ground competition was not determined, so it is only possible to comment on above ground competition, showing that the sharing of light by the component crops was important for better utilization of resources, resulting in higher productivity of the intercropping system. The maize canopy alone could not utilize all incoming radiation during the growing period, and the remaining solar radiation was captured by the legumes when grown under maize, showing complementarity in use of resources. Maize + blackgram paired row in 2:2 row ratio (30/90 cm wide beds) followed by wheat having a B : C ratio of 2.03 had increased contents of organic carbon in soil as 0.63% as compared to initial values of 0.54%. Nitrogen balance under different intercropping treatments was better. The buildup of soil N (121 kg N/ha) was observed under this intercropping. Thus, blackgram intercropping with maize in paired rows in 2:2 row ratios (30/90 cm wide beds) followed by wheat was found to be the most sustainable for semi-arid region of western Uttar Pradesh, India.

## ACKNOWLEDGEMENTS

The authors are grateful to the Director Research of the Sardar Vallabhbhai Patel University of Agriculture and Technology, Meerut U.P., India for providing facilities and encouragement. The financial support received from the Uttar Pradesh council of Agricultural Research, Lucknow during the course of the study is gratefully acknowledged.

### REFERENCES

Anil L, Park RH, Pand Miller FA (1998). Temperate intercropping of cereals for forage: A review of the potential for growth and utilization with particular reference to the UK. Grass Forage Sci. 53:301-317.

Dhima KV, Lithourgidis AA, Vasilakoglou IB, Dordas CA (2007). Competition indices of common vetch and cereal intercrops in two seeding ratio. Field Crops Res. 100:249-256.

Donald CM (1985). The interaction of competition for light and for nutrients. Aust. J. Agric. Res. 9:421-435.

Ghanbari A, Dahmardeh M, Siahsar BA, Ramroudi M (2010). Effect of Maize (*Zea mays* L.) – Cowpea (*Vigna unguiculata* L.) Intercropping on Light Distribution, Soil Temperature and Soil Moisture in Arid Environment. J. Food Agric. Environ. 8(1):102-108.

Hauggaard-Nielsen H, Andersen MK, Jørnsgaard B, Jensen ES (2006). Competitive dynamics in two- and three-component intercrops. Field Crops Res. 95:256-267.

Li L, Yang SC, Li XL, Zhang FS, Christie P (1999). Interspecific complementary and competitive interaction between intercropped maize and faba bean. Plant Soil 212:105–114.

Montgomery DR (2007). Soil erosion and agricultural sustainability. PNAS 104:13268-13272.

Naresh RK, Gupta RK, Singh B, Kumar A, Singh A, Gaurav SS, Singh M, Shahi UP, Ghyanshyam K (2011). Enhancing farmers' income and livelihoods through integrated crop and resource management in the crop diversification in North India. J. Sci. Appl. Res. 2(1):1-19.

Saleem R, Ahmed ZI, Ashraf M, Arif M, Malik MA, Munir M, Khan MA (2011). Response of Maize-Legume Intercropping System to Different Fertility Sources under Rainfed conditions.Sarhad J. Agric. 27(4):503-511.

Shah MH, Koul PK, Khanday BA, Kachroo D (1991). Production potential and monitetary advantage index of maize intercropped with different grain legumes. Indian J. Agron. 36:23-28.

Shackle KA, Hall AE (1984). Effect of intercropping on the water relations of sorghum and cowpea. Field Crops Res. 8:381-387.

Solanki NS, Singh D, Sumeriya HK (2011). Resources Utilization in Maize (*Zea mays*)-based Inter-cropping System under Rainfed Condition. Indian J. Agric. Sci. 81(6):511-515.

Swaminathan MS (1998). Crop production and sustainable food security. In Chopra, V. L., Singh, R. B. and Verma, A. (eds). Crop Productivity and Sustainability—Shaping the Future. Proceedings of the Second International Crop Science Congress, New Delhi, India, pp. 3-18.

Thomsen A (1994). Program AUTOTDR for making automated TDR measurements of soil water content. Users Guide, Version 01. Report 38. Danish Institute of Plant and Soil Science, Tjele, 29p.

Undie UL, Uwah DF, Attoe EE (2012). Growth and Development of Late Season Maize/Soybean Intercropping in Response to Nitrogen and Crop Arrangement in the Forest Agro-ecology of South Southern Nigeria. Int. J. Agric. Res. 7(1):1-16.

Willey RW (1979). Intercropping - Its importance and research needs.Part 1. Competition and yield advantages. Field Crops Abstr. 32:1-10.

Zhu Y, Chen H, Fan J, Wang Y, Li Y, Chen J, Fan J, Yang S, Hu L, Leung H, Mew TW, Teng PS, Wang Z, Mundt CC (2000). Genetic diversity and disease control in rice. Nat. 406:718-722.

# A preliminary approach to halo sensitivity of sorghum cultivars

**Ayaz Khan[1], Muhammad Ibrar[1] and Ishtiaq Ahmad[2]**

[1]Department of Botany, University of Peshawar, Peshawar, Pakistan.
[2]Govt. Degree College, Tangi, Charsadda, Pakistan.

**Effect of five salt concentrations viz, 0, 130, 160, 190, 220 and 250 mM on germination and seedling growth of sorghum cultivars (*Asian, Indian, Mycho* and *Summer graze*) was evaluated under laboratory condition. Analysis of variance (ANOVA) revealed significant differences for germination percentage, plumule and radicle growth, fresh and dry weight for different salt concentrations except moisture contents. Inter cultivars genetic variation and concentration x cultivars interaction showed significant differences for all the parameters studied. Data for germination percentage, plumule and radicle growth, fresh and dry weight and moisture contents clearly demonstrated varietal differences. Though results clearly showed that salinity stress decreased all the parameters studied for sorghum cultivars significantly, tolerance of sorghum cultivars to such high levels of salt stress is worth mentioning and it can be suggested that sorghum can be tried for cultivation on moderately saline areas.**

**Key words:** germination percentage, plumule and radicle growth, fresh and dry weight, moisture contents.

## INTRODUCTION

Salinity is one of the fore most important abiotic factors that retard the activity of embryo in arid and semi arid regions. Most part of the land is under the influence of salinity and water logging. Crops growing under these saline regimes are stunted and weak resulting in low productivity (Ahmed, 2009; Ahmad et al., 2012). Although Pakistan is an agrarian economy yet we are facing scarcity of cereals and forages because of the afore-mentioned problems. Salt stress adversely affects plants at all stages of their life cycle.

Salinity affects seed germination by creating an external osmotic potential that prevents water uptake due to the toxic effects of sodium and chloride ions on the germinating seed (Khajeh-Hosseni, 2003; Kandil et al., 2012). Plants differ in their ability, to develop under saline conditions (Greenway and Munns, 1980; Kandil et al., 2012). Salinity compacts root length and plant height as the level of salt increases (Bashir et al., 2011; Saberi et al., 2011; Kandil et al., 2012). Plant scientists are confronting a challenge of salinity and water logging and they are trying to find out salinity resistant valuable plants. The aim of the present study was to search out salinity stress tolerant sorghum cultivars for vast barren salt effected land of Pakistan. Various researchers studied the effect of salinity on sorghum (Asfaw, 2011; Bashir et al., 2011; Kafi et al., 2011; Kandil et al., 2012; El-Naim et al., 2012; Rani et al., 2012). Though the study is preliminary, yet it will provide important clues and guide line for researchers to study further in this regard.

## MATERIALS AND METHODS

Seeds of sorghum cultivars (*Asian, Indian, Mycho* and *Summer graze*) obtained from open market were subjected to 4 salt concentrations (130, 190, 220 and 250 mM). Tap water was used

**Table 1.** Mean squares analysis of the varience table for plumule and radicle growth, germination percentage (%), fresh and dry weight and moisture contents of sorghum cultivars.

| Source | Degree of freedom | Plumule growth | Radicle growth | Germination (%) | Fresh weight | Dry weight | Moisture contents |
|---|---|---|---|---|---|---|---|
| Concentration | 4 | $0.0000^{S}$ | $0.0000^{S}$ | $0.0000^{S}$ | $0.0000^{S}$ | $0.0000^{S}$ | ****** |
| Cultivars | 3 | $0.0001^{S}$ | $0.0000^{S}$ | $0.0000^{S}$ | $0.0000^{S}$ | $0.0000^{S}$ | $0.0053^{S}$ |
| Concentration x cultivars | 12 | $0.0000^{S}$ | $0.0048^{S}$ | $0.0325^{S}$ | $0.0000^{S}$ | $0.0000^{S}$ | $0.0011^{S}$ |
| Error | 76 | | | | | | |
| Total | 99 | | | | | | |

S = Significant NS = non significant.

**Table 2.** Effect of salinity on plumule growth (cm) of sorghum cultivars.

| Concentration (milli moles) | *Asian* | *Indian* | *Mycho* | *Summer graze* | Concentration means |
|---|---|---|---|---|---|
| Control | $7.51^{ab}$ | $7.68^{a}$ | $6.80^{b}$ | $5.84^{c}$ | $6.96^{a}$ |
| 160 | $5.00^{c}$ | $1.05^{fgh}$ | $3.71^{d}$ | $1.99^{e}$ | $2.94^{b}$ |
| 190 | $0.64^{gh}$ | $0.71^{gh}$ | $1.81^{ef}$ | $1.49^{efg}$ | $1.16^{c}$ |
| 220 | $0.60^{h}$ | $0.48^{h}$ | $0.49^{h}$ | $0.58^{h}$ | $0.54^{d}$ |
| 250 | $0.53^{h}$ | $0.48^{h}$ | $0.32^{h}$ | $0.38^{h}$ | $0.43^{d}$ |
| Cultivars means | $2.86^{a}$ | $2.08^{b}$ | $2.63^{a}$ | $2.06^{b}$ | |

Lsd value at 5% level of significance for concentration = 0.4286, cultivars = 0.3833 and concentration x cultivars interaction = 0.8571. Values bearing similar letters in rows and columns are statistically non significant at 5% level of significance.

as control. Seeds viability was checked prior to the start of the experiment. Seeds were placed equidistantly on two folded Whatman # 1 filter paper as seed bed in Petri dishes of equal size (5 cm). Each treatment was replicated 5 times with 10 seeds in each replicate. Equal volume of solutions was added to the Petri dishes from respective salt concentrations. The glass wares were properly washed and sterilized at 65°C for 24 h in oven prior to use. The dishes were incubated at 25°C for 72 h. After 72 h, data for germination percentage (%), plumule and radicle growth (cm) and fresh weight (gm) were collected. Dry weight (g) and moisture contents (%) were determined after the seedlings were dried in oven at 65°C for 72 h following Hussain (1989). Fisher analysis of variance technique (1985) and LSD test at 5% probability was applied on the data to compare the differences among treatment means (Steel and Torie, 1980).

## RESULTS AND DISCUSSION

### Effect on plumule growth (cm)

ANOVA for plumule growth revealed significant differences for concentrations, cultivars and concentrations x cultivars interaction (Table 1). Maximum mean value (6.96 cm) for plumule growth was recorded in control, followed by 160 (2.94 cm), 190 (1.16 cm), 220 (0.45 cm) and 250 (0.43 cm) mM salt concentration, respectively. Among cultivars, *Asian* showed maximum mean value (2.86 cm) for plumule growth, while cultivar *Summer graze* showed minimum mean value (2.06 cm) for plumule growth. Differences between cultivar *Asian* and cultivar *Mycho* was not significant for plumule growth. Similarly, cultivar *Indian* and cultivar *Summer graze* showed non-significant differences. In concentrations x cultivars interaction, maximum mean value (7.68 cm) was recorded in control for cultivar *Indian,* while minimum (0.32 cm) for cultivar *Mycho* in 250 mM salt concentration (Table 2). Concentration dependent decrease was observed in plumule growth (cm) of sorghum cultivars (Figure 1). Jamil et al. (2006) reported that root and shoot lengths are important traits in salt stress sensitivity evaluation. Arrest of plumule growth of sorghum cultivars may be seen due to the toxic or injurious effects of different ions on embryo or may be due to osmotic disturbance uptake of ions was effected resulting in abnormal metabolism (Farooq and Azam, 2006; Maghsoudi-Moudi and Maghsoudi, 2008; Kazemi and Eskandari, 2011; Ahmad et al., 2012).

One may also conclude that ions of different charges bind with the domain of DNA or enzymes responsible for normal metabolism resulting in retarded plumule growth. Mechanisms of plant adaptation under salt stress are complex. The initial sites of the cell injury by any environmental stress firstly appears in cell membranes (Ashraf and Ali, 2008; Kafi et al., 2011). The imbibition of water was reduced due to the increased osmolality of salt

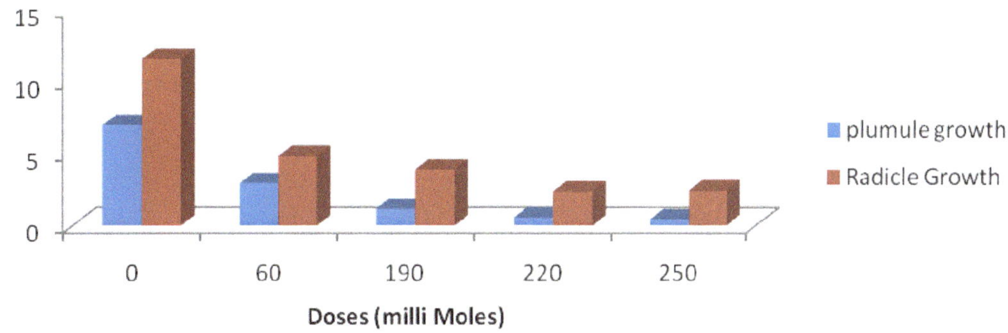


**Figure 1.** Effect of salinity on plumule and radicle growth (cm) of sorghum.

**Table 3.** Effect of salinity on radicle growth (cm) of sorghum cultivars.

| Concentration (milli moles) | *Asian* | *Indian* | *Mycho* | *Summer graze* | Concentration means |
|---|---|---|---|---|---|
| Control | 10.98$^{b}$ | 10.97$^{b}$ | 13.63$^{a}$ | 10.57$^{b}$ | 11.54$^{a}$ |
| 160 | 5.54$^{cd}$ | 2.57$^{fhg}$ | 7.09$^{c}$ | 4.08$^{def}$ | 4.82$^{b}$ |
| 190 | 2.23$^{fgh}$ | 2.78$^{fgh}$ | 5.45$^{cd}$ | 5.07$^{de}$ | 3.88$^{b}$ |
| 220 | 2.61$^{fgh}$ | 1.09$^{h}$ | 3.49$^{efg}$ | 2.39$^{fgh}$ | 2.37$^{c}$ |
| 250 | 2.02$^{gh}$ | 2.70$^{fgh}$ | 1.98$^{gh}$ | 3.05$^{fg}$ | 2.44$^{c}$ |
| Cultivars means | 4.67$^{bc}$ | 4.02$^{c}$ | 6.31$^{a}$ | 5.03$^{b}$ | |

Lsd value at 5% level of significance for concentration = 0.9550, cultivars = 0.8541 and concentration x cultivars interaction = 1.910. Values bearing similar letters in rows and columns are statistically non significant at 5% level of significance.

solutions (Wahid et al., 1998). Among cultivars, *Asian* was more tolerable to salinity and showed maximum plumule growth (Figure 4). It may be concluded that cellular membrane permeability of cultivar *Asian* is less as compared to other cultivars (Ismail, 2003). Naseer (2001) reported that the reduction in shoot length is due to excessive accumulation of salt in the cell wall which modifies the metabolic activities and limits the cell wall elasticity. Further, secondary cell appears sooner and cell wall becomes rigid, as a consequence the turgor pressure efficiency in cell enlargement decreases. These processes may cause the shoot to remain small. Our findings are parallel to those of Jamil et al. (2006) on vegetables species, Rahman et al. (2008) on wheat, Baybordi and Tabatabaei (2009), Keshavarzi (2012) on *Artemesia* and Moradi and Zavareh (2013) on chickpea, thus confirmed our findings for plumule growth.

## Effect on radicle growth (cm)

ANOVA for radicle growth exhibited significant differences for concentrations, cultivars and concentrations x cultivars interaction (Table 1). Maximum mean value (11.54 cm) for radicle growth was recorded in control which was significantly different from the rest of all the salt concentrations. Differences between 160 and 190 mM salt concentrations for radicle growth were non-significant. Similarly, differences between 220 and 250 mM salt concentrations were also non significant for radicle growth. Cultivar *Mycho* showed maximum mean value (6.31 cm) for radicle growth, followed by *Summer graze* (5.03 cm), *Asian* (4.67 cm) and *Indian* (4.02 cm), respectively. In concentrations x cultivars interaction, maximum mean value (13.36 cm) for radicle growth was recorded in control for cultivar *Mycho* while minimum mean value (1.09 cm) was observed in 220 mM salt concentration for cultivar *Indian* (Table 3). Water stress condition caused marked reduction in radicle growth (Ahmad et al., 2012) (Figure 1). Salinity affects the seedling growth of plant by slow or less mobilization of reserved food, suspending the cell division, enlargement and injuring hypocotyls (Tezara et al., 2003).

The data on the average length of radicle and plumule shows that decrease in length of plumule was more pronounced as compared to radicle in all NaCl salt treatment. This indicated that plumule is more sensitive than radicle. It may be due to ability of the radicle system to control entry of ions as compared to plumule (Moradi and Zavareh, 2013). Stress tolerance is dependent upon the genetic and biochemical characteristics of the cultivars (Barakat, 2003).

Our study shows that cultivar *Mycho* is genetically well adapted to salt stress than other sorghum cultivars (Figure 4). Our findings are in complete accordance with the Keshavarzi (2012), Saberi et al. (2012) and Moradi

**Table 4.** Effect of salinity on germination percentage (%) of sorghum cultivars.

| Concentration (milli moles) | *Asian* | *Indian* | *Mycho* | *Summer graze* | Concentration means |
|---|---|---|---|---|---|
| Control | $86.00^{ab}$ | $82.00^{abc}$ | $96.00^{a}$ | $76.00^{bcd}$ | $85.00^{a}$ |
| 160 | $68.00^{cde}$ | $44.00^{h\text{-}k}$ | $68.00^{cde}$ | $40.00^{i\text{-}l}$ | $55.00^{b}$ |
| 190 | $64.00^{d\text{-}g}$ | $54.00^{e\text{-}i}$ | $58.00^{e\text{-}h}$ | $44.00^{h\text{-}k}$ | $55.00^{b}$ |
| 220 | $66.00^{c\text{-}f}$ | $36.00^{j\text{-}m}$ | $48.00^{g\text{-}j}$ | $20.00^{mn}$ | $42.50^{c}$ |
| 250 | $28.00^{k\text{-}n}$ | $12.00^{n}$ | $50.00^{f\text{-}j}$ | $26.00l^{mn}$ | $29.00^{d}$ |
| Cultivars means | $62.40^{a}$ | $45.60^{b}$ | $64.00^{a}$ | $41.20^{b}$ | |

Lsd value at 5% level of significance for concentration = 8.858, cultivars = 7.923 and concentration x cultivars interaction = 17.72. Values bearing similar letters in rows and columns are statistically non significant at 5% level of significance.

and Zavareh (2013) whom studied effect of salinity on *Trifolium*, *Artemesia* and chick pea, respectively, thus confirming our results regarding radicle growth.

## Effect on germination (%)

ANOVA for germination percentage (%) exhibited significant differences for concentrations, cultivars and concentrations x cultivars interaction (Table 1). Maximum mean value (85.00%) for germination percentage (%) was observed in control while minimum in 250 mM salt concentration. Control showed significant differences for all the salt concentrations. Cultivar *Mycho* showed maximum mean value (64.00%) for germination percent-tage (%) while cultivar *Summer graze* showed minimum mean value (41.20%).

Differences between cultivar *Mycho* and cultivar *Asian* was non-significant for germination percentage (%). Similarly cultivar *Indian* and cultivar *Summer graze* was non-significantly different. In interaction of concentrations x cultivars, maximum mean value (96.00%) for germination percentage (%) was recorded in control for cultivar *Mycho* while minimum mean value (12.00%) was observed for cultivar *Indian* in 250 mM salt concentration (Table 4). Salinity concentrations are inhibitory for germination percentage of sorghum cultivars (Figure 3). Salinity exerts its undesirable effects through osmotic inhibition and ionic toxicity (Munns et al., 2006). The differences in germination percentage of forage sorghum cultivars may be due the genetic factors and heredity variation among the four sorghum cultivars (Figure 6). Our results regarding germination percentage of sorghum cultivars are confirmed by Kausar et al. (2012), Keshawarzi (2012), Saberi et al. (2012) and Moradi and Zavareh (2013) who studied ill effects of salinity on germination of sorghum, *Artemesia*, *Trifolium* and chick pea, respectively. Furthermore Anwar et al. (2001), Zia and Khan (2002) and Muhammad and Hussain (2010) also reported reduced germination under saline conditions in some medicinal plants that also strengthen our findings.

## Effect on fresh weight (g)

ANOVA for fresh weight was significantly different for concentrations, cultivars and concentration x cultivars interaction (Table 1). Maximum mean value (1.29 g) for fresh weight was observed in control which was significantly different from all the salt concentrations. Differences among the salt concentrations were significant for fresh weight. Cultivar *Indian* showed maximum mean value (0.75 gm) for fresh weight. While cultivar *Summer graze* showed minimum mean value (0.32 g). In interaction of concentrations x cultivars, maximum mean value (1.77 g) for fresh weight was observed for cultivar *Indian* in control which was higher than control of salt concentrations while minimum mean value (0.11 g) was recorded for cultivar *Indian* in 250 mM salt concentrations (Table 5). Increase in salinity level caused simultaneous reduction of seedling fresh weights in all sorghum cultivars (Figure 2).

This is in line with previous reports in wheat (Afzal et al., 2005), sugar beet, cabbage, amaranth and pak-choi (Jamil et al., 2006) and in sorghum (Asfaw, 2011). Thus our findings regarding fresh biomass are confirmed. Reduction in fresh biomass at higher concentration might be due to poor absorption of water from the growth medium due to physiological drought (Muhammad and Hussain, 2010). Salt stress disturbed the biochemical machinery resulting in reduced biomass. Cultivar *Indian* has got good genetic variations for salinity tolerance thus showed increase in fresh biomass (Figure 4).

## Effect on dry weight (g)

ANOVA for dry weight exhibited significant differences for concentrations, cultivars and concentrations x cultivars interaction (Table 1). Maximum mean value (0.53 g) for dry weight was recorded in control while minimum in 250 mM salt concentration. Among cultivars, *Asian* showed maximum mean value (0.31 g) for dry weight while minimum mean value (0.16 g) was recorded for *Summer graze*. In concentration x cultivars interaction, control of

**Table 5.** Effect of salinity on fresh weight/seedling (gram) of sorghum cultivars.

| Concentration (milli moles) | *Asian* | *Indian* | *Mycho* | *Summer graze* | Concentration means |
|---|---|---|---|---|---|
| Control | 1.44$^{b}$ | 1.77$^{a}$ | 1.37$^{b}$ | 0.56$^{ef}$ | 1.29$^{a}$ |
| 160 | 0.84$^{c}$ | 0.81$^{cd}$ | 0.59$^{de}$ | 0.27$^{ghi}$ | 0.63$^{b}$ |
| 190 | 0.57$^{ef}$ | 0.56$^{ef}$ | 0.52$^{ef}$ | 0.39$^{e-h}$ | 0.51$^{c}$ |
| 220 | 0.35$^{fgh}$ | 0.50$^{efg}$ | 0.27$^{ghi}$ | 0.19$^{hi}$ | 0.33$^{d}$ |
| 250 | 0.26$^{hi}$ | 0.11$^{i}$ | 0.22$^{hi}$ | 0.18$^{hi}$ | 0.19$^{e}$ |
| Cultivars means | 0.69$^{ab}$ | 0.75$^{a}$ | 0.60$^{b}$ | 0.32$^{c}$ | |

Lsd value at 5% level of significance for concentration = 0.1161, cultivars = 0.1039 and concentration x cultivars interaction = 0.2323. Values bearing similar letters in rows and columns are statistically non significant at 5% level of significance.

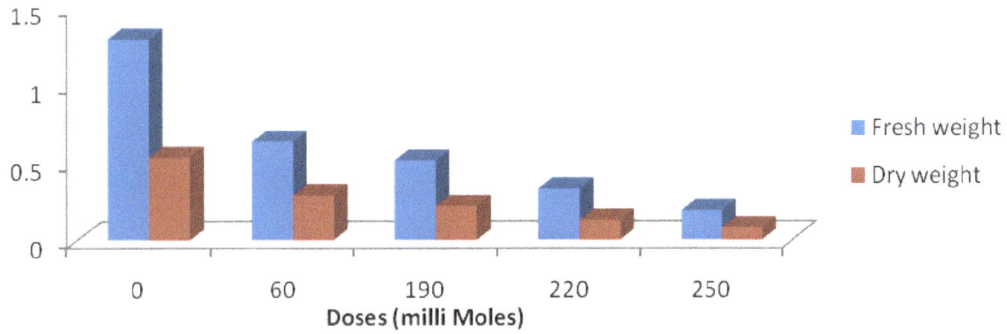


**Figure 2.** Effect of salinity on fresh and dry weight/seedling (gram) of sorghum.

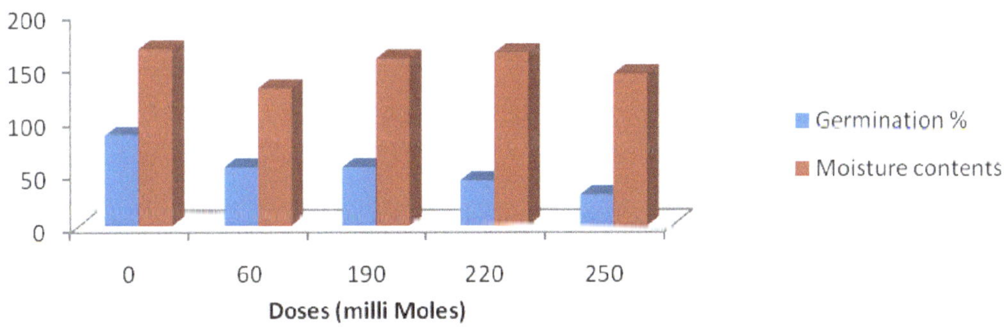


**Figure 3.** Effect of salinity on germination and moisture contents (%) of sorghum.

cultivar *Mycho* represented maximum mean value (0.71 g) which was higher than control of salt concentrations while minimum mean value (0.03 gm) was observed for cultivar *Indian* in 250 mM salt concentration (Table 6). Increased salinity level caused simultaneous reduction of seedling dry weights in all the cultivars (Figure 2). Salt stress disturbed the biochemical machinery resulting in reduced dry biomass. Increase dry biomass of cultivar *Asian* indicated good genetic variations for salinity (Figure 4). This is in line with previous reports in wheat (Afzal et al., 2005), sugar beet, cabbage, amaranth and pak-choi (Jamil et al., 2006) and in sorghum (Asfaw, 2011) and we reported the same. The decrease in seedling dry weight with increasing salinity level is in contrast with the findings of Muhammad and Hussain (2010) who reported increase in seedling dry weight of *Lepidium sativum*, *Linum usitatissimum*, *Plantago ovata* and *Trigonella foenum-graecum* (Figure 5).

## Effect on moisture contents (%)

ANOVA for moisture contents revealed non-significant differences for concentrations while differences were

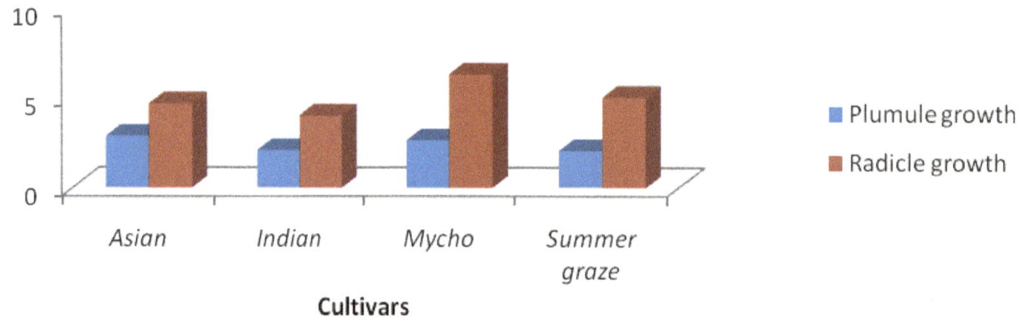


**Figure 4.** Effect of salinity on plumule and radicle growth (cm) of sorghum cultivars.

**Table 6.** Effect of salinity on dry weight/seedling (gram) of sorghum cultivars.

| Concentration (milli moles) | *Asian* | *Indian* | *Mycho* | *Summer graze* | Concentration means |
|---|---|---|---|---|---|
| Control | $0.60^{b}$ | $0.43^{c}$ | $0.71^{a}$ | $0.36^{cd}$ | $0.53^{a}$ |
| 160 | $0.42^{c}$ | $0.32^{de}$ | $0.29^{def}$ | $0.12^{h\text{-}k}$ | $0.29^{b}$ |
| 190 | $0.23^{fg}$ | $0.25^{efg}$ | $0.23^{fg}$ | $0.16^{ghi}$ | $0.22^{c}$ |
| 220 | $0.18^{gh}$ | $0.11^{h\text{-}k}$ | $0.14^{hij}$ | $0.07^{jk}$ | $0.13^{d}$ |
| 250 | $0.11^{h\text{-}k}$ | $0.03^{k}$ | $0.09^{h\text{-}k}$ | $0.08^{ijk}$ | $0.08^{d}$ |
| Cultivars Means | $0.31^{a}$ | $0.23^{b}$ | $0.29^{a}$ | $0.16^{c}$ | |

Lsd value at 5% level of significance for concentration = 0.04454, cultivars = 0.03983 and concentration x cultivars interaction = 0.08907. Values bearing similar letters in rows and columns are statistically non significant at 5% level of significance.

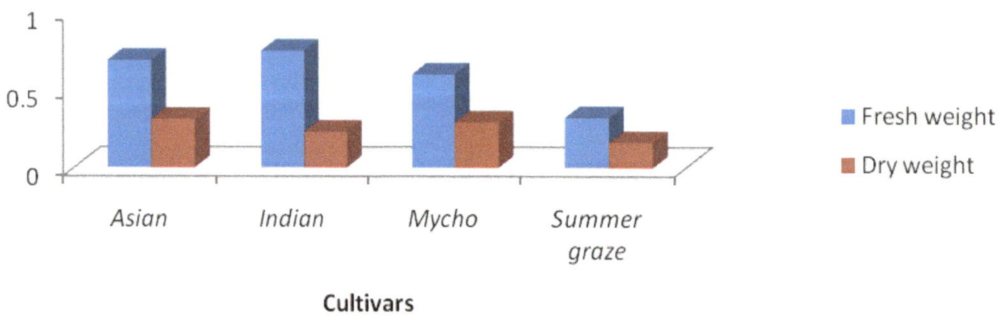


**Figure 5.** Effect of salinity on fresh and dry weight/seedling (gram) of sorghum cultivars.

significant for cultivars and between concentrations and cultivars interaction (Table 1). Maximum mean value (165.9 g) for moisture contents (%) was recorded in control. Cultivar *Indian* showed maximum mean value (195.4%) for moisture contents. While minimum mean value (128.9 g) was observed for cultivar *Asian*. In concentrations x cultivars interaction, maximum mean value (432.1%) for moisture contents was observed for cultivar *Indian* in control while minimum (98.8%) in 220 salt concentration for cultivar *Mycho* (Table 7). Moisture contents of seedling play an important role in various physiological processes including growth (Muhammad and Hussain, 2010; Ahmad et al., 2012). In the present study, decrease in moisture contents was recorded with salinity treatments (Figures 3 and 6). Shadded and Zaidan (1989), Naqvi and Mahmood (1994) and Ahmad et al. (2012) recorded reduction in moisture contents with increase salt stress in *Trigonella*, *Sesbania sesban* and canola, respectively, thus confirmed our findings. The decrease in moisture contents with increasing salinity level is in contrast with the findings of Ibrar and Hussain (2003), Akhtar and Hussain (2008) and Muhammad and Hussain (2010) who reported increase in moisture contents with increase in salinity concentrations while working on *Medicago*, grasses and medicinal plants, respectively.

**Table 7.** Effect of salinity on moisture contents (%) of sorghum cultivars.

| Concentration (milli moles) | *Asian* | *Indian* | *Mycho* | *Summer graze* | Concentration Means |
|---|---|---|---|---|---|
| Control | 146.1[cd] | 324.1[a] | 101.1[d] | 92.2[d] | 165.9 |
| 160 | 93.0[d] | 155.0[bcd] | 119.0[cd] | 147.9[cd] | 128.7 |
| 190 | 150.7[cd] | 163.3[bcd] | 178.0[bcd] | 136.2[cd] | 157.1 |
| 220 | 155.0[cd] | 195.0[bc] | 98.8[d] | 241.3[ab] | 162.5 |
| 250 | 140.0[cd] | 139.5[cd] | 164.0[bcd] | 126.0[cd] | 142.4 |
| Cultivars Means | 128.9[b] | 195.4[a] | 132.2[b] | 148.7[b] | |

Lsd value at 5% level of significance for Cultivars = 40.31 and Concentration x Cultivars interaction = 90.15. Values bearing similar letters in rows and columns are statistically non significant at 5% level of significance.

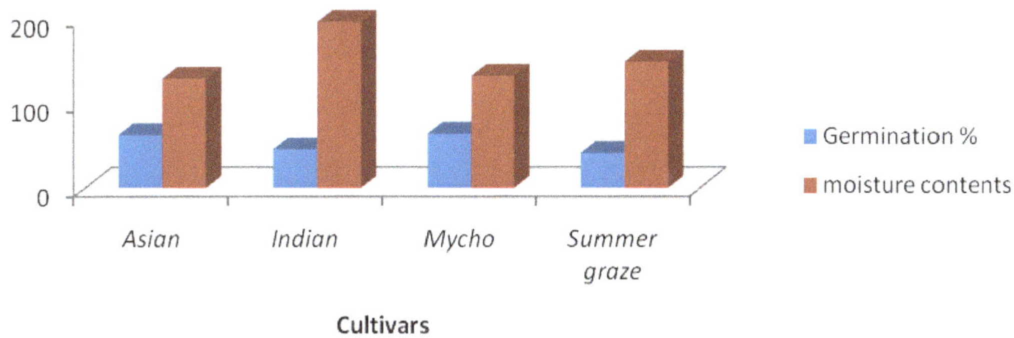


**Figure 6.** Effect of salinity on germination and moisture contents (%) of sorghum cultivars.

This preliminary laboratory study suggests that the tested sorghum cultivars could be grown on marginally saline habitats due to their tolerance to moderate salinity at germination and seedling stage. However, further study is needed to test their salt tolerance under field conditions to assess the possibility of cultivation of sorghum cultivars on otherwise unproductive lands.

## REFERENCES

Afzal I, Shahzad MAB, Iqbal A (2005). The effects of seed soaking with plant growth regulators on seedling vigor of wheat under salinity stress. J. Stress Physio. Biochem. 1(1):6-14.

Ahmad I, Ibrar M, Muhammad Z, Barkatullah (2012). Effect of salinity on canola. J. Pl. Sci. 18(1):85-91.

Ahmed S (2009). Effect of soil salinity on the yield and yield components of mung bean. Pak. J. Bot. 41(1):263-268.

Akhtar P, Hussain F (2008). Salinity tolerance of three range grasses at germination and early growth stages. Pak. J. Bot. 40(6):2437-2441.

Anwar M, Hussain I, Alam SS, Baig F (2001). Effects of NaCl salinity on seed germination, growth and yield of two varieties of Chick pea (*Cicer arietinum* L.) Pak. J. Bio. Sci. 4(2):124-127.

Asfaw KG (2011). Effects of salinity on seedling biomass production and relative water content of twenty sorghum (*Sorghum biolor* L. Moench) accessions. Asi. J. Agri. Sci. 3(3):242-249.

Ashraf M, Ali Q (2008). Relative membrane permeability and activities of some antioxidant enzymes as the key determinants of salt tolerance in canola (*Brassica napus* L.). Environ. Exp. Bot. 63:266-273.

Barakat H (2003). Interactive effects of salinity and certain vitamins on gene expression and cell division. Int. J. Agri. Biol. 5(3):219-225.

Bashir F, Ali M, Hussain K, Majeed A, Nawaz K (2011). Morphological variations in sorghum (*Sorghum bicolor* L.) under different levels of Na2SO4 salinity. BRI 4(1):01-03.

Baybordi A, Tabatabaei J (2009). Effect of salinity stress on germination and seedling properties in canola cultivars. Not. Bot. Hort. Agrobot. Cluj. 37(1):71-76.

El Naim AM, Mohammed KE, Ibrahim EA, Suleiman NN (2012). Impact of salinity on seed germination and early seedling growth on three sorghum (*Sorghum Bicolor* L.Moench ) cultivar. Sci. Tech. 2(2):16-20.

Farooq S, Azam F (2006). The use of cell membrane stability (CMC) technique to screen for salt tolerant wheat varieties. J. Plant Physiol., 163(6):629-637.

Greenway H, Munas R (1980). Mechanisms of salt tolerance in non halophytes. Ann. Rev. Plant Physiol. 31:141-190.

Hussain F (1989). Field and Laboratory Manual of Plant Ecology. University Grants Commission, Islamabad.

Ibrar M, Hussain F (2003). The effect of salinity on the growth of *Medicago polymorpha* Linn. J. Sci. Tech. Univ. Pesh. 27:35-38.

Ismail AM (2003). Physiological studies on the influence of cytokinin or $GA_3$ in the alleviation of salt stress in sorghum plants. Acta Agro. Hung. 51:371-380.

Jamil M, Lee DB, Jung KY, Ashraf M, Lee SC, Rha SE (2006). Effect of salt (NaCl) stress on germination and early seedling growth of four vegetable species. J. Central Eur. Agric. 7(2):273-282.

Kafi M, Nabati J, Masoumi A, Mehrgerdi MZ (2011). Effect of salinity and silicon application on oxidative damage of sorghum [*Sorghum bicolor* (L.) Moench.]. Pak. J. Bot. 43(5):2457-2462.

Kandil AA, Sharief AE, Abido WAE, Ibrahim MM (2012). Effect of salinity on seed germination and seedling characters of some forage sorghum cultivars. Int. J. Agri. Sci., 4(7): 306-311.

Kausar A, Ashraf MY, Ali I, Niaz M, Abbass Q (2012). Evaluation of sorghum varieties/lines for salt tolerance using physiological indices as screening tool. Pak. J. Bio. 44(1):47-52.

Kazemi K, Eskandari H (2011). Effects of salt stress on germination and early seedling growth of rice (*Oryza sativa*) cultivars in Iran. Afr. J. Biotech. 10(77):17789-17792.

Keshavarzi MHB (2012). The effect of different NaCl concentration on germination and early seedling growth of *Artemisia annua* L. Int. J. of Agri. Res. Rev. 2:135-140.

Khaje-hosseini M, Powell AA, Bingham IJ (2003). The interaction between salinity stress and seed vigor during germination of soybean seeds. Seed Sci. Technol. 31:715-725.

Maghsoudi-Moudi A, Maghsoudi K (2008). Salt stress effects on respiration and growth of germinated seeds of different wheat cultivars. W. J. Agric. Sci. 4(3):351-358.

Moradi P, Zavareh M (2013). Effects of salinity on germination and early seedling growth of chickpea (*Cicer arietinum* L.) cultivars. Int. J. Far. All. Sci. 2-3:70-74.

Muhammad Z, Hussain F (2010). Effect of NaCl salinity on the germination and seedling growth of some medicinal plants. Pak. J. Bot. 42(2):889-897.

Munns R, James RA, Lauchli A (2006). Approaches to increasing the salt tolerance of wheat and other cereals. J. Exp. Bot. 57:1025.

Naqvi SAR, Mahmood A (1994). Effect of NaCl salinity on growth, nodulation and nitrogen fixation in *Sesbania sesban L.* International. Symp. on high salinity tolerant plants. Department of Botany, University of Karachi. December 12-16. p. 69.

Naseer SH, Nisar A, Ashraf M (2001). Effect of salt stress on germination and seedling growth of barley (*Hordeum vulgare* L.). Pak. J. Bio. Sci. 4(3):359-360.

Rahman MU, Soomro UA, Haq MZ, Gul S (2008). Effects of NaCl Salinity on Wheat (*Triticum aestivum* L.) cultivars. World J. Agric. Sci. 4(3):398-403.

Rani CR, Reema C, Alka S, Singh PK (2012). Salt tolerance of *sorghum bicolor* cultivars during germination and seedling growth. Res. J. Rec Sci. 1(3):1-10.

Saberi AR, Aishah HS, Halim RA, Zaharah AR (2011). Morphological responses of forage sorghums to salinity and irrigation frequency. Afr. J. Biotechnol. 10(47):9647-9656.

Saberi AR, Kiani AR, Mosavat SA, Halim RA (2012). The effect of different sowing patterns and deficit irrigation management on yield and agronomic characteristics of sweet corn. Afr. J. Biotechnol. 11(74):13882-13887.

Shadded MA, Zaidan MA (1989). Effect of NaCl salinity on the rate of germination, seedling growth and some metabolic changes in *Raphanus sativa* L. and *Trigonella foenum-graecum* L. Beitrage Zur Tropischen Landwirtschaft Und Veterinarmedizin. 27(2):187-194.

Steel RGD, Torrie JH (1984). Principles and Procedures of Statistics. A Biometrical Approach. 2nd Ed. McGraw Hill Book Co. Inc., Singapore. pp. 172-177.

Tezara W, Martinez D. Rengifo E, Herrera A (2003). Photosynthetic response of the tropical spiny shrub *Lycium nodosum* (Solanaceae) to drought, soil salinity and saline spray. Ann. Bot. 92:757-765.

Wahid A, Javed IH, Ali I, Baig A, Rasul E (1998). Short term incubation of sorghum caryopsis in sodium chloride levels: changes in pre- and post-germination physiological parameters. Plant Sci. 139:223-232.

Zia S, Khan M (2002). Comparative effect of NaCl and sea water on seed germination on *Limonium stocksii*. Pak. J. Bot. 34:345-350.

# Release of cowpea line IT99K-494-6 for yield and resistance to the parasitic weed, *Alectra vogelii* Benth. in Malawi

**V. H. Kabambe[1], E. D. L. Mazuma[2], J. Bokosi[1] and E. Kazira[3]**

[1]Bunda College of Agriculture, P. O. Box 219, Lilongwe, Malawi.
[2]Chitedze Research Station, P. O. Box 158, Lilongwe, Malawi.
[3]Lilongwe Agriculture Development Division, P.O Box 259, Lilongwe, Malawi.

**Field trials were conducted between 2008/09 and 2009/10 seasons to evaluate four new cowpea lines, selected in preliminary studies, for yield, adaptation and resistance to the parasitic weed *Alectra vogelii* in Malawi. The design was randomised complete block design with 4 replicates at research stations and 5 at on-farm sites. For on-farm, villages were the sites and each farmer hosted one replicate. The trial sites were Lisasadzi, Mngwangwa, Bunda College, Chitedze Research Station, Rivirivi and Mpokwa in 2008/09 season and in 2009/10 the sites at Chitedze and Mpokwa were replaced by Mkwinda and Chitala Research Station. Four promising cowpea lines, IT98K-503-1, IT97K-825-15, IT99K-7-21-2-2-1 and IT99K-494-6, all originally from IITA and selected for resistance to parasitic weed *Alectra vogelii* in screen house studies, were compared with released varieties Sudan 1, IT82E-16 and a farmers local entry. The following parameters were evaluated: cowpea grain yield, plants/m$^2$, seeds/pod, seed weight (g/100 seeds), days to mid-flower, days to maturity, disease scores (scale 1 [clean] – 9 [most severe]) and *A. vogelii* emergence (plants/m$^2$). The line IT99K-494-6 was found to have comparable or better agronomic traits as compared to control varieties and was eventually released in Malawi in January 2011. The line which is medium maturing has brown grain texture with yield potential of 1-2.0 t/ha. Such a resistant variety is important in *A. vogelii* hot spots to reduce build up in soil seed bank and a viable option for farmers to rotate with cowpeas in order to manage other pests such as *Striga asiatica*, an important parasitic weed, or improve soil fertility through the nitrogen fixation of cowpeas.**

**Key words:** Legumes, witchweeds, parasitic weeds, cowpea, *Alectra vogelii, Vigna unguiculata.*

## INTRODUCTION

Cowpea is amongst the important food legumes in Malawi as it is drought tolerant and can do well in warm areas where beans cannot. In 2009/10 growing season, the yields of cowpeas under smallholder farmers in Malawi averaged 378 kg/ha from an area of 61,082 ha (MoAIWD, 2012). Cowpeas are an important source of

**Table 1.** Test sites, global positioning systems (GPS) and elevation in 2008/09 and '09/10 seasons.

| Site | Season | GPS | Masl | On-farm or station | AE⁺ Zone |
|---|---|---|---|---|---|
| Bunda | 2008/09; 2009/10 | S14°11.096′; E33°46.340′ | 1194 | Station | Mid |
| Mkwinda | 2009/10 | S 14°9.066′; E 33°47.312′ | 1153 | Farm | Mid |
| Lisasadzi | 2008/09; 2009/10 | S13°15.587′; E33°26.344′ | 1045 | Farm | Mid |
| Mngwangwa | 2008/09; 2009/10 | S14°53.284′; E3°42.194′ | 1133 | Farm | Mid |
| Chitedze | 2008/09 | S14°0.246′; E33°38.993′ | 1145 | Station | Mid |
| Mpokwa | 2008/09 | S15°01.761′; E34° 56.386′ | 939 | Farm | Low |
| Rivirivi | 2008/09; 2009/10 | S15°02.704′; E34°54.145′ | 649 | Farm | Low |
| Chitala | 2009/10 | S13°41.206′; 34°14.606′ | 643 | Station | Low |

AE⁺ = Agro-Ecological; Masl = meter above sea level.

protein averaging 22 - 34% (Elias et al., 1964; Neil and Brits, 1992; Mahe et al., 1994) and cash for smallholder farmers. The main constraints to its production are insect pests, diseases and lack of suitable varieties (MoAFS, 2005). In 2009/10 growing season, most farmers planted improved cultivar Sudan 1, variety IT82E-16 and local seeds which are preserved and recycled. Potential yields of cowpea is 2,000 kg/ha (MoAFS, 2005). However, *Alectra vogelii*, a parasitic weed for legumes, is also of growing concern particularly as legumes have been widely promoted for some time (ICRISAT/MAI, 2000; Ngwira et al., 2012). The Malawi Government, through the Farm Input Subsidy Program (FISP) also promotes legumes (MoAIFS, 2007). The weed *A. vogelii* is widely seen in Lilongwe and Kasungu plains and many parts of southern region (Riches and Shaxson, 1993; Kabambe et al., 2005; Mainjeni, 1999; Kabambe et al., 2008a, b). In Tanzania yield losses of up to 50% have been reported (Mbwaga et al., 2000). In Kenya, Bagnall-Oakley et al. (1991) reported total crop loss in some parts. In Botswana, Riches (1989) reported losses of 80 - 100% in a susceptible cultivar. Yield losses of up to 15% have been reported in groundnut in Nigeria (Salako, 1984), while in South Africa 30 - 50% reductions in yield of bambara were reported (Beck, 1987). In northern Nigeria, Lagoke (1989) reported that late-sown crops of soybean may be completely destroyed by the parasite. The witchweed *A. vogelii* may be controlled through trap crops such as dolichos bean (*Lablab purpureus* [L.] Sweet) cotton (*Gossypium hirsutum* L.), okra (*Hibiscus esculentus* L.), sunflower (*Helianthus annuus* L.) pea (*Pisum sativum* L.) and faba bean (*Vicia faba* L.) (Parker and Riches, 1993). There is generally greater availability of resistant genotypes against *A. vogelii* (Riches, 2001; Mainjeni, 1999; Rubiales et al., 2006; Omoigui, 2012). Thus, production is threatened by *A. vogelii* and screening for resistance among existing legumes crops or varieties can help to avoid working with the most susceptible varieties. The parasitic weed *Striga asiatica* is also a serious problem in Malawi (Kabambe et al., 2002, 2005), such that the introduction of legume varieties with resistance *to A. vogelii* would be useful to encourage farmers to adopt legumes as trap crops for it or soil fertility replenishment. Also, resistant varieties would slow down the seed build-up of the pest in soils. This study was therefore conducted to evaluate some selected cowpea lines for yield adaptability and resistance to the parasitic weed *A. vogelii*.

## METHODOLOGY

### Trials design, sites, test materials and management

Four cowpea lines, IT98K-503-1, IT97K-825-15, IT99K-7-21-2-2-1 and IT99K-494-6 were evaluated for yield and resistance to the parasitic weed species *A. vogelii* under field conditions in 2008/09 and 2009/10 in Malawi. These lines were ex-IITA (International Institute of Tropical Agriculture) and were selected for evaluation based on earlier screen house screening for resistance to *A. vogelii* and preliminary yield trials. The lines were compared to released varieties Sudan 1, IT82E-16 and farmers local entry. In both seasons, there were four mid-altitude sites and two low altitude sites. The names, geographic position and altitude of the sites are given in Table 1. The trial design was randomized complete block with four replications at station sites while on-farm there were five replicates, with one farmer hosting one replicate.

Experimental plots had 5 with ridges each 4 m long and 0.75 m apart (gross plot size 15 $m^2$). No fertilizers were applied to plots. Two seeds of cowpeas were planted at 20 cm apart. Planting was done with first rains in low altitude sites and in mid-January for mid-altitude sites according to standard recommendations (MoAIFS, 2005). There were 4-5 farmers per village or site, making one complete trial.

### Data collection and analysis

All crop plant or *A. vogelii* data was collected from the three middle rows (net plot size 12 $m^2$). Data were collected for evaluation of plant cowpea grain yield (adjusted to 12.0% moisture), plant count/$m^2$, *A. vogelii* counts/$m^2$, seed weight, seeds/pod, pod length (cm) days mid-flower and maturity. Counts of *A. vogelii* were taken at time of flowering of cowpeas. Assessment for disease was done at Chitedze Research Station only, where the project pathologist is based. Entries were scored for *Aschochyta* blight (*Ascochyta phaseolorum*), Scab (*Elsinoe phaseoli*) and *Cercospora* leaf spot infection. Disease scores were based on 1-9 scale where 1 was clean and 9 was severe. For infection, the score was the percentage of plants infected. Data were analyzed on site basis

**Table 2.** Mean monthly rainfall (mm) for the some trial sites in 2008/09 and '09/10 seasons.

| Month | Site and season | | | | | | | |
|---|---|---|---|---|---|---|---|---|
| | Bunda | | Mngwangwa | | Rivirivi | | Chitedze | Chitala |
| | 2008/09 | 2009/10 | 2008/09 | 2009/10 | 2008/09 | 2009/10 | 2008/09 | 2009/10 |
| Oct. '09 | 18.5 | 0.9 | 0 | 0 | 3.0 | 16.5 | 45 | 0 |
| Nov. '09 | 119.2 | 60.5 | 51 | 83 | 113.4 | 55.4 | 112 | 5 |
| Dec. '09 | 177.1 | 114.5 | 147 | 184 | 242.1 | 113.5 | 116.4 | 88.1 |
| Jan. '10 | 291.9 | 125.1 | 198 | 249.5 | 300.5 | 222.7 | 227.8 | 220.9 |
| Feb. '10 | 174.0 | 337.6 | 152 | 306 | 145.1 | 320.9 | 121 | 476.5 |
| Mar. '10 | 218.3 | 118.4 | 134 | 71 | 125.1 | 72.6 | 223.7 | 156.2 |
| Apr. '10 | 28.0 | 5.4 | 71 | 36.5 | 6.3 | 34.6 | 18.1 | 0 |
| May '10 | 0 | 4.7 | 0 | 0 | 0 | 0 | 0 | 0 |
| Total | 1026.7 | 767 | 753 | 930 | 932.5 | 836.2 | 864 | 946.7 |

**Table 3.** Effect of genotype on *Alectra* counts m$^{-2}$ taken after flowering at all the trial sites in 2008/09 season.

| Entry name | Bunda* | Lisasadzi | Mngwangwa | Mpokwa | Rivirivi | Chitedze |
|---|---|---|---|---|---|---|
| IT98K-503-1 | 1.35bc | 0.50b | 3.33 | 0.6 | 0.0 | 0.05 |
| IT97K-825-15 | 0.08c | 0.70b | 0.60 | 0.03 | 0.08 | 0.25 |
| IT99K-7-21-2-2-1 | 0.25c | 1.38b | 1.46 | 0.11 | 0.0 | 0.17 |
| IT99K-494-6 | 0.00c | 0.60b | 1.02 | 0.04 | 0.03 | 0.0 |
| Sudan-1 | 5.82a | 3.56ab | 4.03 | 0.86 | 0.0 | 0.58 |
| IT82E-16 | 6.25a | 6.70a | 2.86 | 0.96 | 0.14 | 0.0 |
| Farmers' local | 2.80b | 1.14b | 3.29 | - | 0.03 | 0.11 |
| Mean | 2.36 | 2.08 | 2.43 | 0.44 | 0.04 | 0.17 |
| F Prob | <0.001 | 0.005 | 0.13 | 0.28 | 0.32 | 0.29 |
| LSD (0.05) | 1.59 | 3.28 | 3.0 | 1.0 | 0.13 | 0.56 |

*Means in column denoted by same letter are not significantly different at 5% using LSD.

according to the analysis of variance procedure using the Genstat package. Comparisons between means was done using the least significant difference, LSD, $t_\alpha = 0.05$.

## RESULTS

### Rainfall

In 2008/09 rainfall was normal and fairly distributed for all sites except Mngwangwa and Chitedze, which had relatively poorly rains in February 2009 (Table 2).

### Alectra emergence

In 2008/09, the sites at Mngwangwa and Lisasadzi had high infestation of *Alectra*, even though yields were good (Table 3). However, within the plots, spots that had high numbers of *Alectra* caused stunted growth and wilting on cowpea plants. At Mngwangwa, where pressure was high, entries IT99K-7-21-2-2-1, IT97K-825-15 and IT99K-494-6 showed fair resistance. At Lisasadzi, with high *Alectra* pressure the same lines showed good resistance (low support of parasite). At Bunda College, with high *Alectra* pressure, the same three lines showed resistance. On the overall, entries IT99K-494-6 and IT99K-7-21-2-2-1 were best for resistance with maximum Alectra of 1.02 and 1.46 plants m$^{-2}$, respectively, as compared to maximum emergence of 6.7 and 5.82 plants m$^{-2}$, respectively for IT82E-16 and Sudan-1, the local checks.

In 2009/2010, entry IT99-494-6 was the most resistant line among the seven entries and was followed by IT97K-825-15 and IT99K-7-21-2-2-1, respectively (Table 4). On the overall, IT99K-494-6 and IT97K-825-15 were the best for resistance with 0.10 and 0.80 plants m$^{-2}$ respectively, and entry IT99K-494-6 have nil *Alectra* at Mkwinda and only 0.04 plants m$^{-2}$ at Lisasadzi. There was no Alectra emergence at Bunda College, Chitala and Riviri.

### Grain yields

Tables 5 and 6 show cowpea grain yields from mid-altitude sites in 2008/09 and 2009/10. In 2008/09, the line

**Table 4.** Effect of genotype on *Alectra* counts $m^{-2}$ taken after flowering at three sites in 2009/10.

| Entry | Lisasadzi | Mngwangwa | Mkwinda |
|---|---|---|---|
| IT98K-503-1 | 2.27 | 2.5 | 1.9 |
| IT97K-825-15 | 0.98 | 1.1 | 0.3 |
| IT99K-7-21-2-2-1 | 0.35 | 3.5 | 0.1 |
| IT99K-494-6 | 0.04 | 0.2 | 0 |
| SUDAN-1 | 2.82 | 2.8 | 0.8 |
| IT82E-16 | 2.04 | 11.9 | 0.7 |
| Farmers local | 2.80 | 18.0 | 1.1 |
| Mean | 1.61 | 5.7 | 0.7 |
| F prob | 0.44 | 0.06 | 0.18 |
| LSD (0.05) | 3.33 | 13.3 | 1.59 |

*Means in column denoted by same letter are not significantly different at 5% using LSD.

**Table 5.** Grain yield kg/ha of cowpea entries at mid-altitude trial sites in 2008/09.

| Entry | Site | | | |
|---|---|---|---|---|
| | Lisasadzi | Mngwangwa | Bunda | Chitedze |
| *IT98K-503-1* | $2034^{ab}$ | 856 | 1496 | $1667^{b}$ |
| *IT97K-825-15* | $1527^{b}$ | 796 | 1276 | $1083^{b}$ |
| *IT97K-7-21-2-2-1* | $1749^{b}$ | 543 | 1665 | $1222^{b}$ |
| IT99K-494-6 | $2345^{a}$ | 1114 | 1846 | $972^{b}$ |
| IT82E-16 | $1936^{ab}$ | 903 | 1631 | $3194^{a}$ |
| SUDAN-1 | $2328^{a}$ | 1070 | 1728 | $2278^{ab}$ |
| LOCAL | - | 536 | 794 | $1861^{ab}$ |
| MEAN | 1936 | 831 | 1492 | 1754 |
| F PROB | <0.025 | 0.058 | 0.208 | 0.03 |
| LSD (0.05) | 523 | 435 | 830 | 1340 |

*Means in column denoted by same letter are not significantly different at 5% LSD.

**Table 6.** Grain yield (kg/ha) of cowpea entries at mid-altitude trial sites in 2009/10.

| Entry | Site | | | |
|---|---|---|---|---|
| | Bunda | Lisasadzi | Mngwangwa | Mkwinda |
| IT98K-503-1 | $1808^{ab}$ | $1421^{bc}$ | $730^{c}$ | $979^{b}$ |
| IT97K-825-15 | $1102^{d}$ | $658^{f}$ | $361^{e}$ | $434^{d}$ |
| IT99K-7-21-2-2-1 | $1266^{d}$ | $1024^{e}$ | $619^{d}$ | $763^{c}$ |
| IT99K-494-6 | $1528^{cd}$ | $1757^{a}$ | $932^{a}$ | $1195^{ab}$ |
| Sudan1 | $1621^{bc}$ | $1376^{cd}$ | $938^{a}$ | $1027^{ab}$ |
| IT82E-16 | $2076^{a}$ | $1513^{b}$ | $861^{a}$ | $1060^{ab}$ |
| Farmers local | 1019 | $1275^{de}$ | $528^{d}$ | $1089^{ab}$ |
| Mean | 1489 | 1289 | 710 | 935 |
| P | <0.001 | <0.001 | <0.001 | <0.001 |
| LSD 5% | 337 | 111 | 99 | 139 |

*Means in column denoted by same letter are not significantly different at 5% LSD.

IT99K-494-6 gave the highest yields at Lisasadzi, Bunda and Mngwangwa, which were sites with highest *Alectra* pressure. In 2009/2010, the line IT99K-494-6 gave highest or joint highest yields at Lisasadzi, Mngwangwa

**Table 7.** Grain yield kg/ha of cowpea entries at two low altitude trial sites in 2008/09 and 2009/10 seasons.

| Entry name | Site and season | | Site and season | |
|---|---|---|---|---|
| | 2008/09 | | 2009/10 | |
| | Rivirivi | Mpokwa | Rivirivi | Chitala |
| IT98K-503-1 | 749$^{b}$ | 461$^{b}$ | 1165$^{b}$ | 849 |
| IT97K-825-15 | 589$^{b}$ | 447$^{b}$ | 1064$^{b}$ | 693 |
| IT97K-7-21-2-2-1 | 903$^{b}$ | 353$^{b}$ | 1328$^{b}$ | 852 |
| IT99K-494-6 | 1372$^{ab}$ | 562$^{b}$ | 1189$^{b}$ | 786 |
| IT82E-16 | 2156$^{a}$ | 896$^{ab}$ | 1255$^{b}$ | 855 |
| Sudan-1 | 2158$^{a}$ | 1402$^{a}$ | 2029$^{a}$ | 980 |
| Local | 964$^{b}$ | 353$^{b}$ | 1814$^{ab}$ | 852 |
| Mean | 1270 | 681 | 1406$^{b}$ | 838 |
| LSD | 878 | 544 | 457 | 242 |
| F Prob | 0.004 | 0.004 | 0.002 | 0.38 |

*Means in column denoted by same letter are not significantly different at 5% LSD.

**Table 8.** Effect of genotype on days to 50% flower at 5 sites, 2008/09.

| Entry | Mngwangwa | Bunda | Lisasadzi | Mpokwa | Balaka |
|---|---|---|---|---|---|
| IT98K-503-1 | 46$^{b}$ | 52$^{b}$ | 57$^{b}$ | 46$^{b}$ | 48 |
| IT97K-825-15 | 47$^{b}$ | 56$^{a}$ | 58$^{b}$ | 53$^{a}$ | 52 |
| IT97K-7-21-2-2-1 | 48$^{ab}$ | 54$^{ab}$ | 59$^{b}$ | 53$^{a}$ | 53 |
| IT99K-494-6 | 46$^{b}$ | 52$^{b}$ | 57$^{b}$ | 46$^{b}$ | 47 |
| 52=IT82E-16 | 44$^{c}$ | 48$^{c}$ | 55$^{b}$ | 39$^{c}$ | 46 |
| 53=Sudan-1 | 46$^{b}$ | 47$^{c}$ | 59$^{b}$ | 41$^{c}$ | 51 |
| 54=Local | 50$^{a}$ | 53$^{a}$ | 89$^{a}$ | - | 52 |
| Mean | 47 | 52 | 62 | 46 | 50 |
| P | <0.001 | <0.001 | <0.001 | <0.001 | 0.80 |
| LSD | 2 | 3 | 6 | 3 | 12 |

and Mkwinda (Table 6). However, the two released varieties also gave similar high yields. In the low altitude sites Sudan 1 and IT82E-16 gave better yields than IT99K-494-6 (Table 7).

## Days to mid-flower and maturity

The number of days to 50% flower in 2008/09 season is shown in Table 8. Plants flowered earliest at low altitude sites as compared to mid-altitude sites. The entry IT82E-16 was the earliest to flower at both mid and low-altitude sites (range 39-55 days, followed by Sudan 1 (range 41-59 days) then entries IT99K-494-6 and IT99K825-15 (range 49-57 days). In general, the maximum difference between flowering dates of the improved varieties was about 10 days. In 2009/10 the trend was similar in terms of variety responses. However, flowering was generally delayed by about 2 days (Table 9). The flowering time can vary depending on planting date which affects temperature exposure during the season. Discussions with farmers revealed that such differences were quite important to them.

The effects of variety on days to maturity are shown in Tables 10 and 11. In 2008/09, plants matured early at Rivirivi and Mpokwa (66 days as compared to the other sites (76-81 days), which was expected as these sites are climatically warmer and drier areas. The entry IT99K-494-6 matured between 71 and 82 days in the mid-altitude sites, as compared to 72-74 days for Sudan-1. In 2009/10 season, the entry IT99K-494-6 matured between 71 and74 days in the mid-altitude sites, as compared to 65 and 68 days for Sudan-1.

## Disease incidence and scores

Assessment for disease was done at Chitedze Research Station only in 2008/09. Entries were evaluated for *Aschochyta* blight (*Ascochyta phaseolorum*), scab (*Elsinoe phaseoli*) and *Cercospora* leaf spot infection. The results showed no significant differences amongst

**Table 9.** Effect of genotype on days to 50% flower in 2009/10.

| Entry | Mid altitude sites | | | | Low altitude sites | |
|---|---|---|---|---|---|---|
| | Mngwangwa | Lisa Sadzi | Bunda | Mkwinda | Chitala | Rivirivi |
| IT98K-503-1 | 56d | 52$^{c}$ | 55$^{b}$ | 56 | 50 | 53 |
| IT97K-825-15 | 63$^{c}$ | 57$^{b}$ | 58$^{a}$ | 59 | 51 | 56 |
| IT97K-7-21-2-2-1 | 58$^{c}$ | 56$^{b}$ | 57$^{a}$ | 56 | 49 | 48 |
| IT99K-494-6 | 58$^{a}$ | 56$^{a}$ | 56$^{a}$ | 56 | 49 | 48 |
| SUDAN 1 | 54$^{d}$ | 52$^{c}$ | 52$^{c}$ | 53 | 47 | 51 |
| IT82E-16 | 55$^{d}$ | 51$^{d}$ | 53$^{c}$ | 44 | 48 | 50 |
| LOCAL | 60$^{b}$ | 51$^{d}$ | 57$^{a}$ | 61 | 48 | 50 |
| MEAN | 58 | 54 | 55 | 55 | 49 | 52 |
| P | <0.001 | <.001 | <.001 | 0.39 | 0.10 | 0.002 |
| LSD (0.05) | 2 | 1 | 2 | 15 | 2.7 | 5.4 |

*Means in column denoted by same letter are not significantly different at 5% LSD.

**Table 10.** Effect of cowpea line on days to maturity in 2008/09 season.

| Entry | Mid-altitude sites | | | low altitude sites | |
|---|---|---|---|---|---|
| | Mngwa Ngwa | Bunda | Lisasadzi | Mpokwa | Balaka |
| IT98K-503-1 | 79$^{b}$ | 82$^{b}$ | 73$^{b}$ | 64$^{b}$ | 61 |
| IT97K-825-15 | 83$^{a}$ | 88$^{a}$ | 72$^{b}$ | 74$^{a}$ | 69 |
| IT97K-7-21-2-2-1 | 80$^{b}$ | 83$^{b}$ | 76$^{b}$ | 73$^{a}$ | 70 |
| IT99K-494-6 | 80$^{b}$ | 82$^{b}$ | 71$^{b}$ | 65$^{b}$ | 60 |
| IT82E-16 | 74$^{c}$ | 75$^{c}$ | 60$^{c}$ | 59$^{c}$ | 66 |
| SUDAN-1 | 74$^{c}$ | 74$^{c}$ | 68$^{bc}$ | 60$^{c}$ | 69 |
| LOCAL | 80$^{b}$ | 83$^{b}$ | 107$^{a}$ | - | 68 |
| MEAN | 78 | 81 | 76 | 66 | 66 |
| P | <0.001 | <0.001 | <0.001 | <0.001 | 0.57 |
| LSD (0.05) | 2 | 2 | 11 | 3 | 13.2 |

**Table 11.** Days to maturity of cowpea entries 2009/10.

| Entry | Mid altitude sites | | | | Low altitude sites | |
|---|---|---|---|---|---|---|
| | Bunda | Lisasadzi | Mngwangwa | Mkwinda | Rivirivi | Chitala |
| IT98K-503-1 | 71$^{d}$ | 68$^{d}$ | 71$^{c}$ | 71$^{c}$ | 62$^{cd}$ | 72 |
| IT97K-825-15 | 79$^{a}$ | 79$^{a}$ | 89$^{a}$ | 80$^{b}$ | 68$^{b}$ | 71 |
| IT97K-7-21-2-2-1 | 76$^{b}$ | 74$^{b}$ | 77$^{b}$ | 73$^{c}$ | 73$^{a}$ | 71 |
| IT99K-494-6 | 74$^{c}$ | 71$^{c}$ | 73$^{bc}$ | 73$^{c}$ | 60$^{d}$ | 71 |
| IT82E-16 | 67$^{e}$ | 64$^{e}$ | 65$^{c}$ | 65$^{d}$ | 62$^{cd}$ | 70 |
| Sudan-1 | 67$^{e}$ | 65$^{e}$ | 68$^{c}$ | 66$^{d}$ | 61$^{cd}$ | 70 |
| Local | 70$^{d}$ | 67$^{d}$ | 73$^{b}$ | 89$^{a}$ | 65$^{bc}$ | 71 |
| Mean | 72 | 70 | 74 | 74 | 64 | 71 |
| P | <0.001 | <.001 | <0.001 | <.001 | <.001 | 0.63 |
| LSD (0.05) | 2 | 3 | 5 | 3 | 5 | 2.4 |

entries for early or late evaluation (Table 12). The scores for *Aschochyta* blight showed medium levels at both early and late stages of assessment while scab scores were quite high at the late scoring stage, averaging 6.0, which are close to severe. The infection of *Cercospora* and scores were also high. The high yields of the released varieties reflect a high degree of tolerance and adaptability.

**Table 12.** Summary of analysis of variance for disease scores at Chitedze Research Station, 2008/09.

| Disease and time of scoring | Mean | F Prob | CV (%) |
|---|---|---|---|
| *Aschochyta* blight early | 2.5 | 0.43 | 24 |
| Aschochyta blight late | 3.4 | 0.6 | 21 |
| Scab early | 3.7 | 0.43 | 44 |
| Scab late | 5.9 | 0.09 | 19 |
| Cercospora leafspot incidence (%) early | 25 | 0.38 | 36 |
| Cercospora leafspot incidence (%) late | 71.2 | 0.80 | 23 |
| Cercospora leafspot infection early | 5.1 | 0.44 | 22 |
| Cercospora leafspot infection late | 6.7 | 0.14 | 11 |

**Table 13.** Trial means and standard deviation (in brackets) of yield components and other agronomic traits, 2009/10.

| Entry | Plants $m^{-2}$ | Pod length (cm) | Seed per pod | Seed weight g/100 |
|---|---|---|---|---|
| IT98K-503-1 | 10.00 (2.47) | 12.96 (1.04) | 8.94 (0.90) | 13.20 (0.27) |
| IT997K-825-15 | 10.07 (2.72) | 13.72 (1.26) | 10.66 (0.81) | 11.69 (0.61) |
| IT97K-7-21-2-2-1 | 11.25 (2.35) | 14.26 (0.74) | 9.96 (0.94) | 17.57 (0.34) |
| IT99K-494-6 | 10.07 (2.35) | 13.08 (0.24) | 9.82 (0.35) | 12.92 (0.55) |
| IT82E-16 | 11.13 (4.38) | 17.00 (0.55) | 14.48 (0.94) | 10.44 (1.00) |
| Sudan 1 | 11.47 (1.79) | 16.16 (1.15) | 13.72 (1.30) | 9.84 (0.21) |
| Farmers local | 10.67 (1.44) | 16.72 (1.10) | 13.94 (1.09) | 13.03 (0.56) |

***Other agronomic traits, yield components and variety description***

Trial means for other agronomic traits and yield components for 2009/10 season are shown in Table 13. All entries seemed to have similar plant establishment with a narrow range of 10-11.4 plants per $m^2$. The released varieties were generally superior on pod length and seeds per pod. The new lines were superior in terms of seed weights per pod.

## DISCUSSION

The results are in agreement with other reports which show wide availability in resistant genotypes against *A. vogelii* (Mainjeni, 1999; Kabambe et al., 2008a, Mbwaga et al., 2000; Riches, 1989, 2001; Rubiales et al., 2006; Omoigui, 2012). The results also generally showed that *A. vogelii* is naturally more prevalent in the mid-altitude sites where all entries are better adapted. The nature of these studies did not allow the researchers to determine yield loss due to *A. vogelii* parasitism, however, it is clear that the two local checks IT82E-16 and Sudan 1 were tolerant to it, owing to their high yields although they supported high numbers of *A. vogelii.* In Botswana, Riches (1989) reported losses of 80 – 100% in a susceptible cowpea cultivar. In northern Nigeria, Lagoke (1989) reported that late-sown crops of soybean may be completely destroyed by the parasite. Although three of the new lines showed fair resistance (low support of Alectra), only the line ITK99K-494-6 was most consistent in yield at most sites. While yield is of interest to farmers, continued growing of susceptible cultivars promotes build-up of the seed bank. As *A. vogelii* is a parasite of other common legume crops in Malawi, such as groundnuts and soybean (Kabambe et al., 2008a), it is therefore sound management to reduce build up through planting varieties which support minimal *Alectra*, especially where yield is not compromised as is the case with IT99K-49-6. The *A. vogelii* may be managed through trap crops (false hosts) such as dolichos bean, cotton, okra, sunflower, pea and faba bean (Parker and Riches, 1993; Kabambe et al., 2008b). The same crops are trap crops for managing *S. asiatica* when grown in rotation or intercrops with cereals (Kabambe et al., 2005; Parker and Riches, 1993; Kabambe et al., 2002; Kabambe et al., 2008b; Parkinson et al., 1987; Carsky et al., 1994). By providing *Alectra*- resistant legume varieties to farmers, their choice to incorporate legumes in the farming systems improves. Legumes, including cowpeas furnish nitrogen through biological nitrogen fixation (Bado et al., 2006) and are important for integrated disease management.

The results of days to flower and maturity showed that the two of the new lines (IT99K-494-6 and IT99K825-15) were slightly late to flower and mature as compared to Sudan-1. There was no clear relationship between maturity period and yield observed. As an example, the variety IT82E-16 and IT99K-494-6 all gave highest yields at Lisasadzi in 2008/09 even though IT99K-494-6 matured at 71 days as compared to 60 days for IT82E-16. In the low altitude sites, however, IT82E-16 gave highest yields in both seasons. Earliness therefore was important in these short season areas. Results on diseases showed that no particular susceptibility or resistance for all entries such that existing pest and disease management practices would be recommended. Thus, the line IT99K-494-6 was recommended and released in January 2011 for the mid-altitude areas of Malawi due to consistence in resistance and high yield.

## ACKNOWLEDGEMENTS

The authors would like to thank the McKnight Foundation for financial support for this work. We are also grateful to the management of Chitedze Research Station and Bunda College of Agricultural for allowing the research team and support staff to undertake in this work, and for different forms of financial and material support. We also sincerely thank the management of Lilongwe ADD, Kasungu ADD and Machinga ADD for allowing their staff in the trial sites to collaborate with the research team. Lastly, we extend our gratitude to collaborating farmers for their time and dedication in the field evaluations.

### REFERENCES

Bado BV, Bationo A, Cescas MP (2006). Assessment of cowpeas and groundnut contributions to soil fertility and succeeding sorghum yields in the Guinean savanna zone of Burkina Faso( West Africa). Biol. Fertil. Soils 43:171-176.

Bagnall-Oakley H, Gibberd V, Nyongesa TE (1991). The incidence and control of *A. vogelii*, in Embu district, Kenya. In: Ransom JK, Musselman LJ, Worsham AD, Parker C (Eds.), Proceedings of the 5th International Symposium on Parasitic Weeds, CYMMIT, Nairobi, Kenya.

Beck BDW (1987). The effect *of Alectra vogelii* Benth. on the yield of njugo beans (*Vigna subterranea* (*L.*) *Verdc.*). In: Chr.Weber H, Forstreuter W (Eds.). Parasitic Flowering Plants. Phillips University, Marburg, Germany. pp. 79-82.

Carsky RJ, Singh L, Ndikawa R (1994). Suppression of Striga hermonthica using a cowpea intercrop. Exp. Agric. 30:349-359.

Elias LG, Bressani R, Colindres R (1964). The nutritive value of eight varieties of cowpea (Vigna sinensis). J. Food Sci. 29:118-112.

ICRISAT/MAI (2000). Cost-effective soil fertility management options for smallholder farmers in Malawi. Bulawayo, Zimbabwe: ICRISAT; and Lilongwe, Malawi: Ministry of Agriculture and Irrigation.

Kabambe VH, Katunga L, Kapewa T, Ngwira AR (2008a). Screening legumes for integrated management of witchweeds (*Alectra vogelii* and *Striga asiatica*) in Malawi. Afr. J. Agric. Res. 3(10):706-713.

Kabambe VH, Nambuzi SC, Kauwa AE (2008b). Integrated management of witchweed (*Striga asiatica* {L} Kunze) by means of maize-legume rotations and intercropping systems in Malawi. Bunda J. Agric. Environ. Sci. Technol. 3(2):35-42.

Kabambe VH, Nambuzi S, Kauwa AE, Ngwira AR (2005). Role of cowpea and silverleaf intercropping on witchweed (*Striga asiatica* L. Kuntze) in maize in Malawi. In: Safalaoh, ACL, Gowela JP, Mtethiwa AH (eds). Proceedings of the Second Bunda College Research Dissemination Conference Lilongwe, Malawi. pp. 22-27.

Kabambe VH, Mloza-Banda HR, Nyandule Phiri GY (2002). Controlling witchweeds in cereals in Malawi. An extension bulletin for field staff. Department of Agricultural Research and Technical Services, Extension Bulletin no. 1/2002. Ministry of Agriculture and Irrigation, Lilongwe, Malawi.

Lagoke STO (1989). *Striga* in Nigeria In: Robson TO, Broad HR (Eds.), *Striga* Improved Management in Africa. FAO Plant Production and Protection Paper No. 96, FAO, Rome, Italy. pp. 68-73.

MoAFS (2010). Ministry of Agriculture and Food Security. 2009/10 Agricultural Crop Estimates. Lilongwe, Malawi.

MoAFS (2007). Ministry of Agriculture and Food Security. Annual Agricultural Statistical Bulletin. Planning Division. Government of Malawi. Lilongwe, Malawi.

MoAFS (2005). Ministry of Agriculture, Irrigation and Food Security. A Guide to Agriculture and Natural Resources Management in Malawi. Agricultural Extension Branch. Lilongwe, Malawi.

Mahe S, Gaussere N, Tome D (1994). Legume protein for human requirements. Grain Leg. 7:15-17.

Mainjeni CE (1999). The host range of *Alectra vogelii* Benth. from Malawi and resistance in common bean and cowpea. MSc Thesis, University of Bath.

Mbwaga AM, Kaswende J, Shayo (2000). A reference manual on *Striga* distribution and control in Tanzania. Kilosa, Tanzania: Ilonga Agricultural Research Institute.

Neil FS, Brits FK (1992). Studies on the Nutritive Value of Cowpeas (*Vigna unguiculata*). South Afr. J. Ani. Sci. 22(5):157-160.

Ngwira AR, Kabambe VH, Kambauwa G, Mhango WG, Mwale C D, Chimphero L, Chimbizi A, Mapfumo P (2012). Scaling out best fit legume technologies for soil fertility enhancement among smallholder farmers in Malawi. Afr. J. Agric. Res. 7(6):918-928.

Omoigui LO, Kamara AY, Ishiyaku MF, Boukar O (2012). Comparative responses of cowpea breeding lines to *Striga* and *Alectra* in the dry savanna of northeast Nigeria. Afr. J. Agric. Res. 7(5):747-754.

Parker C, Riches CR (2000). Parasitic Weeds of the World. Biology and Control. CAB International. Oxon, UK.

Parkinson V, Efron Y, Bello L, Dashiel K (1978). Trap crops as cultural measure in Striga control in Africa. FAO Plant Protect. Bull.. 35:51-54.

Riches CR (2001). *Alectra vogelii* Benth. – a constraint to cowpea production in southern Africa. In: Mgonja MA (Ed.), *Striga* Research in Southern Africa and Strategies for Regionalized Control Options. International Crops Research Institute for the Semi-arid Tropics, Bulawayo, Zimbabwe. pp. 89-91.

Riches CR, Shaxson LJ (1993). Parasitic weed problems in southern Malawi and use of farmer knowledge in the design of control measures. In: Munthali DC, Kumwenda JDT, Kisyombe F (eds.). Conference on Agricultural Research for Development organised by the Soil Pests Project (University of Malawi) and the Malawi Maize Research Team, Ministry of Agriculture, Mangochi, Malawi, 7-11 June 1993. Munthali, D.C., J.D.T. Kumwenda and F. Kisyombe (eds.). pp. 165-174.

Riches CR (1989). The biology and control of *Alectra vogelii* Benth. (Scrophulariaceae) in Botswana. PhD thesis, Reading, UK: University of Reading.

Rubiales D, Perez de Luque A, Fernandez-Aparica M, Sillero JC, Roman B, Kharrat M, Khalil S, Joel DM, Riches CR (2006). Screening techniques and sources of resistance to parasitic weeds in grain legumes. Euphytica 147:187-199.

Salako EA (1984). Observations on the effect of *A. vogelii* infestation on the yield of groundnut. Trop. Pest Manag. 30:209-211.

# Comparison of qualitative and quantitative traits of some Advanced breeding lines of tomato (*Lycopersicon esculentum* L.)

**Olakojo, S. A.[1] and Adetula, O. A.[2]**

[1]Institute of Agricultural Research and Training Obafemi Awolowo University, P. M. B. 5029 Moor Plantation, Ibadan.
[2]National Horticultural Research Institute Idi-Isin, Ibadan.

**Field trials and laboratory experiments were conducted on six advanced breeding lines of tomato (*Lycopersicon esculentum* L.) between 2004 and 2006 to genetically assess the organoleptic properties of the tomato lines, determine the relationships among tomato traits using correlation analysis and to estimate the fruit yield potential of the tomato lines. The trial was laid out in randomized complete block design with three replicates. Planting was done on four-row plots of 8 x 5 m using intra row spacing of 60 x 50 cm to obtain a plant population of 33,333 per ha. The results show that agronomic traits such as plant height, fruit weight and fruit yield differed significantly from one line to another, while significant differences exist in their organoleptic properties such as percentage titratable acid, organic acids, sugars and dry matter contents. Fruit yield ranged from 11.0 Mg/ha (CLN 1462B) to 30.0 Mg/ha (Manuella). The highest correlation coefficient (r=0.99) was found in the relationship between fruit weight and fruit yield. Titratable acids was highly and positively correlated with dry matter content, citric acid and malic acid, while malic acid among others was negatively and significantly correlated with fructose, glucose and pH.**

**Key words:** Advanced tomato lines, Organoleptic properties, qualitative traits, correlation coefficients biochemical properties.

## INTRODUCTION

Tomato (*Lycopersicon esculentum* L) is one of the most important fruit vegetables globally. In Nigeria, the most acceptable variety is Roma VF and generally adaptable to Nigerian Savanna ecology (Ibrahim and Dadari, 2002). Genetic improvement of tomato is a major thrust of many research Institutes in different parts of the globe. This research effort focused mainly on development of cultivars for higher yield potential, adaptation, and market value acceptability and enhanced organoleptic properties. Agong et al. *(*2000), used multiple correlation analysis to study relationship between some biochemical parameters and fruit weight in tomato; their results showed that most of the biochemical characters were negatively correlated with fresh tomato yields, suggesting that breeding programme may have to sacrifice larger fruit size to obtain or retain better quality. Mitchell et al. (1991) and

**Table 1.** Mean squares (MS) for the agronomic traits and proximate analysis of the Tomato lines.

| Source of variation | Df | Plant height (cm) | Fruit weight (kg/plot) | Yield (Mg/ha) | Titratable acid (%) | Dry matter content (%) | Citric acid (%) | Malic acid (%) | Fructose (%) | Glucose (%) | pH |
|---|---|---|---|---|---|---|---|---|---|---|---|
| Replicate | 2 | 1.19 | 0.0007 | 0.0002 | 0.00 | 0.003 | 0.0007 | 0.0005 | 0.0005 | 0.009 | 0.027 |
| Variety | 5 | 329.15** | 0.006** | 0.007** | 0.12** | 0.31** | 0.17** | 0.15** | 0.04** | 0.013** | 0.10** |
| Error | 10 | 67.08 | 0.0005 | 0.0007 | 0.001 | 0.001 | 0.002 | 0.002 | 0.002 | 0.0026 | 0.017 |
| Total | 17 | | | | | | | | | | |

**Significant at P<0.01.

Agong et al. (1997) confirmed the use of biochemical characteristics such as electric conductivity brix %, pH value and total titratable acidity as major criteria to judge the organoleptic and processing quality of tomato. Therefore, these properties should be of the selection criteria for improving tomato for better organoleptic properties.

In Nigeria, the major tomato production constraint is bacterial wilt caused by *Pseudomoans solanacearum*. This disease is a serious yield depressing factor in the endemic zones reducing fruit yield by as much as 70%, and in critical situations absolute crop failure have been reported (Olakojo et al., 2005). Breeding efforts at the Institute of Agricultural Research and Training (IAR&T) Obafemi Awolowo University Moor Plantation and National Horticultural Research Institute Ibadan have resulted in selection of some genotypes with considerable level of tolerance to bacterial wilt. The need to improve the fruit size therefore became important for it to be acceptable to end users. On the other hand, fruit sizes have been found to be negatively correlated with important organoleptic and biochemical properties of tomato (Agong et al., 2000). In order to strike a balance between fruit yield and quality traits, this study aimed at assessing the organoleptic properties of some advanced tomato breeding lines identified to be useful as parent materials to improve the fruit size of bacterial wilt tolerant materials.

The objectives of the study therefore were: (i) to assess the organoleptic properties of the tomato lines (ii) determine relationships among the agronomic and biochemical properties and (iii) to ascertain the possibility of using them to improve both the fruit size and organoleptic quality of bacterial wilt resistant genotypes in our national breeding programme.

## MATERIALS AND METHODS

Six advanced breeding lines of tomato (*L. esculentum* L) were evaluated in 2006 at the Institute of Agricultural Research and Training Obafemi Awolowo University Moor Plantation, Ibadan for yield potential and biochemical properties. These advanced breeding lines of tomato were collected from The National Horticultural Research Institute Idi-Ishin, Ibadan and Asian Vegetable Development Centre, China.

Filed was prepared mechanically by ploughing twice harrowing once during the early season of year 2006. Transplanting was done on flat surface of soil in four–row plots measuring 5 x 8 m with a spacing of 60 x 50 cm to obtain a population density of 33,333 plants/ha. Three weeks after transplanting, 150 kg $ha^{-1}$ of (NPK: 20-10-10) fertilizer was applied in granular form round the seedlings. The experimental design was a randomized complete block design with three replications. Weeding was done twice but as at when due. The seedlings were however not staked so as to conform to the farming system practice of Nigerian tomato farmers.

From the laboratory analyses, using the procedure of Food Chemistry Practical Manual, Bruce et al. (2006) (A University of Queensland Publication), vitamin C, titratable acid (TTA) which are acids that can be neutralized with any available base, citric acid, fructose, glucose, dry matter content and percentage moisture were determined. Other data obtained included plant height (cm), fruit weight (kg/plot) and fruit yield Mg $ha^{-1}$

Data were analysed using SAS software package for analysis of variance; correlation coefficients (r) were also computed to determine relationship among the tomato biochemical characters and yield related traits at P<0.05 and P<0.01.

## RESULTS

The results of the analysis of variance (ANOVA) are presented in Table 1. Mean square (MS) for all agronomic characters such as plant height, fruit weight, fruit yield as well as biochemical properties including percentage of titratable acid, dry matter content, citric acid, malic acid, fructose, glucose and pH values were highly significant at p<0.01. This great variability suggests a wide diversity in the tomato lines and their potential for meaningful varietal development that will enhance desirable biochemical properties. These varieties also lend themselves to supporting a broad gene pool for tomato genetic conservation for varied nutritional quality and greater utilization potential.

Mean agronomic characteristics of tomato are presented in Table 2. The plant height varied from 29.0-50.0 (for P3 – 12B) to 55.13 cm (for CLN 1462B). Mean across varieties for plant height was 42.97 cm with standard error of 1.93 while P3-12B was a dwarf tomato, TAS 1-12A, DT95/7401 and DT/95/3023 were of moderate height. Manuella and CLN 1462B (temperate

**Table 2.** Mean values of agronomic characteristics of the tomato lines.

| Tomato line | Plant height (cm) | Fruit weight (kg/plot) | Fruit yield (Mg/ha) |
|---|---|---|---|
| DT/95/8023 | 31.40 | 0.23 | 15.0 |
| P3 - 12B | 29.50 | 0.20 | 13.0 |
| Manuella | 51.40 | 0.38 | 30.0 |
| TAS 1-12A | 46.66 | 0.22 | 15.00 |
| CLN 1462B | 55.13 | 0.16 | 11.0 |
| DT 95/7401 | 43.75 | 0.28 | 20.0 |
| Mean | 42.97 | 0.24 | 16.0 |
| S.E. | 1.93 | 0.003 | 2.00 |

**Table 3.** Percentage of acid and sugar by proximate analysis for the Tomato lines (*Lycopersicon esculentum*).

| Tomato line | pH | Dry matter content | Titratable acid | Citric acid | Malic acid | Fructose | Glucose |
|---|---|---|---|---|---|---|---|
| DT/95/3023 | 4.06 | 5.71 | 1.10 | 1.29 | 1.24 | 0.69 | 0.89 |
| P3-12B | 4.36 | 4.97 | 1.02 | 1.19 | 1.14 | 0.63 | 0.95 |
| Manuella | 4.16 | 5.59 | 0.99 | 1.16 | 1.11 | 0.75 | 0.96 |
| TAS 1 - 12A | 4.46 | 5.05 | 0.60 | 0.70 | 0.67 | 0.81 | 1.08 |
| CLN 1462B | 4.56 | 4.92 | 0.69 | 0.81 | 0.77 | 0.73 | 0.99 |
| DT95/7401 | 4.30 | 5.11 | 0.81 | 0.95 | 0.90 | 0.80 | 1.04 |
| Mean | 4.32 | 5.23 | 0.87 | 1.02 | 0.97 | 0.74 | 0.98 |
| S. E. (0.05) | 0.03 | 0.01 | 0.09 | 0.11 | 0.01 | 0.01 | 0.01 |

origin) appear taller with a height of 50 cm. Fruit weight per plot ranged from 0.16 to 0.38 kg/plot. Mean weight across varieties were 0.24 kg/plot. Fruit yield varied from 11.0 Mg/ha (for CLN 1462B) to 30.0 Mg/ha (for Manuella). Although, Manuella recorded the highest fruit yield, DT95/7401 was equally good with 20 Mg/ha (Table 2).

Percentage of neutralizable acids and sugar from the proximate analysis are presented in Table 3. The pH of the tomato lines though varied significantly from one line to another, it ranged from 4.06(DT/95/3023) to 4.56% (CLN 1462B). DT/95/3023 had the most acidic material with pH of 4.06. These levels of acidity though very mild are able to preserve the tomato paste for some time before further processing. The dry matter content of the tomato lines was quite high. It varied significantly between 4.92 (for CLN 1462B) to 5.59 (Manuella). Titratable acid was slightly high in DT 95/3023 and P3-12B lines with the values of 1.10 and 1.02%, respectively. The values for other lines were generally low with less than 1.0%. Significant differences also exist in citric acid content. The values ranged from 0.70 to 1.29 while DT/95/3023, P3 - 12B and Manuella were high in citric acid, TAS1 - 12A, CLN 1462B and DT 95/7601 had moderate citric acids with less than 1% (Table 3). Malic acid in this study assumed similar trend as in citric acid for all the tomato line.

Proximate analysis of sugar in this study revealed significant differences from one tomato line to another at $p<0.05$. Fructose content of the tomato was generally less than 1% in all the lines with a range of 0.63 (for P-12B) to 0.81% (for TAS1-12A). TAS 1-12A and DT 95/7401 had 0.81 and 0.80%, respectively for % fructose content (Table 3). Glucose content was high in TAS 1-12A and DT95/7401 (over 1.0%), however it reduced in other lines below 1.0%.

Based on results of chemical compositions, all tomato lines are safe for raw consumption, domestically in soup, and are suitable for tomato puree and paste production because the acid content are not injurious to human health. The results of correlation analysis are presented in Table 4. Correlation coefficient (r) among both agronomic and chemical properties were determined. Plant height was significant and positively correlated with percentage fructose and pH with coefficients (r) of 0.54 and 0.50 respectively. On the other hand, negative and significant correlations were obtained for plant height and percentage of titratable acid (-0.60, $p<0.01$) plant height and citric acid (-0.54, $p<0.05$) plant height and % malic acid (-0.56, $p<0.05$) respectively.

Fruit weight was positively and highly ($p<0.01$) correlated with fruit yield (0.99) and percentage of dry matter content (0.62), however, it was negatively ($r=0.50$) correlated with pH. Similarly, fruit yield was positively and

**Table 4.** Correlation Coefficient (r) among tomato agronomic and quality characters.

| Parameter | Plant height | Fruit weight | Fruit yield | % Titratable acid | % Dry matter content | % Citric acid | % Malic acid | % Fructose | % Glucose | PH |
|---|---|---|---|---|---|---|---|---|---|---|
| Plant height | - | 0.18 | 0.l9 | 0.60** | -0.20 | -0.54* | -0.56* | 0.54* | 0.35 | 0.50* |
| Fruit weight | | - | 0.99** | 0.33 | 0.62** | 0.34 | 0.34 | 0.26 | -0.01 | -0.50* |
| Fruit yield | | | - | 0.32 | 0.60** | 0.34 | 0.34 | 0.26 | -0.01 | -0.50* |
| % Titrable acid | | | | - | 0.67** | 0.97** | 0.98** | -0.59** | -0.67** | -0.72** |
| % Dry matter | | | | | - | 0.66** | 0.67** | -.006 | -0.45 | -0.78** |
| % Citric acid | | | | | | - | 0.95** | -0.57* | -0.63** | -0.68** |
| % Malic acid | | | | | | | - | -0.62** | -0.71** | -0.68** |
| % Fructose | | | | | | | | - | 0.61** | 0.14 |
| % Glucose | | | | | | | | | - | 0.48* |
| pH | | | | | | | | | | - |

* and ** Significance at P<0.05 and 0.01, respectively.

highly (r<0.60, p<0.01) correlated with dry matter content, but negatively (r=-0.49, p<.05) correlated with pH. Percentage titratable acid was highly and positively correlated with dry matter content (r=0.67, p, <0.01), citric acid (r=0.97-, p<0.01) and malic acid (r=0.98-, p<0.01), but was negatively correlated with fructose percentage (-0.59-, p<0.01), glucose percentage (-0.67, p<0.01) and pH (-0.72, p<0.01), respectively (Table 4).

Percentage of dry matter content was positively and highly correlated with % citric acid (0.66) and malic acid (0.67), but negatively correlated with pH (-0.78). Percentage citric acid was only positive and significantly correlated with malic acid (0.95), but negatively correlated with percentage of fructose (-0.57) glucose (-0.63) and pH (-0.68) respectively. Malic acid was negatively correlated with tomato sugar such as fructose (-0.62) and glucose percentage (-0.71, p<0.01) and pH (-0.68). Percentage of fructose on the other hand was positively and significantly correlated with glucose percentage (0.61) but glucose percentage was significantly and positively correlated with pH (0.48). While traits with strong and positive relationships may be jointly selected by breeder to enhance quality of tomato for such traits, negative correlation coefficients among tomato characters is an indication that breeder may need to adopt other breeding approach in selecting for such traits in tomato improvement programme.

## DISCUSSION

This study reveals the organoleptic properties of the advanced breeding lines of the tomato. High variability of the agronomic and organoleptic properties will no doubt provide breeders with opportunity to select desirable genotypes when breeding for higher fruit yield and quality organoleptic properties. Fulton et al. (2002) have identified a total of 222 qualitative trait loci (QTL) for 15 tomato traits including sugar, glutamic acid and flavour suggesting the possibility of increasing fruit size without necessarily sacrificing the quality organoleptic traits using molecular genetics.

Percentage of glucose and fructose in this study were very close to the values obtained by Agong et al. (2000), in their study using other tomato varieties. Similarly, sucrose synthase activity and fruit size in tomato had been found to be linearly related by Jia-Hao et al. (1992); this further suggests that sucrose synthase can serve as an indicator of sink strength in growing tomato fruits. In another study by Causse et al. (2001), overall aroma intensity was positively correlated with sweetness and sourness as well as lemon, candy and citrus fruit aroma. This implies that positive and significant correlations among glucose and fructose (r=0.61, p<0.01) as found in this study will no doubt promote good aroma in resultant genotypes.

The significant importance of tomato organoleptic properties is currently attracting international research effort. Contreras et al. (2006) studied the effects of NPK fertilization and irrigation with quality on tomato fruit quality. They observed that increase in concentration of the fertilization solution increased marketable fruits and fruit citric acid in tomato, while high acidity water produced fruit with higher total soluble solid brix, titratable acid and fruit dry matter content.

In the same vein, Baldwin et al. (1991) reported that a complex mixture of sugars, acids, amino acids, minerals and volatile compounds contributes to the characteristic flavour of fresh tomato fruits. In fact, recent scientific findings regarding tomato fruit component have encouraged effort to improve these traits genetically (Paolo et al., 1009). The report of these workers have shown that a strong interconnected nodes among tomato organoleptic properties exist for dry matter, pH and Brix, while a strong association was reported between some metabolic and sensory traits such as citric acid with tomato small, glycine with tomato smell and granularity with dry matter content. Dry matter content also showed a link with ascorbic acid and pH.

## Conclusion

The organoleptic properties of the evaluated tomato lines, relationships among quality traits and their yield potential showed that they are good parental candidates that may be used in improving bacterial wilt tolerant varieties for enhanced fruit-size. In the same vein, the tomato breeding lines could serve as sources of gene for improving tomato crop for dry matter yield and various chemical composition. For example, dry matter content can be improved in tomato using DT/95/3023 and Manuella as parent lines, while glucose and fructose can be enhanced using parent lines TAS1-12A and DT95/7401 as sources of gene for breeding program specially designed to improve these tomato sugars.

## Conflict of Interest

The author(s) have not declared any conflict of interests.

### REFERENCES


Agong SG, Schittenhelm S, Friedt W (1997). Assessment of salt tolerance in Kenyan tomato germplasm. Euphytica 95:57-66.

Agong SG, Schittenhelm S, Friedt W (2000). Genotypic variation of Kenyan tomato (*Lycopersicon esculentum* L.) germplasm. Plant Genet. Resour. Newsl. 123:61-67.

Baldwin EA, Nisperos-Carriedo MO, Baker R, Scott JN (1991). Qualitative analysis of flavour parameters in Six Florida tomato cultivars. J. Agric. Food Chem. 39:1135-1140.

Bruce RDÁ, Geoff H, Ian B (2006). Food Chemistry Practical Manual: A Publication of University of Queensland

Causse V, Saliba-Colombari I, Lesschave MB (2001). Genetic analysis of organoleptic quality in fresh market tomato 2: Mapping QTLS for sensory attributes. Theor. Appl. Genet 102:273-283.

Contreras JI, Segura L, Pascual I, Catela JJ (2006). Effect of NPK fertilization and irrigation water quality on the quality of tomato fruit. Acta Hortic. 747:60.

Fulton TM, Bucheli P, Voirol E, Lopez J, Petiard V, Tanksley SD (2002). Qualitative trait loci (QTL) affecting sugars, organic acids and other biochemical properties. Possibly contributing to flavor, identified in four advanced backcross populations of tomato. Euphytica 127:63-177.

Ibrahim R, Dadan SA (2002). Effect of age of seedling at transplanting and variety in yield and componentsof tomato (*Lycopersicon esculentum*). Niger. J. Hortic. Sci. 6:2-6

Jia-Hao S, Taddusz L, Shi-Jean SS, Clanton CB (1992). Sucrose synthate in wild tomato (*Lycapersicon ehmielwskil*) and tomato fruit sink strength. Plant Physiol. 98:1163-1169.

Mitchell JP, Shonnen C, Grattan SR, May DM (1991).Tomato fruit yields and quality under water deficit and salinity. J. Amer. Soc. Horti. Sci. 116:215-221.

Olakojo SA, Alonge VO, Obmo PM (2005). Evaluation of tomato accessions for resistance to *Pseudomonas solanacearum* (Smith) causal organism of bacterial wilt and yield in a tropical climate. Niger. J. Genet. 19:58-70.

Paolo C, Serena A, Vincenzo F, Luca T, Luigi F, Maria RE (2009). Use of network analysis to capture key traits affecting tomato organoleptic quality. Oxford J. Exp. Bot. 60(12):3379-3386.

# Upland rice response to nutrient application in Uganda

**K. C. Kaizzi[1], J. Byalebeka[1], O. Semalulu[1], I. Alou[1], W. Zimwanguyizza[1], A. Nansamba[1], E. Odama[1] and C. S. Wortmann[2]**

[1]National Agricultural Research Laboratories (NARL), Kawanda, P. O. Box 7065 Kampala, Uganda.
[2]Department of Agronomy and Horticulture, University of Nebraska-Lincoln, Lincoln, NE USA.

**Upland rice (*Oryza* spp.) yields are low in Uganda, partly because of little fertilizer use. Three on-station trials and two clusters of on-farm trials were conducted in Uganda at approximately 1000 m elevation to determine: yield response to N, P and K application; economically optimal nutrient rates (EOR); and N use efficiency components. Mean grain yield, with hulls, was 1.3 and 3.7 Mg ha$^{-1}$ with 0 and 100 kg ha$^{-1}$ N applied, respectively. Grain yield response to applied P when compared with N was less, and mean yield was not increased with K application. Depending on fertilizer cost relative to grain price (CP), mean EOR ranged from 54 to 92 kg ha$^{-1}$ N and 17 to 30 kg ha$^{-1}$ P. Equations were determined for yield response, estimation of EOR, and the benefit: cost ratio (BC) for fertilizer N and P use. Grain N concentration and N harvest index at EOR were 1.55 and 0.55 kg kg$^{-1}$, respectively. Mean recovery efficiency, partial factor productivity and agronomic efficiency declined with increasing N rate and were 0.75, 41 and 28 kg kg$^{-1}$, respectively, at the EOR. Fertilizer N and P use can be highly and moderately profitable, respectively, for upland rice production in Uganda with high N recovery and agronomic efficiency. In maximizing net return on finance-constrained investment in fertilizer use, CP and investment capacity needs to be considered.**

**Key words:** Economic, fertilizer use, nitrogen, phosphorus, potassium, use efficiency.

## INTRODUCTION

Upland rice production is less important in Uganda when compared with maize (*Zea mays* L.) and sorghum (*Sorghum bicolor* L. Moench), but rainfed rice production has doubled (FAOSTAT, 2013) during the past decade due to high market value through expansion of area sown. Smallholders are more likely to apply fertilizer to rice when compared with other cereals because of the high market value of rice. However, levels of nutrient application are low, at least partly to high costs of fertilizer use relative to the value of rice, and mean grain yield in Uganda which is estimated to be 1.5 Mg ha$^{-1}$ (Otsuka and Kalirajan, 2006). Inadequate control of numerous constraints may contribute to the low yield as found in Tanzania with biotic constraints, low input use and low availability of soil N and P constraining productivity (Mghase et al., 2010).

Currently there are no fertilizer recommendations for upland rice production in Uganda. Upland rice is produced across diverse situations in Africa and research findings for nutrient application to upland rice are mostly quite recent and indicate situation specificity but also some opportunity for generalization. Research in Uganda showed that upland rice grain yield can often be increased by more than 100% with application of N and P, and the crop is responsive to *Azolla* spp. and to a preceding green manure crop of *Mucuna pruriens* L. (Kaizzi, 2002; Kaizzi et al., 2007). Yield in Uganda was increased by 2 to 3.5 Mg ha$^{-1}$ in response to 120 kg N ha$^{-1}$ (Onaga et al., 2012). Miyamoto et al. (2012) found that paddy yield could be increased by 46 kg kg$^{-1}$ of applied N. In Benin, yields were less with no N when compared with N applied in diagnostic trials (Koné et al., 2009). In the Ivory Coast, yield was maximized with just 50 kg ha$^{-1}$ of NPK blended fertilizer 12:24:18 or with 12 kg ha$^{-1}$ urea-N applied with the low responsiveness attributed to soil water deficits during grain fill (Galabi et al., 2011). In Burkina Faso, Ethiopia, Ghana and Nigeria, upland rice response to applied N was generally curvilinear to plateau with, on average, about 90% of the grain yield increase with 100 kg N ha$^{-1}$ occurring with the first 50 kg N ha$^{-1}$ (Apaseku et al., 2013; Habtegebrial et al., 2013; Oikeh et al., 2008; Okonji et al., 2012). On the dry savannah land of northern Nigeria, however, rice grain yield increased linearly with N rates up to 90 kg N ha$^{-1}$ (Kamara et al., 2010). Koné et al. (2011) attributed situations of reduced grain yield and root development with N application to mid-season soil water deficits. Oikeh et al. (2008) determined 60 kg N ha$^{-1}$ to be optimal for upland rice production by smallholders in Nigerian forest agro-ecosystems.

A curvilinear to plateau response to applied P is also common with about 40 and 65% of the response to application of 40 kg P ha$^{-1}$ occurring, on average, with the 10 and 20 kg P ha$^{-1}$ rates (Bationo, 2008; Apaseku et al., 2013; Okonji et al., 2012). Upland rice yields were increased from 1.7 to 2.3 Mg ha$^{-1}$ with a Bray-1 soil test P of 4 mg kg$^{-1}$ with P application (Oikeh et al., 2010), and from 0.98 to 1.27 Mg ha$^{-1}$ with Bray-1 of 2 to 3 mg kg$^{-1}$ (Sahrawat, 2000) with 45 kg P ha$^{-1}$. Oikeh et al. (2008) determined 26 kg P ha$^{-1}$ to be optimal for upland rice production in Nigerian forest agro-ecosystems.

Given the inadequate information base for maximizing profit from fertilizer use on upland rice for production in Uganda, research was conducted to: 1) quantify the yield response of upland rice to N, P, and K; 2) determine economically optimal nutrient rates for N, P and K (EONR, EOPR and EOKR) at different CP; and 3) evaluate efficiency of N use by upland rice in Uganda.

## MATERIALS AND METHODS

Five N, P and K response trials were conducted in western Uganda across three cropping seasons from 2009 to 2010 (Table 1). The location and season combinations are referred to as site-seasons. Three site-seasons were conducted at the Bulindi Zonal Agricultural Research and Development Institute and two site-seasons were clusters of single sets of treatments evaluated on-farm with four or five farms per cluster providing replication. Research was carried out in the Western Mid-Altitude Farmlands (Wortmann and Eledu, 1999). Bulindi trial sites received >50 mm rainfall in the two weeks before sowing and received 430 mm or more of rainfall by 100 days after sowing (Figure 1). Rainfall was less in season 2010B when compared with other seasons with only 55 mm from 23 to 55 days after sowing.

The soils were Acric Ferralsols except for pre-dominantly Petric Plinthsol at the Kwera on-farm location (Table 1). Surface soil samples for the 0- to 20-cm depth consisting of 10 cores per site-season were collected before planting and fertilizer application to determine basic soil properties. Sand, clay and soil organic matter content ranged from 154 to 503, 298 to 506 (Bouyoucos, 1936), and 36 to 54 (Walkley and Black, 1934) g kg$^{-1}$ soil, respectively. Mehlich-3 P (Mehlich, 1984) ranged from 3.7 to 7.9 mg kg$^{-1}$ soil. Exchangeable K was always >130 mg kg$^{-1}$.

Site preparation at Bulindi included disk plowing at 15 to 20 cm depth followed by secondary disk tillage at 10 cm depth to reduce soil surface roughness. Land preparation for on-farm trials varied according to the practices of cooperating farmers but always included tillage. The previous crop and sowing dates varied (Table 1). Seeding rates were selected to achieve final plant populations of 50 plant m$^{-2}$, with two plants per point and a spacing of 20 by 20 cm. Weed control was done using hand hoes twice or thrice depending on weed intensity and labour availability. During the season, chloropyrifos 5% (Dursban$^{TM}$) was applied for control of the stem borer complex and the African rice gall midge (*Orseolia oryzivora* Harris & Gagné (Diptera: Cecidomyiidae).

The experimental design was a randomized complete block design. The nutrient rates evaluated were the following: 0 ($N_0$), 50, 100, and 150 kg N ha$^{-1}$; 0, 12.5, 25.0, and 37.5 kg P ha$^{-1}$; and 0, 30, 60, and 90 kg K ha$^{-1}$. The N-P-K treatments were 0-0-0, 50-0-0, 100-0-0, 150-0-0, 50-12.5-0, 100-12.5-0, 150-12.5-0, 50-25-0, 100-25-0, 150-25-0, 50-37.5-0, 100-37.5-0, 150-37.5-0, 50-12.5-30, 100-12.5-30, 150-12.5-30, 50-25-60, 100-25-60, 150-25-60, 50-37.5-90, 100-37.5-90, and 150-37.5-90. The incomplete factorial arrangement limited the number of treatment in consideration of Liebig's law of the minimum, proposed by J. von Liebig in 1840, expecting N and P to be the most and least limiting of the three nutrient deficiencies, respectively. The $N_0$ treatment occurred only with $P_0$ and $K_0$, and P and K effects were tested only with N applied. Similarly, no K was applied for the $P_0$ treatment. The K rates varied with P rates and the K effect was determined by subtracting at the plot level the P rate minus K treatments from the corresponding K-plus treatments after verifying that the P × K interaction was not significant.

Varieties were a sub-plot factor including Nerica-4 and Superica-1. Nerica-4 is a genotype derived from the interspecific hybridization of WAB 56-104 (*Oryza sativa*, tropical japonica type) and CG 14 [*Oryza glaberrima*]). Superica-1 (*O. sativa*) is a Ugandan release. These Uganda releases were not hybrids and each variety had a maturity period of 120 days. The plot size was 4 by 6 m.

The N, P and K sources were urea, triple super phosphate and potassium chloride, respectively. Fertilizer P was applied pre-plant. Fertilizer N and K were applied with 25% pre-plant, 25% at tiller formation and 50% at panicle initiation. The fertilizers were surface broadcast applied at planting and incorporated. The side dress application of N and K was band-applied to the side of the row and covered with soil.

The plants were cut at ground-level from the inner rows in a 1.5 x 2.0 m area and air dried for at least 3 days. The panicles were threshed and the harvested grain weight was determined. After adding the panicle remnants, the straw was weighed to determine the straw yield. The harvested grain was weighed and grain yield calculated. Grain yield was adjusted to 140 g kg$^{-1}$ water content.

**Table 1.** Characteristics of research sites at Bulindi for three seasons, and for clusters of on-farm trials conducted at Kwera and Kiziranfumbi, to determine upland rice response to applied N, P and K in Uganda.

| Site-season† | Soil properties | | | | | | Previous crop | Sowing date | Harvest date |
|---|---|---|---|---|---|---|---|---|---|
| | Sand | Clay | OM | pH | P | K | | | |
| | $g\ kg^{-1}$ | | | | $mg\ kg^{-1}$ | | | | |
| Bulindi 2009B | 154 | 506 | 46 | 5.9 | 4.9 | 418 | FL‡ | Sept. 18 -20 | Jan. 12 - 15 |
| Bulindi 2010A | 326 | 505 | 44 | 5.9 | 4.6 | 235 | CR | Mar. 17 - 19 | June 8 - 21 |
| Bulindi 2010B | 168 | 550 | 49 | 6.1 | 7.9 | 385 | CR | Aug. 23 - 25 | Jan. 9 - 11 |
| Kwera | 405 | 406 | 36 | 6.0 | 6.2 | 231 | CR | Sept. 1 - 10 | Jan 15 - 30 |
| Kiziranfumbi | 503 | 298 | 54 | 7.1 | 3.7 | 132 | CR | Sept. 1 - 10 | Jan 15 - 30 |

†The rainfall is bimodal, with planting for seasons A and B occurring in March and April and in late August and September, respectively. The latitude, longitude and elevation of the research locations, respectively, were: Bulindi, 1°30'N, 31°29'E, 1021 m, Acric Ferralsol; Kwera, 1°49'N, 32°58'E, Petric Plinthsol; and Kiziranfumbi, 1°21'N, 31°12'E, Acric Ferralsol. CR = cereal, FL = fallow.

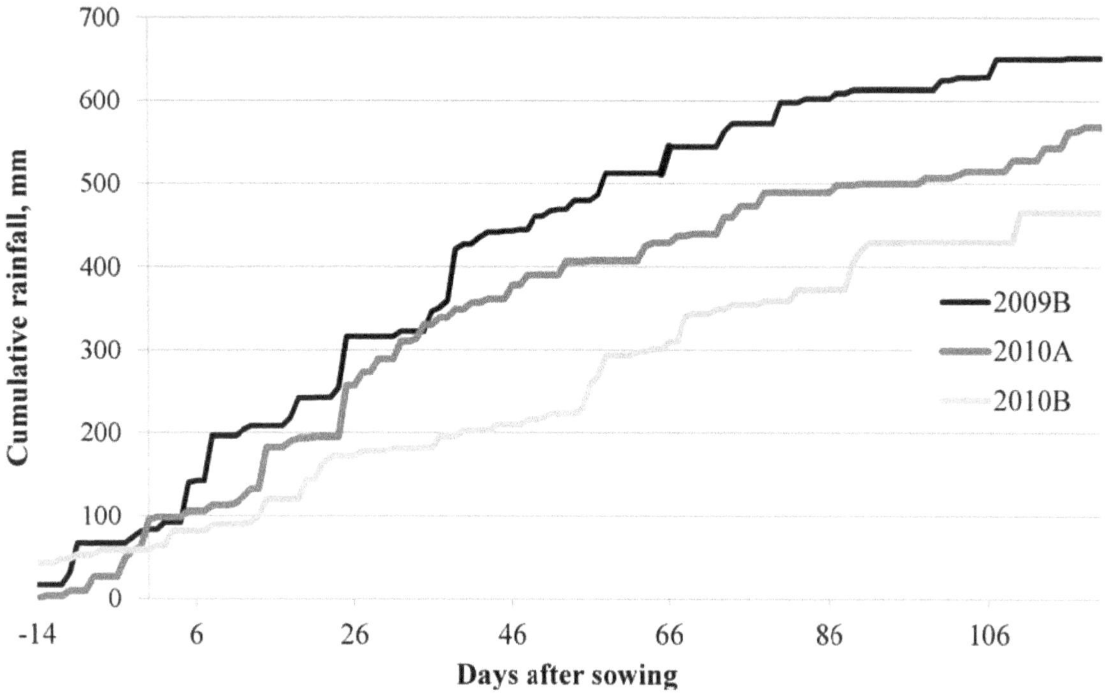


**Figure 1.** Cumulative rainfall for three cropping seasons at Bulindi, Uganda.

Grain and straw samples were oven-dried at 60°C, ground to pass a 0.5-mm sieve, and analyzed for total N in a single digest by a wet-ashing technique with colorimetric determination (Anderson and Ingram, 1993; Okalebo et al., 2002).

The data analyses were done by site-season using Statistix 9 (Analytical Software, Tallahassee, FL) with replications as random variables and varieties and nutrient rates as fixed variables. Analysis of variance combined across the Bulindi seasons was done to test for nutrient rate interactions with seasons after verifying homogeneity of variance. When significant nutrient rate effects occurred for a site-season, an asymptotic yield function was determined: grain yield ($Mg\ ha^{-1}$) = $a - bc^n$, where $a$ was near maximum grain yield, $b$ was the yield increase due to nutrient application, and $c^n$ determined the shape of the curvilinear response where $c$ was a curvature coefficient and $n$ was the nutrient rate. Regression analyses by site-season and combined across site-seasons were done with plot data. Upland rice response to applied N was determined across all P levels after verifying a lack of N x P rate interaction, and response to applied P was determined with the zero N treatment omitted from the analysis.

The EONR and EOPR, or the nutrient application rates that gave the greatest net return $ha^{-1}$ to fertilizer use, were calculated for a range of CP. A grain price of US\$0.40 $kg^{-1}$ (Uganda Sh. 2400 US$^{-1}$) was used for the economic analysis. Equations were developed using non-linear regression analysis to relate EOR to CP.

The BC was considered to be the value of increased yield relative to cost of fertilizer use for the given application rate. Polynomial functions were determined for each crop-nutrient

**Table 2.** Nitrogen application effect on upland rice grain and straw yield in Uganda. The results are the means of two varieties as there was no variety by N rate or P x N rate interaction.

| Site-season | N rate, kg ha$^{-1}$ | | | | | N rate, kg ha$^{-1}$ | | | | |
|---|---|---|---|---|---|---|---|---|---|---|
| | 0 | 50 | 100 | 150 | Pr | 0 | 50 | 100 | 150 | Pr |
| | ------- Grain yield, Mg ha$^{-1}$ ------- | | | | | ------- Straw yield, Mg ha$^{-1}$ ------- | | | | |
| Bulindi 2009B | 0.75$^{c\dagger}$ | 2.19$^{b}$ | 2.45a$^{b}$ | 2.48$^{a}$ | * | 2.74$^{c}$ | 5.26$^{b}$ | 6.06$^{b}$ | 7.04$^{a}$ | *** |
| Bulindi 2010A | 1.00$^{b}$ | 2.56$^{a}$ | 2.55$^{a}$ | 2.62$^{a}$ | * | 3.37$^{b}$ | 5.98$^{a}$ | 5.99$^{a}$ | 5.78$^{a}$ | ** |
| Bulindi 2010B | 0.99$^{b}$ | 3.27$^{a}$ | 3.56$^{a}$ | 3.36$^{a}$ | *** | 4.01$^{c}$ | 7.20$^{b}$ | 8.25$^{a}$ | 8.20$^{a}$ | ** |
| Kwera | 1.79$^{c}$ | 4.59$^{b}$ | 5.02$^{ab}$ | 5.18$^{a}$ | *** | | | | | |
| Kiziranfumbi | 1.87$^{b}$ | 4.32$^{a}$ | 4.97$^{a}$ | 4.45$^{a}$ | *** | | | | | |
| Mean$^{\ddagger}$ | 1.42$^{b}$ | 3.39$^{a}$ | 3.71$^{a}$ | 3.62$^{a}$ | *** | 3.37$^{c}$ | 6.15$^{b}$ | 6.77$^{a}$ | 7.01$^{a}$ | *** |
| SE | 0.26 | 0.53 | 0.63 | 0.59 | | 0.49 | 0.17 | 0.18 | 0.17 | |

$^{\dagger}$ Different letters in a row under grain or straw yield indicate statistically significant differences at $\alpha \leq 0.05$. *, **, *** indicate statistical significance with P = 0.95, 0.99, and 0.999, respectively. The N rate x site-season interaction was significant for straw yield due to a relatively greater increase in straw yield in season 2010B compared with the other seasons.

**Table 3.** The coefficients for the upland rice grain yield response functions to applied N and the economically optimal N rates (EONR) for different cost of fertilizer N use to grain value ratios (CP).

| Site-season | Response coefficients | | | EONR at five N:grain price ratios | | | | |
|---|---|---|---|---|---|---|---|---|
| | --- Mg ha$^{-1}$ --- | | | ------------- \$ kg$^{-1}$ (\$ kg$^{-1}$)$^{-1}$--------------- | | | | |
| | a$^{\dagger}$ | b | c | 2 | 4 | 6 | 8 | 10 |
| Bulindi, 2009B | 2.65 | 1.90 | 0.971 | 113 | 96 | 76 | 66 | 58 |
| Bulindi, 2010A | 2.59 | 1.56 | 0.945 | 67 | 55 | 47 | 43 | 38 |
| Bulindi, 2010B | 3.57 | 2.40 | 0.971 | 118 | 95 | 81 | 71 | 64 |
| Kwera | 5.59 | 3.44 | 0.979 | 113 | 106 | 92 | 85 | 80 |
| Kiziranfumbi | 4.85 | 2.81 | 0.956 | 150 | 132 | 114 | 101 | 91 |
| Combined | 3.67 | 2.40 | 0.958 | 92 | 76 | 66 | 60 | 54 |
| SE | 0.36 | 0.59 | 0.054 | | | | | |

$^{\dagger}$The response coefficients a, b, c are of the asymptotic function with yield = a -b*c$^{n}$; a is estimated yield at near plateau (Mg ha$^{-1}$), b is the yield (Mg ha$^{-1}$) increase due to nutrient application, c determines the shape of the response curve, and *n* (kg ha$^{-1}$) is the rate of applied nutrient.

combination to estimate BC with application rate and CP as independent variables. Differences and relationships were considered significant at $P \leq 0.05$.

Nonlinear functions that related total N in the aboveground biomass at harvest (UN) to the N rate and grain yield were determined. Asymptotic regression analysis, using individual plot data, related NUE properties to N rate. Exceptions were for straw N concentration and uptake which had linear and quadratic relationships to N rate, respectively, and for RE and agronomic efficiency of N use (AE) which had linear and quadratic relationships to N rate, respectively.

The NUE parameters included grain N concentration and content, HI, NHI, internal efficiency (IE) of total plant N taken up from soil and fertilizer, partial factor productivity (PFP), and physiological efficiency (PE), RE and AE for fertilizer N use (Cassman et al., 2002). The NUE components were calculated as follows: IE = *Y*/UN (kg kg$^{-1}$) where *Y* is grain yield (kg ha$^{-1}$); PFP = *Y*/N rate; NHI = grain N/UN; RE = $(UN_{+N} - UN_{N0})$/N rate; PE = $(Y_{+N} - Y_{N0})/(UN_{+N} - UN_{N0})$; and AE = $(Y_{+N} - Y_{N0})$/N rate. The units for IE, PE, AE, and PFP were kg grain kg$^{-1}$ N and kg N kg$^{-1}$ N for NHI. The effects of NHI and grain N concentration on IE were determined using linear regression analysis.

## RESULTS

The mean upland rice paddy grain yield at $N_0$ was 1.42 Mg ha$^{-1}$ and was more in on-farm trials than at Bulindi (Table 2). The predicted mean maximum grain yield was 3.67 Mg ha$^{-1}$ (Table 3), with no significant yield increase for >100 kg ha$^{-1}$ N applied. The N × P rate and N rate × site-season interactions, and the interactions with variety, were not significant for grain yield but there was a relatively greater increase in straw yield in season 2010B due to N application when compared with the other seasons.

Upland rice grain yield increased in response to N application for all site-seasons (Tables 2 and 3). The predicted overall mean grain yield increase was 2.40 Mg ha$^{-1}$. The results indicate a very high probability of response to applied N but the magnitude of the response varies. Mean yield response was greater for on-farm when compared with on-station trials but the mean curvature coefficients were similar. Grain yield was significantly increased by application of >50 kg ha$^{-1}$ N for only two of the five site-seasons but this was not sufficient to result in an N rate x site-season interaction. Combined across all site-seasons, the grain yield response to applied N was

$$Y = 3.67 - 2.40(0.958^{N}) \quad (1)$$

$$Y = 3.39 - 2.12(0.968^{N}) \text{ with no P or K applied} \quad (2)$$

The b and c coefficients of Equations 1 and 2 were not significantly different at α = 0.05 using a z-test. Upland rice straw yield was increased by 50 kg N ha$^{-1}$ for the three site-seasons where measured, with an additional increase by applying more N for two site-seasons.

Net returns to N application were positive for all site-seasons and all CPs, and the site-season EONR ranged from 38 to 150 kg ha$^{-1}$ depending on the CP (Table 3). The mean EONRs determined from the analysis combined across all site-seasons ranged from 54 and 92 kg ha$^{-1}$ with CPs of 10 and 2, respectively (Figure 2). Net returns to applied N were more sensitive to N rate as the CP increased. The mean EONR can be estimated from the CP according to:

$$EONR, kg\,ha^{-1} = 107.2 - 8.78CP + 0.339CP^{2} \quad (3)$$

The P × K rate interaction was not significant but the P rate × site-season interaction was significant for grain yield with P application, in the presence of applied N, resulting in increased upland rice grain yield for three of the five site-seasons (Table 4). Yield increased with up to 25 kg P ha$^{-1}$ for two site-seasons. In the combined analysis, grain yield was increased by 0.38 Mg ha$^{-1}$ with 12.5 kg P ha$^{-1}$. The yield response function from the combined analysis was:

$$Y = 3.79 - 0.556(0.947^{P}) \quad (4)$$

Applying Equation 4, EOPR was determined to be related to CP, with CP for fertilizer P ranging from 4 to 12, as:

$$EOPR, kg\,ha^{-1} = 53.4 - 4.69CP + 0.134CP^{2} \quad (5)$$

Straw yield was increased with 12.5 kg ha$^{-1}$ P applied at Bulindi in the 2010B season only (Table 4). Mehlich-3 P was low, ranging from 4 to 8 mg kg$^{-1}$, but there was no indication of a relationship between soil test P and grain yield response to applied P. Grain and straw yield were not affected by K application for any site-season.

The BC of fertilizer use was related to nutrient application rate and CP for ≤100 kg N ha$^{-1}$ and ≤40 kg P ha$^{-1}$ as follows:

$$Nrate \le 100\,kg\,ha^{-1} : BC_{N} = 56.52 - 0.541N - 9.983CP + 1.83E10^{-4}N^{2} + 0.425CP^{2} + 0.0326NCP \quad (6)$$

$$Prate \le 40\,kg\,ha^{-1} : BC_{P} = 10.67 - 0.172P - 0.383CP + 1.01E10^{-3}P^{2} + 0.0513CP^{2} + 0.00854PCP \quad (7)$$

The BC of fertilizer use was greater with N compared with P application, and decreased as application rate and CP increased.

Plant UN ranged from 31 to 89 kg ha$^{-1}$ for $N_0$ and with 150 kg N ha$^{-1}$ and was 79 kg ha$^{-1}$ at EONR (Table 5). Variation in UN accounted for 82 and 74% of the variation in biomass and grain yield, respectively. Mean grain yield was 14.7 kg kg$^{-1}$ UN for $N_0$ and 27.6 kg kg$^{-1}$ UN across all N rates.

Internal efficiency (IE), or the efficiency of converting UN to grain yield, is a function of NHI and grain N concentration. The linear effects of NHI and grain N concentration accounted for 72 and 11% of the variation in IE, respectively. Grain N concentration, NHI and IE at an EONR = 66 kg ha$^{-1}$ for CP = 6 were 15.5 g kg$^{-1}$, 55% and 35.4 kg grain (kg UN)$^{-1}$, respectively, which were higher than for the 0N rate (Table 5). The IE decreased with increased N rates.

Mean PFP declined with increased N rate and was 41 kg kg$^{-1}$ at EONR. Mean AE decreased with N rate and was estimated to be 28 kg kg$^{-1}$ at EONR when compared with 52 and 64 kg kg$^{-1}$ for sorghum and maize, respectively (Kaizzi et al., 2012a, b). Mean PE of fertilizer N was not affected by N rate which is consistent with the results of the sorghum and maize studies. The following equations, determined from plot data of the three site-seasons at Bulindi, represent the N rate effect on various components of NUE for upland rice in Uganda.

$$UN, kg\,ha^{-1} = 91.82 - 60.89(0.977^{N}) \quad (8)$$

$$GrainN\,concentraion, kg\,kg^{-1} = 17.50 - 3.47(0.992^{N}) \quad (9)$$

$$GrainUN, kg\,ha^{-1} = 47.60 - 34.42(0.968^{N}) \quad (10)$$

$$StoverN\,concentraton, kg\,kg^{-1} = 5.32 - 0.0060N \quad (11)$$

$$StoverUN, kg\,ha^{-1} = 17.73 - 0.356N - 0.00129N^{2} \quad (12)$$

$$HI, kg\,kg^{-1} = 0.303 - 0.691(0.613^{N}) \quad (13)$$

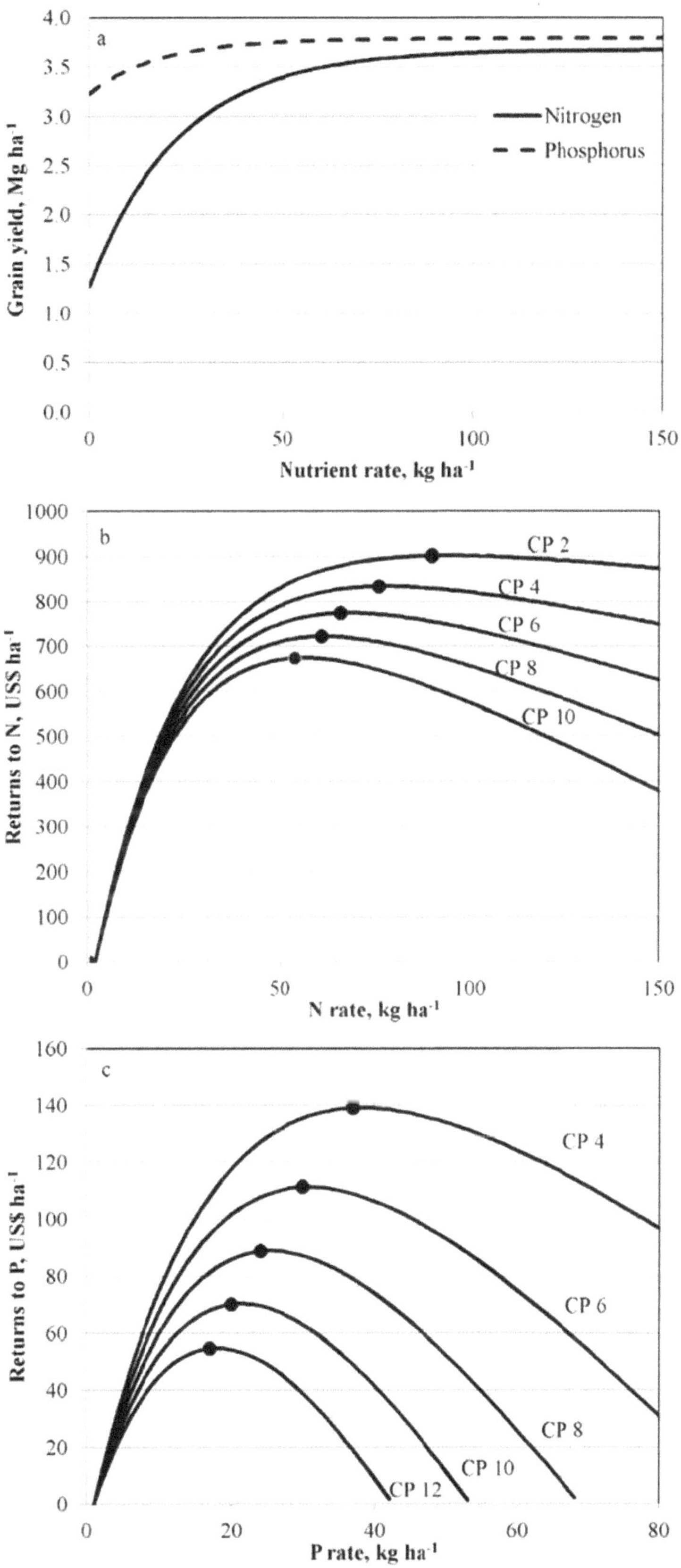


**Figure 2.** Upland rice yield response to applied N and P in Uganda with economically optimal N and P rates for different ratios of fertilizer use cost to grain price ratios (CP).

**Table 4.** Phosphorus application effect on upland rice grain and straw yield in Uganda. The results are the means of two varieties as there was no variety by N rate or P x N rate interaction.

| Site-season | P rate (kg ha$^{-1}$) | | | | | P rate (kg ha$^{-1}$) | | | | |
|---|---|---|---|---|---|---|---|---|---|---|
| | 0 | 12.5 | 25 | 37.5 | Pr | 0 | 12.5 | 25 | 37.5 | Pr |
| | ------- Grain yield, Mg ha$^{-1}$------- | | | | | ------- Straw yield, Mg ha$^{-1}$ ------- | | | | |
| Bulindi, 2009B | 2.21† | 2.25 | 2.41 | 2.55 | ns | 5.81 | 6.04 | 6.22 | 6.23 | ns |
| Bulindi, 2010A | 2.36b | 2.53b | 2.50ab | 2.80a | * | 5.68 | 5.92 | 5.86 | 6.09 | ns |
| Bulindi, 2010B | 3.03b | 3.29b | 3.34ab | 3.70a | * | 7.14b | 7.62ab | 8.18ab | 8.55a | * |
| Kwera | 4.53 | 4.73 | 4.94 | 5.22 | ns | | | | | |
| Kiziranfumbi | 4.44b | 5.66a | 4.51b | 4.75ab | * | | | | | |
| Mean‡ | 3.31c | 3.69ab | 3.67bc | 3.88a | ** | 6.21 | 6.53 | 6.75 | 6.97 | ns |
| SE | 0.50 | 0.65 | 0.51 | 0.52 | | | | | | |

†Different letters in a row under grain or straw yield indicate statistically significant differences at $\alpha \le 0.05$. ‡The P rate x site-season interaction was significant for grain yield with greater response to P rate for some site-seasons when compared with others. The grain yield response function to applied P was: Yield = $3.79 - 0.556 \times 0.947^{P}$.

**Table 5.** Mean effect of N rate on components of N use efficiency by upland rice averaged for 5 site-seasons in Uganda.

| Component | Units | N rate, kg ha$^{-1}$ | | | | Pr | EONR† |
|---|---|---|---|---|---|---|---|
| | | 0 | 50 | 100 | 150 | | |
| Grain N concentration | g kg$^{-1}$ | 14.2 | 15.6 | 16.1 | 16.8 | *** | 15.5 |
| Straw N concentration | g kg$^{-1}$ | 5.60 | 5.32 | 6.08 | 6.05 | *** | 5.6 |
| Grain N content | kg ha$^{-1}$ | 13.2 | 41.4 | 46.7 | 47.9 | *** | 43.6 |
| Straw N content | kg ha$^{-1}$ | 18.0 | 32.3 | 41.6 | 41.2 | *** | 35.6 |
| Plant N content | kg ha$^{-1}$ | 31.2 | 73.7 | 88.3 | 89.2 | *** | 78.7 |
| Harvest index | kg kg$^{-1}$ | 0.24 | 0.31 | 0.30 | 0.30 | *** | 0.30 |
| N harvest index | kg kg$^{-1}$ | 0.44 | 0.57 | 0.54 | 0.54 | *** | 0.55 |
| Recovery efficiency | kg kg$^{-1}$ | | 0.85 | 0.57 | 0.38 | *** | 0.75 |
| Agronomic efficiency | kg kg$^{-1}$ | | 34.4 | 19.7 | 12.7 | *** | 28.1 |
| Internal efficiency | kg kg$^{-1}$ | 31.0 | 36.5 | 33.6 | 32.6 | *** | 35.4 |
| Partial factor productivity | kg kg$^{-1}$ | | 52.8 | 28.9 | 18.9 | *** | 41.4 |
| Physiological efficiency | kg kg$^{-1}$ | | 40.3 | 37.5 | 32.6 | ns | |

†The EONR was 66 kg ha$^{-1}$ for a fertilizer N use cost to farm-gate grain price ratio of 6. ns, no significant effect at $\alpha \le 0.05$. *** Significant effect at $\alpha \le 0.001$.

$$NHI, kg\,kg^{-1} = 0.547 + 0.106 - 0.106(0.609^{N}) \quad (14)$$

$$IE, kg\,kg^{-1} = 23.63 + 0.0797N - 0.000570N^{2} \quad (15)$$

$$AE, kg\,kg^{-1} = 57.6 - 0.564N + 0.00177N^{2} \quad (16)$$

$$RE, kg\,kg^{-1} = 104.7 - 0.446N \quad (17)$$

$$PFP, kg\,kg^{-1} = 12.9 + 101(0.981^{N}) \quad (18)$$

## DISCUSSION

The mean yield of 3.67 Mg ha$^{-1}$ is not high relative to genetic potential of upland rice indicating the importance of other abiotic and biotic constraints in addition to deficiencies of N, P and K. Inadequate rainfall was not an apparent constraint to grain yield in the Bulindi trials as yield was highest in the 2010B season when rainfall was least and less well-distributed when compared with 2009B and 2010A. Variety performance across N rates was similar. This was in agreement with Onaga et al. (2012) who evaluated more varieties and reported a variety x N rate interaction but which accounted for less than 1% of the N rate main effect on variation in yield.

The indigenous soil N supply was apparently low as indicated by a mean grain yield of 1.42 Mg ha$^{-1}$ and 31 kg ha$^{-1}$ UN with $N_0$, even though SOM was always >36 g kg$^{-1}$. In comparison, mean grain yield was about 2 Mg ha$^{-1}$ across several other studies in Africa (Apaseku et al., 2013; Bationo, 2008; Habtegebrial et al., 2013; Oikeh et al., 2008; Okonji et al., 2012). For further consideration of

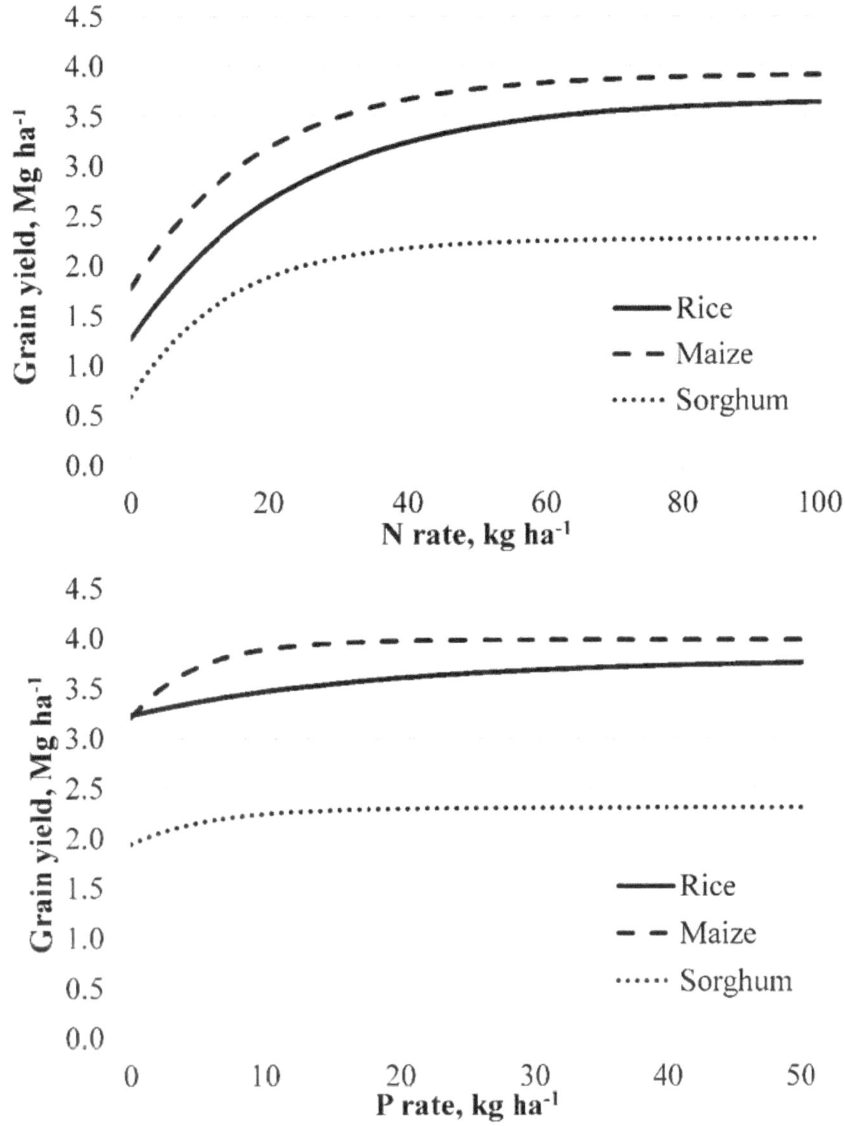


**Figure 3.** The N and P grain yield response functions of upland rice when compared with maize and grain sorghum in Uganda (Kaizzi et al., 2012a, b).

these agricultural soils of Uganda to supply N, UN with $N_0$ was 31 and 46 kg ha$^{-1}$ by sorghum and maize, respectively (Kaizzi et al., 2012a, b).

The large response of upland rice grain yield to N application was consistent with earlier results from Uganda (Kaizzi, 2002; Kaizzi et al., 2007; Onaga et al., 2012), and with the responses of maize and sorghum in Uganda (Figure 3) (Kaizzi et al., 2012a, b). The response to applied N was about twice the mean for that of several studies cited above (Apaseku et al., 2013; Habtegebrial et al., 2013; Oikeh et al., 2008; Okonji et al., 2012). Yield with N applied and at $N_0$ was more and less for maize and sorghum, respectively, when compared with upland rice. The coefficient values for *b* of 2.14 and 2.40 and for *c* of 0.94 and 0.96 for maize and upland rice, respectively, were not significantly different resulting in similar curve shape although a higher yield plateau for maize when compared with upland rice.

Upland rice response to applied P, with N applied, was near linear and to a higher P rate when compared with maize and sorghum which had little response beyond 10 kg ha$^{-1}$ P. The significant but small increase in upland rice yield to P application, with N applied, was generally consistent with the responses reported by Oikeh et al. (2008) and Sahrawat (2000) but low and high when compared with results of Bationo (2008) and Apaseku et al. (2013), respectively.

Depending on CP, the mean EONR and EOPR for upland rice ranged from 54 to 92 and 17 to 37 kg ha$^{-1}$, respectively. Finance-constrained farmers commonly do not have enough money to apply fertilizer at EOR to maximize net returns ha$^{-1}$ for all their cropland. Optimizing the choice of crop-nutrient-rate combinations, in consideration of CPs, is needed to maximize net returns on their constrained investment. Considering 14 other crop nutrient combinations in Uganda using common CP values, Kaizzi et al. (2012c) found that the decreasing order of BC for fertilizer nutrients applied at EOR was bean (*Phaseolus vulgaris* L.) N > groundnut (*Arachis hypogaea* L.) P = maize N = soybean (*Glycine max* L.) P > sorghum N > groundnut K > maize P > bean P > soybean K and > K applied to maize or sorghum. The BC for N applied to upland rice at EOR is greater than for any of the above and the BC for applied P to upland rice is less and more when compared with soybean P and sorghum N. This order could change with changes in relative CP due to changed grain values or nutrient costs. The upland rice EONR allows for BC >2, but BC >2 at EOPR only if CP ≤4 for fertilizer P. The results demonstrate that N application to upland rice can increase farm productivity with high profitability and should have priority over the 14 other crop-nutrient response functions evaluated here and in Kaizzi et al. (2012a, b, c) when fertilizer use is financially constrained. Phosphorus application for upland rice can be profitable when CP is not too high and/or when applied as less than EOPR.

Recovery of applied N in the aboveground upland rice biomass was 75% at EONR which is intermediate between the RE reported for maize and sorghum (Kaizzi et al., 2012a, b). The RE of sorghum was >100% as sorghum performance was poor at $N_0$ and applied N apparently boosted plant vigor and root growth to recover nitrate-N that was otherwise lost to leaching beyond the root zone. Other fertilizer N use efficiency components were low at EONR for upland rice as compared to maize and sorghum including AE and PFP, but this is largely due to the higher EONR of upland rice associated with the higher market value of rice when compared with maize and sorghum. Agronomy efficiency for upland rice from this Uganda study with 50 kg N $ha^{-1}$ applied was 34.4 when compared with a mean of 18 kg $kg^{-1}$ calculated from several other studies (Apaseku et al., 2013; Habtegebrial et al., 2013; Oikeh et al., 2008; Okonji et al., 2012) implying relatively high efficiency and potential for high net returns to N applied to upland rice in Uganda.

## Conclusion

Upland rice yield increased by 178% with N applied at the EONR for a CP of 6. Yield can be further increased with P application but the results demonstrate that N deficiency is much more limiting than P or K deficiency. The results indicate that N application for upland rice production is highly profitable and a priority fertilizer application option relative to 14 other crop-nutrient options of finance-constrained smallholder farmers in Uganda. Application of P is also likely to be profitable when applied at EOR, but less so compared to N, while K application is not likely to be profitable. The recovery and agronomic efficiencies of applied N are high for EONR or lower rates, implying little residual effect for the following crop but also little loss to the environment.

## Conflict of Interests

The author(s) have not declared any conflict of interests.


## ACKNOWLEDGEMENTS

The authors are grateful to the Alliance for a Green Revolution in Africa (AGRA) and the Government of Uganda for funding the study. The research was also partly supported by the Hatch Act and the U.S. Agency for International Development under the terms of Grant No. LAG-G-00-96-900009-00. The excellent cooperation of participating farmers, field assistants and village based facilitators was essential for good and efficient implementation of the many trials.


**Abbreviations: AE,** Agronomic efficiency of N use; **BC,** benefit/cost ratio; **CP,** fertilizer use cost to grain price ratio; **EONR,** economically optimal nitrogen rate; **EOPR,** economically optimal P rate; **EOKR,** economically optimal K rate; **EOR,** economically optimal rate; **HI,** harvest index; **IE,** internal efficiency; **$K_0$,** no K applied; **NHI,** N harvest index; **NUE,** N use efficiency; **$N_0$,** no N applied; **PE,** physiological efficiency; **PFP,** partial factor productivity; **$P_0$,** no P applied; **RE,** recovery efficiency; **SOM,** soil organic matter; **UN,** total N in the aboveground biomass at harvest.

## REFERENCES


Anderson JM, Ingram JSI (1993). Tropical soil biology and fertility: A handbook of methods. CAB Int. Wallingford, UK.

Apaseku JA, Mbatchou VC, Dogbe W, Dawda S (2013). Lowland NERICA and some sativa line rice responses to nitrogen and phosphorus in Guinea Savanna agroecoystem. Int. J. Res. Chem. Environ. 3:295-307.

Bationo A (2008). Integrated soil fertility management options for agricultural intensification in the Sudano-Sahelian Zone of West Africa. p. 114.

Bouyoucos GJ (1936). Directions for making mechanical analysis of soils by the hydrometer method. Soil Sci. 42:225-230. doi:10.1097/00010694-193609000-00007.

Cassman KG, Dobermann A, Walters DT (2002). Agroecosystems, nitrogen use efficiency, and nitrogen management. Ambio. 31:132-140.

FAOSTAT (2013). Food and Agriculture Organization, Rome. Available at http://faostat.fao.org (verified 4 March 2013).

Galabi TJ, Camara M, Yao-Kouame A, Keli ZJ (2011). Profitability of mineral fertilizers in upland rice tray in Gagnoa area in central western Ivory Coast. J. Appl. Biosci. 46:3153-3162.

Habtegebrial K, Mersha S, Habtu S (2013). Nitrogen and sulphur fertilizers effects on yield, nitrogen uptake and nitrogen use efficiency of upland rice variety on irrigated Fulvisols of the Afar region, Ethiopia. J. Soil Sci. Environ. Manag. 4:62-70.

Kaizzi CK (2002). The potential benefit of green manures and inorganic fertilizers in cereal production on contrasting soils in eastern Uganda. Ph. D Thesis. Centre for Development Research. Ecology and Development Series No. 4. Cuvillier Verlag Gottingen.

Kaizzi CK, Byalebeka J, Semalulu O, Alou I, Zimwanguyizza W, Nansamba A, Musinguzi P, Ebanyat P, Hyuha T, Wortmann CS (2012c). Optimizing smallholder returns to fertilizer use: bean, soybean and groundnut. Field Crops Res. 127:109-119.

Kaizzi CK, Ssali H, Nansamba A, Vlek PLG (2007). The potential benefits of Azolla, velvet bean (*Mucuna pruriens*) and N fertilizers in rice production under contrasting systems of eastern Uganda. In: Bationo A, Kihara J, Kimetu J, Waswa B (eds) Advances in Integrated Soil Fertility Research in Sub-Saharan Africa: Challenges and Opportunities. Springer, The Netherlands, pp. 423-433.

Kaizzi KC, Byalebeka J, Semalulu O, Alou I, Zimwanguyizza W, Nansamba A, Musinguzi P, Ebanyat P, Hyuha T, Wortmann CS (2012a) Maize response to fertilizer and nitrogen use efficiency in Uganda. Agron. J. 104:73-82.

Kaizzi KC, Byalebeka J, Semalulu O, Alou I, Zimwanguyizza W, Nansamba A, Musinguzi P, Ebanyat P, Hyuha T, Wortmann CS (2012b). Sorghum response to fertilizer and nitrogen use efficiency in Uganda. Agron. J. 104:83-90.

Kamara AY, Ekeleme F, Omoigui LO, Oikeh SO, Chikoye D, Tegbaru A (2010). Response of upland rice cultivars to nitrogen fertilizer in the savannas of Nigeria. Agron. J. 102:333-339.

Koné B, Amadji GL, Aliou S, Diatta S, Akakpo C (2011). Nutrient constraint and yield potential of rice on upland soil in the south of the Dahoumey gap of West Africa. Arch. Agron. Soil Sci. 57:763-774.

Koné B, Amadji GL, Igue AM, Ogunbayo A (2009). Rainfed upland rice production on a derived savannah soil in West Africa. J. Anim. Plant Sci. 3:156-162.

Mehlich A (1984). Mehlich 3 soil test extractant: A modification of Mehlich 2 extractant. Commun. Soil Sci. Plant Anal. 15:1409-1416. doi:10.1080/00103628409367568

Mghase JJ, Shiwachi H, Nakasone K, Takahashi H (2010). Agronomic and socio-economic constraints to high yield of upland rice in Tanzania. Afr. J. Agric. Res. 5:150-158.

Miyamoto K, Maruyama A, Haneishi Y, Matsumoto S, Tsuboi T, Asea G, Okello S, Takagaki M, Kikuchi M (2012). NERICA cultivation and its yield determinants: the case of upland rice farmers in Namulonge, Central Uganda. J. Agric. Sci. 4:120-135.

Oikeh SO, Nwilene F, Diatta S, Osiname O, Touré A, Okeleye KA (2008). Responses of upland NERICA rice to nitrogen and phosphorus in forest agroecosystems. Agron. J. 100:735-741.

Oikeh SO, Somado EA, Sahrawat KL, Toure A, Diatta S (2010). Integrated soil fertility management involving promiscuous dual-purpose soybean and upland NERICA enhanced rice productivity in the savannas. Nutr. Cycl. Agroecosyst. 88:29-38.

Okalebo JR, Gathua KW, Woomer PL (2002). Laboratory methods of soil and plant analysis: a working manual. 2nd ed. TSBF-CIAT, Nairobi, Kenya.

Okonji CJ, Okeleye KA, Oikeh SO, Aderibigbe SG, Nwilene FE, Ajayi O, Oyekanmi AA (2012). Rice yield and yield components as influenced by phosphorus and nitrogen application rates in the Moist Savanna of West Africa. J. Agric. Sci. Tech 2:1214-1223.

Onaga G, Asea G, Lammo J, Kikafunda J, Bigirwa G (2012). Comparison of response to nitrogen between upland NERICAs and ITA (*Oryza sativa*) rice varieties. J. Agric. Sci. 4:197-205.

Otsuka K, Kalirajan K (2006). Rice Green Revolution in Asia and its transferability to Africa. Dev. Econ. 44:107-122. http://dx.doi.org/10.1111/j.1746-1049.2006.00009.x

Sahrawat KL (2000). Determining fertilizer phosphorous requirement for upland rice. Commun Soil Sci. Plant Anal. 31:1195-1208.

Walkley A, Black IA (1934). An examination of the Degtjareff method for determining soil organic matter, and a proposed modification of the chromic acid titration method. Soil Sci. 37:29-37. doi:10.1097/00010694-193401000-00003

Wortmann CS, Eledu CA (1999). Uganda's agroecological zones: a guide for planners and policy makers. CIAT, Kampala, Uganda.

# Acute toxicity effects of dichloromethane fraction of ethanol extract of stem bark of *Piliostigma reticulatum* on rats

**Kassim DOSSO[1] , Benoit Banga N'GUESSAN[1,2], Boua Narcisse GNANGORAN[1] and Angoué Paul Yapo[1]**

[1]Laboratory of Physiology, Pharmacology and Phytotherapy, UFR SN, University of Abobo-Adjamé, 02 BP 801 Abidjan 02, Côte d'Ivoire.
[2]Department of Pharmacology and Toxicology, School of Pharmacy, College of Health Sciences, University of Ghana, P.O Box KB 52, Korle-Bu, Accra, Ghana.

**The aim of this study was to investigate the oral acute toxicity effects of dichloromethane fraction of ethanol extract of stem bark of *Piliostigma reticulatum* on rats using conventional methods. The rats were divided in nine groups. The control group received normal saline. The others groups received the order of doses of 800, 1000, 2000, 3000, 4000, 5000, 6000 and 7000 mg/kg body weight. However, the rats showed signs of immobility with the dichloromethane fraction at doses of 5000 mg/kg body weight and mortality at doses of 2000 mg/kg body weight. The LD50 was 3000 mg/kg body weight and the quotient LD5/LD95 was 0.17. This value shows that the toxic dose and the therapeutic dose were not distant. Phytochemical screening revealed the presence of tannins, flavonoids, polyphenols and reducing sugars in the stem bark of *P. reticulatum*. These results suggest that dichloromethane fraction of the ethanol extract of stem bark of *P. reticulatum* could be used with some degree of safety especially by oral route.**

**Key words:** Acute, toxicity, stem, bark, *Piliostigma reticulatum*, rats.

## INTRODUCTION

Hundreds of years ago, plants were known to play mainly nutritional roles. Today they play an additional important function in treatment of diverse pathologies. Plants, although pharmacologically active, are used without knowledge of toxicological parameters. For this reason, the World Health Organisation approved the use of herbal products for national policies and drug regulatory measures in order to strengthen research and evaluation of the toxicity of these products (Saxena, 2001). Thus, the toxicity of certain plants has been shown in many studies. For example, the acute toxicity of *Anogeissus leiocarpus, Daphnia magna* and *Cansjera rheedii* has been studied respectively by Agaie et al. (2007), Altindag et al. (2008) and Mounnissamy et al. (2010).

*Piliostigma reticulatum* which is the object of our study is traditionally used in Côte d'Ivoire for diarrhoea including bacterial diarrhoea (Yelemou et al., 2007; Dosso et al., 2012). Thus Babajide et al. (2008) have shown that the piliostigmol (6-C-methyl-2-p-hydroxyphenyloxychromonol), a substance isolated in *P. reticulatum,* inhibited *Escherichia coli* with a MIC equal to 2.57 µg/mL. In our previous study, results showed that an ethanol extract of the stem bark of *P. reticulatum* significantly reduced the gastrointestinal transit, the number, volume and weight of faeces of rats (Dosso et al., 2012). Also, the extract was not toxic in rats. In the present study, the aim of the current investigation was to determinate the toxic effects of the dichloromethane fraction of a crude ethanol extract of the stem bark of *P. reticulatum* on rats as well as to screen its phytochemical constituents.

## MATERIALS AND METHODS

### Plant collection

Stem barks of *P. reticulatum* (DC.) Horscht (Ceasalpiniaceae) were collected in Abidjan (South region of Côte d'Ivoire) in October 2007. The plant was identified and authenticated by Pr AKE-ASSI Laurent. A voucher specimen (No. 18033) of the plant has been deposited in the herbarium of the National Centre of Floristic of University of Cocody-Abidjan.

### Preparation of dichloromethane fraction

Stem barks of *P. reticulatum* were washed with distilled water, cleaned, cut into smaller pieces and kept at room temperature for two weeks. Then they were ground into a fine powder. The powder (100 g) was extracted with two liters of a solution of ethanol (96%) / water (80:20) for 24 h under constant stirring (this operation was repeated twice).

The extract was filtered twice through cotton wool, then through Whattman filter paper (No 1). The filtrate was evaporated to dryness using a rotavapor (Buchi R110/NKE6540/2) at 45°C, and dried under reduced pressure. Percentage yield was found to be 13.6%.

After successive liquid-liquid fractionations, five fractions (heptane, dicloromethane, ethyle acetate, butanol and water) were obtained from the crude ethanol extract (Harborn, 1984; Samsam-Shariat, 1992).

### Animals

Healthy young adult albino Wistar rats (weighing 150-200 g) of both sexes that were provided by UFR Biosciences (University of Cocody-Abidjan, Côte d'Ivoire) were housed in standard metal cages. They were kept under standard laboratory temperature conditions one week before the experiments for acclimation. The animals were fed with a diet of commercial pellets (Ivograin®, Abidjan, Côte d'Ivoire) and were given water *ad libitum*. They were deprived of food for at least 18 h prior to experiments but allowed free access to drinking water. The equipment usage and handling and sacrificing of the animals were in accordance with the European Council legislation 87/609/EEC for the protection of experimental animals (Mitjans, 2008).

### Acute toxicity studies

Fifty forth albino rats were divided into nine groups of six rats each and were given graded doses (800, 1000, 2000, 3000, 4000, 5000, 6000 and 7000 mg/kg body weight) of the fraction by oral route. The rats were observed for signs of toxicity and death over a period of 72 h as described by Lorke (1983). The first group received single oral dose of 2 ml of normal saline through the same route. The $LD_5$, $LD_{50}$ and $LD_{95}$ of the fraction were calculated using the arithmetic method of Karber as modified by Aliu and Nwude (1982).

### Phytochemical analysis of the fraction

The dichloromethane fraction was screened for the presence of tannins, flavonoids, alkaloids, sterols, saponins, polyphenols, polyterpenes and anthraquinones. Detection of these constituents was performed according to Bekro et al. (2007).

## RESULTS

### Phytochemical analysis of the fraction

Phytochemical screening tests of dichloromethane fraction for various constituents revealed the presence of major components such as tannins, polyphenols and flavonoids. Reducing sugars were present, and anthraquinones, alkaloids, coumarins, polyterpenes and sterols were absent.

### Acute toxicity studies

The results in the Table 1 show that with the dichloromethane fraction at doses of 800 to 4000 mg/kg body weight, the survival animals are mobile. At the dose of 5000 mg/kg body weight, we observed the immobility of rats. Also the animals began to die at dose of 2000 mg/kg body weight. At dose of 1000 mg/kg body weight, we had 0% of mortality. The $LD_{50}$ (50% of mortality) is 3000 mg/kg body weight. Also, at the dose of 7000 mg/kg body weight, 100% of animals did not survive with dichloromethane fraction (Tables 1 and 2). The $DL_5$ and $DL_{95}$ are respectively 1118 and 6425 mg/kg body weight (Figure 1). The quotient $DL_5$ / $DL_{95}$ reports 0.17. This result is far from 1.

## DISCUSSION

The study of the acute toxicity of dichloromethane fraction showed that the $LD_{50}$ is equal to 3000 mg/mL body weight. Animals showed toxicity signs in a dose-dependent manner. Also, at a dose of 7000 mg/mL, there were no survivors. The results obtained with the dichloromethane fraction, are consistent with those of Diallo and

**Table 1.** Acute toxicity signs observed in rats after treatment with dichloromethane fraction.

| Dose (mg/kg body weight ) | Acute toxicity signs | |
|---|---|---|
| | mobility | Death |
| 800 | + | – |
| 1000 | + | – |
| 2000 | + | + |
| 3000 | + | + |
| 4000 | + | + |
| 5000 | – | + |
| 6000 | – | + |
| 7000 | – | + |

**Table 2.** Percentage of mortality in rats treated with dichloromethane fraction.

| Group | Number of animals | Dose (mg/kg body weight) | Number of death | Mortality (%) |
|---|---|---|---|---|
| 1 | 6 | 800 | 0 | 0 |
| 2 | 6 | 1000 | 0 | 0 |
| 3 | 6 | 2000 | 2 | 33.33 |
| 4 | 6 | 3000 | 3 | 50 |
| 5 | 6 | 4000 | 4 | 66.67 |
| 6 | 6 | 5000 | 4 | 66.67 |
| 7 | 6 | 6000 | 5 | 83.33 |
| 8 | 6 | 7000 | 6 | 100 |

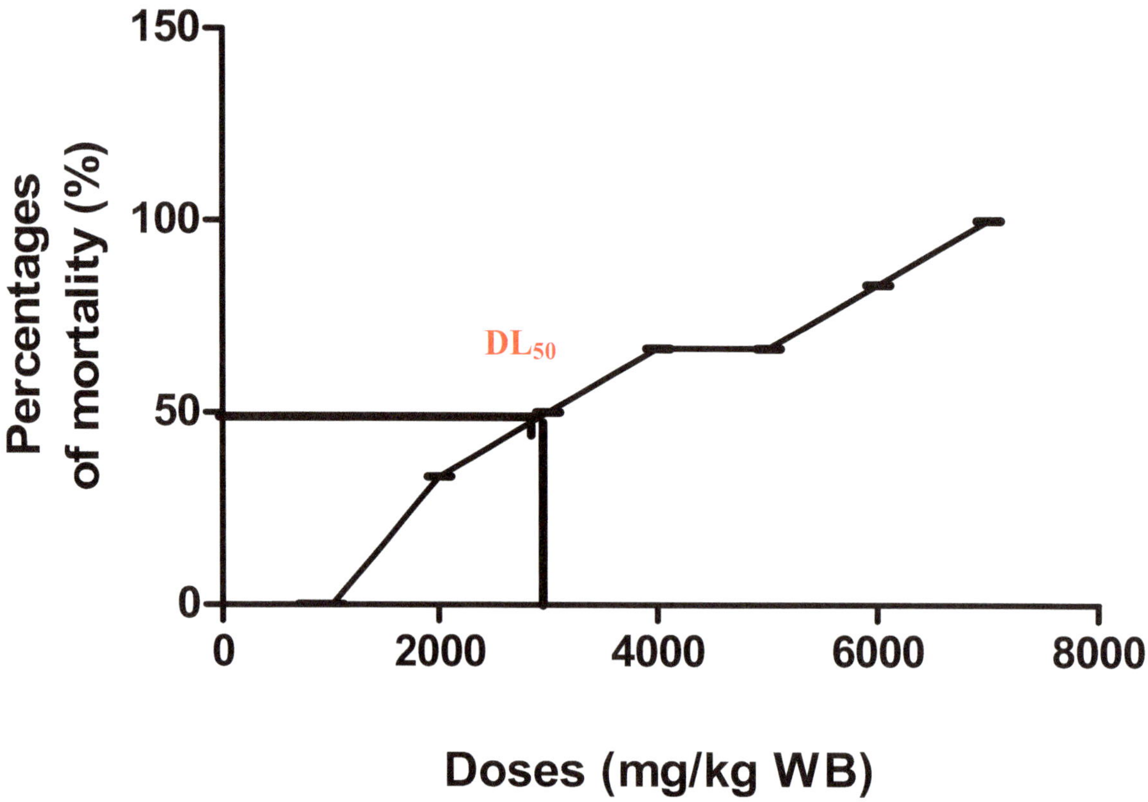


**Figure 1.** Evolution of mortality percentage in rats treated with dichloromethane fraction.

Diouf (2002) which showed that the $LD_{50}$ of the aqueous extract of *Piliostigma reticulatum* leaves in rats was 1700 mg/mL body weight. The leaves of this plant therefore center more toxic substances in the leaves than stem bark. The $LD_{50}$ obtained with our active fraction is higher than that obtained with the aqueous extract of the leaves. Agaie et al. (2007) studied the acute toxicity of aqueous extract of leaves of *Anogeissus leiocarpus* rats and was near 1200 mg/kg body weight like $LD_{50}$. This plant is more toxic than that used in our studies with a low $LD_{50}$. The study of the toxicological properties of the latex of *Calotropis procera* yielded approximately a $LD_{50}$ like 2611.75 mg/kg body weight (Lohoues et al., 2006). The latex of *Calotropis procera* focuses more toxic substances than the dichloromethane fraction.

The quotient $DL_5$ / $LD_{95}$ (0.17) is far from 1. These results indicate that the use of this fraction is not safe in its therapeutic use with the understanding that the therapeutic dose is not distinct from its toxic dose (Tamboura et al., 2005). These results are contrary to those obtained by Lohoues et al. (2006). According to these authors, the $DL_5$ / $DL_{95}$ quotient in the study of the acute toxicity of the latex of *C. procera* was 0.71 - 1. The therapeutic dose of this product is distinct from the toxic dose. The latex of this plant is probably non toxic.

After administration of the fraction of plant, immobility and death of animals are observed at high doses. These clinical manifestations had been also reported by Lohoues et al. (2006) in the study of the acute toxicity of the latex of *C. procera* in Côte d'Ivoire.

The phytochemical screening of dichloromethane fraction showed the presence of flavonoids, tannins, polyphenols and reducing sugars. Our tests have not revealed the presence of alkaloids, quinones, coumarins, saponins, sterols and polyterpenes. Phytochemical studies by Kubmarawa et al. (2007) on the aqueous extract of the roots of *P. reticulatum* showed the presence of chemical compounds such as saponins, tannins and flavonoids. Contrastingly, chemical compounds in the same plant could be explained by the different organs used in the studies or the difference of natural environments. Thus, we observed a variation of secondary metabolites in the same plant according the living environment (Sofowora, 1996).

The active compounds observed in the stem bark of *P. reticulatum* could explain the toxicity signs in rats. Thus, when in excess in the body, certain chemical compounds may exceed their therapeutic activities, incite some malfunctions or lethal disorders (Lohoues et al., 2006). For example, astringent tannins have a role in reducing foods in animals (Alldredge, 1993; Agaie et al., 2007). This causes weakness in animals leading to immobility. Polyphenols are endowed with surfactant and hemolytic properties (Lohoues et al., 2006).

Thus, according to the same authors, compounds that are swallowed or even inhaled, are known to cause, among others, digestive tract burns, cyanosis, hypoxia and seizures.

Schultz and Riggin in 1985 stated in this regard that the phenol poisoning lead to death by acute respiratory failure. These could have caused immobility and death of the animals in our study.

## Conclusion

Our results show that the stem bark of *P. reticulatum* contains toxic natural substances. These could be due to chemical constituents observed in dichloromethane fraction like tannins, flavonoids, reducing sugars and polyphenols.

It will be necessary to eliminate the probable toxic substances in dichloromethane fraction. Thus, the fraction could be used to treat diarrhea without risk. Also, the use of aerial parts of the plant could represent an alternative to the utilization of its roots, therefore limiting the biodiversity degradation.

## Conflict of Interests

The author(s) have not declared any conflict of interests.

## ACKNOWLEDGEMENTS

This project was supported in part by the Non-Governmental Organisation for the Promotion of Scientific Research in African Traditional Medicine (NGO "PRORESMAT"). The authors wish to thank Professor Ake-Assi Laurent of the National Centre of Floristic, University of Cocody-Abidjan for botanical identification of the plant, Doctor AMOATENG Patrick of Department of Pharmacology and Toxicology, University of Accra (Ghana), for arranging the graph.

## REFERENCES

Agaie BM, Onyeyili PA, Muhammad BY, Ladan MJ (2007). Acute toxicity effects of the aqueous leaf extract of *Anogeissus leiocarpus* in rats. Afr. J. Biotech. 6(7):886-889.

Alldredge J (1993). The effect of condensed tannins on browsers and grazers: Quantitative and Qualitative defense? Colorado State University, Fort Collins. Colorado, p.7.

Aliu YO, Nwude N (1982). Phytochemistry and preliminary toxicity studies of the methanol extract of the stem bark of *Garcinia kola* (Heckel). Vet. Pharmacol. Toxicol. Exp. 1st ed. pp. 104-110.

Altında A, Ergönül MB, Yigit S, Baykan Ö (2008). The acute toxicity of lead nitrate on *Daphnia magna* Straus. Afr. J. Biotech. 7(23):4298-4300.

Babajide OJ, Babajide OO, Daramola AO, Mabusela WT (2008). Flavonols and an oxychromonol from *Piliostigma reticulatum.* Phytochem. 69(11):2245-2250.

Bekro Y, Bekro J, Boua BB, Tra Bi F, Ehile EE (2007). Etude ethnobotanique et screening phytochimique de *Caesalpinia benthamiana* (Baill.) Herend et Zarrucchi (Caesalpiniaceae). Sci. Nat. 4(2):217-225.

Diallo B, Diouf A (2002). Odonto-stomatologie. Tropicale. 92:5-11.

Dosso K, N'guessan BB, Bidie AP, Gnangoran BN, Méité S, N'guessan D, Yapo AP, Ehilé EE (2012). Antidiarrhoeal activity of an ethanol extract of the stem bark of *Piliostigma reticulatum* (Caesalpiniaceae) in rats. Afr. J. Tradit. Complement. Altern. Med. 9(2):242-249.

Harborn JB Phytochemical methods (1984). In: A Guide to ModernTechniques of Plant Analysis. 2nd ed. Chapman and Hall. London pp. 4-7.

Kubmarawa D, Ajoku GA, Enwerem NM, Okorie DA (2007). Preliminary phytochemical and antimicrobial screening of 50 medicinal plants from Nigeria. Afr. J. Biotech. 6(14):1690-1696.

Lorke D (1983). A new approach to practical acute toxicity testing. Arch. Toxicol. 54: 275-287.

Lohoues EE, Tiarou GG, Monde AA, Djinhi J, Sess ED (2006). Proprietes toxicologiques aigues du latex de *Calotropis procera* chez le rat. Pharm. Méd. Trad. Afr. 14:187-199.

Mitjans M, Garcia L, Marrero E,Vinardell MP (2008). Study of ligmed-A, an antidiarrheal drug based on liguin, on rat small intestine enzyme activity and morphometry. J. Vet. Pharmacol. Ther. 24:349-351.

Mounnissamy VM, Kavimani S, Sankari G, Quine DS. Subramani K (2010). Evaluation of acute and sub-acute toxicity of ethanol extracts of *Cansjera rheedii* J. Gmelin (Opiliaceae). J. Brewing Distillin. 1(1):011-014.

Samsam-Shariat SH (1992). Qualitative and Quantitative Evaluation of the Active Constituents and Control Methods for Medical Plants. Mani Publications, Isfahan. pp. 23–30 (in Persian).

Schultz TW, Riggin GW (1985). Prédictive correlations for the toxicity of alkyl and halogen substituted phenols. Toxicol. Lett. 25:47-54.

Saxena MJ (2001). Relevance of herbs in improving health index of livestock animals. Proceedings of 38th congress of Nigeria. Vet. Med. Assoc. pp.14-16.

Sofowora A (1996). Plantes médécinales et medecine traditionnelle d'Afrique. Acad.Suisse des Sc. Nat et Ed. Kharthala, co-édit, Berne/Paris. 378p.

Tamboura HH, Bayala B, Lompo M, Guissou P (2005). Distribution écologique, caractéristiques morphologiques et toxicité aiguë des extraits aqueux de *Holarrena floribunda* (G. MET) Durand et Schinz, *Leptadenia hastata* (PERS.) DECNE et *Casse sieberiana* (C.C) employée par les guérisseurs vétérinaires au Burkina Faso. Afr. J. Tradit. Complement. Altern. Med. 2(1):13-24.

Yelemou B, Bationo B, Yameogo G, Millogo-Rasolodimby J (2007). Bois et Forêts des tropiques. Coden. 291:55-66.

# Assessment of drought stress tolerance in root and tuber crops

**Placide Rukundo[1], Hirut Getinet Betaw[2], Stephan Ngailo[3] and Fekadu Balcha[4]**

[1]Rwanda Agriculture Board (RAB), P.O. Box 5016, Kigali, Rwanda.
[2]Amhara Agricultural Research Institute, P.O. Box 527 code 100, Bahidar, Ethiopia.
[3]Sugarcane Research Institute, P.O. Box 30031, Kibaha, Coast Region, Tanzania.
[4]Southern Agricultural Research Institute, Hawassa Research Centre, P.O. Box 6, Hawassa, Ethiopia.

**Plant growth and productivity is adversely affected by environmental abiotic and biotic stresses. Drought is one of abiotic stresses that cause serious damage to plants. Therefore, understanding the extent of drought stress and assessing the mechanisms of drought tolerance in plants is very crucial to devise different coping mechanisms with the stresses. Brief discussions have been given on some of the terminologies such as drought tolerance vs. resistance; potential yield vs. yield potential, and water use efficiency. Mechanisms of adaptation to drought stress, inheritance of drought tolerance and the contribution of biotechnology to develop methods of coping up with the problems have been discussed. Finally, methods of screening for drought tolerance have been outlined. The integration of both agricultural practices and conventional and contemporary breeding strategies are necessary in developing crop varieties that are tolerant to abiotic stresses including drought.**

**Key words:** Drought stress, resistance, root and tuber crops, tolerance.

## INTRODUCTION

Plant growth and productivity is adversely affected by various abiotic and biotic stresses. Some of the abiotic factors such as low temperature, salt, flooding, drought, heat, oxidative stress and heavy metal toxicity (these stresses) induce various biochemical and physiological responses in plants, which respond and adapt in order to survive and reproduce even under sub-optimal conditions (Barnabás et al., 2008; Mahajan and Tuteja 2005; Seki et al., 2003).

Drought is an extended period of dry weather characterized by a shortage of water supply to plants (Acquaah, 2007). Drought begins when the readily available soil water in the root zone is exhausted (Taiz and Zeiger, 2006). Drought can be permanent, periodic, or random, occurring early, late, or in the middle of the crop season. Drought can also be cumulative or specific and short. Drought in conjunction with coincident high temperature and radiation poses the most important environmental constraints to plant survival and to crop productivity (Chaves et al., 2003; Neumann 2008; Zheng et al., 2009). Further, they are a menace for plants and prevent them from reaching their full genetic potential (Mahajan and Tuteja, 2005). Drought can either be terminal or intermittent (Neumann, 2008). While terminal drought, the availability of soil water decreases progressively and can lead to premature plant death, intermittent drought results from finite periods of inadequate irrigation occurring at one or more intervals during the growing season and is not necessarily lethal. The plant may adjust to short-term water stress by closing stomata and thereby reducing water loss through the leaves. Drought is an extended period of dry weather characterized by a shortage of water supply to plants (Acquaah, 2007). In drought conditions, water potential ($\Psi$w) of soil becomes very negative due to concentration increase of soil solutes. The movement of cell water is determined by the water potential gradient ($\Delta\Psi$w) that acts as a driving force for transport through a permeable cell membrane (Taiz and Zeiger, 2006). A plant can continue to absorb water only if its $\Psi$w is lower than that of the soil. Indeed, drought stress causes changes in plant cells and tissues to adapt to drought stress condition and continue to acquire little available water from the soil (Bartels and Sunkar, 2005). Symptoms of drought stress start when crop has used between 50 and 80% of available soil moisture (Acquaah, 2007). In fact, the failure of plant to absorb the soil water has severe consequences.

Water plays a crucial role in the life of plant and its availability is a main factor that determines the plant population in the environment (Coley et al., 2009). Water is the main constituent of plant tissues but its quantity varies within plant tissues and plant species. The water content was estimated at 80 to 95% in mass of growing tissues, 85 to 95% in vegetative tissues, 35 to 75% in wood with dead cells, and at 5 to 15% in dried seeds (Taiz and Zeiger, 2006). The distribution of plant species in the environment is associated with their tolerance to environmental stresses (Brenes et al., 2009). Most widespread plant species are drought tolerant (Baltzer et al., 2008; Brenes et al., 2009). A low temperature was suspected to be the main limiting factor of life in the Antarctic environment; however, it was found that the water deficit is the major cause and a positive correlation was observed between the soil moisture and the abundance of plant species (Kennedy, 1993).

Drought is the primary abiotic stress that affects crop production and food availability. In many countries, agriculture depends on rain which in many cases does not meet the crop need (Ober, 2008). The fluctuation of rain affects the growth and production of crop and this causes famines in many semi-arid countries (Acquaah, 2007). Drought can cause the big loss of the production as compared to other isolated biotic or abiotic stress (Boyer, 1982; Ober, 2008). It affects production by reducing the genetic production potential of a plant (Mitra, 2001). Consequently, it is responsible for the difference between the mean yield and the yield potential of a crop and the cause of yield instability in time (Sorrells et al., 2000).

Drought induces physiological, biochemical and molecular changes that have consequences on a plant growth and production (Reddy et al., 2004). The drought osmotic stress causes the removal of water from the cytoplasm to the extracellular space and cell dehydration (Bartels and Sunkar, 2005). Water deficit affects the photosynthetic ability of plants by changing the content and components of chlorophyll, reducing the net $CO_2$ uptake by leaves, or by decreasing activities of enzymes in the Calvin cycle (Becana et al., 1998; Cornic, 2000; Gong et al., 2005; Lawlor and Tezara, 2009). The osmotic stress of water deficit inhibits strongly the growth of leaves and stem of plants. This has negative effects on the crop yield potential (Westgate and Boyer, 1985). However, the degree of growth inhibition and yield potential reduction depends on the duration and intensity of drought stress, and the genotype of plant species (Bartels and Sunkar, 2005; Monakhova and Chernyad'ev, 2002).

The major cause of reduction of photosynthesis ability and plant growth under drought stress is the disequilibrium between the production of reactive oxygen species (ROS) and their scavenging systems (Becana et al., 1998). Plants under abiotic stresses generate ROS that cause oxidative reactions (Lin et al., 2006). The main ROS are hydrogen peroxide ($H_2O_2$) and superoxide ($O_2^{\cdot-}$). These ROS are routinely produced in different cellular reactions catalyzed by various enzymes such as lipoxygenase, peroxidase, NADPH oxidase and xanthine oxidase, but the main source of these molecules is the Fenton and Haber-Weiss reactions (Blokhina et al., 2003; Debarry et al., 2005; Lin et al., 2006). The ROS damage lipids, carbohydrates and proteins of cell membrane and cell nucleic acids (Blokhina et al., 2003; Fu and Huang, 2001; Sairam et al., 1997; Zhang and Kirkham, 1996). When a plant is under a serious stress condition, there is an accumulation of ROS because its ROS scavenging and repairing mechanisms of ROS damages are surpassed (Lin et al., 2006). Therefore, a plant must have efficient mechanisms of defense against ROS to survive a severe drought osmotic stress and adapt to drought condition. This paper review concepts of drought stress and methods for assessing drought tolerance in root and tuber crops.

## DROUGHT RESISTANCE VERSUS DROUGHT TOLERANCE

Drought resistance refers to the ability of a plant to maintain favorable water balance and turgidity even when exposed to drought condition and its consequences.

When a genotype yields better than another under a severe strain of drought, it is relatively more drought resistant (Blum, 2002). The strain of drought is developed when crop demand for water is not met by the supply, and plant water status is reduced. Plants can resist drought by either dehydration avoidance or dehydration tolerance. Physiological responses of drought resistance depend on magnitude of stress and plant growth stage. For example, drought resistance in seedlings grown in a pot has nothing to do with drought resistance during grain filling in the field (Blum, 2011). Drought tolerance is defined as the relative capacity of plants to maintain functional growth under low leaf water status.

A genotype is drought resistant when it produces an economic crop, within the limits of its production potential under conditions of limited water availability. A genotype can be drought resistant due to the following mechanisms: drought escape, drought tolerance, drought avoidance and drought recovery (Barnabás et al., 2008; Chaves et al., 2003; Ekanayake, 1990). These mechanisms are not mutually exclusive and provide the crop with the ability to resist drought at any given period during its growth cycle (Chaves et al., 2003; Ekanayake, 1990). Plant escape drought by exhibiting high degree of developmental plasticity by being able to complete their life cycle before physiological water deficit occurs; this relies on successful reproduction prior to the onset of severe water stress such as by short life cycle, high rate of growth or the efficient storage and use of reserves for seed production (Barnabás et al., 2008; Chaves et al., 2003). Also, plants endure drought conditions by avoiding dehydration while maintaining tissue water potential as high as possible or tolerating low tissue water potential. This can be achieved either by minimizing water loss or maximizing water intake (Chaves et al., 2003). Water loss can be minimized by closing stomata, reducing light absorbance through leaf rolling, increasing reflectance by dense trichome layer, steep leaf angles or by decreasing canopy leaf area through reduced growth and shedding older leaves (Ehleringer and Cooper, 1992; Larcher, 2000).

## Water use efficiency

Water use efficiency is defined by the relationship between grams of water transpired by a crop per gram of dry matter produced (Zangerl and Bazzaz, 1984). Therefore, crops that use water more efficiently produce more dry matter per gram of water transpired, so the efficient use of water is directly correlated to time of stomata opening and stomatal conductance (Zangerl and Bazzaz, 1984). As the plant absorbs $CO_2$ for photosynthesis, water is lost by transpiration, with variable intensity depending on the potential gradient between leaf surface and the atmosphere, following a stream of water potentials (Shwani et al., 2010).

## Yield potential and potential yield

Yield potential is defined as the yield of a cultivar when grown in environments to which it is adapted, with nutrients and water non-limiting and with pests, diseases, weeds, lodging and other stresses effectively controlled (Acquaah, 2007). As such, it is distinguished from potential yield, which is defined as the maximum yield which could be reached by crop in given environments, as determined, for example, by simulation models with plausible physiological and agronomic assumptions (Evans and Fischer, 1999).

## Mechanisms of adaptation to drought stress

The interaction and adaptation of plants to environmental signals and stresses is a complex network model (Shwani et al., 2010). Figure 1 highlights the probable physiological, biochemical and molecular responses to drought-stress in higher plants. Plants can withhold the drought stress by dehydration tolerance, dehydration avoidance or drought escape (Ludlow, 1989). In the strategy of dehydration tolerance, it was reported resurrection plants that can survive an extreme internal water deficit. These plants can be still alive when there is 95% of leaf water loss (Scott, 2000). The dehydration avoidance consists of maximizing the water absorption and minimizing the water loss under water deficit conditions. This model is mainly observed on succulent and C4 plant species. The strategy of drought escape is based on a short life cycle. Plants grow and reproduce before appearance of a drought season (Mckay et al., 2003; Mitra, 2001; Passioura, 1996; Richards, 1996). Even though there are these strategies, plants present various changes to resist drought stress. Perennial and deciduous plants reduce their foliage in drought seasons. Plants that are always green present sometimes thick leaves with solid cuticle, deep roots, highly sclerophyllous and reduced size leaves, and effective water use and control of evapotranspiration (Ain-Lhout et al., 2001; Lebreton et al., 1995; Sanguineti et al., 1999; Sorrells et al., 2000; Taiz and Zeiger, 2006). These strategies are usually assisted by others such as accumulation of compatible solutes and increased expression and production of antioxidants. Accumulation of compatible solutes is one of biochemical processes that help plants to survive under drought condition (McCue and Hanson, 1990). Compatible solutes play an adaptive role by osmotic adjustment and protection of cellular compounds (Ain-Lhout et al., 2001; Hare et al., 1998). The compatible solutes are mainly nitrogen containing molecules such as amino acids and polyamines, and hydroxyl compounds. Types of these compatible solutes and the level of their accumulation vary with plant species (McCue and Hanson, 1990). These molecules work together with antioxidants which intervene to eliminate ROS and repair

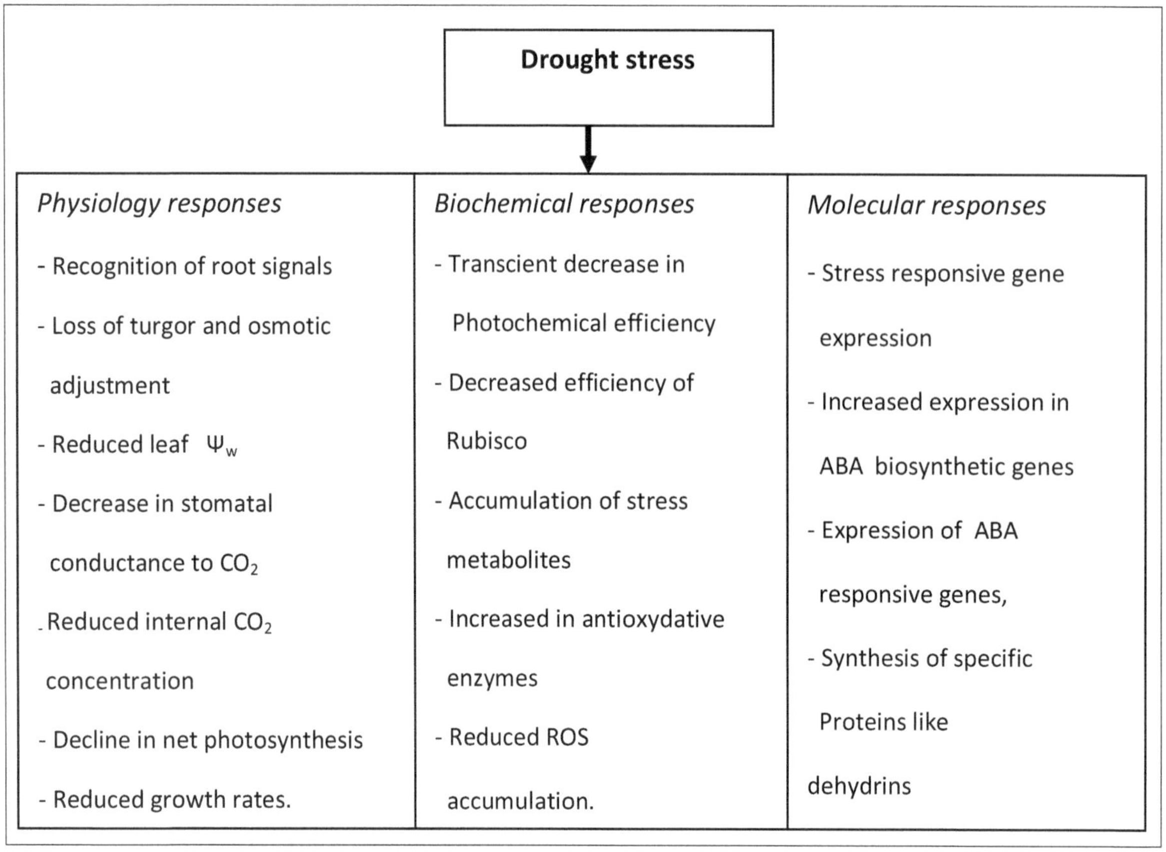


**Figure 1.** Physiological, biochemical and molecular responses to drought stress in plants (Reddy et al., 2004).

damages of ROS.

Plants produce different antioxidants that have abilities to scavenge ROS. Antioxidants are molecules with small molecular mass such as ascorbic acid, glutathione, tocopherols, phenolic compounds, ROS-interacting enzymes such as superoxide dismutase (SOD), ascorbate peroxidise (APX) and catalase (CAT) (Blokhina et al., 2003; Brosché et al., 2010). These molecules play an important role in controlling the equilibrium between the production and the elimination of free radicals. Moreover, they work in cohesive network reactions and use mainly redox reactions (Lin et al., 2006). Plant varieties that are drought resistant or tolerant express a higher quantity of antioxidants than sensitive varieties (Herbinger et al., 2002; Lin et al., 2006). The quality and the quantity of these molecules are plant species dependent and their expression is affected by plant environmental conditions (Blokhina et al., 2003; Herbinger et al., 2002; Lin et al., 2006). Therefore, understanding the expression mechanisms of these molecules and their function models can assist to identify and develop of drought tolerant varieties.

Adaptation mechanisms to drought stress incriminate plants. Varieties of short life cycle can escape a drought period; but produce low yield (Acquaah, 2007). Mechanisms of dehydration avoidance such as stomatal closure and a decrease of leaf area reduce the assimilation of light and atmospheric $CO_2$ necessary for photosynthesis (Cornic, 2000; Lawlor and Tezara, 2009). Dehydration tolerance with the accumulation of compatible solutes, the synthesis of antioxidants and the process of ROS scavenging requires too much plant energy. Consequently, these mechanisms reduce the ability of plants to synthesize organic products for stock organs (Mitra, 2001). Thus, the development of a drought tolerant variety such as sweet potato needs to balance all drought tolerance mechanisms without sacrificing the crop productivity (Mckay et al., 2003; Mitra 2001; Passioura, 1996; Richards, 1996).

## Inheritance of drought tolerance

Drought tolerance is a complex trait because of the heterogeneity of drought stress in the time and space, and unexpected characteristics of drought stress (Sorrells et al., 2000). The drought tolerance involves actions and interactions of various biochemical, morphological and physiological mechanisms that are controlled by products expressed by different genes (Acquaah, 2007; Mitra, 2001). Moreover, it is difficult to study isolated single gene and to understand its role of drought tolerance in plant (Mitra, 2001).

Contradiction cases of quantitative nature of drought tolerance were reported in various studies. Ekanayake et al. (1985) have observed that root characteristics are controlled by a qualitative inheritance under a drought condition. The leaf rolling, the osmotic adjustment and the number of roots were identified to be qualitative traits (Mitra, 2001). Study on water deficit mediator genes has indicated that plant species vary in symptoms and reactions to water deficit (Sorrells et al., 2000). The genes responsible for earliness of stem reserves, leaf persistence and dwarfing were identified to be associated with drought tolerance (Foulkes et al., 2007). A drought resistance gene linked with genes for plant height and pigmentation that has pleiotropic effects on a root system was identified in rice and crowpea (Agbicodo et al., 2009; Mitra, 2001; Morgan, 1995). Other proposed candidate genes that are involved in drought tolerance are genes coding for dehydrin proteins that protect cellular components under dehydration condition (Shinozaki and Yamaguchi-Shinozaki, 2007), proteins controlling the equilibrium and damages of ROS (Foyer and Noctor, 2005), proteins involving in osmotic adjustment and plant morphology (Moinuddin et al., 2005; Ober, 2008) and enzymes involved in the accumulation of compatible solutes (Mitra 2001). Indeed, drought tolerance involves many genes which code for products working in a highly coordinated network.

## DROUGHT AND BIOTECHNOLOGY

Good genetic maps based on molecular marker technologies are now available for major crops. Many of the traits determining abiotic stress tolerance and the quality and quantity of yield are controlled by a large number of genes, which have only minor individual effects but which act together (quantitative trait loci, QTL) (Barnabás et al., 2008). In crop species with large, complex genomes such as sweet potato, QTL analysis is an important tool in the identification of genetic markers to assist breeding efforts. Barnabas et al. (2008) suggested that correlating genetic information with physiological and morphological traits related to drought tolerance and other traits such as yield will allow the development of new varieties with improved yield safety under water-limited conditions using molecular marker-assisted breeding. The same author highlighted the use of proteomics in investigating the effect of drought and/or heat stress on protein composition as an important step towards understanding the link between environmental factors and plant development. On the other hand, the introduction of isolated individual genes into the germplasm through genetic engineering offers a variety of opportunities to increase environmental stress tolerance in crops (Barnabás et al., 2008). Wahid et al. (2007) applauded the significant contribution of biotechnology in a better understanding of the genetic basis of heat tolerance.

Recent advances in plant genomics have led to the identification of a vast number of potentially beneficial water-stress-related genes, plus technologies for gene overexpression or silencing. Moreover, these can be introduced into transgenic plants under the control of appropriate promoters and are transmitted to subsequent generations (Neumann, 2008).

### Screening for drought tolerance

#### *Environmental conditions of study*

Study of drought tolerance in plants can be carried out under field or controlled environmental conditions (Acquaah, 2007). The field condition consists of conducting trials under natural conditions. These trials are carried out in the real environments of a plant but it has some limitations of fluctuation of water availability caused by unexpected rainfall. Moreover, environmental factors such as temperature, air humidity and light are variable. Therefore, the screening for drought tolerance is complicated by difficulties of field management (Lafitte et al., 2004). The rainout shelter and *in vitro* techniques were proposed to overcome the limitations of selection for drought tolerance under field condition (Acquaah, 2007). The rainout shelter is a mobile infrastructure that protects plants under experiment from rain. This method controls the uniformity of water supply to plants (Blum, 2002). The *in vitro* approach consists of growing cells or tissues of plant or plantlets on a defined drought stressing culture media under an aseptic and controlled environment (Ahloowalia et al., 2004; Wang et al., 1999). The *in vitro* technique provides precise results but the working environment differs from the natural environment of plant. Therefore, the combination of *in vitro* screening with selection under the natural condition or under the rainout shelter could improve the quality of results.

### Screening methods for drought tolerance

Drought stress tolerance is a complex quantitatively inherited and controlled trait. These stresses affect the plant in different ways and induce different genetic responses. Therefore, very careful strategies and powerful methods are required to assess the level of genetic resistance and/or tolerance.

Drought tolerance can be identified by quantifying phenological, morphological, physiological and biochemical characteristics and using molecular tools (Blum, 2002). Phenological and morphological characteristics are mostly used in breeding for drought tolerance. In these approaches data collection consists of measurement of plant growth (size of roots, stem and leaf

**Table 1.** A scale of visual symptoms of leaf wilting.

| Score | Estimation (in %) of turgid leaf area | Description of the symptoms |
|---|---|---|
| 9 | > 95 | All leaves turgid |
| 8 | 80 | Lower leaves wilted |
| 7 | 70 | |
| 6 | 60 | Lower half of the plant wilted |
| 5 | 50 | |
| 4 | 40 | |
| 3 | 30 | Top leaves still turgid |
| 1 | <5 | Plant completely wilted; leaves show necrosis |
| 0 | 0 | |

area, gain of fresh and dry weights and yield loss), growth stage (days to flowering and maturity), senescence and leaf rolling (Cheema and Sadaqat, 2004; Spitters and Schaapendonk, 1990). The water content and water potential of plant are indicators used to identify drought tolerant varieties. A variety that maintains its internal water status under a drought stress is considered as drought tolerant. Drought tolerance is also determined by quantifying plant biochemical products such as compatible solutes, chlorophyll, antioxidants and other proteins produced by plant as responses to drought stress (Kasukabe et al., 2006; Reddy et al., 2004; Wang et al., 1999). Diffusion porometry for leaf water conductance, root penetration, distribution and density in the field and infrared aerial photography for dehydration and leaf temperature are used commonly in studies for drought tolerance (Mitra, 2001). Practical methods to measure the drought tolerance are mainly based on phenological, morphological and physiological traits and on biotechnological approaches.

## Phenological traits

Growth rate and yield are main phonological traits for determining drought tolerance in crops. The yield reduction due to drought stress, the number of surviving plants and/or a count of plants with tubers are often practiced to study drought tolerance in root crops (Ekanayake, 1990). Data of growth and yield reduction due to drought stress are collected from trial carried out under control and drought stress conditions. These data are transformed into drought tolerance indices such as drought resistance indix (DRI) = (Ys/Yn)/(Ms/Mn), mean productivity (MP) = (Ys+Yn)/2, stress tolerance index (STI) = (Ys)(Yn)/(Mp), stress susceptible index (SSI) = (1-(Ys/Yn))/(1- (Ms/Mp)) and tolerance (TOL) = Yn-Ys. Where Yn: yield of a given genotype in non-stressed environment, Ys: yield of a given genotype in drought stressed condition, Ms: mean yield over all genotypes in stressed environment, Mn: mean yield over all genotype in non-stressed environment (Golabadi et al., 2006; Talebi et al., 2009). These drought tolerances indices assist to identify drought tolerant genotypes within germplasm collections

## Morphological traits

### *Leaf wilting*

Visual symptoms of leaf wilting (Table 1) are recorded on a scale base and the measurement unit is expressed in cm $s^{-1}$. But this might be very subjective since it is observation from visual symptoms.

### *Senescence*

The slower the process of leaf senescence, the better is the genotype. It is measured using scores on a scale from 0 to 10, dividing the percentage of estimated total leaf area that is dead. The description of this scoring is presented in Table 2.

### *Leaf rolling*

The leaf rolling is a good indicator of dehydration avoidance mechanism. The best genotypes show the low scores. It is measured using scores from 1 to 5 (Table 3), when leaves are still more upright at the hottest time of the day.

## Physiological traits

### *Canopy temperature*

Canopy temperature measurements have been widely used to study the drought response of various crops. This approach is based upon the close, inverse relationship between leaf temperature and transpirational cooling. When stress develops and water status in low

**Table 2.** A scale of visual symptoms of leaf senescence.

| Score | Description | Score | Description |
|---|---|---|---|
| 1 | 10% dead leaf area | 6 | 60% dead leaf area |
| 2 | 20% dead leaf area | 7 | 70% dead leaf area |
| 3 | 30% dead leaf area | 8 | 80% dead leaf area |
| 4 | 40% dead leaf area | 9 | 90% dead leaf area |
| 5 | 50% dead leaf area | 10 | 100% dead leaf area |

**Table 3.** A scale of visual symptoms of leaf rolling.

| Score | Description |
|---|---|
| 1 | Unrolled, turgid |
| 2 | Leaf rim starts to roll |
| 3 | Leaf has a the shape of a V |
| 4 | Rolled leaf rim covers part of leaf blade |
| 5 | Leaf is rolled like an onion |

transpiration is reduced and canopy temperature rises. Blum et al. (1989) found a positive correlation between drought susceptibility of wheat genotypes and canopy temperature in stressed environments. Genotypes that suffered greater relative yield losses under drought stress tended to have warmer canopies at midday. Stark et al. (1991) also reported that the most drought-resistant potato genotypes usually had the lowest canopy temperatures during periods of drought. Canopy temperature is measured by a hand infrared thermometer (IRT) which is designed to sense long-wave infrared radiation emitted from its target and convert it to average temperature display which can be related to transpiration. Measurements (°C) will be made from 9:00 to 10:00 ideally on clear (cloudless), windless and sunny days.

### *Relative leaf water content (RWC)*

Relative water content (RWC) is determined by weighing the entire leaves or leaf discs immediately after cutting, following prolonged immersion in water and after oven drying by the method described by Okono (2010). Three topmost fully expanded leaves were sampled from a plant. Six random replications were taken from a single plot. Each sample was represent a different plant within the genotype. The leaf sampling was taken in the afternoon after 2:00 pm, when the change in atmospheric condition and solar radiation slow in order to reduce change in water relation with time. Each sample was be placed in plastic bag. The plastic bag was immediately be placed in a cooler (at around 10-15°C). In the laboratory, each leaf in a plastic bag was weighed to obtain the leaf sample weight (fresh weight, FW), after which the sample was immediately placed hydrated to full turgidity for four hours under normal room light and temperature conditions. Samples were re-hydrated by floating on deionised water in a covered petri dish. After four hours, the samples were taken out of the water and will be thoroughly dried of any surface moisture quickly and lightly with paper towel and immediately weighed to obtain the fully turgid weight (TW). Samples are then oven dried at 80°C for 24 h and weighed (after being cooled down in a desiccator) to determine the dry weight (DW). All weighing was done to the nearest mg. The relative water content is calculated as followings:

$$RWC\ (\%) = \left(\frac{FW - DW}{TW - DW}\right) X\ 100\ ,$$

where FW: = sample fresh weight, TW: sample turgid weight; and DW: sample dry weight.

### *Stomatal conductance*

Stomatal responses to environmental drought have a substantial influence on plant adaptation to arid climates (Bates and Hall, 1982). Stomata regulate water use and the development of water stress, and they influence plant growth rates through effects on carbon dioxide assimilation. Stomatal conductance is measured by using diffusion porometer. It can be calculated by the following empirical formula by Ball-Berry et al. (1987) model as modified by Xu and Baldocchi (2003):

$$g = g_0 + m \frac{ARH}{C_a}$$

Where, g is stomatal conductance, RH is the relative humidity at the leaf surface, $C_a$ is the $CO_2$ concentration at the leaf surface, and $g_0$ and $m$ are the intercept and slope, respectively.

Tuzet et al. (2003) suggested another model for determining stomatal conductance which incorporated the stomatal response to leaf water potential as follows:

$$g_{CO2} = g_0 + \frac{aA}{ci - \Gamma} . f\psi v$$

Where, $g_{CO2}$ is the stomatal conductance for $CO_2$, $g_o$ is the residual conductance, A is assimilation rate, $ci$ is the $CO_2$ concentration in the intercellular spaces, $\Gamma$ is the $CO_2$ compensation point and $a$ is an empirical coefficient. Stomatal conductance varies with leaf irradiance, leaf temperature, atmospheric water vapour pressure deficit and $CO_2$ concentration (Tuzet et al., 2003). On the other hand, stomatal conductance depends on guard cell and epidermal turgor.

### *Cell membrane thermo-stability (CMS)*

The CMS is highly correlated with yield under drought stress, therefore, this trait can assist to identify drought tolerance in crops (Blum, 2005). Identification of cell membrane stability consists of measurement of electrolyte using the conductimetric method as described by Blum (2011). 15 to 20 $cm^2$ leaf materials are sampled from 5 plants of each genotype grown on control and drought stress conditions and placed in tubes. These leaves samples are washed three times with 20 ml desionised water. Clean samples are soaked in 20 ml of desionised water and incubated at 10°C for 24 h. The conductivity (T1) of the liquid of each tube is measured after 1 h of equilibration at room temperature. Each tube is covered and autoclaved for 15 min to kill the leaf tissues. The conductivity (T2) of each autoclaved sample is measured again after equilibration of one hour at room temperature. The CMS is determined with the following formula:

$$CMS\,(\%) = \frac{1-(1-T_1/T_2)}{1-C_1/C_2} \times 100$$

Where T1 and T2: Treatment conductivities before and after autoclaving, and C1 and C2 = control conductivities before and after autoclaving.

### *Leaf chlorophyll content*

Strong correlations between higher chlorophyll and carotenoids contents and stress tolerance in plants were reported. It was observed that the chlorophyll content is different between drought resistant and susceptible genotypes. Drought resistant genotypes showed the highest chlorophyll content. This characteristic was suggested to be used as screening tool for drought tolerance (Ganji Arjenaki et al., 2012). Houman et al. (2011) reported that genotypes with high chlorophyll index under drought stress conditions revealed the highest yield. It was observed that drought stress had a negative effect on yields of different genotypes but the genotypes that were able to maintain their chlorophyll content were also able to maintain its yield potential (Khayatnezhad and Gholamin, 2012). Drought stress imposed during vegetative growth or anthesis significantly decreased chlorophyll a, chlorophyll b and total chlorophyll content. The results showed that mesophyll resistance is the basic determinate of rate of photosynthesis under drought stress conditions. Under drought conditions, the drought tolerant variety gave the highest yield whereas the drought sensitive variety gave the lowest yield (Mafakheri et al., 2010). Based on the above findings, the measurement of chlorophyll fluorescent was used as new tool to investigate effects of different stresses such as drought, salinity and heat on crop yield (Ort, 2002). Nowadays, handheld instruments such CCI-200 device and chlorophyll meter SPAD 502 Plus that measure the leaf chlorophyll content are available. These instruments determine the relative chlorophyll present in leaf by measuring the absorbance of leaf.

## BIOTECHNOLOGICAL APPROACHES

Many biotechnological studies based on molecular biology have the aim to show the gene function and gene expression profile. To achieve this goal, different molecular approaches such as analysis of genome, transcriptome, metablome and proteome were developed (Carpentier, 2007). The analysis of the genome is a complex study of an organism's genome. This discipline known as genomics, is based on a complete genome analysis and involves DNA sequencing, assembly of sequences, annotation and mapping of genes (Arabidopsis, 2000).

The study of gene expression and its regulation is another approach for understanding the gene function. This approach known as transcriptomics uses efficient tools such as microarray analysis, cDNA fragment fingerprinting and serial analysis of gene expression (SAGE) to carry out the transcriptome analysis (Brown and Botstein, 1999; Schena et al., 1998). Metabolome represents a collection of all metabolites in a biological organism at a specific time and under specific conditions. These metabolites are the end products of genes expression of the organism.

The study of metablome (metabolomics) is a comprehensive, qualitative and quantitative study of all small molecules (less than or equal to 1500 daltons) participating in important metabolic functions and fulfilling critical roles such as drought tolerance, signaling molecules or secondary metabolites in an organism (Oliver et al., 1998). The main methods for metablome analysis are metabolite profiling and metabolite fingerprinting (Hall, 2006). Proteomics is another approach to understand the gene function and expression profile. It focuses on the characterisation of the cellular proteome which is defined as a set of protein species present in a biological unit at a specific developmental stage and under determined external biotic and abiotic conditions (Jorrín et al., 2006; Klug et al., 2000; Prescott et al., 2005). Proteomics involves protein biochemistry and bioinformatics to determine the spatial and temporal expression of proteins in cells and tissues of a living thing (Karr, 2007). The main tool of proteome analysis is a two dimensional gel electrophoresis (2-DE).

All these approaches (genomics, transcriptomics, metabolomics and proteomics) are powerful tools for massive screening of several genes and aim to reveal the changes of what might be happening in a cell. They can also serve as tools to identify candidates with desired traits such as drought tolerance in sweet potato germplasm.

## CONCLUSION

The effect of drought stress on yield is highly complex and involves diverse processes in plants. Tolerance to a combination of different stress conditions, particularly those that mimic the field environment, should be the focus of future research programmes aimed at developing transgenic crops with an enhanced tolerance to naturally occurring environmental conditions. Modern genetic and molecular tools will significantly contribute to understanding physiological mechanisms related to drought and forge the way forward to develop new crop varieties with drought tolerance. The advent of biotechnology will contribute significantly to survey drought tolerance in various crop species, including sweet potato, cassava and potato. Further, advancements in marker technology and genetic transformation are expected to contribute significantly to the development of plants with various tolerances in the future. Integration of both agricultural practices, conventional and contemporary breeding strategies are necessary in developing crop varieties tolerant to abiotic and biotic stresses to insure food security of the worldwide population.

### Conflict of Interests

The author(s) have not declared any conflict of interests.

## ACKNOWLEDGEMENTS

The Alliance for a Green Revolution in Africa (AGRA) is apprciated for providing financial support. The authors thank African Center for Crop Improvement (ACCI) for organising a writing sessions.

### REFERENCES

Acquaah G (2007). Principles of plant genetics and breeding. Wiley-Blackwell, Oxford.

Agbicodo E, Fatokun C, Muranaka S, Visser RGF, Linden van der C (2009). Breeding drought tolerant cowpea: Constraints, accomplishments and future prospects. Euphytica 167:353-370.

Ahloowalia BS, Prakash J, Savangikar VAC, Savangikar C (2004). Plant tissue culture, In: IAEA (Ed.), Low cost options for tissue culture technology in developing countries. International Atomic Energy Agency, Vienna. pp. 3-11.

Ain-Lhout F, Zunzunegui M, Diaz Barradas M, Tirado R, Clavijo A, Garcia Novo F (2001). Comparison of proline accumulation in two Mediterranean shrubs subjected to natural and experimental water deficit. Plant Soil 230:175-183.

Arabidopsis GI (2000). Analysis of the genome sequence of the flowering plant *Arabidopsis thaliana*. Nature 408:796.

Baltzer JL, Davies SJ, Bunyavejchewin S, Noor N (2008). The role of desiccation tolerance in determining tree species distributions along the Malay-Thai Peninsula. Functional Ecol. 22:221-231.

Barnabás B, Jäger K, Fehér A (2008). The effect of drought and heat stress on reproductive processes in cereals. Plant Cell Environ. 31:11-38.

Bartels D, Sunkar R (2005). Drought and salt tolerance in plants. Critical Rev. Plant Sci. 24:23-58.

Becana M, Moran J, Iturbe-Ormaetxe I (1998). Iron-dependent oxygen free radical generation in plants subjected to environmental stress: Toxicity and antioxidant protection. Plant Soil 201:137-147.

Blokhina O, Virolainen E, Fagerstedt KV (2003). Antioxidants, oxidative damage and oxygen deprivation stress. Ann. Bot. 91:179-194.

Blum A (2002). Drought tolerance is it a complex trait?, In: Saxena N P and O'Toole JC (Eds.), Proceedings of an international workshop on field screening for drought tolerance in rice. Rockefeller Foundation, New York. pp. 24-59.

Blum A (2005). Drought resistance, water-use efficiency, and yield potential are they compatible, dissonant, or mutually exclusive? Australian J. Agric. Res. 56:1159-1168.

Blum A (2011). Plant breeding for water-limited environments. Springer, London.

Blum A, Shpiler L, Golan G, Mayer J (1989). Yield stability and canopy temperature of wheat genotypes under drought-stess. Field Crops Res. 22:289-296.

Boyer JS (1982). Plant productivity and environment. Sci. 218:443-448.

Brenes AT, Coley PD, Kursar TA (2009). Pests vs. drought as determinants of plant distribution along a tropical rainfall gradient. Ecol. 90:1751-1761.

Brosché M, Overmyer K, Wrzaczek M, Kangasjarvi (2010). Stress signaling III: Reactive oxigen species (ROS), In: Pareek A (Ed.), Abiotic stress adaptation in plant. Springer, Dordrecht. pp. 91-102.

Brown PO, Botstein D (1999). Exploring the new world of the genome with DNA microarrays. Nature Genet. 21:33-37.

Carpentier S (2007). Optimized proteomic methods to unravel biochemical processes in banana meristems during in vitro osmotic stress acclimation. PhD Thesis, Leuven.

Chaves MM, Maroco JP, Pereira JS (2003). Understanding plant responses to drought from genes to the whole plant. Functional Plant Biol. 30:239-264.

Cheema KL, Sadaqat HA (2004). Potential and genetic basis of drought tolerance in canola (*Brassica napus*): Generation mean analysis for some phenological and yield components. Int. J. Agric.Biol. 6:74-81.

Coley PD, Kursar TA, Brenes-Arguedas T (2009). Pests vs. drought as determinants of plant distribution along a tropical rainfall gradient. Ecol. 90:1751-1761.

Cornic G (2000). Drought stress inhibits photosynthesis by decreasing stomatal aperture not by affecting ATP synthesis. Trends in Plant Sci. 5:187-188.

Debarry M, Marten I, Ngezahayo A, Kolb HA (2005). Differential defense responses in sweetpotato suspension culture. Plant Sci. 168:1171-1179.

Ehleringer J, Cooper T (1992). On the role of orientation in reducing photoinhibitory damage in photosynthetic-twig desert shrubs. Plant Cell Environ.15:301-306.

Ekanayake I, Garrity J, Masajo D (1985). Inheritance of root characters and their relations to drought resistance in rice. Crop Sci. 25:927-933.

Ekanayake IJ (1990). Evaluation of potato and sweetpotato genotypes for drought resistance CIP, Lima.

Evans LT, Fischer RA (1999). Yield potential: Its definition, measurement, and significance. Crop Sci. 39:1544-1551.

Foulkes M, Sylvester-Bradley R, Weightman R, Snape J (2007). Identifying physiological traits associated with improved drought resistance in winter wheat. Field Crops Res. 103:11-24.

Foyer CH, Noctor G (2005). Redox homeostasis and antioxidant signaling: A metabolic interface between stress perception and physiological responses. The Plant Cell Online 17:1866-1875.

Fu J, Huang B (2001). Involvement of antioxidants and lipid peroxidation in the adaptation of two cool season grasses to localized drought stress. Environ. Exp. Bot. 45:105-114.

Ganji AF, Jabbari R, Morshedi A (2012). Evaluation of drought stress on relative water content, chlorophyll content and mineral elements of wheat (*Triticum aestivum L.*) varieties. Int. J. Agric. Crop Sci. 11:726-729.

Golabadi M, Arzani A, Maibody SAM (2006). Assessment of drought tolerance in segregating populations in durum wheat. Afr. J. Agric. Res. 5:162-171.

Gong H, Zhu X, Chen K, Wang S, Zhang C (2005). Silicon alleviates oxidative damage of wheat plants in pots under drought.

Plant Sci. 169:313-321.
Hall RD (2006). Plant metabolomics: From holistic hope, to hype, to hot topic. New Phytol. 169:453-468.
Hare P, Cress W, Van Staden J (1998). Dissecting the roles of osmolyte accumulation during stress. Plant Cell Environ. 21:535-553.
Herbinger K, Tausz M, Wonisch A, Soja G, Sorger A, Grill D (2002). Complex interactive effects of drought and ozone stress on the antioxidant defence systems of two wheat cultivars. Plant Physiol. Biochem. 40:691-696.
Houman H, Morteza SD, Parisa M (2011). Effect of drought stress on leaf chlorophyll in corn cultivars (*Zea mays*). Middle-East J. Sci. Res. 3:418-420.
Jorrín J, Rubiales D, Dumas-Gaudot E, Recorbet G, Maldonado A, Castillejo M, Curto M (2006). Proteomics: A promising approach to study biotic interaction in legumes. A review. Euphytica 147:37-47.
Karr T (2007). Application of proteomics to ecology and population biology. Heredity 100:200-206.
Kasukabe Y, He L, Watakabe Y, Otani M, Shimada T, Tachibana S (2006). Improvement of environmental stress tolerance of sweetpotato by introduction of genes for spermidine synthase. Plant Biotechnol. 23:75-83.
Kennedy AD (1993). Water as a limiting factor in the Antarctic terrestrial environment: a biogeographical synthesis. Arctic Alpine Res. pp. 308-315.
Khayatnezhad M, Gholamin R (2012). The effect of drought stress on leaf chlorophyll content and stress resistance in maize cultivars (*Zea mays*). Afr. J. Microbiol. Res. 12:2844-2848.
Klug WS, Cummings MR, Spencer CA, Ward SM (2000). Concepts of genetics. Prentice Hall Englewood Cliffs, USA.
Lafitte HR, Price AH, Courtois B (2004). Yield response to water deficit in an upland rice mapping population: Associations among traits and genetic markers. Theoret. Appl. Genet. 109:1237-1246.
Larcher W (2000). Temperature stress and survival ability of Mediterranean sclerophyllous plants. Plant Biosyst. 134:279-295.
Lawlor DW, Tezara W (2009). Causes of decreased photosynthetic rate and metabolic capacity in water deficient leaf cells: A critical evaluation of mechanisms and integration of processes. Ann. Bot. 103:561-579.
Lebreton C, Lazi-Jani V, Steed A, Peki S, Quarrie S (1995). Identification of QTL for drought responses in maize and their use in testing causal relationships between traits. J. Exp. Bot. 46:853-865.
Lin KH, Tsou CC, Hwang SY, Chen LFO, Lo HF (2006). Paclobutrazol pre-treatment enhanced flooding tolerance of sweetpotato. J. Plant Physiol. 163:750-760.
Ludlow MM (1989). Strategies of response to water stress, in: Kreeb KH (Ed.), Structural and functional responses to environmental stresses. SPB Academic, The Hague. pp. 269-281.
Mafakheri A, Siosemardeh A, Bahramnejad B, Struik PC, Sohrabi Y (2010). Effect of drought stress on yield, proline and chlorophyll contents in three chickpea cultivars. Auatralian J. Crop Sci. 8:580-585.
Mahajan S, Tuteja N (2005). Cold, salinity and drought stresses: An overview. Archives of Biochemistry and Biophys. 444:139-158.
McCue KF, Hanson AD (1990). Drought and salt tolerance: Towards understanding and application. Trends Biotechnol. 8:358-362.
Mckay JK, Richards J, Mitchell-Olds T (2003). Genetics of drought adaptation in *Arabidopsis thaliana*: Pleiotropy contributes to genetic correlations among ecological traits. Mol. Ecol. 12:1137-1151.
Mitra J (2001). Genetics and genetic improvement of drought resistance in crop plants. Current Sci. 80:758-763.
Moinuddin F, Sayre R, Reynolds K (2005). Osmotic adjustment in wheat in relation to grain yield under water deficit environments. Agron. J. 97:1062-1071.
Monakhova O, Chernyad'ev I (2002). Protective role of kartolin-4 in wheat plants exposed to soil draught. Appl. Biochem. Microbiol. 38:373-380.
Morgan J (1995). Growth and yield of wheat lines with differing osmoregulative capacity at high soil water deficit in seasons of varying evaporative demand. Field Crops Res. 40:143-152.
Neumann PM (2008). Coping mechanisms for crop plants in drought-prone environments. Ann. Bot. 101:901-907.
Ober ES (2008). Breeding for improved drought tolerance and water use efficiency, in: HGCA (Ed.), Arable cropping in a changing climate. HGCA, Higham. pp. 28-37.
Okono R (2010). Practical measurement of generic drought adaptation-related traits, In: Monneveux P and Ribaut J M (Eds.), Drought phenotyping in crops: From theory to practice. Generation Challenge Programme, Cornell, USA, pp. 451-457.
Oliver S G, Winson M K, Kell D B, Baganz F (1998). Systematic functional analysis of the yeast genome. Trends Biotechnol. 16:373-378.
Ort D (2002). Chilling-induced limitations on photosynthesis in warm climate plants: Contrasting mechanisms. Environ. Contr. Biol. 40:7-18.
Passioura JB (1996). Drought and drought tolerance. Plant Growth Regul. 20:79-83.
Prescott L, Harley J, Klein D (2005). Microbial growth: Microbiology. 6$^{th}$ edion, McGraw-Hill, New York, NY pp.109-132.
Reddy AR, Chaitanya KV, Vivekanandan M (2004). Drought induced responses of photosynthesis and antioxidant metabolism in higher plants. J. Plant Physiol. 161:1189-1202.
Richards R (1996). Defining selection criteria to improve yield under drought. Plant Growth Regul. 20:157-166.
Sairam R, Shukla D, Saxena D (1997). Stress induced injury and antioxidant enzymes in relation to drought tolerance in wheat genotypes. Biologia Plantarum 40:357-364.
Sanguineti M, Tuberosa R, Landi P, Salvi S, Maccaferri M, Casarini E, Conti S (1999). QTL analysis of drought-related traits and grain yield in relation to genetic variation for leaf abscisic acid concentration in field grown maize. J. Exp. Bot. 50:1289-1297.
Schena M, Heller RA, Thenault TP, Konrad K, Lachenmeier E, Davis RW (1998). Microarrays: Biotechnology's discovery platform for functional genomics. Trends Biotechnol. 16:301-306.
Scott P (2000). Resurrection plants and the secrets of eternal leaf. Ann. Bot. 85:159-166.
Seki M, Kamei A, Yamaguchi-Shinozaki K, Shinozaki K (2003). Molecular responses to drought, salinity and frost: Common and different paths for plant protection. Curr. Opin. Biotechnol. 14:194-199.
Shinozaki K, Yamaguchi-Shinozaki K (2007). Gene networks involved in drought stress response and tolerance. J. Exp. Bot. 58:221-227.
Shwani P, Sudhir KS, Hans JB, Govindjee (2010). Abiotic stress adaptation in plant: Phyisiology, molecular and genomic foundation. Springer, University of Illinois, Ubana, USA.
Sorrells M, Diab A, Nachit M (2000). Comparative genetics of drought tolerance. Options Méditerranéennes 40:191-201.
Spitters CJT, Schaapendonk AHCM (1990). Evaluation of breeding strategies for drought tolerance in potato by means of crop growth asimulation. Plant Soil 123:193-203.
Stark JC, Pavek JJ, McCann IR (1991). Using canopy temperature measurements to evaluate drought tolerance of potato genotypes. J. Am. Soc. Hortic. Sci. 116:412-415.
Taiz L, Zeiger E (2006). Stress physiology, In: Taiz L and Zeiger E (Eds.), Plant physiology. Sinauer Associates, Inc., Sunderland, MA. pp. 671-681.
Talebi R, Fayaz F, Naji A (2009). Effective selection criteria for assessing drought stress tolerance in drum wheat (*Triticum durum* Desf.). Gen. Appl. Plant Physiol. 35:64-74.
Tuzet A, Perrier A, Leuning R (2003). A coupled model of stomatal conductance, photosynthesis and transpiration. Plant Cell Environ. 26:1097-1116.
Wahid A, Gelani S, Ashraf M, Foolad MR (2007). Heat tolerance in plants: An overview. Environ. Exp. Bot. 61:199-223.
Wang HL, Lee PD, Liu LF, Su JC (1999). Effect of sorbitol induced osmotic stress on the changes of carbohydrate and free amino acid pools in sweetpotato cell suspension cultures. Bot. Bull. Acad. Sin. 40:219-225.
Westgate M, Boyer J (1985). Osmotic adjustment and the inhibition of leaf, root, stem and silk growth at low water potentials in maize. Planta 164:540-549.
Xu L, Baldocchi DD (2003). Seasonal trends in photosynthetic parameters and stomatal conductance of blue oak (*Quercus douglasii*) under prolonged summer drought and high temperature. Tree Physiol. 23:865-877.

Zangerl AR, Bazzaz FA (1984). Effects of short-term selection along environmental gradients on variation in populations of *Amaranthus retroflexus* and *Abutilon theophrasti*. Ecol. 65:207-217.
Zhang J, Kirkham M (1996). Antioxidant responses to drought in sunflower and sorghum seedlings. New Phytol. 132:361-373.
Zheng X, Chen B, Lu G, Han B (2009). Overexpression of a NAC transcription factor enhances rice drought and salt tolerance. Biochem. Biophys. Res. Commun. 379:985-989.

# *In vitro* and *in vivo* production of polygalacturonase, polymethylgalacturonase and cellulase enzymes by *Alternaria solani* at different incubation periods

**Amit Kumar Chaurasia[1], Shubha Chaurasia[2], Sushmita Chaurasia[3] and Shridha Chaurasia[1]**

[1]Department of Botany, Government P.G. College, Tikamgarh (M.P.) 472001, India.
[2]Department of Boatany, Government College, Palera District, Tikamgarh (M.P.) 472221, India.
[3]Department of Botany, Government College, Jatara District, Tikamgarh (M.P.) 472118, India.

**The production of polygalacturonase (PG), polymethylgalacturonase (PMG) and cellulase (Cx) enzymes have been studied at different incubation period. The fruit rot pathogen of tomato *Alternaria solani* was cultured at 28°C on semi ripe tomato fruit broth medium for 2, 4, 6, 8, 10 and 12 days. Isolated pathogen *A. solani* has a capability of producing PG, PMG and Cx enzyme within short incubation period, that is, 2 days incubation. Production of these enzymes gradually increased by increasing the length of incubation period up to 6 days and further increase in length of incubation up to 12 days, did not show any effect on the production of PG, PMG and Cx. The 6 days incubation was found to be the best incubation period for the production of all these three enzymes. The production of PG and Cx was also more in comparison with PMG enzyme. The enzyme activity in healthy and diseased semi ripe tomato fruits was also assayed at different incubation periods and it was found to be maximum in diseased tomato fruits in comparison with healthy tomato fruits. It was also found that PG and PMG enzymes were produced in higher concentration than the Cx enzyme. Among six incubation periods, 6 days incubation period was found best for maximum production of all these three enzymes.**

**Key words:** *Alternaria solani*, semi ripe tomato fruit medium, semi ripe tomato fruit, incubation period.

## INTRODUCTION

Fungal diseases of fruit and vegetable plants are known to cause great damages all over the world. Tomato (*Lycopersicon esculentum* Mill.) is the most ancient among the vegetable fruits. Among the fungal diseases, fruit rot is the most severe disease of tomato, which is caused by *Alternaria solani* (Mehta, 1973; Chaurasia, 2001; Chaurasia and Chaurasia, 2010 and Chaurasia et.al., 2013a).

Pectin is present in the middle lamella of cell wall, in the form of magnesium and calcium pectate. They are complex and colloidal in nature and mostly comprises of anhydroglacturonic acid units, linked together by α-1,4 glycosidic linkages to form a chain like structure of poly-galacturonic acid units. The middle lamella is degraded by pectolytic enzymes secreted by the phytopathogens. However, the production and activity of pectolytic enzymes by fungal pathogens is an important process of pathogenesis (Gupta, 1956; Wood, 1960; Brown, 1965; Bateman and Millar, 1966; Mehrotra et al., 1988).

Next to pectin, the cellulose is the major component and basic unit of structural framework of plant cell walls, which act as barrier. The pathogens feed successfully only after breaking this barrier with the help of cellulolytic enzyme. Cellulolytic enzymes secreted by pathogen play on important role on living plant tissues in softening them and may further participate indirectly by hydrolysing cellulose into soluble saccharides which serve as food for pathogen. The production of cellulolytic enzymes and degradation of cellulose by several fungi has been studied by many workers like Reese (1956, 1963), Basu and Ghose (1960), Gascoigne and Gascoigne (1960), Bateman (1964), Norkrans (1967), Singh and Singh (1988), Kaur et al. (1989) and Sohail et al. (2011). Production of cellulolytic enzymes by several pathogenic fungi and its role in the development of diseases has also been reported (Lucas and Sherwood, 1966; Hasija and Batra, 1982; Sanghi and Rajak, 1987; Singh et al., 1991; Singh and Shukla, 1999; Sharma, 2000; Chaurasia, 2001; Shukla and Dwivedi, 2012; Jat et al., 2013, Chaurasia et al., 2013b, c).

Keeping the above facts in mind, in the present paper, the production and activity of pectolytic and cellulolytic enzymes by *A. solani* in culture and in the healthy and diseased tomato fruits at different incubation periods have been investigated.

## MATERIALS AND METHODS

### Organism

*A. solani* (Ellis & Marts) Jones & Grout was isolated from the diseased fruits of tomato (Chaurasia and Chaurasia, 2010). The organism was maintained on potato dextrose agar slants.

### Production of enzymes *in vitro*

The semi ripe tomato fruit medium tested by Chaurasia et al. (2013b) was used to study polygalacturonase (PG), polymethyl galacturonase (PMG) and cellulase (Cx) enzyme production *in vitro* at different incubation periods. The pathogen *A. solani* was grown in 150 ml Erlenmeyer flask containing 25 ml of semi ripe tomato fruit medium. The flasks were sterilized at 15 lb/sq in pressure for 15 min. After sterilization, each flask was inoculated by a 8.0 mm disc taken from the periphery of four day old colony of the pathogen growing on potato dextrose agar medium. The inoculated flasks were incubated for 2, 4, 6, 8, 10 and 12 days under stationary conditions at 28°C. Three replicates were taken in each case. After desired incubation, fungal mat was removed from the medium and the culture fluid was diluted with 35 ml of distilled water. The filtrates thus obtained was centrifuged for 15 min at 10,000 rpm and the supernatant was used as crude enzyme extract.

### Production of enzymes *in vivo*

Semi ripe tomato fruits were used for this purpose. Healthy semi ripe tomato fruits were first surface sterilized with 0.01% mercuric chloride and then washed several times with sterilized distilled water. The surface sterilized fruits were inoculated with *A. solani* by cavity method (Granger and Hornes, 1924; Chaurasia et al., 2009; Chaurasia and Chaurasia, 2010). The inoculated fruits were kept in sterilized moist chambers having 98-100% relative humidity. These chambers were incubated at 28°C. Healthy surface sterilized fruits were kept as control. After 2, 4, 6, 8, 10 and 12 days of incubation, the infected tissue was taken out with the help of a spatula. Thirty grams of diseased tissue was mixed with 30 ml distilled water and the mixture was homogenized in Waring blender for 10 min. The homogenate thus obtained was centrifused for 20 min at 10,000 rpm. The supernatant was used as enzyme extract. Enzyme extract from uninoculated healthy tissue was also prepared in a similar manner.

### Assay of enzyme activity

Enzyme activity was measured by using standard viscometric method (Hancock and Millar, 1965; Capellini, 1966; Chaurasia et.al., 2013c).

Oswald viscometers were clamped in stands which were fixed vertically in water bath, with temperature adjusted to 28°C. For the assaying of PG, PMG and $C_X$ enzymes, the following freshly prepared substrate components were used:

1. Polygalacturonase (PG): 1.2% sodium polypectate - 3.5 ml; distilled water- 1.5 ml; citrate phosphate buffer (pH 4.6) - 1.5 ml;
2. Polymethylgalacturonase (PMG): 1.2% citrus pectin - 3.5 ml; distilled water- 1.5 ml; citrate phosphate buffer (pH4.6) - 1.5 ml.
3. Cellulase ($C_X$): 1.2% carboxymethyl cellulose- 3.5 ml; distilled water- 1.5 ml; citrate phosphate buffer (pH 5.5) - 1.5 ml.

At the time of determination of PG, PMG and $C_X$ enzyme activity, desired substrate component was taken into the stalk bulb of viscometer. Then, 1.5 ml of freshly prepared enzyme extract was poured into viscometer and then efflux time of the enzyme reaction mixture was determined at the intervals of 0, 20, 40, 60 and 80 min. Efflux time for 8.0 ml of distilled water was also noted in each viscometer.

### Determination of percent loss in viscosity

Percent loss in viscosity was calculated with the help of the following formula (Capellini, 1966; Chaurasia et.al. 2013c): where, $ET_0$ = Efflux time in seconds at zero time/control. $ET_t$ = efflux time in seconds at any specific interval of time. $ET_w$ = efflux time

$$\text{Percent loss in viscosity} = \frac{ET_0 - ET_t}{ET_0 - ET_W} \times 100$$

**Table 1.** Production of PG, PMG and Cx enzymes by *A. solani* in semi ripe tomato fruit medium at different incubation periods.

| Days after incubation | Polygalacturonase (PG) | | | | | Polymethylgalacturonase (PMG) | | | | | Cellulase ($C_X$) | | | | |
|---|---|---|---|---|---|---|---|---|---|---|---|---|---|---|---|
| | Enzyme activity (%Loss in viscosity) | | | | | Enzyme activity (%Loss in viscosity) | | | | | Enzyme activity (%Loss in viscosity) | | | | |
| | Reaction time (min) | | | | | Reaction time (min) | | | | | Reaction time (min) | | | | |
| | 20 | 40 | 60 | 80 | REA | 20 | 40 | 60 | 80 | REA | 20 | 40 | 60 | 80 | REA |
| 2 | 32.5 | 43.4 | 50.3 | 53.2 | 65.01 | 7.2 | 15.3 | 19.4 | 21.2 | 0.00 | 42.5 | 52.3 | 58.2 | 62.2 | 85.03 |
| 4 | 53.4 | 66.3 | 76.0 | 78.3 | 106.83 | 9.3 | 19.3 | 25.4 | 30.0 | 16.93 | 50.4 | 62.3 | 68.2 | 71.4 | 100.80 |
| 6 | 56.3 | 68.2 | 77.4 | 80.2 | 112.61 | 10.4 | 22.4 | 30.2 | 40.3 | 20.13 | 66.6 | 78.4 | 84.4 | 88.3 | 133.33 |
| 8 | 37.5 | 47.3 | 52.1 | 55.1 | 75.01 | 8.2 | 13.4 | 16.3 | 18.3 | 0.00 | 31.2 | 41.3 | 47.5 | 50.2 | 62.42 |
| 10 | 32.3 | 42.4 | 48.3 | 51.4 | 64.64 | 8.0 | 11.3 | 14.6 | 16.2 | 0.00 | 20.3 | 29.4 | 35.4 | 38.2 | 29.40 |
| 12 | 28.0 | 32.2 | 35.3 | 37.3 | 56.02 | 7.4 | 9.8 | 11.2 | 13.4 | 0.00 | 18.0 | 26.2 | 31.3 | 33.6 | 26.20 |

in seconds for distilled water.

#### Determination of relative enzyme activity (REA)

Values for percent loss in viscosity were determined for 0, 20, 40, 60 and 80 min reaction time. These values were then plotted against the reaction time, thus a curve was obtained and from this curve, the time to bring a 25% loss in viscosity was determined. REA was then calculated using the following formula:

$$REA = \frac{1000}{t}$$

Where, t = represent the time in min to reach 25% loss in viscosity. Thus:

$$\mathbf{REA} = \frac{\mathbf{1000}}{\mathbf{25}}$$

## RESULTS AND DISCUSSION

### Production of PG, PMGand Cx enzymes *in vitro*

The pathogen *Alternaria solani* was cultured in the semi ripe tomato fruits medium and incubated for 2, 4, 6, 8, 10 and 12 days at 28°C. In obtained culture filtrates, the activity of PG, PMG and Cx enzymes was assayed and results are presented in Table 1 and Figure 1. It is clear that pathogen *A. solani* was able to produce PG, PMG and Cx enzymes with a short period of incubation, that is, within 2 days. The gradual increase in incubation period up to 6 days, the activity of PG, PMG and Cx enzymes significantly increased in the culture filtrates. The 6 day of incubation period was found to be the best for the maximum production of PG, PMG and Cx enzymes as 80.2 (REA 112.61), 40.3 (REA 20.13) and 88.3 (REA 133.33) percent loss in viscosity has been recorded respectively at 80 min of reaction time. After 6 days, further increase in incubation period up to 12 days, has no effect and production of PG, PMG and Cx enzymes declined gradually. Mehta et al. (1974) has reported the maximum production of PG, PMG and Cx enzymes in between 4 to 12 days, in culture filtrates of *A. solani* and *Alternaria tenuis*. Mehta and Mehta (1985) also reported maximum PG, PMG and Cx enzyme activity in 6 days old culture of *Fusarium oxysporium*. The gradual fall in production of PG, PMG and Cx enzymes in long incubation period could be due to the slow inactivation of PG, PMG and Cx enzymes by the appearance of oxidized phenols in the semi ripe tomato fruit medium. The same explanation could hold for the observation of Harter and Weimer (1921) on the production of pectic enzyme by *Rhizopus nigricans* and *Rhizopus artocarpi*, in sweet potato broth. Similar observations were made by Balasubramanian and Srivastava (1973).

### Production of PG, PMG and Cx enzymes *in vivo*

Data (Table 2 and Figure 2), indicates the activity of PG, PMG and Cx enzymes in the diseased and healthy tissue of semi ripe tomato fruits, in different incubation periods. From the results, it is evident that PG, PMG and Cx enzymes present in healthy tomato fruits and maximum activity of these enzymes were recorded in 6 days incubition period. After 6 day incubation, with the age, the activity of PG, PMG and Cx enzymes was decreased gradually in healthy tissues. It was also observed that the activity of PG and Cx enzymes has completely disappeared in healthy tomato fruits kept for 12 day incubation.

From the above result, it is concluded that the presence of PG, PMG as well as Cx enzyme in healthy tissues shows their constitutive nature. Comparatively, PG enzyme has been found to be

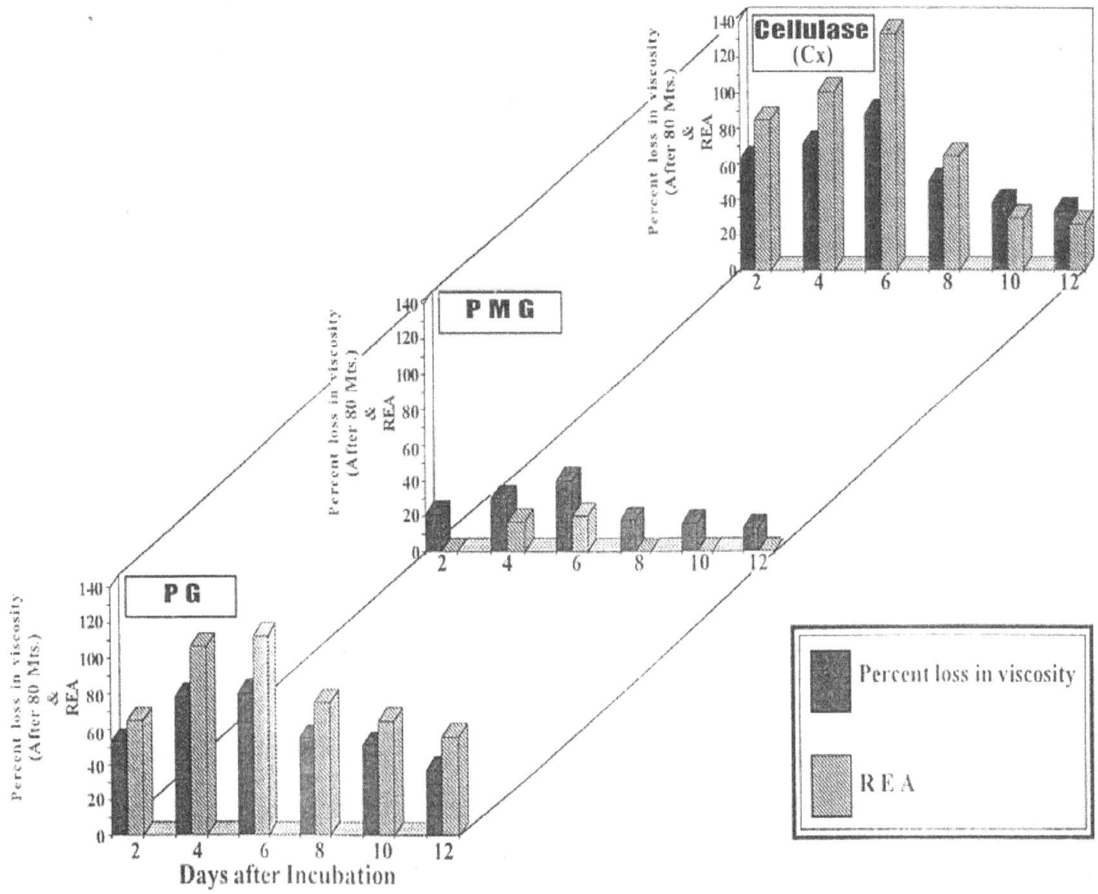


**Figure 1.** Production of PG, PMG and Cx enzymes by *A. solani* in semi ripe tomato fruit medium at different incubation periods.

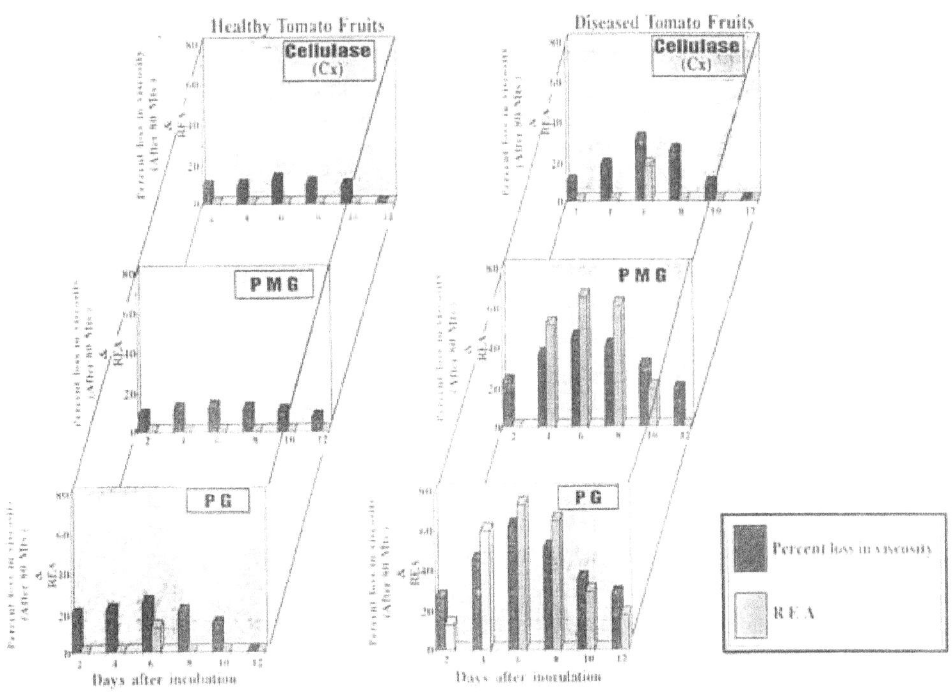


**Figure 2.** Production of PG, PMG and Cx enzymes in healthy and diseased semi ripe tomato fruits at different incubation periods.

**Table 2.** Production of PG, PMG and Cx enzymes by *A. solani* in healthy and diseased semi ripe tomato fruits at different incubation periods.

| Days after incubation | Polygalacturonase (PG) | | | | | Polymethylgalacturonase (PMG) | | | | | Cellulase (Cx) | | | | |
|---|---|---|---|---|---|---|---|---|---|---|---|---|---|---|---|
| | Enzyme activity (%Loss in viscosity) | | | | | Enzyme activity (%Loss in viscosity) | | | | | Enzyme activity (%Loss in viscosity) | | | | |
| | Reaction time (min) | | | | | Reaction time (min) | | | | | Reaction time (min) | | | | |
| | 20 | 40 | 60 | 80 | REA | 20 | 40 | 60 | 80 | REA | 20 | 40 | 60 | 80 | REA |
| | | | | | | Healthy tomato fruits | | | | | | | | | |
| 2 | 8.00 | 14.30 | 18.00 | 19.00 | 0.00 | 6.00 | 7.00 | 8.00 | 8.00 | 0.00 | 4.60 | 6.00 | 7.50 | 8.50 | 0.00 |
| 4 | 8.00 | 16.00 | 19.20 | 21.10 | 0.00 | 6.20 | 8.00 | 10.00 | 11.00 | 0.00 | 5.00 | 7.5. | 8.60 | 9.00 | 0.00 |
| 6 | 8.20 | 17.40 | 22.80 | 25.20 | 12.60 | 6.40 | 8.60 | 10.20 | 12.10 | 0.00 | 7.20 | 9.30 | 11.40 | 12.00 | 0.00 |
| 8 | 8.00 | 15.10 | 18.30 | 20.30 | 0.00 | 6.10 | 8.00 | 10.00 | 11.20 | 0.00 | 7.00 | 8.00 | 9.50 | 10.00 | 0.00 |
| 10 | 6.20 | 10.20 | 13.40 | 14.00 | 0.00 | 5.40 | 7.20 | 9.50 | 10.00 | 0.00 | 6.50 | 7.00 | 8.00 | 9.00 | 0.00 |
| 12 | 0.00 | 0.00 | 0.00 | 0.00 | 0.00 | 4.50 | 6.00 | 7.00 | 7.00 | 0.00 | 0.00 | 0.00 | 0.00 | 0.00 | 0.00 |
| **Days after inoculation** | | | | | | Diseased tomato fruits | | | | | | | | | |
| 2 | 16.40 | 21.40 | 24.10 | 26.30 | 13.15 | 12.20 | 17.50 | 20.80 | 22.00 | 0.00 | 5.40 | 7.00 | 8.00 | 9.00 | 0.00 |
| 4 | 30.00 | 39.60 | 43.30 | 45.20 | 7.02 | 25.60 | 31.40 | 34.60 | 36.10 | 51.20 | 6.00 | 11.30 | 15.30 | 17.50 | 0.00 |
| 6 | 36.50 | 48.20 | 57.70 | 62.20 | 73.04 | 32.60 | 38.30 | 42.50 | 44.20 | 65.23 | 13.30 | 21.60 | 26.50 | 30.30 | 17.66 |
| 8 | 32.60 | 43.30 | 48.30 | 51.30 | 65.23 | 30.50 | 35.00 | 38.30 | 40.00 | 61.01 | 11.70 | 18.40 | 22.30 | 24.60 | 0.00 |
| 10 | 19.00 | 29.40 | 33.40 | 35.40 | 29.40 | 18.60 | 24.60 | 28.40 | 30.30 | 18.93 | 4.50 | 6.50 | 7.80 | 8.00 | 0.00 |
| 12 | 12.40 | 22.10 | 26.50 | 28.10 | 17.66 | 9.40 | 14.00 | 17.00 | 18.00 | 0.00 | 0.00 | 0.00 | 0.00 | 0.00 | 0.00 |

more active in healthy tissues than the PMG and Cx enzyme. Ulrich (1958), Mc Cready et al. (1955) has also reported the higher activity of PG enzyme in healthy tissues of various fruits.

In inoculated diseased semi ripe tomato fruit, the activity of PG, PMG and Cx enzymes were found to be higher in comparison with healthy tissue at various given incubation period. Within a short incubation period, that is, within 2 day, activity of PG, PMG and Cx enzymes has been noted in diseased tissues, which increased with increased in incubation period up to 6 days. The 6 days incubation period was found to be the best for the production of PG, PMG and Cx enzymes in diseased tissues, in which 73.04, 65.23 and 0.0 REA were observed respectively. After 6 day incubation, further increase in incubation period had no effect on the production of PG, PMG and Cx enzymes. In 12 day old infected tissues, the Cx activity has completely disappeared which indicates the inactivation of Cx enzyme in a very long incubation period.

Comparatively, the activity of PG and PMG enzymes were found to be always higher than the Cx enzyme in diseased tissues of various incubition periods, which indicated that PG and PMG enzymes played a significant role in pathogenesis of *A. solani*. To some extent, these results are similar and in agreement with the work of Mehta et al. (1974), Agarwal and Gupta (1978), Hasija and Chawdhury (1979), Sanghi and Rajak (1987) and Sharma (2000).

In general, production of PG, PMG and Cx was higher *in vitro* than *in vivo*. This could be due to the negligible inactivation of these enzymes by phenols present in the semi ripe tomato fruit medium or perhaps the fungus is forced to secrete the enzymes in large quantities to spread in the semi ripe tomato fruit medium. This situation does not prevail when the fungus is grown in host tissue. The ability of the fungus to secrete PG, PMG and Cx *in vitro* and *in vivo* indicates the importance of these enzymes in the pathogenesis of the fungus.

## Conflict of Interests

The author(s) have not declared any conflict of interests.

### REFERENCES


Agarwal GP, Gupta S (1978). Production of pectic enzymes by *Corynespora cassicola* causing fruit rot of papaya. Curr. Sci. 45(5):161-163.

Balasubramanian KA, Srivastava DN (1973). In vitro and in vivo production of pectic enzymes by *Rhizopus stolonifer*. Indian J. Exp. Biol. 11:127-128

Basu SN, Ghosh SN (1960). The production of cellulase by fungi on mixed cellulosic substances. Can. J. Microbiol 6:265-288.

Bateman DF (1964). Cellulase and *Rhizoctonia* disease of bean. Phytopathology 54(11):1372-1377.

Bateman DF, Millar RL (1966). Pectic enzymes in tissue disintegration. Ann. Rev. Phytopath. 4:119-146.

Brown W (1965). Toxin and cell wall dissolving enzymes in relation of plant disease. Ann. Rev. Phytopath. 3:1-18.

Capellini RA (1966). Growth and polygalacturonase producting by *Rhizopus stolonifer*. Phytopathol. 56:734-737.

Chaurasia Amit Kumar (2001). Studies on the fruit rot disease of tomato (*Lycopersicon esculentum* Mill.) caused by *Alternaria solani* (Ellis & Mart.) Jones & Grout. M.Sc. thesis, Dr. H.S. Gaur Univ. Sagar (M.P.) India pp. 68.

Chaurasia AK, Chaurasia Shirdha (2010). Effect of different inoculation methods and temperatures on tomato fruit rot pathogen *Alternaria solani* and its host range. Biozone Int. J. Life Sci. 2(1&2):236-243.

Chaurasia AK, Chaurasia Shridha, Chaurasia SC, Yadav VK (2009). Studies on the host range of apple isolate of *Botryodiplodia theobromae* Pat. Biozone Int. J. Life Sci. 1(1):34-38

Chaurasia Amit Kumar, Chaurasia Shridha, Chaurasia Sushmita, Chaurasia Shubha (2013a). Studies on the development of fruit rot of tomato caused by *Alternaria solani* (Ellis & Mart.) Jones & Grout. Int. J. Pharm. Life Sci. 4(6):2713-2716.

Chaurasia Amit Kumar, Chaurasia Shridha, Chaurasia Shubha, Chaurasia Sushmita (2013b). Production of polygalacturonase, polymethylgalacturonase and cellulase enzymes by *Alternaria solani*. Nat. J. Life Sci. in press.

Chaurasia Amit Kumar, Chaurasia Sushmita, Chaurasia Shubha, Chaurasia Shridha, Chaurasia SC (2013c). Influence of temperature on production of polygalacturonase, polymethylgalacturonase and cellulase enzymes by *Alternaria solani in vitro*. Int. J. Curr. Res. 5(7):1648-1650.

Gascoigne JA, Gascoigne MM (1960). Biological degradation of cellulase. Butler worths & Co. London.

Granger K, Horne AS (1924). A method of inoculating the apples. Ann. Bot. 38:212.

Gupta SC (1956). Studies in the physiology of parasitism XXII. The production of pectolytic enzymes by *pythium debaryanum* Hasse. Ann. Bot. 20:179-190.

Hancock JB, Millar RL (1965). Association of cellulotytic, Proteolytic and xylolytic enzymes with southern anthracnose, spraying block stem and *Stomphylium* leaf spot of alfalfa. Phytopathology 55:356-360.

Hasija SK, Batra S (1982). Production of cellulolytic enzymes by *Phoma destructiva*. Indian Phytopath. 35(3):384-387.

Hasija SK, Chawdhury SR (1979). Pectic and cellulolytic enzymes by *Phomopsis vexans*. Acta biol. Acad. Sci. Hung. 30(1-2):161-163.

Jat BL, Sharma P, Gaur HN (2013). Production of enzymes and culture filtrates by *Colletotrichum gloeosparioides* penz. causing banana fruit rot. Proc. Nat. Acad. Sci., India 83B(2):177-180.

Kaur G, Gupta AK, Mehrotra RS (1989). Effect of different carbon sources on the activity of pectolytic and cellulolytic enzymes *Phtophthora drechsleri* var. *cajani*. Plant Dis. Res. 4(2):198.

Lucas GB, Sherwood RT (1966). Role of pectic enzymes of *Alternaria tenuis* in brown spot disease of tobacco. *Proc. IVth Inter.,* Tobacco scientific cong. Athens. pp. 410-416

Mc Cready RM, Mac Comb EA, Jansen EF (1955). The action of tomato and avocado polygalacturonase. Food. Res. 20:186-191.

Mehrotra RS, Agarwal A, Kaur G (1988). Enzymic degradation of host cell walls by different *Phytophthora* sp. with special reference to the work done in india. In current trends in physiological plant pathology. Association of plant pathologists of India, Lucknow. pp. 89-111.

Mehta A, Mehta P (1985). Production of pectolytic and cellulolytic enzymes by *Fusarium oxysporum* and *F. Moniliformae* under different cultivation conditions. Folia Microbiol. 30:42-50.

Mehta P (1973). Biochemical investigations of fruit rot diseases of tomatose (with special reference to the pathogenesis caused by *Alternaria solani* (Ellis & Mart.), Jones & Grout and *Alternaria tenuis* Auct.) Ph.D. Thesis, Saugar Univ. Sagar (M.P.) India pp. 413.

Mehta P, Vyas KM, Saksena SB (1974). Production of pectolytic and cellulolytic enzymes by *Alternaria* sp. during pathogenesis of tomato fruits. Hindustan Antibiot. Bull. 16(4):210-214.

Norkrans B (1967). Cellulase and cellulolysis. Adv. Appl. Microbiol. 9:91-130.

Reese ET (1956). A microbiological progress report enzymatic hydrolysis of cellulose. Appl. Microbiol. 4:39-45.

Reese ET (1963). Advances in enzyme hydrolysis of cellulose and related materials. Mcmillan & Co. New York.

Sanghi N, Rajak RC (1987). Production of pectic and cellulolytic enzymes by two strains of *Myrothecium roridum* causing fruit rot of tomato. Prof. G.P. Agarwal Festschrift Volume. Today and Tomorrow's printers and publishers, New Delhi -*5* pp. 111-116.

Sharma RL (2000). "*In vivo*" and "*in vitro*" production of cellulolytic enzymes by *Monilinia* sp. causing brown rot of apple. Indian J. Mycol. Plant Pathol. 30(1):31-33.

Singh PL, Gupta MN, Singh AL (1991). Cell wall degrading enzymes production by seed brone fungi isolated from chilgoza. Indian J. Mycol. Plant Pathol. 21(3):263-267.

Singh RS, Singh OS (1988). Quantification of cellulase produced by *Macrophomina phaseolina* and Pythium butleri. Plant Dis. Res. 3(2):229-231.

Singh SN, Shukla P (1999). Variation in the production of enzymes by different isolates of *Myrothecium roridum* caused leaf spot of mungbean. Indian Phytopath. 52(1):63-65.

Sohail M, Ahmad A, Khan SA (2011). Production of cellulases from *Alternaria* sp. MS28 and their partial characterization. Pak. J. Bot. 43(6):3001-3006.

Shukla A, Dwivedi SK (2012). Pathogenic action of Cx, PG and PMG enzymes of *Fusarium udum* and *Fusrium oxysporum f. sp. ciceri*. Int. J. Curr. Res. 4(6):111-113

Ulrich R (1958). Post harvest physiology of fruits. Ann. Rev. Plant Pathology. 9:384-416.

Wood RKS (1960). Pectic and cellulolytic enzymes in plant disease. Ann. Rev. Plant Physiol. 11:299-322.

# 31

# Cloning and expression of anthocyanidin synthase (ANS) gene from peel of mango (*Mangifera indica* Linn)

**Zhao Zhichang, Chen Yeyuan, Gao Aiping, Huang Jianfeng and Dang Zhiguo**

Tropical Crops Genetic Resources Institute, Chinese Academy of Tropical Agricultural Sciences/Key Laboratory of Crop Gene Resources and Germplasm Enhancement in Southern China, Ministry of Agriculture, Danzhou 571737, P. R. China.

**Fruit coloring is a major source of anthocyanin and carotenoid in mango. The anthocyanidin synthase (ANS) is a key enzyme in the anthocyanin biosynthesis. The *ANS* gene was cloned from mango peel by homologous cloning method. This manuscript reported characterization of *ANS* from mango peel that comprised 1,262 bp full-length cDNA with open reading frame (ORF) of 1,056 bp and encoding a protein of 351 amino acids. The theoretical molecular weight of the deduced amino sequence of *ANS* was 39.8 kDa. It was found that the gene encoding for the protein had close relationship with mountain grape (*Vitis vinifera*), cocoa (*Theobroma cacao*), mulberry (*Morus alba*), litchi (*Litchi chinensis*), and other fruit trees through phylogenetic analysis. Expression of *ANS* was maximum in the green rather than the yellow and red fruit peel. Expression was down-regulated in mature fruit, there was no response to fruit coloration which was affected by anthocyanin.**

**Key words:** Cloning, expression, anthocyanidin synthase (*ANS*), mango.

## INTRODUCTION

The anthocyanins are one of the most widespread studied members of the tricyclic flavonoid family of secondary metabolites in plants (Grotewold, 2006). The anthocyanins contribute to some of the most important pigments in flowers and fruits for flower pollination or fruit and seed dispersal, and also the precursors of tannin oligomers present in tea and red wine, which have long-established biomedicinal properties, including inhibition of cell proliferation and antimutagenic, antimicrobial, anti-inflammatory, antioxidant and antihypertensive properties (Pool-Zobel et al., 1999; Harborne and Williams, 2000; Akihisa et al., 2003; Parejo et al., 2004). The structural genes that encode enzymes involved in the anthocyanin pathway and many regulatory genes for transcriptional regulation of the structural genes have been cloned and characterized from a wide variety of plants (Mol et al., 1998; Holton and Cornish, 1995; Lesnick and Chandler, 1998).

Anthocyanidin synthase (ANS), an enzyme of the biosynthetic pathway to anthocyanin, catalyzes the reaction(s) from the colorless leucoanthocyanidins to the colored anthocyanidins. The ANS proteins belong to the 2-oxoglutarate iron-dependent oxygenases and were cloned first from *Perilla frutescens* (Saito et al., 1999). In a series of studies on recombinant ANS from *Arabidopsis thaliana* evidence was presented for the initial oxidation of the substrate at C-3 (Welford et al., 2001).

Furthermore, the ANS formed predominantly quercetin

and cis and trans-dihydroquercetin (DHQ) with cyanidin being a minor product only (Welford et al., 2001; Turnbull et al., 2003), and the product pattern from (2R, 3S, 4S) cis-leucocyanidin vs. that from (2R, 3S, 4R)-trans-leucocyanidin implied that cis-DHQ, trans-DHQ and cyanidin resulted mostly from the unnatural (2R, 3S, 4R)-trans-leucocyanidin (Turnbull et al., 2003). Moreover, co-crystallization of the ANS with $Fe^{2+}$, 2-oxoglutarate and racemic trans-DHQ or enantiomerically pure (2R, 3R)-DHQ as a substrate analogue in the absence or presence of molecular oxygen supported the stereoselective C-3 hydroxylation of the substrate and surprisingly revealed two molecules of the substrate analogue in the active site with (2R, 3R)-trans-DHQ closest to the iron atom, whereas either enantiomer was bound at the other location (Wilmouth et al., 2003).

Clearly, additional data are required to define the substrate specificity of ANS. Mango (*Mangifera indica* Linn) is one of the delicious seasonal fruits grow in the tropics, and cultivated in many regions of India and now distributed wide across the world in many continents. Mango fruit is one of the most popular, nutritionally rich fruits with unique flavor, fragrance, taste, and heath promoting qualities making it a common ingredient in new functional foods often labeled "The king of the fruits." Internally, it has juicy flesh features, orange-yellow in color with numerous soft fibrils radiating from the husk. Its flavor is pleasant and rich, and tastes sweet with mild tartness. Outer skin of mango is smooth and green in un-ripe, but turns into golden yellow, bright yellow or orange-red when ripen depending upon the cultivar.

In the present study, we reported the isolation of cDNA clone coding for an ANS gene from mango. This is the first report on ANS gene from mango peel, these studies add to the understanding of the biosynthesis of anthocyanin in tropical fruit, with significant practical implications for enhancing the specific biomedical anthocyanin contents in mango by genetic engineering.

## MATERIALS AND METHODS

### Plant materials

Mango (*M. indica* Linn) cv "Guifei" cultivar was grown at Tropical Crops Genetic Resources Institute, Chinese Academy of Tropical Agricultural Sciences in Danzhou, China. Before the commercially mature stage, green, yellow and red peels were taken from at least 20-30 fruit at a month interval throughout the growing season. Fresh fruit samples were taken to laboratory and immediately frozen in liquid nitrogen and stored at -80°C.

### cDNA of *ANS* cloning and expression

Total RNA was isolated from mango peel sample according to the method of Wan and Wilkins (1994). Reverse transcription was achieved with oligo-d $(T)_{18}$ primers by using 3 µg of total RNA from peel of mango litchi and the primeScript II 1st strand cDNA synthesis kit (TaKaRa). To isolate partial cDNA clone, two oligonucleotides designed by on the basis of conserved amino acid sequences of several available ANSs were used: forward primer (5' AAGTACATCCATCCAGTCATYTT '3); reverse primer (5' GAATCCAAGGATCCYGAGAAYGA 3'). The PCR reaction was carried out using the following conditions: 94°C, 3 min (1 cycle); 94°C, 30 s; 55°C, 30 s; 72°C, 1 min (30 cycles); 72°C, 5 min (1 cycle). Product of 280 bp, whose identity was confirmed by sequencing was amplified. The 3'RACE and 5'RACE of *ANS* gene was gotten according to 3'-Full RACE Core Set ver.2.0 and 5'-Full RACE Kit (TaKaRa in Dalian, China). For the analysis of *ANS* expression in three phages of mango peel, RT-PCR was performed for 30 cycles to determine the linearity of the PCR. The thermal cycling parameters used for the RT-PCR for all genes were as follows: 94°C for 30 s, 56°C for 30 s, 72°C for 100 s; followed by 72°C for 10 min. The cDNA was amplified from 30 ng of total RNA, using specific primers sets for, *ANS*-F 5' ATTATGGCAGTGTTATCAATCGGGT'3, *ANS*-R 5' GGGAGGAGCGTGGAAAGTCGTCGTA'3. As a positive control, *actin7* fragment was amplified under the same RT-PCR conditions, using the primer pair: *Actin7*-F 5'AATGGAACTGGAATGGTCAAGGC'3, *Actin7*-R5' TGCCAGATCTTCTCCATGTCATCCCA '3.

### Sequence analysis

Nucleotide and amino acid sequence comparisons were performed using the LASERGENE DNA software package (DNAstar, Madison, WI). Sequence similarity searching was performed using the BLASTN and BLASTX, nucleic and protein databases at NCBI (http://www.ncbi.nlm.nih.gov). The phylogenetic analysis of ANS from mango and other species was carried out by alignment with the bioEdit software (http://www.mbio.ncsu.edu/BioEdit/bioedit.html). The phylogenetic and molecular evolutionary analysis used MEGA (molecular evolutionary genetics analysis) version 5.0 (Kumar et al., 2004).

## RESULTS

### Isolation of the anthocyanidin synthase (ANS) gene from mango

To obtain *ANS* from mango, the 280 bp cDNA *ANS* fragments were isolated with two oligonucleotides primers by RT-PCR. The entire length of the cDNA *ANS* was gotten with the rapid amplification of cDNA ends (RACE) reactions. The 920 and 530 bp cDNA *ANS* fragments were gotten by the 3'-RACE and 5'-RACE reactions.

The isolated *ANS* cDNA had a 1,262 bp full-length with ORF of 1,056 bp, and encoding a protein of 351 amino acids (Figure 1). The theoretical molecular weight and isoelectric point of the deduced amino sequence of *ANS* were 39.8 kDa and 5.68, respectively.

### Sequence analysis of anthocyanidin synthase (ANS) from mango

A search for amino acid sequence homology revealed high sequence identity (*Arabidopsis thaliana*, GenBank number NM_118417.1, *Vitis vinifera*, GenBank Accession number ABV82967.1, *Pyrus pyrifolia*, GenBank

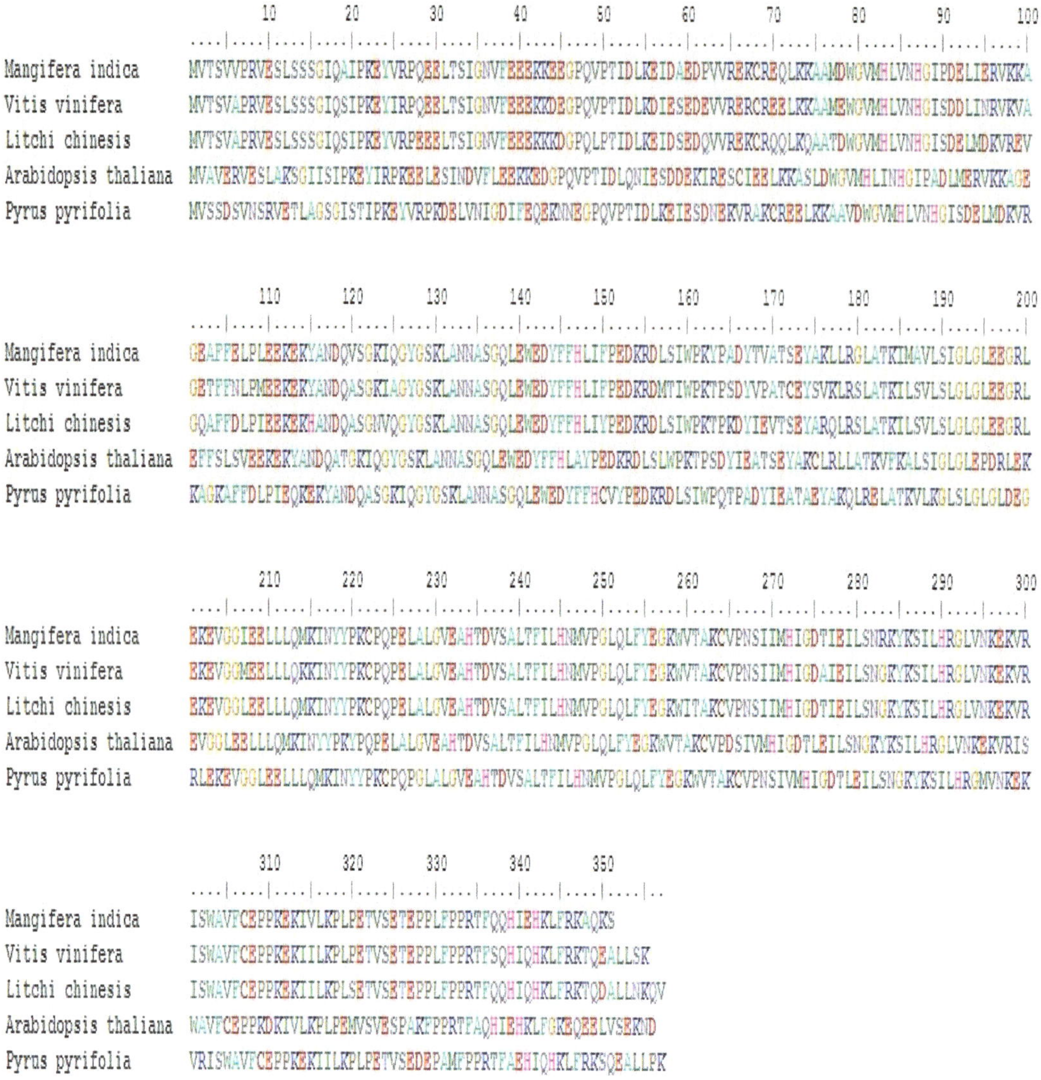


**Figure 1.** Amino acid sequence of ANS comparison between mango and other species. Sequence analysis of amino acid region proposed to determine specificity.

Accession number ADP09379.1, *Litchi chinesis* GenBank Accession number HQ402913.1) between ANS and its closest sequence homolog (Figure 1). An alignment of the deduced amino acid sequences of all these proteins indicated that ANS contain a conserved sequence with other plant in the same domain. Phylogenetic analysis was performed on ANS protein sequences. The tree confirms the conclusions that ANS of mango was most closely related to *Morus alba* (GenBank Accession number AEN55613.2), *V. vinifera* (GenBank Accession number ABV82967.1), and *Litchi chinensis* (GenBank Accession number HQ402913.1). ANS of mango from *Medicago truncatula* (GenBank Accession number ABU40983.1), and *Anthurium andraeanum* (GenBank Accession number ABK76317.1) was more distant related in this work clusters. It formed two separate branches, one branch mainly included fruit and crops, for example *Vitis*, *Litchi*, *Malus* and *Fragopyrum*, interesting, other branch mainly included flower of plants, for example, *Glycine max*, *M. truncatula* and *Ipomoea trifida* (Figure 2).

## Expression of *ANS* and measurement of anthocyaninin different stages in mango peel

Expression of *ANS* was the highest in the green stage, and the lowest in the yellow stage (Figure 3). A close correlation was also observed between the *ANS* gene expression and enzyme activity (data not shown). In this study, ANS activity was detected, and changed over the growth period. Anthocyanin was measured in the three phenotypes for peel (green, yellow and red). The green stage examined contains lowest detectable amounts of

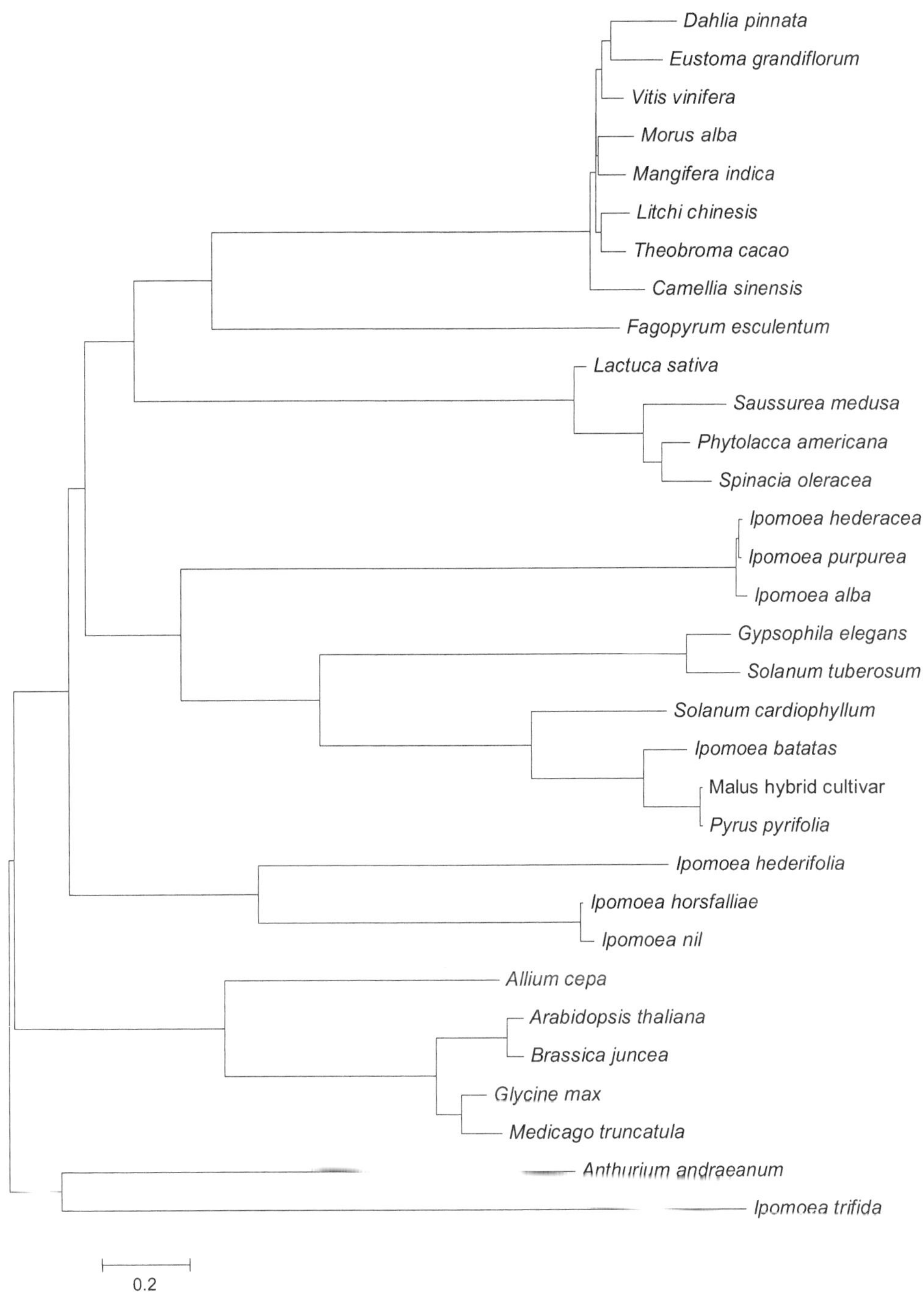


**Figure 2.** Phylogenetic relationship of ANS. The tree was constructed using MEGA (version 5.0). *Vitis vinifera* (ABV82967.1), *Gypsophila elegans* (AAP13054.1), *Glycine max* (AAR26525.1), *Ipomoea batatas* (ACT31916.1), *Arabidopsis thaliana* (NM_118417.1), *Anthurium andraeanum* (ABK76317.1), *Camellia sinensis* (AAV88087.1), *Allium cepa* (ABR24157.1), *Ipomoea batatas* (ABM63373.1), *Dahlia pinnata* (BAJ21536.1), *Brassica juncea* (ACH58398.1), *Ipomoea hederifolia* (BAK78919.1), *Solanum tuberosum* (AEJ90548.1), *Fagopyrum esculentum* (ADT63066.1), *Pyrus pyrifolia* (ADP09379.1), *Ipomoea nil* (BAB71811.1), *Malus* hybrid cultivar (ACP30363.1), *Solanum tuberosum (*ADP37440.1), *Eustoma grandiflorum* (BAJ08929.2), *Ipomoea trifida* (ACF59729.1), *Lactuca sativa* (BAJ10383.1), *Saussurea medusa* (AAS48200.1), *Ipomoea batatas* (ADE08370.1), *Theobroma cacao* (ADD51356.1), *Ipomoea purpurea* (AAP82030.1), *Glycine max* (ABY51685.1), *Medicago truncatula* (ABU40983.1), *Allium cepa* (ABM66367.1), *Phytolacca americana* (BAE54521.1), *Spinacia oleracea* (BAE54520.1), *Brassica juncea* (ACH58397.1), *Solanum cardiophyllum* (AEJ90547.1), *Ipomoea horsfalliae* (ACS71531.1), *Ipomoea alba* (AAP82018.1), *Morus alba* (AEN55613.2) and *Litchi chinensis (*HQ402913.1).

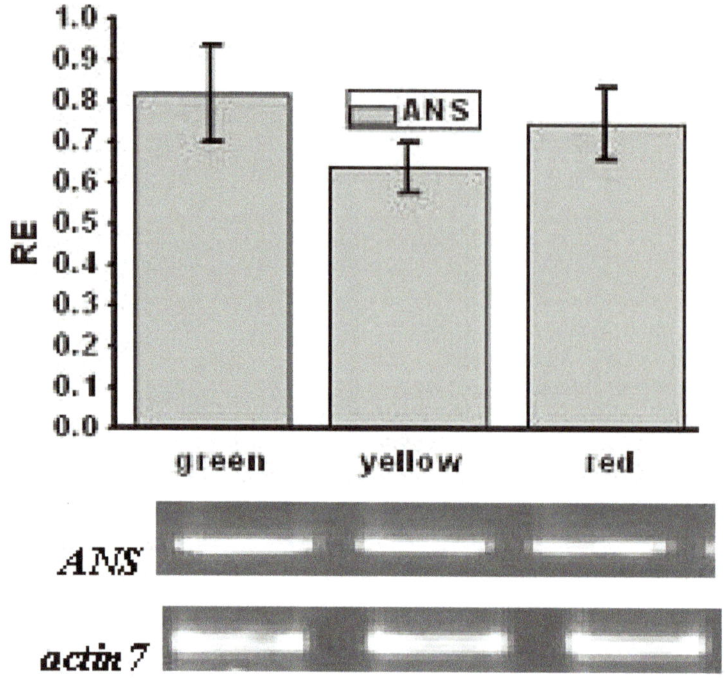


**Figure 3.** Expression of *ANS* was measured by RT-PCR at three developmental stages. Expression of *ANS* was the highest in the green stage, and the lowest in the yellow stage. Values are the average ±SD of three replicant.

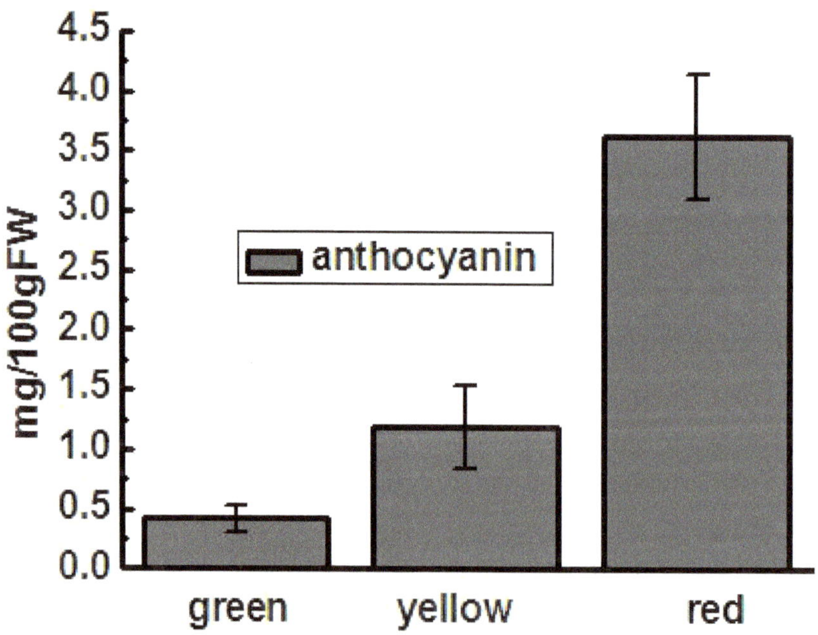


**Figure 4.** Total anthocyanin concentration in the three stages of mango peel.

anthocyanin in the peel. The accumulation of anthocyanin in the peel started at the yellow stage and was 1.12 mg/100 g. The anthocyanin was 8-fold higher in red peel as compared to green peel (Figure 4).

## DISCUSSION

Some mango cultivars do not have anthocyanin in their shaded skin, however, anthocyanin was tested in "Guifei" mango. Although several structural genes as well as some regulatory genes in the anthocyanin pathway of mango had been isolated, there are currently no reports on the *ANS* gene in anthocyanin pathway of mango. In this experiment, we initially investigated the sequence characterization, expression patterns of ANS from mango and analyzed the contents of total anthocyanin. The results indicated that the ANS is one of the key enzymes in anthocyanin-pigmentation pathway of mango, and it is responsible for the formation of anthocyanin in the outer mango peel. Molecular evolutionary trees deduced from amino acid sequences indicated that the enzymes in anthocyanin biosynthesis form a distinct group in the supper family of 2-oxoglutarate-dependent enzymes. ANS had close relationship with mountain grape (*V. vinifera*), cocoa (*Theobroma cacao*), mulberry (*M. alba*) litchi (*Litchi chinensis*), and other fruit trees through phylogenetic analysis. ANS, a key enzyme of the biosynthetic pathway to anthocyanin, catalyzes the reaction(s) from the colorless leucoanthocyanidin to the colored anthocyanidin. So the present research on isolation and characterization analysis of ANS from mango will provide an alternative to controlling the overall metabolic flux to the target products by appropriate genetic engineering strategies.

## Conflict of Interests

The author(s) have not declared any conflict of interests.

## ACKNOWLEDGEMENTS

We wish to thank He Junhu, Zhu Min and Chen Huarui in Tropical Crops Genetic Resources Institute, Chinese Academy of Tropical Agricultural Sciences. This work was supported by "the Fundamental Scientific Research Funds for CATAS-TCGRI" (Project No. 1630032013003), and "The Non-Profit Scientific Research Institution Reform Started Funding Projects for CATAS" (pzs-201225).

## REFERENCES

Akihisa T, Tokuda H, Ukiya M, Iizuka M, Schneider S, Ogasawara K, Mukainaka T, Iwatsuki K, Suzuki T, Nishino H (2003). Chalcones, coumarins, and flavanones from the exudate of Angelica keiskei and their chemopreventive effects. Cancer Lett. 201:133-137.

Grotewold E (2006). The genetics and biochemistry of floral pigments. Annu. Rev. Plant Biol. 57:761-780.

Harborne JB, Williams CA (2000). Advances in flavonoid research since 1992. Phytochem. 55:481-504.

Holton TA, Cornish EC (1995). Genetics and Biochemistry of anthocyanin biosynthesis. Plant Cell 7:1070-1083.

Lesnick ML, Chandler VL (1998). Activation of the maize anthocyanin gene a2 is mediated by an element conserved in many anthocyanin promoters. Plant Physiol. 117:437-445.

Mol J, Grotewold E, Koes R (1998). How genes paint flowers and seeds. Trends Plant Sci. 3:212-217.

Parejo I, Viladomat F, Bastida J, Schmeda-Hirschmann G, Burillo J, Codina C (2004). Bioguided isolation and identification of the nonvolatile antioxidant compounds from fennel (*Foeniculum vulgare* Mill.) waste. J. Agric. Food Chem. 52:1890-1897.

Pool-Zobel BL, Bub A, Schro¨ der N, Rechkemmer G (1999). Anthocyanins are potent antioxidants in model systems but do not reduce endogenous oxidative DNA damage in human colon cells. Eur. J. Nutr. 38:227-234.

Saito K, Kobayashi M, Gong Z, Tanaka Y, Yamazaki M (1999). Direct evidence for anthocyanidin synthase as a 2-oxoglutarate-dependent oxygenase: molecular cloning and functional expression of cDNA from a red forma of Perilla frutescens. Plant J. 17:181-189.

Turnbull JJ, Nagle MJ, Seibel JF, Welford RWD, Grant GH, Schofield CJ (2003). The C-4 stereochemistry of leucocyanidin substrates for anthocyanidin synthase affects product selectivity. Bioorg. Med. Chem. Lett. 13:3853-3857.

Welford RWD, Turnbull JJ, Claridge TDW, Schofield CJ, Prescott AG (2001). Evidence for oxidation at C-3 of the flavonoid C-ring during anthocyanin biosynthesis. Chem. Commun. 12:1828-1829.

Wilmouth RC, Turnbull JJ, Welford RWD, Clifton IJ, Prescott AG, Schofield CJ (2002). Structure and mechanism of anthocyanidin synthase from Arabidopsis thaliana. Structure. 10:93-103.

# Influence of medium type and growth regulators on *in vitro* micropropagation of pineapple (*Ananas comosus* (L.), var. Smooth cayenne)

**Pamela Eloho Akin-Idowu, Sunday Oluseyi Solomon Akinyemi and Dorcas Olubunmi Ibitoye**

Fruits and Biotechnology Programme, National Horticultural Research Institute, P.M.B. 5432, Idi-Ishin, Ibadan, Oyo State, Nigeria.

**Aseptic cultures of pineapple 'smooth cayenne' were established from shoot tip explants. They were initiated on Murashige and Skoog (MS) basal medium with vitamins supplemented with various combinations of 6-benzylaminopurine (BAP) (0, 0.1, 0.2, 0.3 and 0.4 mg $L^{-1}$) and Gelrite [1.0 g $L^{-1}$ (semi-liquid), 1.5 g $L^{-1}$ (semi-solid) and 2.0 g $L^{-1}$ (solid)]. Explants were transferred five weeks after *in vitro* initiation to semi-liquid medium containing higher concentrations of BAP (0, 0.5, 1.0, 1.5 and 2.0 mg $L^{-1}$) for proliferation. Four sub-cultures were made at five weeks interval for twenty weeks in the proliferation medium. Established shoots were introduced into rooting medium containing either full (4.4 g $L^{-1}$) or half (2.2 g $L^{-1}$) strength MS basal medium with vitamins supplemented with 0, 0.5 mg $L^{-1}$ BAP alone or in combination with 0, 0.9 and 1.8 mg $L^{-1}$ α-naphthaleneacetic acid (NAA). Semi-liquid MS basal medium supplemented with 0.1 and 0.3 mg $L^{-1}$ BAP gave the best regeneration results, producing the highest average shoot length (34.6 mm) and average number of shoot buds (2.4) per explants. Significant difference ($p \leq 0.05$) in average shoot number per explant was observed with semi-liquid MS basal medium supplemented with 1.5 mg $L^{-1}$ BAP, producing the highest average shoot number (6.1) per explant at 5 weeks and this increased to 167.7 after 20 weeks. The half-strength MS basal medium without growth regulators or with 0.9 mg $L^{-1}$ NAA gave the highest average root length (29.3 mm) and root number per shoot (7.9). The semi-liquid MS basal medium supplemented with low BAP (1.5 mg $L^{-1}$) is a cost-effective method for *in vitro* propagation of pineapple. Half strength MS basal medium without hormones or with low NAA are also cost-effective methods for inducing roots in pineapple.**

**Key words:** Gelrite, growth regulators, medium type, pineapple, propagation.

## INTRODUCTION

Pineapple [*Ananas comosus* (L.) Merr] is a tropical fruit of great economic importance with a lot of health benefits (Duval et al. 2001). It is considered to be an exotic fruit, used for dessert due to its attractive flavor and nutritive value (Khan et al., 2004). Pineapples are commonly consumed as fresh fruits or processed into canned fruit, juice and jam for export (Roostika and Mariska, 2003). Nigeria is the largest producer of pineapples in Africa and the 8th largest producer in the world; with 'smooth cayenne' being the most commonly cultivated variety in Nigeria (Odusote, 2013).

Large number of healthy pineapple planting materials are required (planting density of 60,000 per hectare) in order to meet the demand of both processing and an expanding fresh-market sector (Danso et al., 2008). Conventionally, pineapple is propagated vegetatively through suckers, slips or crowns (Khan et al., 2004), but this has limitations such as low multiplication rate, transmission of diseases and lack of uniformity (Lieu et al. 2004). *In vitro* propagation of pineapple shoots is proposed as a means of addressing these problems because it allows for efficient and rapid multiplication of disease-free pineapple plantlets in a relatively shorter period independent of the season (Firoozabady and Gutterson, 2003; Sether et al., 2001). The success of an *in vitro* or a micropropagation procedure with regards to the survival rate and performance of the plants depends on several factors including explants source, type of growth hormones, gelling agent, carbon source, pH of the medium and condition of the growth room during the *in vitro* growth process (Zuraida et al., 2011).

Micropropagation of shoot tips and buds from crown have been reported in pineapple (Al-Saif et al., 2011; Hammad and Taha, 2008; Kiss et al., 1995). A commonly used method was a combination of 6-benzylaminopurine (BAP) and naphthalene acetic acid (NAA) (Firoozabady and Gutterson, 2003), indole acetic acid (IAA) (Hamad and Taha, 2008), indole butyric acid (IBA) (Boxus et al. 1991) or 2,4-dichlorophenoxy acetic acid (2,4-D) (Liu et al., 1989). Although *in vitro* micropropagation of pineapple with BAP alone has been reported (Almeida et al., 2002; Be and Debergh, 2006), the use of BAP alone or at low concentrations for rapid *in vitro* multiplication of pineapple will be a cost-effective method.

Liquid culture medium is known to promote faster rates of growth of *in vitro* plantlets because there is rapid absorption of the nutrients by the plantlets during continuous agitation of the medium on a shaker (Danso et al., 2008; Firoozabady and Gutterson, 2003). The cost of production using this technique can be high particularly in developing countries like Nigeria with unstable power supply, which depend on alternative power backup. Testing the culture medium for regeneration and proliferation of pineapple using different of levels Gelrite (gelling agent) in combination with BAP and or NAA may result in the production of a cost-effective and efficient method for *in vitro* multiplication of pineapple. This study was therefore carried out to determine the concentration(s) of Gelrite and BAP that promotes maximum rate of multiplication of pineapple var. smooth cayenne with minimal input of chemical materials.

The effect of full and half strength MS supplemented with or without selected growth regulators on root induction was also studied.

## MATERIALS AND METHODS

### Plant material and sterilization

Crowns from freshly harvested pineapple 'smooth cayenne' were used as the explant source. Leaves were removed from crowns by gentle peeling, leaving about five primordial leaves and the base intact (Khan et al., 2004). Afterwards, the peeled crowns were thoroughly washed under running tap water for 60 min before transferring to the Laminar Flow Hood for surface sterilization. Surface sterilization was done by sequential immersion in 70% (v/v) ethanol for 5 min with gentle swirling, 20% sodium hypochlorite solution (v/v) is containing three drops of Tween 20 for 10 min and finally, 15% sodium hypochlorite solution (v/v) containing 3 drops of Tween 20 for 15 min. This was followed by three rinses in sterile distilled water to remove all traces of sodium hypochlorite and other chemicals.

### Regeneration medium and culture conditions

Under aseptic conditions, the shoot tip with one to two leaf primordial (approximately 1 $cm^3$), was inoculated into freshly prepared Murashige and Skoog (MS), (1962) basal medium with vitamins in a test tube. This medium was supplemented with 30.0 g $L^{-1}$ sucrose, BAP at 0, 0.1, 0.2, 0.3, and 0.4 mg $L^{-1}$ and Gelrite at 1.0, 1.5, and 2.0 g $L^{-1}$; which constitutes semi-liquid, semi-solid and solid media; respectively. Prior to use of Gelrite, the pH of the medium was adjusted to 5.8 using 0.1 M sodium hydroxide (NaOH) or 0.1 M hydrochloric acid (HCl). The medium was dispensed at 15 ml to each test tube of size (24 x 145 mm) and sterilized by autoclaving at 15 psi and 121°C for 15 min. The cultures were maintained in the growth room at 26 ± 2°C under white fluorescent bulbs controlled with automatic timer to supply 16 h photoperiod (40 µmoles photons $m^{-2}$ $s^{-1}$). A treatment consisted of eight test tubes arranged in a completely randomized design with three replications according to Snedercor and Cochran (1980). The best regeneration medium was used in subsequent studies.

Data were collected on the number of shoot buds, and the shoot length (mm) per tube for a successive period of 5 weeks.

### Multiplication medium

Regenerated shoots obtained from the shoot-tip cultures were used as explants for this experiment. These were planted on freshly prepared multiplication medium. The multiplication medium consisted of MS basal salt with vitamins supplemented with 3% sucrose, BAP (0.5, 1.0, 1.5 and 2.0 mg $L^{-1}$) and Gelrite (1.0 g $L^{-1}$). A treatment consisted of eight culture jars arranged in a completely randomized design with three replications. Successive subcultures were carried out for up to 4 cycles with 5-weeks interval per cycle. The cultures were maintained in the growth room under the same condition as the regeneration stage. The number of shoots produced per initial explants in each jar was counted at 5 weeks interval for 20 weeks.

### Rooting medium

Proliferated shoots were separated, and each single shoot was

**Table 1.** Effect of different combinations of Gelrite and BAP concentrations on *in vitro* regeneration of shoot-tips of pineapple after 5 weeks of initiation[a].

| Gelrite (g $L^{-1}$) | BAP (mg $L^{-1}$) | Average no. of shoot buds/explants | Average shoot length (mm) explants |
|---|---|---|---|
| | 0.0 | 1.0 ± 0.0 | 24.3 ± 2.2 |
| | 0.1 | 2.0 ± 0.3 | 34.6 ± 2.7 |
| 1.00 | 0.2 | 1.8 ± 0.3 | 30.4 ± 1.6 |
| | 0.3 | 2.4 ± 0.3 | 32.3 ± 2.4 |
| | 0.4 | 1.9 ± 0.3 | 28.0 ± 2.7 |
| | 0.0 | 1.0 ± 0.0 | 21.0 ± 1.6 |
| | 0.1 | 1.8 ± 0.3 | 23.5 ± 1.6 |
| 1.50 | 0.2 | 1.8 ± 0.3 | 24.8 ± 2.8 |
| | 0.3 | 1.5 ± 0.2 | 27.6 ± 1.4 |
| | 0.4 | 1.6 ± 0.3 | 26.8 ± 2.3 |
| | 0.0 | 1.0 ± 0.0 | 20.9 ± 1.8 |
| | 0.1 | 1.0 ± 0.0 | 23.4 ± 1.7 |
| 2.00 | 0.2 | 1.0 ± 0.0 | 28.1 ± 2.9 |
| | 0.3 | 1.0 ± 0.0 | 24.5 ± 2.4 |
| | 0.4 | 1.0 ± 0.0 | 29.9 ± 1.1 |
| LSD | | 0.6 | 5.4 |

[a]Results are mean values ± standard error.

transferred to a rooting medium in test tube, for root induction and elongation. The rooting medium was composed of full-strength (4.4 g $L^{-1}$) or half-strength (2.2 g $L^{-1}$) MS basal medium with vitamins, supplemented with or without growth hormones. Medium supplemented with hormones had varying concentrations; BAP (0, and 0.5 mg $L^{-1}$) and NAA (0, 0.9 and 1.8 mg $L^{-1}$). The hormones were added alone and in combination.

Data on number of roots per shoot, number of days to first root emergence and root length were recorded for six weeks. Rooted plantlets were subsequently rinsed free of culture medium and transferred to black polythene bags containing sterile top soil. These were maintained under shade for 40 days before they were transferred to the field.

#### Statistical analysis

Data were subjected to analysis of variance (ANOVA) using the GLM procedure of SAS (Statistical Analysis Software, 2008). Means were separated using Fisher's LSD at $p \leq 0.05$.

## RESULTS AND DISCUSSION

### Shoot regeneration

Experiments were carried out to determine the best medium composition for shoot regeneration from shoot-tip cultures under controlled environment. The results showed that all the medium solidification types supplemented with BAP promoted regeneration of shoot buds but the semi-liquid medium (1.0 g $L^{-1}$ Gelrite) gave the best regeneration in terms of the average number of shoot buds produced per explant (2.4) and average shoot length (34.6 mm) (Table 1). The highest average shoot length (34.6 mm) was observed in the medium with 0.1 mg $L^{-1}$ BAP and the highest average number of shoot buds (2.4) was observed in 0.3 mg $L^{-1}$ BAP. This result agrees with previous report (Firoozabady and Gutterson, 2003), which shows that BAP (0.1 to 0.5 mg $L^{-1}$) added to MS medium is essential for the regeneration of plants from shoot apices of pineapple. Paiva et al. (1998) obtained best results in the shoot induction of pineapple, cv. Skay, with either 1.0 mg $L^{-1}$ BAP or 0.1 mg $L^{-1}$ TDZ. This is different from results of this study in which lower concentrations of BAP (0.1 and 0.3 mg $L^{-1}$) gave best results for shoot induction.

The differences between both results may be due to genotype effect. Al-Saif et al. (2011) reported that 2.0 mg $L^{-1}$ BAP in solid medium (7.0 g $L^{-1}$ solidified with agar) gave shoot length of 9.5 mm. Result of this study showed that lower concentrations of BAP (0.1 to 0.4 mg $L^{-1}$) in solid medium produced higher shoot length (20.9 to 29.9 mm) (Table 1). With the exception of 0.3 mg $L^{-1}$ BAP, an increase in BAP concentrations from 0 to 0.4 mg $L^{-1}$ in the solid medium (2.0 g $L^{-1}$ gelrite) resulted in increased shoot length. This is not in agreement with report of Hamad and Taha (2009) who observed a decrease in shoot length as BAP concentration increased up to 0.5 mg $L^{-1}$ in solid medium. The highest average number of shoot buds per explant and shoot length obtained in

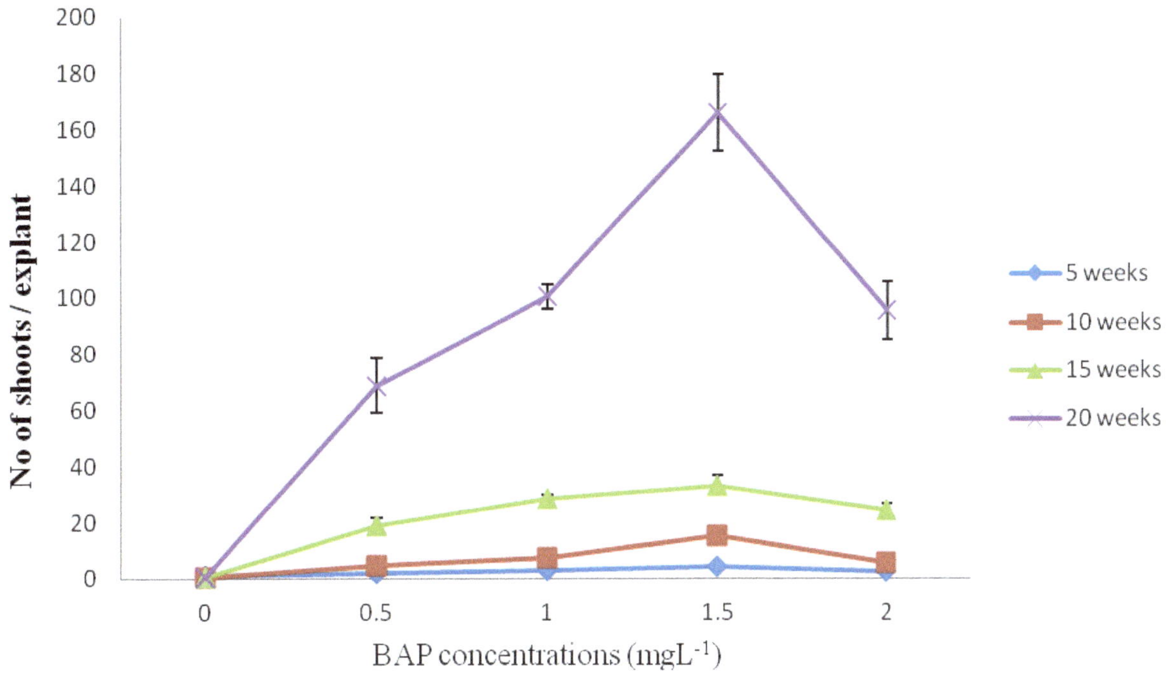


**Figure 1.** Effect of BAP concentrations on average shoot number of pineapple produced in a semi-liquid MS basal medium with vitamins ±SE.

semi-liquid medium (solidified with 1.0 g $L^{-1}$ Gelrite) in this study maybe due to increased intake of the nutrients made possible by free movement of compounds or nutrients from the culture medium into the plant tissues. Gupta et al. (1981) showed that the use of liquid medium facilitates the nutrient uptake by the plant due to better distribution in the culture medium. Similar higher regeneration rates in liquid medium when compared with solid medium have been reported by Alvard et al. (1993).

## Shoot proliferation

After the determination of the optimal Gelrite concentration for regeneration, the BAP concentration was increased from 0.5 to 2.0 mg $L^{-1}$ with the aim of inducing more shoots in the semi-liquid medium (Figure 1). After five weeks of culture, the basal portions of pineapples consisting of proliferating buds (clumped explant) were observed to produce more shoots (Plate 1A and B). Medium supplemented with 1.5 mg $L^{-1}$ BAP produced most average number of shoots (6.1) and this was significantly higher ($p \leq 0.05$) than that of other BAP concentrations evaluated (Figure 1). Zuraida et al. (2011) reported that explants cultured on liquid medium supplemented with 1.0 mg $L^{-1}$ BAP produced the highest number of shoots (31) after 4 weeks when compared with 5 mg $L^{-1}$ BAP which produced 14 shoots after four weeks of culture. This confirms the assertion that liquid culture explants usually display a higher frequency of growth rate when compared with solid or semi-liquid cultures (Gupta et al., 1981), and maybe as a result of continuous shaking of liquid cultures thus promoting increased intake of the nutrients in the medium. After a second sub-culture, a 3-fold increase in the average number of shoots (18.3) was recorded for 1.5 mg $L^{-1}$ BAP at 10 weeks; and at 20 weeks a 28-fold increase was observed (167.6) (Figure 1, Plate 1C). Result of this study is similar to report of Almeida et al. (2002) in which MS medium supplemented with 1.5 mg $L^{-1}$ BAP gave the best treatment at proliferation stage, producing an average of 701.1 shoots per explant in liquid medium after five subcultures. Hamad and Taha (2008) also reported that in 4 cycles of culture, shoot number was highest in BAP at 1.5 mg $L^{-1}$.

The proliferation induced by BAP corro-borates report of Kyte and Kleyn (1996) that a cytokinin is very crucial for cell division and axillary bud multiplication in plants. In this study, BAP (1.5 mg $L^{-1}$) alone produced average shoot number of 18.3 per clumped explant. This is higher than the average shoot number of 12.0 per explant observed in solid medium supplemented with BAP (2.0 mg $L^{-1}$) (Al-Saif et al., 2011); 9.0 shoots per explant obtained in response to 1.0 mg $L^{-1}$ BAP (Be and Debergh, 2006). Barboza and Caldas (2001) used etiolated nodal segments for micropropagation of the pineapple hybrid PE X SC-52, and observed that BAP (2.0 mg $L^{-1}$) promoted the highest number of plants per

**Plate 1.** Different stages of *in vitro* propagation of smooth cayenne in semi-liquid medium. A = Regeneration stage with bud formation. B = Initial multiplication stage; C = advanced multiplication stage; D = rooting in half-strength MS basal medium with 0.9 NAA mg $L^{-1}$. E = rooting in half-strength MS basal medium without growth hormones.

shoot culture (10.4) and per nodal segment, when compared with kinetin (5.0 mg $L^{-1}$) or a combination of BAP (2.0 mg $L^{-1}$) and NAA (1.86 mg $L^{-1}$). Alvard et al. (1993) and George and Sherrington (1984) both reported higher rates of growth in liquid cultures when compared with solid medium. This may be attributed to the exposure to greater surface of the explants to a medium with uniform distribution of nutrients, and enhanced nutrient uptake. Medium containing BAP 1.0 mg $L^{-1}$ which gave the second highest average number of shoots per explants (4.6) was observed to have a 2-fold increase in the average number of shoots per explant (9.5) after 10 weeks and a 25-fold increase in the average number of shoots (102.4) per explant after 20 weeks (Figure 1). Further increase in BAP concentration (2.0 mg $L^{-1}$) did not result in appreciable increase in the number of shoots. This may be due to inherent cytokinnin already present in the plant resulting to habituation decline.

## Root induction

Medium with half-strength MS basal medium with vitamins (2.2 g $L^{-1}$) supplemented with 0.5 mg $L^{-1}$ BAP + 0.9 mg $L^{-1}$ NAA gave the lowest average days to root emergence (1.2) and the lowest average number of roots per shoot (approx. 0.3) (Table 2). This was followed by half- strength MS basal medium with vitamins supplemented with 0.9 mg $L^{-1}$ NAA alone which gave average days to root emergence of (approximately 7.0) and the highest mean number of roots per shoot (approximately 7.9). There was no significant difference ($p \leq 0.05$) in

**Table 2.** Influence of MS strength and different growth regulator concentrations on root induction of pineapple cultured for 6 weeks[a].

| MS strength | Plant growth regulators (mg $L^{-1}$) | | Average days to root emergence | Average no of roots/shoot | Average root length (mm) | Rooted plantlet (%) |
|---|---|---|---|---|---|---|
| | BAP | NAA | | | | |
| Half strength | 0.00 | 0.00 | 7.33[b] ± 0.23 | 6.25[b] ± 0.29 | 29.3[a] ± 1.0 | 100[a] ± 0.0 |
| | 0.00 | 0.90 | 6.96[b] ± 0.19 | 7.88[a] ± 0.54 | 26.5[b] ± 0.8 | 100[a] ± 0.0 |
| | 0.00 | 1.80 | 7.46[b] ± 0.19 | 7.79[a] ± 0.40 | 28.7[a] ± 1.1 | 100[a] ± 0.0 |
| | 0.50 | 0.00 | - | - | - | - |
| | 0.50 | 0.90 | 1.17[c] ± 0.81 | 0.29[d] ± 0.20 | 1.0[d] ± 0.7 | 8[c] ± 0.06 |
| | 0.50 | 1.80 | - | - | - | - |
| Full strength | 0.00 | 0.00 | - | - | - | - |
| | 0.00 | 0.90 | - | - | - | - |
| | 0.00 | 1.80 | 10.71 [a] ± 0.3 | 3.46c ± 0.42 | 6.4[c] ± 0.6 | 92[b] ± 0.06 |
| | 0.50 | 0.00 | - | - | - | - |
| | 0.50 | 0.90 | - | - | - | - |
| | 0.50 | 1.80 | - | - | - | - |

[a]Means with same letter in the same column are not significantly different at $P \leq 0.05$ according to L.S.D.

the number of roots induced when NAA concentration was doubled from 0.9 NAA (7.9) to 1.80 mg $L^{-1}$ NAA (7.8) in half strength MS medium. Contrary to result of this study, Hamad et al. (2013) reported that in liquid medium, shoots failed to form roots in half-strength MS basal medium containing NAA. Danso et al. (2008) reported that shoots cultured on 7.5 mg $L^{-1}$ BAP and NAA concentrations (7.5 to 15.0 mg $L^{-1}$) did not result in any root formation in MD2 pineapple per shoot. Firoozabady and Gutterson (2003) obtained roots from liquid cultures of pineapple cultured on MS medium supplemented with 0.5 mg $L^{-1}$ NAA and 0.5 mg $L^{-1}$ IBA. NAA and IBA are root inducing growth regulators and have been used either alone or in combination for root induction in many cultures (Be and Debergh, 2006; Gupta et al., 1981). The average number of roots per shoot obtained with medium supplemented with 0.9 mg $L^{-1}$ NAA (7.9) (Plate 1D) or 1.8 mg $L^{-1}$ NAA (7.8) are higher than those obtained in medium supplemented with 1.0 mg $L^{-1}$ IBA (5.0) (Khan et al, 2004); and when IBA was substituted by 1.0 mg $L^{-1}$ NAA (3.6); this took between 8-15 days for induction of root.

The full or half-strength MS basal medium supplemented with 0.5 mg $L^{-1}$ BAP did not produce any roots. This result agrees with the reports of Be and Debergh (2006), Danso et al. (2008) and Gupta et al. (1981) which showed that NAA, as an auxin, is important for root induction and regulation in plants. The half-strength MS basal medium containing no growth regulators had the highest root length (29.3 mm) (Table 2, Plate 1E) and the third highest mean number of roots (6.3). Almeida et al. (2002) also recorded success in rooting when shoots were transferred to MS medium with half the concentration of salts and no growth regulators for 30 days. The half-strength MS supplemented with 1.8 mg $L^{-1}$ NAA gave the second highest average roots' length (28.7 mm, Table 2). Shoots cultured on half strength MS basal medium with vitamins and no growth regulators, 0.9 mg $L^{-1}$ NAA or 1.8 mg $L^{-1}$ NAA recorded 100% rooting. The full-strength MS medium with no growth hormones did not produce any root. Only full-strength MS supplemented with 1.8 mg $L^{-1}$ NAA produced roots; the mean number of roots per explants and root lengths were 3.46 and 6.4 mm, respectively. These were significantly lower ($p \leq 0.05$) than results obtained from shoots cultured on half-strength MS medium.

Nearly all plantlets were successfully hardened in black polythene bags (survival rate of 85%) containing sterile top soil and placed under shade.

The results of this study demonstrates the efficient use of semi-liquid medium with low concentrations of BAP for *in vitro* regeneration and multiplication of pineapple, as well as the use of half-strength MS basal medium with vitamins and no growth hormones or low NAA concentration for root induction. The minimal use of materials and chemicals in tissue culturing of pineapple, directly translates to low cost of production. This economic approach to micro propagation enhances the availability and affordability of *in vitro* derived pineapple cv. smooth cayenne as quality planting materials to meet the ever increasing demand by farmers.

## Conflict of interest

Authors declare that there is no conflict of interest.

whether financial or relevant interest that has influenced this study.

## Conflict of Interests

The author(s) have not declared any conflict of interests.

## ACKNOWLEDGEMENTS

The authors gratefully acknowledge National Horticultural Research Institute, Ibadan, Nigeria for financial support. They also wish to thank Mrs Funmi Amoran, and all other staff of the biotechnology laboratory for their technical assistance.

## REFERENCES

Almeida WAB, Santana GS, Rodriguez APM, Costa MAP (2002). Optimization of a protocol for the micropropagation of pineapple. Rev. Bras. Frutic. 24(2):296-300.

Al-Saif AM, Hossain ABMS, Taha RM (2011). Effects of benzylaminopurine and naphthalene acetic acid on proliferation and shoot growth of pineapple (*Ananas comosus* L. Merr) *in vitro*. Afr. J. Biotech. 10(27):5591-5595.

Alvard D, Cote F, Teisson C (1993). Comparison of methods of liquid medium cultures for banana micropropagation. Effect of temporary immersion of explants. Plant Cell. Tiss. Organ. Cult. 32:555-600.

Barboza SBSC, Caldas LS (2001). Estiolamento e regeneração na multiplicação *in vitro* do abacaxizeiro híbrido PE x SC-52. Pesq Agropec Bras. 36(3):417-423.

Be LV, Debergh PC (2006). Potential low-cost micropropagation of pineapple (*Ananas comosus*). South Afr. J. Bot. 71:191-194.

Boxus P, Terzi JM, Lieves C, Pylyser M, Ngaboyamahina P, Duhem K (1991). Improvement and perspectives of micropropagation techniques applied to some hot climate plants. Acta. Hort. 289:55-59.

Danso KE, Ayeh KO, Oduro V, Amiteye S, Amoatey HM (2008). Effect of 6-benzylaminopurine and naphthalene acetic acid on *in vitro* production of MD2 pineapple planting materials. World. Appl. Sci. J. 3(4):614-619.

Duval MF, Noyer JL, Perrier X, Coppens d'Eeckenbrugge G, Hamon P (2001). Molecular diversity in pineapple assessed by RFLP markers. Theor. Appl. Genet. 102:83-90.

Firoozabady E, Gutterson N (2003). Cost effective *in vitro* propagation methods for pineapple. Plant Cell Rep. 21:844-850.

George F, Sherrington PD (1984). Plant propagation by tissue culture. Exegetics Limited, pp. 284-330.

Gupta PK, Mascarenhas AF, Jagannathan V (1981). Tissue Culture of Trees: clonal propagation of mature trees of *Eucalyptus citriodora* Hook, by tissue culture. Plant Sci. Lett. 20:195-201.

Hamad AM, Taha RM (2008). Effect of sequential subcultures on *in vitro* proliferation capacity and shoot formation pattern of pineapple (*Ananas comosus* L. Merr.) over different incubation periods. Sci. Hort. 117:329-334.

Hamad AM, Taha RM (2009). Effect of explants density on the *in vitro* proliferation and growth of separated and cluster shoots of smooth cayenne pineapple (*Ananas comosus* L. Merr.). Asian J. Plant Sci. 8:313-317.

Hamad AM, Taha RM, Mohajer S (2013). In vitro induction and proliferation of adventitious roots in pineapple (Ananas comosus L.) cultivars of smooth cayenne. Aust J. Crop. Sci. 7(7):1038-1045.

Khan S, Nasib A, Saeed BA (2004). Employment of *in vitro* technology for large scale multiplication of pineapples (*Ananas comosus*). Pak. J. Bot. 36(3):611-615.

Kiss E, Kiss J, Gyulai G, Heszky LE (1995). A novel method for rapid micropropagation of pineapple. Hort Sci. 30(1):127-129.

Kyte L, Kleyn J (1996). Plants from test tubes: An introduction to micropropagation. Third edition, Daquinta J.H., Gonzalez Y., Desjardins C.G, pp. 82.

Lieu P, Tinh NN, Vui PV, Khai TP, Teisson C (2004). Study of multiplication rate of conventional propagation *in vivo* of Cayenne (*A. comosus* L.). Impact de 10 annees de cooperation francaise sur l'amelioration des productions fruitieres au Vietnam – SOFRI. Conference, Mai, pp. 1-11.

Liu LJ, Rosa-Marquez E, Lazardi E (1989). Smooth leaf (spineless) red spanish pineapple (*Ananas comosus* (L.) Merr) propagated *in vitro*. J. Agric. Univ. Puerto Rico. 73:301-311.

Murashige T, Skoog F (1962). A revised medium for rapid growth and bioassays with tobacco tissue cultures. Physiol. Plant. 9:8-11.

Odusote I (2013). Tomato, PineApple: We are largest producers in Africa. Naija AgroNet. http://www.naijaagronet.com.ng/2013/08/tomato-pineapple-were-largest-producers.html. Accessed 10th march, 2014.

Paiva PD de O, Mayer MDB, Kawamura MI, Pasqual M, Paiva R (1999). Efeito de BAP, Thidiazuron e Sulfato de Adenina na Propagação *in vitro* de Abacaxi. Revista Ceres, Viçosa. 46(265):231-237.

Roostika IT, Matiska I (2003). *In Vitro* Culture of Pineapple by Organogenesis and Somatic Embryogenesis: Its Utilization and Prospect. Bull. AgroBio. 6(1):34-40.

Sether DM, Karasev AV, Okumura C, Arakawa C, Zee F, Kislan MM, Busto JL, Hu JS (2001). Differentiation, distribution and elimination of two different pineapple mealybug wilt-associated viruses found in pineapple. Plant Disease. 85:856-864.

Snedercor WG, Cochran GW (1980). "Statistical Methods" 6th Ed. Iowa State University Press, Ames, Iowa, U.S.A.

Statistical Analysis Software (SAS) (2008). Statistical Analysis Software (SAS) Institute Inc., Cary, NC, USA, version 9.3.

Zuraida AR, Nurul Shahnadz AH, Harteeni A, Roowi S, Che Radziah CMZ, Sreeramanan S (2011). A novel approach for rapid micropropagation of maspine pineapple (*Ananas comosus* L.) shoots using liquid shake culture system. Afr. J. Biotech. 10(19):3859-3866.

# Permissions

All chapters in this book were first published in AJPS, by Academic Journals; hereby published with permission under the Creative Commons Attribution License or equivalent. Every chapter published in this book has been scrutinized by our experts. Their significance has been extensively debated. The topics covered herein carry significant findings which will fuel the growth of the discipline. They may even be implemented as practical applications or may be referred to as a beginning point for another development.

The contributors of this book come from diverse backgrounds, making this book a truly international effort. This book will bring forth new frontiers with its revolutionizing research information and detailed analysis of the nascent developments around the world.

We would like to thank all the contributing authors for lending their expertise to make the book truly unique. They have played a crucial role in the development of this book. Without their invaluable contributions this book wouldn't have been possible. They have made vital efforts to compile up to date information on the varied aspects of this subject to make this book a valuable addition to the collection of many professionals and students.

This book was conceptualized with the vision of imparting up-to-date information and advanced data in this field. To ensure the same, a matchless editorial board was set up. Every individual on the board went through rigorous rounds of assessment to prove their worth. After which they invested a large part of their time researching and compiling the most relevant data for our readers.

The editorial board has been involved in producing this book since its inception. They have spent rigorous hours researching and exploring the diverse topics which have resulted in the successful publishing of this book. They have passed on their knowledge of decades through this book. To expedite this challenging task, the publisher supported the team at every step. A small team of assistant editors was also appointed to further simplify the editing procedure and attain best results for the readers.

Apart from the editorial board, the designing team has also invested a significant amount of their time in understanding the subject and creating the most relevant covers. They scrutinized every image to scout for the most suitable representation of the subject and create an appropriate cover for the book.

The publishing team has been an ardent support to the editorial, designing and production team. Their endless efforts to recruit the best for this project, has resulted in the accomplishment of this book. They are a veteran in the field of academics and their pool of knowledge is as vast as their experience in printing. Their expertise and guidance has proved useful at every step. Their uncompromising quality standards have made this book an exceptional effort. Their encouragement from time to time has been an inspiration for everyone.

The publisher and the editorial board hope that this book will prove to be a valuable piece of knowledge for researchers, students, practitioners and scholars across the globe.

# List of Contributors


**Harnet Abrha**
Alamata Agricultural Research Center, Alamata, Ethiopia

**Derbew Belew**
Jimma University, Jimma, Ethiopia

**Gebremedhin Woldegiorgis**
Holleta Agricultural Research Center, Addis Ababa, Ethiopia

**S. T. Anjorin**
Department of Crop Science, University of Abuja, Abuja, Nigeria

**M. Mohammed**
Department of Crop Production, Federal University of Technology, Minna, Niger State, Nigeria

**R. L. Bhardwaj**
Krishi Vigyan Kendra, Sirohi, Agriculture University Jodhpur-342304, Rajasthan, India

**N. Amadi**
Department of Plant Science and Biotechnology, College of Natural and Applied Sciences, University of Port Harcourt, Port Harcourt, Rivers State, Nigeria

**F. B. G. Tanee**
Department of Plant Science and Biotechnology, College of Natural and Applied Sciences, University of Port Harcourt, Port Harcourt, Rivers State, Nigeria

**J. Kubiriba**
National Agricultural Research Organization, P.O. Box 7065, Kampala, Uganda

**W. K. Tushemereirwe**
National Agricultural Research Organization, P.O. Box 7065, Kampala, Uganda

**Otitoloju Kekere**
Department of Plant Science and Biotechnology, Adekunle Ajasin University, Akungba Akoko, Ondo State Nigeria

**Joseph Femi Bamidele**
Department of Plant Biology and Biotechnology, University of Benin, Benin City, Edo State, Nigeria

**Grace Koech**
World Agroforestry Centre, United Nations Avenue, Gigiri, P. O. Box 30677-00100, Nairobi, Kenya
Jomo Kenyatta University of Agriculture and Technology, P. O. BOX 62000-00100, Nairobi, Kenya

**Daniel Ofori**
World Agroforestry Centre, United Nations Avenue, Gigiri, P. O. Box 30677-00100, Nairobi, Kenya

**Anne W.T. Muigai**
Jomo Kenyatta University of Agriculture and Technology, P. O. BOX 62000-00100, Nairobi, Kenya

**Martha Makobe**
Jomo Kenyatta University of Agriculture and Technology, P. O. BOX 62000-00100, Nairobi, Kenya

**Muriuki, Jonathan**
World Agroforestry Centre, United Nations Avenue, Gigiri, P. O. Box 30677-00100, Nairobi, Kenya

**G. Jeremias Mowo**
World Agroforestry Centre, United Nations Avenue, Gigiri, P. O. Box 30677-00100, Nairobi, Kenya

**Ramni Jamnadass**
World Agroforestry Centre, United Nations Avenue, Gigiri, P. O. Box 30677-00100, Nairobi, Kenya

**Naser Arabzadeh**
Agricultural and Natural Resources Researches Center, Sodoughi Avenue, P.O. Box 76175-538, Kerman, Iran

**Denre Manas**
Department of Agricultural Biochemistry, Bidhan Chandra Krishi Viswavidyalaya, Mohanpur, Nadia-741252, West Bengal, India

**P. K. Bandopadhyay**
Department of Plant Physiology, Bidhan Chandra Krishi Viswavidyalaya, Mohanpur, Nadia-741252, West Bengal, India

**A. Chakravarty**
Department of Agricultural Biochemistry, Bidhan Chandra Krishi Viswavidyalaya, Mohanpur, Nadia-741252, West Bengal, India

**S. Pal**
Department of Agricultural Biochemistry, Bidhan Chandra Krishi Viswavidyalaya, Mohanpur, Nadia-741252, West Bengal, India

**A. Bhattacharya**
Department of Agricultural Biochemistry, Bidhan Chandra Krishi Viswavidyalaya, Mohanpur, Nadia-741252, West Bengal, India

**Misheck Chandiposha**
Department of Agronomy, Midlands State University, P Bag 9055, Gweru, Zimbabwe

**Fanuel Chivende**
Department of Agronomy, Midlands State University, P Bag 9055, Gweru, Zimbabwe

**R. Dhivya**
Centre for Plant Breeding and Genetics, Tamil Nadu Agricultural University, Coimbatore-641 003, India

**P. Amalabalu**
Centre for Plant Breeding and Genetics, Tamil Nadu Agricultural University, Coimbatore-641 003, India

**R. Pushpa**
Centre for Plant Breeding and Genetics, Tamil Nadu Agricultural University, Coimbatore-641 003, India

**D. Kavithamani**
Centre for Plant Breeding and Genetics, Tamil Nadu Agricultural University, Coimbatore-641 003, India

**Uday Chand Basak**
Regional Plant Resource Centre, Nayapalli Bhubaneswar-751015, Odisha, India

**Dipika Dash**
Regional Plant Resource Centre, Nayapalli Bhubaneswar-751015, Odisha, India

**Gourisankar J. P. Jena**
Regional Plant Resource Centre, Nayapalli Bhubaneswar-751015, Odisha, India

**Ajay Kumar Mahapatra**
Regional Plant Resource Centre, Nayapalli Bhubaneswar-751015, Odisha, India

**Tewodros Mesfin**
Melkassa Agriculture Research Center-EARI, P. O. Box 436, Adama Ethiopia

**Jibril Mohammed**
Melkassa Agriculture Research Center-EARI, P. O. Box 436, Adama Ethiopia

**Abuhay Taklete**
Melkassa Agriculture Research Center-EARI, P. O. Box 436, Adama Ethiopia

**Feyera Merga**
Melkassa Agriculture Research Center-EARI, P. O. Box 436, Adama Ethiopia

**Charles Wortmann**
Melkassa Agriculture Research Center-EARI, P. O. Box 436, Adama Ethiopia
University of Nebraska-Lincoln, 369 Keim Hall, Lincoln EN 68583-0915, Ethiopia

**Ezekiel Edward**
Faculty of Forestry and Nature Conservation, Sokoine University of Agriculture,P. O. Box 3010, Morogoro, Tanzania

**Shabani A. O. Chamshama**
Faculty of Forestry and Nature Conservation, Sokoine University of Agriculture,P. O. Box 3010, Morogoro, Tanzania

**Yonika M. Ngaga**
Faculty of Forestry and Nature Conservation, Sokoine University of Agriculture,P. O. Box 3010, Morogoro, Tanzania

**Mathew A. Mndolwa**
Tanzania Forestry Research Institute-Lushoto Centre, P. O. Box 95, Lushoto, Tanzania

**E. Nafie**
Department of Botany, College for Women (Arts, Science and Education), Ain Shams University, Cairo, Egypt

**K. B.Adejobi**
Cocoa Research Institute of Nigeria, P. M. B. 5244, Ibadan, Nigeria

**O. S. Akanbi**
Cocoa Research Institute of Nigeria, P. M. B. 5244, Ibadan, Nigeria

**O. Ugioro**
Cocoa Research Institute of Nigeria, P. M. B. 5244, Ibadan, Nigeria

**S. A. Adeosun**
Cocoa Research Institute of Nigeria, P. M. B. 5244, Ibadan, Nigeria

**I. Mohammed**
Cocoa Research Institute of Nigeria, P. M. B. 5244, Ibadan, Nigeria

**B. A. Nduka**
Cocoa Research Institute of Nigeria, P. M. B. 5244, Ibadan, Nigeria

**D. O. Adeniyi**
Cocoa Research Institute of Nigeria, P. M. B. 5244, Ibadan, Nigeria

**Berhanu Abraha Tsegay**
Department of Biology, Bahir Dar University, Ethiopia

**Berhane Gebreslassie**
Department of Biology, Woldia University, Ethiopia

**A. R. Saberi**
Scientific Board of Agricultural and Natural Science Research Center of Golestan Province, Gorgan, Iran

**Benaouf Zohra**
Faculty of Biology, Laboratory of Ecology, Houari Boumediene University of Sciences and Technology (U.S.T.H.B), El Alia, BP 32, Bab Ezzouar, 16111 Algiers, Algeria

**Miloudi Ali**
Laboratory of Plant Biology - Faculty of Science, University of Mascara 29000 Algeria

**Belkhodja Moulay**
Laboratory of Plant physiology - Faculty of Science, University of Oran 31000 Algeria

**Fikru Mekonnen**
School of Plant Sciences, College of Agriculture and Environmental Sciences, Haramaya University, P.O. Box, 138, Dire Dawa, Ethiopia

**Firew Mekbib**
School of Plant Sciences, College of Agriculture and Environmental Sciences, Haramaya University, P.O. Box, 138, Dire Dawa, Ethiopia

**Shiv Kumar**
International Center for Agricultural Research in the Dry Areas (ICARDA), Rabat, Morocco

**Seid Ahmed**
International Center for Agricultural Research in the Dry Areas (ICARDA), P O Box 5689, Addis Ababa, Ethiopia

**T. R. Sharma**
Department of Agricultural Biotechnology, Chaudhary Sarwan Kumar Himachal Pradesh Agricultural University, Palampur 176 062, India

**Amartuvshin NARANTSETSEG**
Institute of Botany, Mongolian Academy of Sciences, Ulaanbaatar 210351, Mongolia

**Habtamu Ashagre**
College of Agriculture and Veterinary Sciences, Department of Plant Sciences, Ambo University, P. O. Box 19, Ambo, Ethiopia

**Ibrahim A. Hamza**
College of Agriculture and Veterinary Sciences, Department of Plant Sciences, Ambo University, P. O. Box 19, Ambo, Ethiopia

**Urgecha Fita**
College of Agriculture and Veterinary Sciences, Department of Plant Sciences, Ambo University, P. O. Box 19, Ambo, Ethiopia

**Worku Nedesa**
College of Agriculture and Veterinary Sciences, Department of Plant Sciences, Ambo University, P. O. Box 19, Ambo, Ethiopia

**R. K. Naresh**
Department of Agronomy, Sardar Vallabhbhai Patel University of Agriculture and Technology, Meerut (U.P), India

**S. S. Tomar**
Department of Agronomy, Sardar Vallabhbhai Patel University of Agriculture and Technology, Meerut (U.P), India

**U. P. Shahi**
Department of Soil Science, Sardar Vallabhbhai Patel University of Agriculture and Technology, Meerut (U.P), India

**S. P. Singh**
Department of Soil Science, Sardar Vallabhbhai Patel University of Agriculture and Technology, Meerut (U.P), India

**Purushottam**
Department of Pathology & Microbiology, Sardar Vallabhbhai Patel University of Agriculture and Technology, Meerut (U.P), India

**B. Singh**
Department of Horticulture, Sardar Vallabhbhai Patel University of Agriculture and Technology, Meerut (U.P), India

**Parvinder Kumar**
Department of Agricultural Meterology C. C. S. H. A. U., Hissar, Haryana, India

**Ayaz Khan**
Department of Botany, University of Peshawar, Peshawar, Pakistan

**Muhammad Ibrar**
Department of Botany, University of Peshawar, Peshawar, Pakistan

**Ishtiaq Ahmad**
Govt. Degree College, Tangi, Charsadda, Pakistan

**V. H. Kabambe**
Bunda College of Agriculture, P. O. Box 219, Lilongwe, Malawi

**E. D. L. Mazuma**
Chitedze Research Station, P. O. Box 158, Lilongwe, Malawi

**J. Bokosi**
Bunda College of Agriculture, P. O. Box 219, Lilongwe, Malawi

**E. Kazira**
Lilongwe Agriculture Development Division, P.O Box 259, Lilongwe, Malawi

**S. A. Olakojo**
Institute of Agricultural Research and Training Obafemi Awolowo University, P. M. B. 5029 Moor Plantation, Ibadan

**O. A. Adetula**
National Horticultural Research Institute Idi-Isin, Ibadan

**K. C. Kaizzi**
National Agricultural Research Laboratories (NARL), Kawanda, P. O. Box 7065 Kampala, Uganda

**J. Byalebeka**
National Agricultural Research Laboratories (NARL), Kawanda, P. O. Box 7065 Kampala, Uganda

**O. Semalulu**
National Agricultural Research Laboratories (NARL), Kawanda, P. O. Box 7065 Kampala, Uganda

**I. Alou**
National Agricultural Research Laboratories (NARL), Kawanda, P. O. Box 7065 Kampala, Uganda

**W. Zimwanguyizza**
National Agricultural Research Laboratories (NARL), Kawanda, P. O. Box 7065 Kampala, Uganda

**A. Nansamba**
National Agricultural Research Laboratories (NARL), Kawanda, P. O. Box 7065 Kampala, Uganda

**E. Odama**
National Agricultural Research Laboratories (NARL), Kawanda, P. O. Box 7065 Kampala, Uganda

**C. S. Wortmann**
Department of Agronomy and Horticulture, University of Nebraska-Lincoln, Lincoln, NE USA

**Kassim DOSSO**
Laboratory of Physiology, Pharmacology and Phytotherapy, UFR SN, University of Abobo-Adjamé, 02 BP 801 Abidjan 02, Côte d'Ivoire

**Benoit Banga N'GUESSAN**
Laboratory of Physiology, Pharmacology and Phytotherapy, UFR SN, University of Abobo-Adjamé, 02 BP 801 Abidjan 02, Côte d'Ivoire

**Boua Narcisse GNANGORAN**
Department of Pharmacology and Toxicology, School of Pharmacy, College of Health Sciences, University of Ghana, P.O Box KB 52, Korle-Bu, Accra, Ghana

**Angoué Paul Yapo**
Laboratory of Physiology, Pharmacology and Phytotherapy, UFR SN, University of Abobo-Adjamé, 02 BP 801 Abidjan 02, Côte d'Ivoire

**Placide Rukundo**
Rwanda Agriculture Board (RAB), P.O. Box 5016, Kigali, Rwanda

**Hirut Getinet Betaw**
Amhara Agricultural Research Institute, P.O. Box 527 code 100, Bahidar, Ethiopia

**Stephan Ngailo**
Sugarcane Research Institute, P.O. Box 30031, Kibaha, Coast Region, Tanzania

**Fekadu Balcha**
Southern Agricultural Research Institute, Hawassa Research Centre, P.O. Box 6, Hawassa, Ethiopia

**Amit Kumar Chaurasia**
Department of Botany, Government P.G. College, Tikamgarh (M.P.) 472001, India

**Shubha Chaurasia**
Department of Boatany, Government College, Palera District, Tikamgarh (M.P.) 472221, India

**Sushmita Chaurasia**
Department of Botany, Government College, Jatara District, Tikamgarh (M.P.) 472118, India

**Shridha Chaurasia**
Department of Botany, Government P.G. College, Tikamgarh (M.P.) 472001, India

**Zhao Zhichang**
Tropical Crops Genetic Resources Institute, Chinese Academy of Tropical Agricultural Sciences/Key Laboratory of Crop Gene Resources and Germplasm Enhancement in Southern China, Ministry of Agriculture, Danzhou 571737, P. R. China

**Chen Yeyuan**
Tropical Crops Genetic Resources Institute, Chinese Academy of Tropical Agricultural Sciences/Key Laboratory of Crop Gene Resources and Germplasm Enhancement in Southern China, Ministry of Agriculture, Danzhou 571737, P. R. China

**Gao Aiping**
Tropical Crops Genetic Resources Institute, Chinese Academy of Tropical Agricultural Sciences/Key Laboratory of Crop Gene Resources and Germplasm Enhancement in Southern China, Ministry of Agriculture, Danzhou 571737, P. R. China

**Huang Jianfeng**
Tropical Crops Genetic Resources Institute, Chinese Academy of Tropical Agricultural Sciences/Key Laboratory of Crop Gene Resources and Germplasm Enhancement in Southern China, Ministry of Agriculture, Danzhou 571737, P. R. China

**Dang Zhiguo**
Tropical Crops Genetic Resources Institute, Chinese Academy of Tropical Agricultural Sciences/Key Laboratory of Crop Gene Resources and Germplasm Enhancement in Southern China, Ministry of Agriculture, Danzhou 571737, P. R. China

**Pamela Eloho Akin-Idowu**
Fruits and Biotechnology Programme, National Horticultural Research Institute, P.M.B. 5432, Idi-Ishin, Ibadan, Oyo State, Nigeria

**Sunday Oluseyi Solomon Akinyemi**
Fruits and Biotechnology Programme, National Horticultural Research Institute, P.M.B. 5432, Idi-Ishin, Ibadan, Oyo State, Nigeria

**Dorcas Olubunmi Ibitoye**
Fruits and Biotechnology Programme, National Horticultural Research Institute, P.M.B. 5432, Idi-Ishin, Ibadan, Oyo State, Nigeria